Geographies of Health, Disease, and Well-being

T0172570

This book is a collection of papers reflecting the latest advances in geographic research on health, disease, and well-being. It spans a wide range of topics, theoretical perspectives, and methodologies – including anti-racism, post-colonialism, spatial statistics, spatiotemporal modeling, political ecology, and social network analysis. Health issues in various regions of the world are addressed by interdisciplinary authors, who include scholars from epidemiology, medicine, public health, demography, and community studies. The book covers the major themes in this field such as health inequalities; environmental health; spatial analysis and modeling of disease; health care provision, access, and utilization; health and well-being; and global/ transnational health and health issues in the global south. There is also a specially commissioned book review in addition to the chapters included in these six sections. Together, these chapters show cogently how geographic perspectives and methods can contribute in significant ways to advancing our understanding of the complex interactions between social and physical environments and health behaviors and outcomes.

This book was published as a special issue of *Annals of the Association of American Geographers*.

Mei-Po Kwan is Professor of Geography and Geographic Information Science at the University of Illinois at Urbana-Champaign. A health/urban geographer with a focus on environmental health, mobility and health, and application of advanced geospatial methods in health research, she has received many prestigious research awards and garnered substantial research support from sources including the National Institutes of Health and the National Science Foundation.

Geographies of Health, Disease, and Well-being

Recent Advances in Theory and Method

Edited by
Mei-Po Kwan

Routledge
Taylor & Francis Group

LONDON AND NEW YORK

First published 2014
by Routledge
2 Park Square, Milton Park, Abingdon, Oxon, OX14 4RN

Simultaneously published in the USA and Canada
by Routledge
711 Third Avenue, New York, NY 10017

Routledge is an imprint of the Taylor & Francis Group, an informa business

British Library Cataloguing in Publication Data
A catalogue record for this book is available from the British Library

ISBN13: 978-0-415-87001-6

Typeset in Garamond
by Taylor & Francis Books

Publisher's Note
The publisher accepts responsibility for any inconsistencies that may have arisen during the conversion of this book from journal articles to book chapters, namely the possible inclusion of journal terminology.

Disclaimer
Every effort has been made to contact copyright holders for their permission to reprint material in this book. The publishers would be grateful to hear from any copyright holder who is not here acknowledged and will undertake to rectify any errors or omissions in future editions of this book.

Contents

CONTENTS

CONTENTS

Citation Information

The chapters in this book were originally published in the *Annals of the Association of American Geographers*, volume 102, issue 5 (September 2012). When citing this material, please use the original page numbering for each article, as follows:

Chapter 1
Geographies of Health
Mei-Po Kwan
Annals of the Association of American Geographers, volume 102, issue 5 (September 2012) pp. 891–892

Chapter 2
Ethnic Density and Maternal and Infant Health Inequalities: Bangladeshi Immigrant Women in New York City in the 1990s
Sara McLafferty, Michael Widener, Ranjana Chakrabarti, and Sue Grady
Annals of the Association of American Geographers, volume 102, issue 5 (September 2012) pp. 893–903

Chapter 3
With Reserves: Colonial Geographies and First Nations Health
Sarah de Leeuw, Sean Maurice, Travis Holyk, Margo Greenwood, and Warner Adam
Annals of the Association of American Geographers, volume 102, issue 5 (September 2012) pp. 904–911

Chapter 4
Smoking, Ethnic Residential Segregation, and Ethnic Diversity: A Spatio-temporal Analysis
Graham Moon, Jamie Pearce, and Ross Barnett
Annals of the Association of American Geographers, volume 102, issue 5 (September 2012) pp. 912–921

Chapter 5
Spatial Methods to Study Local Racial Residential Segregation and Infant Health in Detroit, Michigan
Sue Grady and Joe Darden
Annals of the Association of American Geographers, volume 102, issue 5 (September 2012) pp. 922–931

Chapter 6
Connecting the Dots Between Health, Poverty, and Place in Accra, Ghana
John R. Weeks, Arthur Getis, Douglas A. Stow, Allan G. Hill, David Rain, Ryan Engstrom, Justin Stoler, Christopher Lippitt, Marta Jankowska, Anna Carla Lopez-Carr, Lloyd Coulter, and Caetlin Ofiesh
Annals of the Association of American Geographers, volume 102, issue 5 (September 2012) pp. 932–941

Chapter 7
Geospatial Methods for Reducing Uncertainties in Environmental Health Risk Assessment: Challenges and Opportunities
Nina Siu-Ngan Lam
Annals of the Association of American Geographers, volume 102, issue 5 (September 2012) pp. 942–950

Please direct any queries you may have about the citations to clsuk.permissions@cengage.com

Geographies of Health, Disease, and Well-being

Mei-Po Kwan

Department of Geography and Geographic Information Science, University of Illinois at Urbana-Champaign

In the past two decades or so, geographers and researchers in cognate disciplines have significantly advanced our understanding of the geographies of health, disease, and well-being in different areas of the world (e.g., Gatrell and Elliott 2009; Brown, McLafferty and Moon 2010; Meade and Emch 2010; Pearce and Witten 2010). Various theoretical perspectives and geographic methods have been applied to study health issues (e.g., Kearns 1993; Dorn and Laws 1994; Elliott et al. 2000; Andrews and Evans 2008; Cromley and McLafferty 2011). Importantly, health scholars have re-asserted the roles of environment, place, context, and neighborhood as significant influences on health behaviors and outcomes (e.g., Jones and Moon 1993; Diez Roux 1998, 2001; Curtis and Jones 1998; Macintyre, Ellaway, and Cummins 2002; Kawachi and Berkman 2003; McLafferty 2008; Nemeth et al. 2012). It is now widely recognized that geographic variations in health, disease, and well-being cannot be explained exclusively in terms of the characteristics of individuals, as specific characteristics of place or neighborhoods (e.g., collective efficacy) also exert significant influence on health.

The fourth annual special issue of the *Annals of the Association of American Geographers* focuses on the geographies of health, disease, and well-being. The call for abstracts was issued in early 2010, and review of full papers and revised submissions lasted from December 2010 to March 2012. Papers were sought to address social, cultural, political, environmental, theoretical, and methodological issues related to health, disease, and well-being, including topics such as access to health care, spatial disparities in health outcomes, the effect of geographic context on health outcomes, mobility and health, environmental health, development and health, space-time modeling and geographic information systems (GIS)-based analysis of health outcomes. The purpose of the annual special issues is to publish articles that reflect the range of research contributions of geographers and scholars in cognate disciplines to contemporary issues of significant social relevance. They seek to highlight the work of geographers around an important global theme and publish articles covering a wide spectrum of the discipline in a format accessible to a wide range of readers.

The articles in the special issue (and this book) reflect some of the latest theoretical and methodological advances in geographic research on health, disease, and well-being.[1] They span a wide range of topics, theoretical approaches, and methodologies (e.g., anti-racism, post-colonialism, spatial statistics, political ecology, nature-society perspectives, biopolitics, spatial cluster analysis, social network analysis, governmentality, and spatial-temporal modeling). Many of them are highly interdisciplinary, involving researchers from disciplines like epidemiology, medicine, public health, demography, and community studies. Health issues in about fourteen countries are addressed (Bangladesh, Belize, Canada, Cuba, Ghana, India, Kenya, Mexico, Netherlands, New Zealand, South Africa, Taiwan, U.K., U.S.)

The thirty-four main chapters in this book are organized in six sections according to their main themes and the nature of their contribution: (1) health inequalities (five articles); (2) environmental health (seven articles); (3) spatial analysis and modeling of disease (seven articles); (4) health care provision, access, and utilization (seven articles); (5) health and well-being (four articles); and (6) global/transnational health and health issues in the Global South (four articles). There is also a specially commissioned book review by Melinda Meade in addition to the thirty-four main chapters. This organization, however, should be understood largely as a heuristically derived scheme, as many of the chapters address issues that cross-cut the themes of several sections and thus cannot be unproblematically assigned to any one section (e.g., McLafferty et al. (Chapter 2) and Weeks et al. (Chapter 6) examine issues of health disparities in transnational contexts; Messina et al. (Chapter 17) investigate environmental health through dynamic spatial models).

Investigations of *health inequalities* and their social and policy implications have been an important research theme in public health for decades. The five chapters on health inequalities highlight how geographic processes and factors affect the health and well-being of various social groups. Using a GIS-based ethnic density measure (kernel density estimation) and spatial data on mothers' residential locations, McLafferty et al. (Chapter 2) examine infant health inequalities in New York City. The study found that in 2000, Bangladeshi immigrant women living in isolated settings and those in the highest ethnic density areas are vulnerable to poor infant health outcomes. Based on interviews with members of First Nations communities in the northern interior region of British Columbia in Canada, de Leeuw et al. (Chapter 3) found that colonial geographies and anti-Indigenous racism are important determinants of the health and well-being of the Indigenous people who live on Indian reserves in the region.

Using census data of New Zealand geocoded to local and urban area levels and multilevel modeling, Moon, Pearce, and Barnett (Chapter 4) observe that individual ethnic status and area-level deprivation are important factors shaping people's smoking behavior. Grady and Darden (Chapter 5) examine the relationship between local racial residential segregation and infant health in Detroit, Michigan. The study found persistent effects of racial residential segregation and socioeconomic neighborhood inequality on the health of black women and their low birth weight infants when compared to white women. Using satellite imagery and census and survey data, Weeks et al. (Chapter 6) investigate the spatial and health characteristics of cities in Accra, Ghana and observe that local levels of health and well-being are closely related to the abundance of vegetation in a neighborhood.

Research on *environmental health* seeks to assess the influence of physical and social environments on health outcomes or behaviors. Seven chapters address this theme. Lam (Chapter 7) addresses the need to reduce errors and uncertainties in environmental health risk assessment. She identifies four major sources of uncertainties (data, methods of analysis, interpretations of the findings, and reactions to the findings) and discusses five groups of geospatial methods that can help reduce these uncertainties. Focusing particularly on studies that examine the effects of area-based attributes on individual behaviors or outcomes, Kwan (Chapter 9) articulates a fundamental methodological problem that can confound research results in significant ways: the *uncertain geographic context problem* (UGCoP). The chapter clarifies the nature and sources of the problem, highlights some of the inferential errors it may cause, and discusses some means for mitigating the problem. Together these two chapters discuss many important methodological issues pertinent to the study of environmental health.

Guthman (Chapter 8) and Mansfield (Chapter 10) assert the importance of the human body as a lens through which new insights about environmental health can be gained. Through a critical political ecology perspective, Guthman argues that investigations of the relationship between urban form and obesity tend to "black-box the biological body as the site where excess calories are putatively metabolized into fat and made unhealthy." The chapter highlights some anomalies not well explained by current geographic approaches and discusses evidence from emerging biomedical research that questions these approaches to examining the relationship between obesity and health. Informed by the Focauldian perspective of biopolitics, Mansfield interrogates current approaches in public health research on contaminated seafood. She argues that the "dominant approach is one in which risk is used to secure the population by calculating the net benefits and hazards of women's seafood consumption," with the intention to influence women's seafood choices.

Using three case studies as evidence, Scott, Robbins, and Comrie (Chapter 11) illustrate the coevolution dynamics of pathogen ecologies and human institutions. The study shows that mutual influence between humans and pathogens is significantly mediated by the motives and responses of individuals and institutions, and investigations of place-based contextual exposure provide only a partial explanation of disease transmission. Using data on birth defects in North Carolina, Root (Chapter 12) examines the geographic relationship between socioeconomic status (SES) and orofacial clefts (cleft lip and cleft palate) at different spatial scales. As an attempt to explore geographic and statistical methods for mitigating the modifiable areal unit problem (MAUP), the study illustrates how researchers can identify the best neighborhood size for examining such relationship using the Brown–Forsythe test. Groenewegen et al. (Chapter 13) report findings from a research program that examines the relationships between greenspace and health, focusing on three particular mechanisms: physical activity, stress reduction, and social cohesion. The results, as the authors elaborate, indicate that both the quantity and quality of greenspace in residential areas have positive impacts on people's health (more because of stress reduction and social cohesion than physical activity).

The seven chapters on *spatial analysis and modeling of disease* showcase important advances in geographic methods for health research, and several of them focus on the spatiotemporal dynamics of disease transmission. Through a study of two bacterial diarrheal diseases in rural Bangladesh, Emch et al. (Chapter 14) explore how disease transmission may be mediated by kinship-based social networks and lead to variations in disease incidence beyond the effects of the local neighborhood context. The chapter concludes that simultaneous social network and spatial analysis can help us better understand disease transmission. Based on simulation of individuals' vulnerability to influenza in an urban area in the Northeastern U.S., Bian et al. (Chapter 15) highlight the role of individuals in disease transmission and explore a model design for representing mobile, heterogeneous, and interacting individuals and their vulnerability to infectious diseases. With a focus on the effects of routine human movement on dengue transmission, Wen, Lin, and Fang (Chapter 16) examine the role of commuters in the transmission of the virus from their homes to workplaces in Tainan City, Taiwan. The chapter concludes that commuting was a significant risk factor contributing to the geographic diffusion of the epidemic and certain local neighborhood characteristics are independent facilitating factors.

Exploring spatiotemporal patterns of disease incidents can help researchers identify high-risk areas and specific disease risk factors. Through a Kenya case study, Messina et al. (Chapter 17) present a modeling framework for combining a temporally and spatially dynamic species distribution model with a dynamically downscaled regional climate model to predict tsetse populations over space and time. Using residential histories data, Wheeler, Ward, and Waller (Chapter 18) explore spatiotemporal clusters of non-Hodgkin lymphoma in four metropolitan regions in the U.S. The study found that genetic factors and exposure to polychlorinated biphenyls (PCBs) did not fully explain previously detected spatiotemporal clusters of the disease. Based on data from a sample of predominately injection drug users in Baltimore City, Tobin et al. (Chapter 19) investigate the spatial pattern of individuals who report sex exchange and who do not exchange. The results indicate spatial clustering of sex exchangers and in particular identify the high density of sex exchangers in one specific housing complex. The last chapter in this section by Beyer, Tiwari, and Rushton (Chapter 20) describes five important properties of disease maps that will lead to maps that are best for supporting public health uses. It presents an approach to implement these properties and demonstrates it with small-area data from a population-based cancer registry.

Seven chapters examine *health care provision, access, and utilization* from different theoretical and methodological perspectives. Drawing upon interview data from a study of long-term home care in Ontario, England and Dyck (Chapter 21) explore the lived experience of care work by migrant workers through the themes of routes, responsibilities, and respect that emphasize the embodied care work relation. The chapter shows that the experience of care workers can be understood in terms of the complex interplay among labor market inequalities, embodiment, and Ontario's regulatory mechanisms of care provision. Based on a study of the siting of residential social service facility ("group home") in central Massachusetts, Pierce et al. (Chapter 22) examine the role of informal development politics in generating landscapes of mental health provision that is highly uneven in socially and economically depressed areas. Results of the study highlight the need to include the social and political processes of siting in geographic analysis of mental health. Mennis, Stahler, and Baron (Chapter 23) examine how accessibility and neighborhood socioeconomic context influenced treatment continuity for a sample of drug-dependent patients at an inner city hospital in Philadelphia. The study found that a high crime rate in the patient's home neighborhood or the hospital neighborhood and longer travel time to treatment suppress treatment continuity.

Wang (Chapter 24) provides a helpful review of recent methodological advances in the measurement, optimization, and impact of health care accessibility. The chapter suggests that the development of simplified and transparent proxy measures of health care accessibility will be particularly relevant to public health professionals, in light of the increasing complexity of accessibility models. The two chapters that follow focus on nationwide analysis of the utilization and demand for cancer screening services in the U.S. Using Medicare data on insured persons aged sixty-five or older, Mobley et al. (Chapter 25) examine breast cancer and colorectal cancer screening behavior in each state of the U.S. The study observed distinctive geographic patterns and racial differences in the utilization of cancer screening services (e.g., Hispanics in six states are significantly more likely to utilize mammography than whites). Shi et al. (Chapter 26) assess demand for cancer screening facilities using a two-step floating catchment area method that takes into account both travel time and facility capacity. The results show distinctive geographic patterns of demand for cancer screening facilities: spatially continuous but relatively low in eastern regions but sporadic and tends to be high in the west. Using patient registration data, Lewis and Longley (Chapter 27) examine access to primary health care in the London borough of Southwark. The study observed that different ethnic groups have different behavioral patterns in accessing general practitioner-run health centers.

Four chapters explore pertinent issue of *health and well-being* in various contexts through feminist, therapeutic landscape, and nature-society perspectives. They emphasize the importance of local-level social and cultural dynamics that shape health behaviors and outcomes. Based on two case studies, Thien and Del Casino (Chapter 28) demonstrate how various sociospatial practices of masculinity affect men's health and their affective relationships with support systems for health. The article concludes that "men's health is not only about the management of their responsibilities as political citizens but as biological citizens with all the attendant emotional geographies." Pope's chapter (Chapter 29) addresses the recent history of HIV and HIV policy in Cuba and Belize through a therapeutic imaginaries framework. Through these two case studies, it shows that countries in the same region can develop different care policies that lead to different biomedical and sociocultural outcomes. The chapter demonstrates how gender dynamics, geopolitics, economic philosophies, and cultural norms intertwine to create different disease outcomes even for countries in the same region.

Drawing upon insights from feminist geographies of well-being and nature–society geographies of health, Sultana (Chapter 30) examines chronic arsenic poisoning and water contamination in rural Bangladesh in order to highlight the role of the complex interactions among economic, political, geological, and social systems in shaping health and well-being in the context of development. The chapter emphasizes that the experiences of health and well-being are often highly complex, and public health crises like slow poisoning can be an outcome of development endeavors and environmental factors at the same time. In the context of governmental promotion of traditional medicine in managing human health, King (Chapter 31) investigates local-level dynamics that influence perceptions of health, health decision-making, and the use of traditional medicine within rural areas in South Africa. The chapter concludes that future research on health and well-being needs to attend to the social and cultural processes that shape health perceptions and decision-making in important ways.

The last section addresses *global/transnational health and health issues in the Global South*. To advance our critical understanding of global health, Brown, Craddock, and Ingram (Chapter 32) propose a "critical geographical approach that entails reflexivity about the processes by which problems are constituted and addressed as issues of global health." The chapter discusses three analytical approaches that offer complementary insights into these processes: governmentality, risk, and assemblage. It emphasizes that "global health problems and responses are not given but are enabled, imagined, and performed via particular knowledges, rationalities, technologies, affects, and practices across a variety of sites, spaces, and relations." Based on in-depth interviews with key informants from urban health posts, Wadhwa (Chapter 34) examines AIDS awareness and attitudes in the community and HIV/AIDS policy efficacy in four slums in two cities in India. The study revealed a largely reactive governmental response and significant socioeconomic and institutional barriers to timely conceptualization and implementation of HIV/AIDS policies. The chapter shows how the structural violence and grief model frameworks can be fruitfully employed to shed light on the slow policy progression in India.

Using GIS, spatial statistics, interactive mapping, and data about where participants lived, worked, bought drugs, and injected drugs, Brouwer et al. (Chapter 33) investigate the spatial epidemiology of HIV among injection drug users in Tijuana, Mexico. The study identified a 16-km^2 hotspot near the Mexico–U. S. border with distinctive social and structural environmental characteristics (e.g., lower homeownership and higher divorce and female-headed household rates). It found that HIV-positive participants most strongly clustered by injection locations when compared to residence or work place. The chapter concludes that targeting only a small area in Tijuana would have considerable effect on HIV incidence because of the observed high geographic concentration of HIV. Using a representative survey from the Mexican 2000 Census and multilevel models, Riosmena et al. (Chapter 35) examine how migrant flows may transform the health and nutritional profile of people in former communities. The study found that, largely through the mediating effects of remittance intensity, community-level migration intensity had a significant and positive effect on individual risk of being overweight and obese.

Together these thirty-four chapters cogently show how geographic perspectives and methods can contribute in significant ways to advancing our understanding of the complex interactions between social, political, economic, cultural, institutional, and physical environments on the one hand and health behaviors and outcomes on the other. They indicate that geographies of health, disease, and well-being are far too complex to be fully deciphered by any single perspective or any one group of explanatory factors (e.g., individual attributes, environmental features, social relations, institutional processes, and cultural norms). Different theoretical and methodological perspectives can often enrich each other and enhance our understanding better than using just one approach. It would thus be helpful to move beyond the binary thinking that treats analytical medical geography and social/cultural health geography as two separate (and even antagonistic) domains (Kwan 2004). Further, as discussed in the chapters by Lam (Chapter 7) and Kwan (Chapter 9), it is important to note that research findings of health studies can be much less certain or reliable than they appear due to uncertainties and errors arising from various sources. To improve the validity or reliability of research findings, much research on the many sources of spatial and temporal uncertainties and means for mitigating their effects on research results is still needed.

Place-based analysis of health behaviors and outcomes need to attend to a vast array of processes and contexts that interact in a highly complex manner. As most people move around in their daily life and over the life course, they are under the influence of many different places at different times besides their residence (Gatrell 2011; Kwan 2009). Health geographers thus need to move beyond the traditional focus on static locations or places (e.g., residential neighborhood) and extend conventional conceptualizations of geographic context to take into account the effects of people's movement and mobility on health (Kwan 2012a,b, 2013). In order to do this, health researchers and geographers have recently begun to deploy advanced geospatial technologies and various location-aware devices (e.g., GPS and mobile phones) to collect high-resolution space-time data about people's activities and trips (e.g., Wiehe et al. 2008; Shoval et al. 2011; Almanza et al. 2012; Rodríguez et al. 2012; Richardson et al. 2013). For instance, it is only through collecting and using GPS and mobile phone data that researchers can now track the mobility of human or animal agents (hosts) and be able to study the complex interactions between the movement of these agents and the spread of the disease (e.g., Wesolowski et al. 2012). Further, it is only through the use of an integrated personal real-time air pollution monitoring system, which integrates a GPS with a personal real-time environmental monitoring unit, that assessment of people's exposure to air pollution can achieve an accuracy level not possible before (Fang and Lu 2012). To better identify modifiable risk factors and to inform intervention measures, many health studies also integrate and use a variety of spatial data to analyze the complex causal pathways between contextual variables and health outcome/behavior variables. The possibility to develop and use new methods to operationalize more robust conceptualizations of geographic context along this line is indeed promising.

Acknowledgments

I thank all the individuals who have helped to make the publication of the special issue possible. I benefited immensely from the experience of Audrey Kobayashi, Richard Aspinall, and Karl Zimmerer, who edited the three preceding special issues (Kobayashi 2009; Aspinall 2010; Zimmerer 2011). I am grateful to all who submitted abstracts and manuscripts for consideration. I deeply appreciate all the authors for their excellent contributions and their efforts in adhering to a tight timetable and the length restriction we are required to impose. I am grateful to Editorial Board members and over one hundred reviewers who provided constructive and timely comments. I especially would like to thank Antoinette Winklerprins for kindly handling the blind review process and making editorial decisions for my article (including the abstract). Peter Muller graciously provided helpful suggestions and invited Melinda Meade to contribute the excellent book review essay that accompanies the special issue and this book. More important, I cannot imagine what editing the special issue would be like without the enormous help of Managing Editors Robin Maier and Miranda Lecea at the AAG Office. Their competent assistance throughout the entire editorial process had been critical in making the special issue possible.

Note

1. This is a revised and expanded version of the original introduction to the special issue published in 2012.

References

Almanza, E., M. Jerrett, G. Dunton, E. Seto, and M. A. Pentz. 2012. A study of community design, greenness, and physical activity in children using satellite, GPS and accelerometer data. *Health & Place* 18: 46–54.

Andrews, G. J., and J. Evans. 2008. Understanding the reproduction of health care: Towards geographies in health care work. *Progress in Human Geography* 32: 759–80.

Aspinall, R. 2010. Geographical perspectives on climate change. *Annals of the Association of American Geographers* 100 (4): 715–18.

Brown, T., S. McLafferty, and G. Moon, eds. 2010. *A companion to health and medical geography*. Malden, MA: Wiley-Blackwell.

Cromley, E. K., and S. L. McLafferty. 2011. *GIS and public health*, 2nd ed. New York: Guildford.

Curtis, S., and I. R. Jones. 1998. Is there a place for geography in the analysis of health inequality? *Sociology of Health & Illness* 20 (5): 645–72.

Diez-Roux, A. V. 1998. Bringing context back into epidemiology: Variables and fallacies in multilevel analysis. *American Journal of Public Health* 88 (2): 216–22.

Diez-Roux, A. V. 2001. Investigating neighborhood and area effects on health. *American Journal of Public Health* 91 (11): 1783–89.

Dorn, M., and G. Laws. 1994. Social theory, body politics, and medical geography: Extending Kearns' invitation. *The Professional Geographer* 46: 106–10.

Elliott, P., J. C. Wakefield, N. G. Best, and D. J. Briggs. 2000. *Spatial epidemiology: Methods and applications*. Oxford: Oxford University Press.

Fang, T. B., and Y. Lu. 2012. Personal real-time air pollution exposure assessment methods promoted by information technological advances. *Annals of GIS* 18 (4): 279–88.

Gatrell, A. C. 2011. *Mobilities and health*. Aldershot: Ashgate.

Gatrell, A. C., and S. J. Elliott. 2009. *Geographies of health: An introduction*. 2nd ed. Malden, MA: Wiley-Blackwell.

Jones, K., and G. Moon. 1993. Medical geography: Taking space seriously. *Progress in Human Geography* 17 (4): 515–24.

Kawachi, I., and L. F. Berkman, eds. 2003. *Neighborhoods and health*. Oxford: Oxford University Press.

Kearns, R. A. 1993. Place and health: Towards a reformed medical geography. *The Professional Geographer* 45 (2): 139–47.

Kobayashi, A. 2009. Geographies of peace and armed conflict: Introduction. *Annals of the Association of American Geographers* 99 (5): 819–26.

Kwan, M.-P. 2004. Beyond difference: From canonical geography to hybrid geographies. *Annals of the Association of American Geographers* 94 (4): 756–63.

Kwan, M.-P. 2009. From place-based to people-based exposure measures. *Social Science & Medicine* 69 (9): 1311–13.

Kwan, M.-P. 2012a. How GIS can help address the uncertain geographic context problem in social science research. *Annals of GIS* 18 (4): 245–55.

Kwan, M.-P. 2012b. The uncertain geographic context problem. *Annals of the Association of American Geographers* 102 (5): 958–68.

Kwan, M.-P. 2013. Beyond space (as we knew it): Toward temporally integrated geographies of segregation, health, and accessibility. *Annals of the Association of American Geographers* 103 (5).

Macintyre, S., A. Ellaway, and S. Cummins. 2002. Place effects on health: How can we conceptualise, operationalise and measure them? *Social Science & Medicine* 55: 125–39.

McLafferty, S. 2008. Placing substance abuse: Geographical perspectives on substance use and addiction. In *Geography and drug addiction*, eds. Y. F. Thomas, D. Richardson, and I. Cheung, 1–16. New York: Springer.

Meade, M. S., and M. Emch. 2010. *Medical geography*. 3rd ed. New York: Guilford.

Nemeth, J. M., S. T. Liu, E. G. Klein, A. K. Ferketich, M.-P. Kwan, and M. E. Wewers. 2012. Factors influencing smokeless tobacco use in rural Ohio Appalachia. *Journal of Community Health* 37 (6): 1208–17.

Pearce, J., and K. Witten, eds. 2010. Geographies of obesity: Environmental understandings of the obesity epidemic. Farnham: Ashgate.

Richardson, D. B., N. D. Volkow, M.-P. Kwan, R. M. Kaplan, M. F. Goodchild, and R. T. Croyle. 2013. Spatial turn in health research. *Science* 339 (6126): 1390–92.

Rodríguez, D. A., G.-H. Cho, D. R. Evenson, T. L. Conway, D. Cohen, B. Ghosh-Dastidar, J. L. Pickrel, S. Veblen-Mortenson, and L. A. Lytle. 2012. Out and about: Association of the built environment with physical activity behaviors of adolescent females. *Health & Place* 18: 55–62.

Shoval, N., H.-W. Wahl, G. Auslander, M. Isaacson, F. Oswald, T. Edry, R. Landau, and J. Heinik. 2011. Use of the global positioning system to measure the out-of-home mobility of older adults with differing cognitive functioning. *Ageing & Society* 31: 849–869.

Wesolowski, A., N. Eagle, A. J. Tatem, D. L. Smith, A. M. Noor, and R. W. Snow. 2012. Quantifying the impact of human mobility on malaria. *Science* 338: 267–70.

Wiehe, S. E., S. C. Hoch, G. C. Liu, A. E. Carroll, J. S. Wilson, and J. D. Fortenberry. 2008. Adolescent travel patterns: Pilot data indicating distance from home varies by time of day and day of week. *Journal of Adolescent Health* 42: 418–20.

Zimmerer, K. S. 2011. New geographies of energy: Introduction to the special issue. *Annals of the Association of American Geographers* 101 (4): 705–11.

Ethnic Density and Maternal and Infant Health Inequalities: Bangladeshi Immigrant Women in New York City in the 1990s

Sara McLafferty,* Michael Widener,[†] Ranjana Chakrabarti,[‡] and Sue Grady[§]

*Department of Geography, University of Illinois at Urbana–Champaign
[†]Department of Geography, University at Buffalo
[‡]Independent Scholar, New York, New York
[§]Department of Geography, Michigan State University

How do the social and material characteristics of residential contexts in the host country affect immigrant maternal and infant health? We examine this question through the lens of the ethnic density hypothesis, a hypothesis that posits beneficial effects on immigrant health of living in areas of high ethnic density; that is, among a socially and linguistically similar population. We analyze the association between infant low birth weight and ethnic density for Bangladeshi immigrant mothers in New York City during a period of rapid and sustained immigration (1990–2000). For Bangladeshi immigrant women, ethnic neighborhoods can provide an important source of social and material support during pregnancy. Geographic information systems (GIS) and spatial analysis methods are used to create a fine-grained indicator of ethnic density. Results show that the relationship between ethnic density and infant low birth weight changed over time. The lack of association in the early years (1990 and 1993) might reflect the fact that the Bangladeshi population had not yet reached a sufficient size, or spatially clustered settlement pattern, to provide dense ethnic neighborhoods and concentrations of social and material resources. In 2000, we observe a U-shaped association between low birth weight and density: Women living in ethnically isolated settings and those living in high-density enclaves are more vulnerable to adverse infant health outcomes. The results suggest the need for a more nuanced understanding of immigrant maternal and infant health and ethnic density that incorporates the dynamism of immigrant experiences and their associations with shifting spatially and socially defined residential environments.

东道国居住环境的社会和物质特性怎样影响移民孕产妇和婴儿的健康？我们通过种族密度的假设来研究这个问题，我们假定生活在族群密度高的地区，即生活在社会和语言类似的人群之间，对移民的健康存在有益的影响。我们分析一个快速和持续的移民时期（1990 至 2000 年），在纽约市的孟加拉移民母亲所生婴儿的低出生体重和种族密度之间的关联。对于孟加拉国的移民妇女，民族聚居区是其怀孕期间的社会和物质支持的一个重要来源。使用地理信息系统（GIS）和空间分析方法创建一个种族密度的细粒度指标。结果表明，民族的密度和婴儿低出生体重之间的关系随着时间的推移而改变。相关性在早年（1990 和 1993）的缺乏，可能反映了孟加拉人口尚未达到足够的规模，或空间聚集的聚落形态不足以提供密集的民族聚居区，社会，以及物质资源的强度。我们观察到在 2000 年，低出生体重和密度之间的 U 形关联：生活在种族分离环境和那些生活在高密度飞地里的妇女更容易受不利于婴儿健康的因素影响。结果表明，我们需要更细致入微地了解移民孕产妇和婴儿的健康和种族密度，综合考虑移民经验的动态以及空间转移和社会性居住环境的相关性。*关键词：孟加拉国，族群密度，移民，低出生体重，纽约市。*

¿Cómo afectan las características sociales y materiales del contexto residencial en el país anfitrión la salud de madre e hijo inmigrantes? Examinamos esta cuestión a través de la lente de la hipótesis de densidad étnica, una hipótesis que propone efectos benéficos para la salud de los inmigrantes que viven en áreas de alta densidad étnica; esto es, que viven entre una población social y lingüísticamente similar. Analizamos la asociación existente entre el bajo peso de los infantes al nacer con la densidad étnica para madres inmigrantes de origen bangladeshí en la Ciudad de Nueva York, durante un período de inmigración rápida y sostenida (1990–2000). Para las mujeres inmigrantes bangladeshí, los vecindarios étnicos pueden representar una fuente importante de apoyo social y material durante la preñez. Se utilizaron métodos de sistemas de información geográfica (SIG) y de análisis espacial para crear un indicador de la densidad étnica de grano fino. Los resultados muestran que la relación entre la densidad étnica y el bajo peso de los infantes al nacer cambió a través del tiempo. La falta de asociación en los años iniciales (1990 y 1993) podría reflejar el hecho de que la población bangladeshí todavía no había alcanzado un tamaño suficiente, o un patrón de asentamiento espacialmente aglomerado, para proveer vecindarios étnicos densos y concentraciones de recursos sociales y materiales. En el 2000 observamos una asociación en forma de U entre bajo peso al nacer y la densidad: las mujeres que vivían en conjuntos étnicamente aislados y la que residían en enclaves de alta densidad eran más vulnerables a tener que enfrentar situaciones adversas de salud en sus infantes. Los resultados sugieren la necesidad de un entendimiento más matizado de la salud materna e infantil de los inmigrantes, y de la densidad étnica, que incorpore el dinamismo de las experiencias del inmigrante y sus asociaciones con los cambiantes entornos residenciales espacial y socialmente definidos.

I mmigrant health is a critically important and relatively neglected concern in the United States. Although the past three decades saw the highest levels of immigration to the United States since the early 1900s (Monger and Yankay 2011), little is known about socioeconomic and environmental influences on immigrant health and how those play out in particular place contexts. Immigrant health has strong transnational dimensions, reflecting migrant experiences in the home and host countries, along with the physical and psychosocial stresses associated with migration and resettlement (Kerner et al. 2001). Researchers theorize that immigrant health in the host country is influenced in part by "ethnic density"—the localized concentration of an immigrant's own ethnic group in her or his residential neighborhood. Ethnic density, it is argued, provides opportunities for social interaction and access to culturally appropriate resources and services that are beneficial for physical and psychosocial well-being. Ethnic density, however, is a dynamic construct that evolves over time as immigrant settlement patterns change and unfold.

For Bangladeshi immigrants in New York City, we analyze changes in the association between immigrant infant and maternal health and ethnic density during a period of rapid and sustained immigration (1990–2000). During the 1990s, the Bangladeshi population in the city grew by over 400 percent, from 4,955 to 28,269, primarily as a result of immigration (New York City Department of City Planning 2004). Rapid growth led to changes in settlement, including the emergence of relatively dense Bangladeshi neighborhoods. We investigate the links between these geographic changes in immigrant settlement and reproductive health inequalities for Bangladeshi immigrant women. To gauge reproductive health, we use infant low birth weight (LBW; <2,500 g), a widely used indicator of the health and well-being of both infants and mothers. Geographic information systems (GIS) and spatial analysis methods are used to create a fine-grained indicator of ethnic density. The results suggest the need for a more nuanced understanding of immigrant health and ethnic density that incorporates the dynamism of immigrant experiences and their associations with shifting spatially and socially defined residential place environments.

Immigrant Health Inequalities and Ethnic Density

Wide inequalities in health among population groups are a persistent concern in the United States. Research on health inequalities typically emphasizes class and racial divisions and neglects the impact of immigration. In health statistics, immigrants are often hidden from view, grouped with nonimmigrants into broad ethnic categories. The "healthy immigrant effect"—the fact that many immigrants are healthier than their counterparts in their home country and U.S.-born citizens—also detracts attention from immigrant health concerns. Although research has confirmed the healthy immigrant effect for a range of health conditions and immigrant populations (Singh and Yu 1996; Baker and Hellerstedt 2006), there are also wide health disparities within and among immigrant groups.

Following contemporary relational perspectives in health geography (Cummins et al. 2008), we can conceptualize immigrant health as reflecting experiences and conditions in the country of origin, the host country, and the process of migration. Regarding experiences in the host country, studies have explored the roles of contextual factors that describe an immigrant's living environment (Lorant, Van Oyen, and Thomas 2008). Place context affects immigrants' everyday negotiations, environmental exposures, and access to health-related resources—all of which affect health and well-being. It influences access to social and economic resources including employment, education, and social and health care services (Smaje 1995; Wang 2007; Thomas 2010) and exposure to stressors in the local environment including crime, traffic, and environmental hazards (Elliott et. al. 2004; De Jesus et al. 2010). At the same time, place context is closely related to social and cultural resources and networks. Immigrants often have strong place-based social ties based on ethnicity and country of origin (Dyck 1995). Dyck (2006) described the "sedimentation of place" for immigrant women through daily routines and activities that revolve around culturally familiar institutions, shops and services, and strong local social networks. Families and communities provide essential social resources for maintaining health in the host country (Cervantes, Keith, and Wyshak 1999). These social and spatial relations, rooted in place, intersect in complex ways to affect immigrant health.

The *ethnic density hypothesis*, first described in the mental health literature, posits beneficial effects on health of living among a socially and linguistically similar population, especially for immigrant and ethnic minority populations. Health improvements for those living in areas of high ethnic density are attributed to reduced exposure to prejudice and increased social support (Halpern and Nazroo 1999; Becares, Nazroo, and Stafford 2009). Ethnic density could foster high levels of social capital that develop through place-based institutions and social networks that provide resources and support during times of stress (Fullilove 1988; Sampson, Morenoff, and Earls 1999). Studies have found improvements in physical health through increased protection from the consequences of stigmatization and racism (Pickett and Wilkinson 2008), enhanced social cohesion (Kawachi 1999), mutual social support (Berkman and Glass 2000), a greater sense of community and belonging (Smaje 1995), and reduced levels of stress (Wilkinson 2005). Studies of specific health outcomes have shown an inverse relationship between ethnic density and self-reported ill health status (Smaje 1995), diet (Osypuk et al. 2009), life satisfaction (Lackland 1998), and heart disease (Franzini and Spears 2003). The protective mechanisms reported in these studies included a positive ethnic identity, civic participation, help from social networks, and positive coping behaviors.

At the same time, other research presents more mixed evidence about the associations between ethnic density and health. Pickett et al. (2009) utilized data from the Millennium Cohort Study in the United Kingdom on infant outcomes among various ethnic groups. This study found that Pakistani and Bangladeshi

mothers living in neighborhoods with 5 to 30 percent same-ethnic density were significantly less likely to experience limited long-standing illness. Indian and Pakistani mothers were also less likely to report "ever being depressed" if they lived in dense ethnic neighborhoods. Other health outcomes did not vary with ethnic density, however. Analyzing the relationship between premature birth and ethnic density for broad racial and ethnic groups in New York City, Mason et al. (2010) found that high ethnic density was protective for whites but detrimental for blacks—a finding observed in studies of racial segregation and health (Grady 2006). Using county-scale data for the United States, Shaw, Pickett, and Wilkinson (2010) reported reduced odds of infant mortality among Hispanic mothers living in areas of high group density. Similarly, a study focusing on infant LBW for ethnic minority populations in Chicago found that ethnic density was protective for Hispanics and whites (Masi, Hawkley, Piotrowski, and Pickett 2007), but the study did not assess the effects of immigration. In general, findings are difficult to compare because of the diversity of health indicators, geographic settings, immigrant race and ethnic populations, and definitions of ethnic density used in testing for density effects.

Although elegant in its simplicity, the ethnic density hypothesis raises several important questions. It is ambiguous about the geographical scale at which density effects might be observed. There might be complex threshold effects—critical density levels needed to sustain social interactions in a particular place. Two recent studies indicate a nonlinear association between density and health (Neeleman, Wilson-Jones, and Wesseley 2001; Fagg et al. 2006). Furthermore, a large literature suggests that living in ethnically or racially segregated communities could be detrimental to health (Williams and Collins 2001; Subramanian, Acevedo-Garcia, and Osypuk 2005; White and Borrell 2011). For immigrants, this can occur if immigrants cluster in economically and environmentally disadvantaged communities or if segregation is associated with feelings of stress, subordination, and disempowerment caused by racial or ethnic discrimination (Osypuk and Acevedo-Garcia 2010). Living in ethnic enclave areas also inhibits social interactions with other groups, limiting development of "bridging" social capital; that is, social capital derived from interactions among dissimilar persons (Kim, Subramanian, and Kawachi 2006). Given the complexity of processes linking ethnic density and health, the association (if any) is likely to change over time as immigrant settlement patterns evolve and as social interactions within and among immigrant and nonimmigrant populations develop and change.

We explore the ethnic density hypothesis for Bangladeshi immigrant women and their infants during the 1990s, a period of very rapid Bangladeshi immigration to New York City. The 1986 Immigration Reform and Control Act, coupled with the diversity visa program aimed at encouraging immigration from underrepresented countries, fueled Bangladeshi immigration to the United States, making Bangladesh one of the top twenty source countries of immigrants in the 1990s (New York City Department of City Planning 2004). Many of these immigrants sought employment in low-income sectors of the economy (e.g., waiters, bartenders, vendors, taxi drivers). This wave of immigration also gave rise to several dense Bangladeshi neighborhoods with visible social, cultural, and economic spaces. A high poverty rate, large household sizes, and unstable employment made Bangladeshis vulnerable to adverse socioeconomic circumstances. Bangladeshi immigrants to New York City during the early 1990s were also very young (with a median age of twenty-three years) and predominately male, with a ratio of 143 males per 100 females (New York City Department of City Planning 1996).

Bangladeshi women immigrants in New York City came primarily as spouses with limited English proficiency and were relatively young and with low income. They had significant needs for health care services, especially reproductive health care. Qualitative research shows that the everyday lives of Bangladeshi women residing in the city are quite proscribed spatially and culturally (Chakrabarti 2010). Being primarily housewives, women shoulder the major burden of household and child care responsibilities and are tied to neighborhood spaces. Some work in part-time, low-paying jobs near home to supplement family income. Women's embeddedness in neighborhood spaces and their interactions with other Bangladeshi women and families characterize their daily sociospatial settings (Chakrabarti 2010), as they do in countries like the United Kingdom (Phillipson, Ahmed, and Latimer 2003). Moreover, Bangladeshi immigrant women in New York City are not homogeneous. Recent low-income immigrants whose everyday socio-spatial spaces are constrained by strict gender roles and a limited knowledge of English are particularly vulnerable (Chakrabarti 2010). These qualitative studies suggest that spaces of high ethnic density present a mix of opportunities and challenges for Bangladeshi immigrant women, with potentially complex implications for health disparities.

Data and Methods

Data for this project come from vital statistics birth records for New York City for 1990, 1993, 1996, and 2000. For each birth, the data report characteristics of the mother such as age, birthplace, and education, along with characteristics of the infant such as birth weight, parity, and gender. Representing a near-complete registry of births among New York City residents, these vital statistics data have been widely used in analyzing maternal and infant health outcomes (Grady 2006; Mason et al. 2010). Despite limitations such as missing values, the data are a rich source of information about changes in health inequalities through space and time. We identified Bangladeshi immigrant mothers by country of birth, and the mothers' locations were geocoded to the residential census tract at the time of birth. Multiple births (i.e., twins, triplets) were excluded because these infants are highly likely to be small (low weight) at birth. The numbers of births ranged from 318 in 1990 to 1,373 in 2000.

The health outcome of interest is LBW, a dichotomous variable indicating whether an infant weighed less than 2,500 g at birth. LBW is a widely used measure of maternal, infant, and community health. It is the most important risk factor for infant mortality and has been linked to a host of physical and developmental problems in infancy and childhood. It is also a key indicator of maternal health, reflecting stress, nutrition, risk behaviors, and access to health care during pregnancy. Independent variables describe maternal characteristics including marital status, age, education, and insurance coverage. Maternal characteristics were defined as dichotomous variables including marital status (MARRIED; 1 = married), high school education (HIGHSCH, 1 = completed twelve years of education), and enrollment in the Medicaid health insurance program (MEDICAID).

In estimating ethnic density, we use the birth data rather than data on total Bangladeshi population. Census counts of total Bangladeshi population were not available for either of the intercensal years (1993, 1996) or for 1990, so we used data on mothers' residential locations to approximate the spatial distribution of Bangladeshi population. In this sense, our measure describes a mother's geographical proximity to other Bangladeshi mothers during the year of birth.

Kernel density estimation (KDE), calculated within a circular window using a quartic kernel function, was used to characterize the uneven density of mothers (per square mile) across the city. Several recent studies employ KDE in computing measures of residential segregation and ethnic concentration (O'Sullivan and Wong 2007; Mason et al. 2010). The resulting KDE output was then used to assign each mother a density value corresponding to her residential location. A critical parameter when computing a KDE is the search bandwidth (radius). We used two different bandwidths: 1.0 and 1.5 miles. The 1.5-mile bandwidth produced a smoother, less spiky map than the 1.0-mile bandwidth and appeared to better represent local neighborhood-scale variation in ethnic density. We only discuss the 1.5-mile results here, although findings were very consistent for both bandwidth values.

This ethnic density measure differs from those used in previous studies in that it focuses only on mothers and not the entire immigrant population. It reflects our assumption that other mothers who give birth in the same year provide an important source of social support during pregnancy, which is supported by qualitative research on Bangladeshi mothers' access to prenatal care (Chakrabarti 2010). Our measure is limited, however, insofar as it omits a large fraction of the Bangladeshi population that also offers social and material support. To check the correspondence between our density measure and one based on total population, we obtained data from the 2000 Census on Bangladeshi population by census tract and computed ethnic density via the same kernel estimation procedures. The two density measures were very highly correlated ($r = 0.988$), indicating that our measure does well in capturing geographic variation in total Bangladeshi population.

Also important is the fact that our ethnic density measure emphasizes absolute density rather than a group's share of the local population. Given the small size of the Bangladeshi population, even in the neighborhoods of highest density, this community represents only a tiny fraction of the total population. Focusing on absolute density highlights the presence of mothers from the same ethnic background in a woman's residential neighborhood, representing the potential for ethnic group–based social interactions and networks, irrespective of the group's share of local population. In contrast, relative density measures, used in many studies of the ethnic density hypothesis (Becares, Nazroo, and Stafford 2009; Mason et al. 2010), describe a group's segregation in an area to the exclusion of other population groups.

To assess whether this absolute measure of ethnic density might be confounded by overall population density, we included population density as an additional independent variable. Decennial census data were used for computing population density values in 1990 and 2000; for 1993 and 1996, values were estimated via linear interpolation. For 2000, the year with the largest sample size, additional analyses were performed to determine the influence of the following contextual variables that describe social and spatial characteristics of the residential census tract: percentage of households with incomes below poverty, median household income, network distance to the closest publicly funded prenatal care center, and the prevailing smoothed tract LBW percentage, excluding births to Bangladeshi mothers. These contextual factors might underpin any observed associations between ethnic density and LBW risk.

Results

Reflecting the rapid growth of Bangladeshi population in New York City, the total number of Bangladeshi mothers in our sample increased by more than 400 percent from 1990 to 2000 (Table 1). Data from the birth certificates present a shifting portrait of Bangladeshi infants and mothers throughout the decade. The percentage of LBW incidents ranges from 7.9 in 1990 to

Table 1. Social and characteristics of Bangladeshi immigrant mothers and births

Year	Number of births	% LBW (n)	% ≥12 years education	% ≥16 years education	% married	% Medicaid	% self pay
			Characteristics of mothers and infants				
1990	318	7.9 (25)	59.8	21.7	78.6	53.8	13.8
1993	592	11.0 (65)	72.1	19.6	68.8	61.0	21,6
1996	726	8.8 (64)	77.2	24.2	55.0	75.5	8.0
2000	1,373	8.5 (117)	72.9	22.9	78.5	79.1	3.1

Note: LBW = low birth weight.

11.0 in 1993. In all years except 1990, the LBW percentage for Bangladeshi mothers exceeded the city-wide value by more than 1.5 percentage points. Bangladeshi women's greater risk of delivering an LBW infant has been observed in other studies (Kelly et al. 2008; Stein et al. 2009), and it is consistent with the high risk seen among women in Bangladesh. Because the vast majority of women in our data set are likely to be recent immigrants, their elevated risk of infant LBW might in part reflect risks associated with the home country; still, the consistently elevated values show that LBW is an important health issue for this vulnerable population.

The mothers' educational characteristics remained relatively stable during the 1990s: Approximately three fourths of mothers reported having a high school education and 20 percent had a college education. The percentage of mothers with Medicaid insurance coverage rose from 53.8 percent in 1990 to 79.1 percent in 2000, reflecting the expansion of Medicaid eligibility in the 1990s (Zavodny and Bitler 2010). At the same time, the proportion of uninsured, "self-pay" mothers declined to just 3 percent in 2000. This indicates Bangladeshi mothers' expanded access to government-funded medical insurance coverage during the study period.

The data on marital status present a more complex picture. The percentage of Bangladeshi mothers listed as married fluctuated greatly from year to year, ranging from 55.0 percent in 1996 to 78.6 percent in 1990. Both the low percentages and fluctuation over time raise concerns about the accuracy of reporting nonmarital status. Unlike most states, New York State does not permit the asking of a direct question to determine marital status for the birth certificate. Instead, marital status is inferred. Births are considered nonmarital if a paternity acknowledgement was filed or the father's name was missing from the birth certificate (New York City Department of Health and Mental Hygiene 2010). It is possible that due to language and cultural barriers, some Bangladeshi immigrant mothers might not have written the father's name on the birth certificate, even if married. During the early and mid-1990s, few hospitals had Bengali language translation services, placing immigrants who did not speak English in a vulnerable position. Thus, the unmarried group could include some married women who did not provide the necessary information to be recorded as "married." In 2008, only 11.3 percent of Bangladeshi mothers were identified as unmarried—a value that is probably closer to the true value for the years we studied (New York City Department of Health and Mental Hygiene 2008).

Table 2. Spatial characteristics of Bangladeshi immigrant births

| | Geographic characteristics | | | |
Year	No. tracts with Bangladeshi birth	Average density of Bangladeshi births	Moran's I	% mothers in enclave
1990	188	7.4	0.301	36.5
1993	272	14.7	0.318	40.4
1996	303	18.1	0.323	41.8
2000	416	42.4	0.537	44.7

During the 1990s, rapid immigration fueled changes in the spatial distribution of Bangladeshi mothers. The number of census tracts reporting a birth to a Bangladeshi mother rose from 188 to 416 (out of approximately 2,216 tracts) between 1990 and 2000, and the average ethnic density score among mothers also increased (Table 2). In addition, Bangladeshi births became more spatially clustered: The Moran's I value for births by tract increased from 0.301 to 0.537. Maps of the ethnic density measure for 1990 and 2000 (Figure 1) show the highly uneven geographic distribution of Bangladeshi mothers and the formation of a large enclave area in northern Queens, in the Astoria and Jackson Heights neighborhoods. The 1990s saw marked increases in density in this enclave area (south-central Astoria and western Jackson Heights), as the percentage of mothers residing in the enclave rose from 36.5 to 44.7 (Table 2). By 2000, several smaller, secondary enclaves had emerged in central Brooklyn and southern Queens. The end result is a strongly nucleated settlement pattern, a pattern typical of Bangladeshi populations in other cities in North America (Ghosh 2007).

To illustrate the association, if any, between ethnic density and LBW, we plotted smoothed LBW as a function of ethnic density for each year, using a local polynomial smoothing algorithm in STATA (Figure 2). The bandwidth ranged from five in 1990 to ten in 2000, to take into account the increasing range of density values. Note that the range of density values is quite small in 1990 and much larger in 2000, reflecting overall population growth. For 1990, no trend is evident between LBW and ethnic density. This makes sense because the small Bangladeshi population size meant that there were no dense concentrations of settlement. The 1993 plot shows an inverse gradient in LBW with increasing density, an indication that LBW outcomes were somewhat better for women living in higher

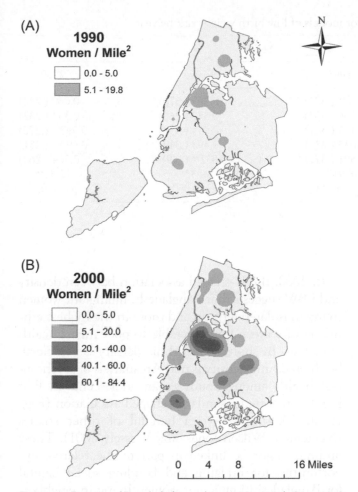

Figure 1. Density of Bangladeshi mothers in New York City, 1990 and 2000. (Color figure available online.)

density neighborhoods. The observed trend might not be statistically significant, however, given the relatively small number of births in that year. The 1996 plot shows a slight U-shaped trend with the highest LBW percentage in high-density areas. A more striking U-shaped

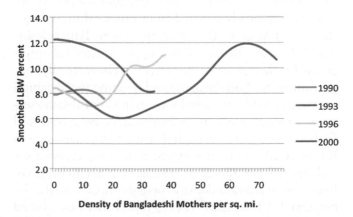

Figure 2. Smoothed low birth weight (LBW) percentage by ethnic density, 1990, 1993, 1996, and 2000. (Color figure available online.)

pattern emerges in 2000, with the lowest rate of infant LBW among women living in neighborhoods with moderate levels of ethnic density (10–40 per square mile): Mothers living in both isolated settings and areas of high Bangladeshi population concentration have a higher risk of infant LBW.

To determine whether the trends observed in these graphs are statistically significant after controlling for maternal risk factors, we estimated a series of logistic regression models for each year. Variables representing the mother's age and health insurance coverage were found to be insignificant in all models and had minimal effect on parameter estimates for ethnic density; therefore, they are excluded from the results presented here. We represented ethnic density by using linear and quadratic terms and by defining binary variables representing high and low density. For each year, low density was defined as fewer than ten mothers per square mile; the cutoff value for high density was thirty in 1993 and 1996 and forty in 2000. High density was omitted for 1990 due to the lack of high-density Bangladeshi areas at that time. For ease of interpretation, we only present the models incorporating the low- and high-ethnic-density designations.

For each of the first three years, the models indicate no statistically significant association between ethnic density and LBW (Table 3), suggesting that ethnic density has little to no relationship with LBW during the first three years. This might be due to small sample sizes that result in insufficient statistical power. It also could be related to a threshold effect. In the early years of immigration, geographic concentrations of Bangladeshis had not reached a sufficient population density threshold for health effects to develop and emerge. In fact, the magnitude of the parameter estimate for the high-density variable becomes large in 1996, pointing to an emerging association between high ethnic density and LBW as Bangladeshi settlement expanded. Associations between other independent variables and LBW also were generally weak during the first three time periods. The consistent inverse association with population density suggests that, independent of ethnic density, living in densely populated urban neighborhoods might be beneficial for Bangladeshi mothers and infants in New York City.

In contrast, the 2000 model shows significant associations between density and LBW (Table 3). The results confirm the U-shaped pattern apparent in Figure 2 in which risk is elevated among infants born to mothers living in low-ethnic-density neighborhoods, as well as among those living in high-density neighborhoods.

Table 3. Logistic regression coefficients for models of low birth weight risk by year

	Year			
	1990	1993	1996	2000
High school	−0.779 (.486)	−0.537* (.303)	0.367 (.370)	−0.060 (.224)
Married	−0.357 (.477)	0.166 (.302)	−0.352 (.266)	−0.334 (.232)
Density<10	0.106 (.525)	0.135 (.311)	0.054 (.328)	0.597** (.270)
High density		0.035 (.435)	0.495 (.339)	0.690** (.251)
Population density (ln)	−0.937 (.512)	−0.501 (.371)	−0.832** (349)	−0.304 (.262)
	4.085	6.915	9.832*	12.82**
Model chi-square	$p = 0.394$	$p = 0.227$	$p = 0.080$	$p = 0.028$

Note: Standard errors in parentheses.
*Statistically significant, $p < 0.10$.
**Statistically significant, $p < 0.05$.

Living in areas of moderate density confers protective effects with respect to LBW risk. Compared to women living in areas of moderate Bangladeshi ethnic density, similar women in low-density areas have an 82 percent higher risk of delivering a low-weight infant, and those in high-density areas have almost double the odds (Table 4).

To check whether these ethnic density associations are an outcome of well-established contextual factors such as poverty and income levels in the local population, we estimated a series of multilevel models that incorporated the four contextual variables mentioned earlier: poverty, income, distance to prenatal care, and prevailing LBW percentage. The variables were included as level-2 fixed effects. None of these variables, either singly or in combination, yielded a statistically significant improvement in model fit or had a statistically significant coefficient. Moreover, including the variables resulted in little change (<10 percent) in the estimates of the LBW–ethnic density relationship. This indicates that economic disadvantage and other measured characteristics of the contexts in which Bangladeshi women live do not account for the observed association between ethnic density and health.

Table 4. Low birth weight (LBW) by ethnic density category in 2000: LBW percentages and odds, adjusted for marital status, population density, and education

Ethnic density	LBW %	Number of births	Adjusted odds of LBW (95% confidence interval)
<10	10.78	269	1.82 (1.22, 3.26)
10–40	6.85	567	1.00
>40	10.24	537	1.99 (1.00, 3.08)

In 2000, the U-shaped association between density and LBW suggests that Bangladeshi immigrant women living in isolated settings and those in the highest ethnic density areas are vulnerable to poor infant health outcomes. In short, some ethnic density is beneficial, but high ethnic density is not. An analysis of psychosocial health among South Asian adolescents in East London reported a similar U-shaped association (Fagg et al. 2006), as have a handful of other studies (Neeleman, Wilson-Jones, and Wessely 2001). These findings might be linked in part to the relative opportunities for bridging and bonding social capital for Bangladeshi immigrant women living in neighborhoods with varying levels of ethnic density. With few Bangladeshi neighbors, women in low-density settings might experience a lack of opportunities for bonding social capital—an outcome of social and spatial isolation from one's own ethnic group. In contrast, those living in areas of high Bangladeshi density might experience low levels of bridging social capital, leading to a different form of isolation. Although in the high-density areas studied here Bangladeshi people are not geographically isolated from other ethnic groups, strong within-group (bonding) ties among Bangladeshis could decrease the need and opportunities for intergroup (bridging) social interaction. Theoretically, opportunities for both bridging and bonding social capital are maximized for women living in medium-density settings. Protective effects on health have been observed for both types of social capital (Kim, Subramanian, and Kawachi 2006). To the extent that these types of social capital are rooted in residential neighborhoods, living in moderate-density neighborhoods might be beneficial. Chakrabarti (2010) found that place-based social interactions with Bangladeshi mothers provided an important source of social and emotional support for

Bangladeshi women during pregnancy; however, bridging social capital was not evaluated. Direct measurement of social capital across a range of neighborhood contexts is needed to assess the validity of these hypothesized links among ethnic density, bridging and bonding social capital, and maternal and infant health.

Our results indicate that living in areas of high ethnic concentration is associated with poor LBW outcomes—a finding that contradicts the ethnic density hypothesis. Data on birth outcomes for Bangladeshi mothers in London showed a similar elevated risk of LBW in high-concentration areas; however, the risk was not statistically significant (Pickett et al. 2009). In addition to the limited development of bridging social capital, other contextual factors might influence the ties between living in areas of high ethnic concentration and adverse infant health outcomes. Our findings show no influence of absolute economic deprivation at the census tract level, but relative deprivation could come into play, especially given the very high rates of poverty and low income among New York City's Bangladeshi immigrants. Living conditions also warrant attention. There is evidence of necessity-based overcrowding among Bangladeshis, as extended families share small living spaces, and even take on renters, to economize on rental payments (Asian American Federation 2005). Living with extended family members in crowded housing units could result in psychosocial stress, which heightens the risk of infant LBW (Chakrabarti 2010). These crowded conditions might be more prevalent in high-density enclave areas, given the areas' sociocultural attractions for Bangladeshi settlement. Additionally, selective migration might be relevant in understanding the elevated risk of infant LBW for mothers living in high-density areas (Oakes 2004). Over time, such areas might attract recent immigrants or other vulnerable Bangladeshi subpopulations. For example, the birth data show an increasing share of unmarried mothers living in high-density areas. Lacking data on length of residence and other sociodemographic characteristics of the Bangladeshi population, we are unable to assess the role of selective migration; however, it is an important topic for future investigation.

This study points to the adverse health effects for Bangladeshi mothers of ethnic isolation. Such a finding is consistent with the ethnic density hypothesis, but it emphasizes isolation as the primary factor influencing immigrant health, rather than high density. The health vulnerability of immigrant women living in isolated settings has important policy implications. Interventions to promote immigrant women's health, such as culturally appropriate clinics and health care providers, are often spatially targeted to immigrant enclave areas to maximize geographical access. Given how geographic barriers intersect with cultural barriers in shaping immigrants' access to health care (Wang 2007), however, such policies disadvantage immigrants living in ethnically isolated settings.

Conclusions

Understanding how the social characteristics of residential spaces in the host country affect immigrant health is an important and relatively neglected topic in health geography. This research suggests that both the health impacts and our ability to detect them are deeply bound up with the changing geographies of immigrant settlement. Although limited to a single health outcome, in a single immigrant population, our findings highlight an important temporal dimension in the development of an association between ethnic density and health. For Bangladeshi mothers in New York City during a period of rapid immigration, an association only became apparent after the Bangladeshi population had reached a sufficient size and a geographically concentrated settlement pattern had emerged. This makes sense: After all, ethnic density can only influence health if the group population size is large enough for density to exist and the settlement pattern is geographically uneven, including some dense concentrations. Links between settlement and ethnic density have been mentioned in recent empirical studies, but none have addressed the temporal dimension (Pickett et al. 2009). We acknowledge that our findings for the earlier years might simply reflect a lack of statistical power due to small numbers of births; however, this is also directly tied to immigrant settlement.

Results for 2000 show a nonlinear, U-shaped association between ethnic density and LBW: Women living in ethnically isolated settings and those living in high-density enclaves are more vulnerable to adverse LBW outcomes. Such outcomes are driven by a combination of effects in which both residential settings are linked to increased vulnerability. These findings support recent studies that show complex and heterogeneous associations between density and health (Subramanian, Acevedo-Garcia, and Osypuk 2005; Osypuk et al. 2009; Mason et al. 2010). Research on ethnic density needs to consider and explicitly model such nonlinear effects. Analyzing whether the U-shaped association persists in the future, or whether continued Bangladeshi immigration and changing settlement patterns alter its form and significance, is an important topic for future investigation. Moreover, research on other immigrant populations, with varying settlement histories and in diverse geographic contexts, would show whether the U-shaped association applies more broadly.

This research also highlights the benefits of linking GIS, spatial analysis methods, and vital statistics data in studying immigrant health in the United States. The dynamism of immigrant populations and their rapidly changing settlement patterns mean that decennial census data are often out of date, and ethnic group data from the American Community Survey are not released for small levels of geography. Our findings suggest that vital statistics information could be used to approximate the residential changing geographies of ethnic populations. Although quantitative assessments of contextual effects on immigrant maternal and infant health provide an important starting point, the many unanswered questions that arise from our research point to the pressing need for qualitative investigations. Bangladeshi immigrant women's experiences of space, access to social and economic resources, and place-based interactions are critically important in shaping their health vulnerabilities in different geographic settings.

References

Asian American Federation. 2005. Census-based profile depicts poverty, language barriers and other hurdles amid rapid growth for New York City's Bangladeshi American population. http://www.afny.org/press/pressrelease.asp?prid=17&y=2005 (last accessed 10 April 2012).

Baker, A. N., and W. L Hellerstedt. 2006. Residential racial concentration and birth outcomes by nativity: Do neighborhoods matter? *Journal of the National Medical Association* 98 (2): 172–80.

Becares, L., J. Nazroo, and M. Stafford. 2009. The buffering effects of ethnic density on experienced racism and health. *Health and Place* 15 (3): 700–08.

Berkman, L., and T. Glass. 2000. Social integration, social networks, social support, and health. In *Social epidemiology*, ed. L. Berkman and I. Kawachi, 137–73. New York: Oxford University Press.

Cervantes, A., L. Keith, and G. Wyshak. 1999. Adverse birth outcomes among native-born and immigrant women: Replicating national evidence regarding Mexicans at the local level. *Maternal and Child Health Journal* 3 (2): 99–109.

Chakrabarti, R. 2010. Therapeutic networks of pregnancy care: Bengali immigrant women in New York City. *Social Science and Medicine* 71 (2): 362–69.

Cummins, S., S. Curtis, A. Diez-Roux, and S. Macintyre. 2008. Understanding and representing "place" in health research: A relational approach. *Social Science and Medicine* 65:1825–38.

De Jesus, M., E. Puleo, R. C. Shelton, and K. M. Emmons. 2010. Associations between perceived social environment and neighborhood safety: Health implications. *Health and Place* 16 (5): 1007–13.

Dyck, I. 1995. Putting chronic illness "in place": Women immigrants' accounts of their health care. *Geoforum* 26 (3): 247–60.

———. 2006. Travelling tales and migratory meanings: South Asian migrant women talk of place, health and healing. *Social & Cultural Geography* 7 (1): 1–18.

Elliott, M. R., Y. Wang, R. A. Lowe, and P. R. Kleindorfer. 2004. Environmental justice: Frequency and severity of US chemical industry accidents and the socioeconomic status of surrounding communities. *Journal of Epidemiology and Community Health* 58 (1): 24–30.

Fagg, J., S. Curtis, S. Stansfeld, and P. Congdon. 2006. Psychological distress among adolescents, and its relationship to individual, family and area characteristics in east London. *Social Science and Medicine* 63 (3): 636–48.

Franzini, L., and W. Spears. 2003. Contributions of social context to inequalities in years of life lost to heart disease in Texas, USA. *Social Science and Medicine* 47 (4): 469–76.

Fullilove, M. T. 1988. Promoting social cohesion to improve health. *Journal of the American Medical Women's Association* 53 (2): 72–76.

Ghosh, S. 2007. Transnational ties and intra-immigrant group settlement experiences: A case study of Indian Bengalis and Bangladeshis in Toronto. *GeoJournal* 68:223–42.

Grady, S. C. 2006. Racial disparities in low birthweight and the contribution of residential segregation: A multilevel analysis. *Social Science and Medicine* 63:3013–29.

Halpern, D., and J. Nazroo. 1999. The ethnic density effect: Results from a national community survey of England and Wales. *International Journal of Social Psychiatry* 46:34–46.

The Immigration Reform and Control Act, Public Law 99-603, 100 Stat. 3359. 1986.

Kawachi, I. 1999. Social capital and community effects on population and individual health. *Annals of the New York Academy of Sciences* 896:120–30.

Kelly, Y., L. Panico, M. Bartley, M. Marmot, J. Nazroo, and A. Sacker. 2008. Why does birthweight vary among ethnic groups: Findings from the Millenium Cohort Study. *Journal of Public Health* 31 (1): 131–37.

Kerner, C., A. Bailey, A. Mountz, I. Miyares, and R. Wright. 2001. Thank God she's not sick: Health and disciplinary practice among Salvadoran women in northern New Jersey. In *Geographies of women's health*, ed. I. Dyck, N. Lewis, and S. McLafferty, 127–42. London and New York: Routledge.

Kim, D., A. Subramanian, and I. Kawachi. 2006. Bonding versus bridging social capital and their associations with self-rated health: A multilevel analysis of 40 U.S. communities. *Journal of Epidemiology and Community Health* 60:116–22.

Lackland, S. D. 1998. Predicting life satisfaction among adolescents from immigrant familes in Norway. *Ethnicity and Health* 3 (1–2): 5–18.

Lorant, V., H. Van Oyen, and I. Thomas. 2008. Contextual factors and immigrants' health status: Double jeopardy. *Health and Place* 14 (4): 678–92.

Masi, C. M., L. C. Hawkley, Z. H. Piotrowski, and K. E. Pickett. 2007. Neighborhood economic disadvantage, violent crime, group density, and pregnancy outcomes in a diverse, urban population. *Social Science & Medicine* 65:2440–57.

Mason, S., J. Kaufman, J. Daniels, M. Emch, V. Hogan, and D. Savitz. 2010. Neighborhood ethnic density and preterm birth across seven ethnic groups in New York City. *Health and Place* 17 (1): 280–88.

Monger, R., and J. Yankay. 2011. U.S. legal permanent residents: 2010. Annual Flow Report, Office of Immigration Statistics, Department of Homeland Security, Washington, DC.

Neeleman, J., C. Wilson-Jones, and S. Wesseley. 2001. Ethnic density and deliberate self-harm: A study in south east London. *Journal of Epidemiology and Community Health* 55:85–90.

New York City Department of City Planning. 1996. *The newest New Yorkers, 1990–1994*. New York: New York City Department of City Planning.

———. 2004. *The newest New Yorkers, 2000*. New York: New York City Department of City Planning.

New York City Department of Health and Mental Hygiene. 2008. Summary of vital statistics for the City of New York, 2008. http://www.nyc.gov/html/doh/downloads/pdf/vs/2008sum.pdf (last accessed 10 April 2012).

———. 2010. Summary of vital statistics 2010 the city of New York, pregnancy outcomes. http://www.nyc.gov/html/doh/downloads/pdf/vs/vs-pregnancy-outcomes-report2010.pdf (last accessed 10 April 2012).

Oakes, J. M. 2004. The (mis)estimation of neighborhood effects: Causal inference for a practicable social epidemiology. *Social Science and Medicine* 58 (10): 1929–52.

O'Sullivan, D., and D. Wong. 2007. A surface-based approach to measuring spatial segregation. *Geographical Analysis* 39 (2): 147–68.

Osypuk, T., and D. Acevedo-Garcia. 2010. Beyond individual neighborhoods: A geography of opportunity perspective for understanding racial/ethnic health disparities. *Health and Place* 16 (6): 1113–23.

Osypuk, T., A. Diez Roux, C. Hadley, and N. Kandula. 2009. Are immigrant enclaves healthy places to live? The multi-ethnic study of atherosclerosis. *Social Science and Medicine* 69:110–20.

Phillipson, C., N. Ahmed, and J. Latimer. 2003. *Women in transition: A study of the experiences of Bangladeshi women living in Tower Hamlets*. Bristol, UK: The Policy Press.

Pickett, K. E., R. J. Shaw, K. Atkin, K. E. Kiernan, and R. G. Wilkinson. 2009. Ethnic density effects on maternal and infant health in the Millennium Cohort Study. *Social Science and Medicine* 69:1476–83.

Pickett, K. E., and R. G. Wilkinson. 2008. People like us: Ethnic group density effects on health. *Ethnicity and Health* 13:321–34.

Sampson, R., J. Morenoff, and F. Earls. 1999. Beyond social capital: Spatial dynamics of collective efficacy for children. *American Sociological Review* 64 (5): 633–60.

Shaw, R. J., K. E. Pickett, and R. G. Wilkinson. 2010. Maternal smoking during pregnancy in the US linked birth and infant death data set. *American Journal of Public Health* 100 (4): 707–13.

Singh, G. K., and S. M. Yu. 1996. Adverse pregnancy outcomes: Differences between US- and foreign-born women in major US racial and ethnic groups. *American Journal of Public Health* 86 (6): 837–43.

Smaje, C. 1995. Ethnic residential concentration and health: Evidence for a positive effect? *Policy and Politics* 23:251–69.

StataCorp. 2007. *Stata statistical software: Release 10*. College Station, TX: StataCorp LP.

Stein, C., D. Savitz, T. Janevic, C. Ananth, J. Kaufman, A. Herring, and S. Engel. 2009. Maternal ethnic ancestry and adverse perinatal outcomes in New York City. *American Journal of Obstetrics and Gynecology* 201 (8): 584.e1–584.e9.

Subramanian, S., D. Acevedo-Garcia, and T. Osypuk. 2005. Racial residential segregation and geographic heterogeneity in black/white disparity in poor self-rated health in the US: A multilevel statistical analysis. *Social Science and Medicine* 60 (8): 1667–79.

Thomas, F. 2010. Transnational health and treatment networks: Meaning, value and place in health seeking amongst southern African migrants in London. *Health and Place* 16 (3): 606–12.

Wang L. 2007. Immigration, ethnicity and accessibility to culturally diverse family physicians. *Health and Place* 13 (3): 656–71.

White, K., and L. Borrell. 2011. Racial/ethnic segregation: Framing the context of health risk and health disparities. *Health and Place* 17 (2): 438–48.

Wilkinson, R. G. 2005. *The impact of inequality: How to make sick societies healthier*. New York: New Press.

Williams, D. R., and C. Collins. 2001. Racial residential segregation: A fundamental cause of racial disparities in health. *Public Health Reports* 116 (5): 404–16.

Zavodny, M., and M. Bitler. 2010. The effect of Medicaid eligibility expansion on fertility. *Social Science and Medicine* 71 (5): 918–24.

With Reserves: Colonial Geographies and First Nations Health

Sarah de Leeuw,* Sean Maurice,* Travis Holyk,† Margo Greenwood,‡ and Warner Adam†

*Northern Medical Program, University of Northern British Columbia, Faculty of Medicine, University of British Columbia
†Carrier Sekani Family Services
‡First Nations Studies Program and the National Collaborating Centre for Aboriginal Health, University of Northern British Columbia

Health disparities between Indigenous and non-Indigenous peoples persist globally. Northern interior British Columbia, where many Indigenous people live on Indian[1] reserves allocated in the late nineteenth century, is no exception. This article reviews findings from fifty-eight interviews with members of thirteen First Nations communities in Carrier, Sekani, Wet'suwet'en, and Babine territories. The results suggest that colonial geographies, both physical and social, along with extant anti-Indigenous racism, are significant determinants of the health and well-being (or lack thereof) of many First Nations in the region.

土著和非土著人民之间的健康差距在全球范围持续。在不列颠哥伦比亚省内北部，生活在十九世纪末期所分配的印第安保留地里的许多土著人民也不例外。本文评估了包含 Carrier, Sekani, Wet' suwet'en, 和 Babine 领土里的第一民族社区的 13 个成员的五十八个访谈的结果。结果表明，殖民地自然和社会地理，与现存的反土著种族主义，是在该地区的许多第一民族的健康和福祉（或缺乏）的重要决定因素。关键词：不列颠哥伦比亚殖民主义，第一民族的健康，社会和地理因素。

La desigualdades por salud entre pueblos indígenas y no indígenas persisten a escala global. El interior septentrional de la Columbia Británica, donde numerosos pueblos indígenas viven en reservaciones indias establecidas a finales del siglo XIX, no es la excepción a este respecto. Este artículo revisa los hallazgos logrados en cincuenta y ocho entrevistas hechas a miembros de trece comunidades de las Primeras Naciones, en los territorios de Carrier, Sekani, Wet'sewet'en y Babine. Los resultados sugieren que las geografías coloniales, tanto físicas como sociales, junto con el racismo anti-indigenista existente, son determinantes significativos de la salud y bienestar (o falta de los mismos) de varias de las Primeras Naciones de la región.

Lee Edmond[2] is a twenty-six-year-old Dakelh (Carrier) First Nation man born and raised in Cheslatta (Grassy Plains). Cheslatta, on the south side of François Lake in northern interior British Columbia, Canada, has fewer than 350 people. Edmond describes himself as "constantly working as much as [he] can. Just kind of living life, pretty much." Mention of health services spurs wide-ranging discussions: "My left eye is a prosthetic. I was shot with a 22 when I was eleven months old. The bullet went right in and came out just above my left ear."

According to 2010 population data, Edmond is one of seventy-six registered Cheslatta-Carrier First Nations men residing in his "home" community, Indian Reserve 620. In 1951, when the Nechako River Valley was flooded to accommodate a hydroelectric dam, Cheslatta people were relocated there from previous reserve lands allocated between 1881 and 1897 without treaty or First Nations agreement. Canada's Department of Indian and Northern Affairs characterizes present-day Cheslatta as a Zone 2 Indian reserve: It is between 50 and 350 km from the nearest service center with year-round road access. Hunting, fishing, and trapping remain crucial aspects of life (Fiske and Patrick 2000) and unique health challenges (e.g., gunshot wounds) persist, including some of the province's highest rates of mortality from unintentional injuries (British Columbia Provincial Health Officer 2009).[3] Edmond cannot access any medical services for his prosthetic eye in Cheslatta or any nearby community:

> Every six months or so I have to catch a bus. It takes quite a while. Fourteen hours down to Vancouver and eighteen and a half hours home. I usually just stay for one night. They're cleaning my eye and resizing it and all that. Sometimes they let other people in ahead of me and stuff. I pretty much know why. . . . I'm Native.

This article is about the ill health and health experienced by Edmond and many other First Nations peoples in northern interior British Columbia.

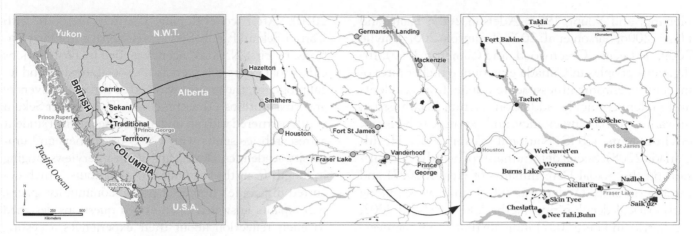

Figure 1. Traditional territories, First Nation communities, present-day Indian reserve spaces and allocations.

Specifically, we argue that First Nations' poor health can only be ameliorated if understood as geographically and historically determined, linked to colonial practices, and associated with dominant systems of social power that spatially and socially (re)produce Indigenous peoples as perpetually othered. Our research is informed by studies about the socially determined nature of Indigenous peoples' health (Kirmayer, Brass, and Tait 2000; Nettleton, Napolitano, and Stephens 2007; de Leeuw, Greenwood, and Cameron 2010) and is meant to dialogue with critical health geography and the emergent subfield of Indigenous geography, both of which value social justice and the politicization of geographic inquiry (Parr 2004; Shaw, Herman, and Dobbs 2006; Buzzelli and Veenstra 2007; Pualani-Louis 2007; Watson and Huntington 2008; Cameron, de Leeuw, and Greenwood 2009; Carmalt and Faubion 2009; Curtis and Riva 2010a, 2010b). Our work privileges partnerships with First Nations and is grounded in qualitative data gathered from fifty-eight participants from thirteen First Nations communities (see Figure 1). Despite its regional context, our conclusions might offer insights into persistent gaps between Indigenous and non-Indigenous peoples' health globally.

Background

British Columbia is the outcome of long-standing and ongoing colonial work, both discursive and material in nature, undertaken in shifting, complex, and asymmetrical ways (Williams-Braun 1997; Harris 2004; Kobayashi and de Leeuw 2010; Oliver 2010). The results are tense geographies of different social, cultural, economic, and political powers and unstable boundaries, borders, laws, policies, and sociocultural protocols that, although continually resisted, still persist (Blomley 1996; Harris 1997, 2002, 2004; Sparke 1998; Clayton 2000, 2001–2002; Rossiter and Wood 2005). Colonial ideologies and government legislation that, literately and figuratively, mapped Indigenous peoples out of British Columbia and onto Indian reserves (Brealey 1995; Harris 2002, 2004) have resulted in contemporary tensions about everything from jurisdictional privileges and laws, a lack of treaties, human rights violations, and unequal allocation of resources (Sterritt et al. 1998; Schouls 2003; Stanger-Ross 2008; Irlbacher-Fox 2009). Eighteen residential schools, focused on transforming children and cultures, operated for over 100 years in the province (de Leeuw 2007, 2009). Indigenous peoples were not able to vote federally until 1960. The federal 1876 Indian Act, remarkably unchanged until 1951, produced taxonomies of being or not being Indian that carry into today.[4] The minimal rights and privileges linked to being a Status Indian include some limited access to extended health benefits, educational support, and resource harvesting rights. These socially engineered categories were, and remain, premised on blood-quantum logics and the places where a person lived (Lawrence 2003, 2004; de Leeuw and Greenwood 2011). Today, Status Indians in British Columbia carry status cards to prove identity when accessing, among other things, health services and, like all other Status Indians across Canada, they are governed differently than their non-Indigenous counterparts (Legal Services Society of British Columbia 2007).

In neocolonial landscapes around the world, colonialism is a factor in health divisions between Indigenous and non-Indigenous peoples (Waldram,

Herring, and Young 1995; Adelson 2005; Baum and Harris 2006; Larson et al. 2007; Tarantola 2007; Smylie 2009; de Leeuw, Greenwood, and Cameron 2010). Indigenous peoples in Canada experience higher rates of morbidity, chronic illness, acute trauma from accidents and violence, suicide, addiction, mental health issues, unwanted teenage pregnancy, and exposure to high-risk environments (RHS National Team 2007; Health Canada 2008). Social determinants of health (Commission on the Social Determinants of Health 2007) and intersectionality (Valentine 2007) frameworks, along with critical decolonizing methodologies (Nash 2002; Gilmartin and Berg 2007) indicate that social engineering, legislated disparities in access of services and resources, and forced colonial education have all directly and negatively impacted Indigenous peoples' contemporary well-being (Kelm 1999; Marmot 2005; Kelly et al. 2007; Marmot et al. 2008; Richmond and Ross 2009; Castleden et al. 2010). Much of this research, however, lacks specific or purposeful geographic analysis.

Research Methods and Methodologies

Undertaking research with First Nations is not a straightforward or apolitical process. Indigenous scholars and community members around the world associate research with colonialism. Research still positions Indigenous peoples as passive subjects from whom information is extracted, to whom results are often not communicated, and for whom the work is often not useful (Smith 1999; Brant-Castellano 2004; Schnarch 2004; Giles and Castleden 2008). Canadian health research is no exception (Smylie 2009). Consequently, in partnership with diverse Aboriginal groups, the Canadian Institutes for Health Research (CIHR) developed comprehensive guidelines for ethical research with Indigenous peoples. Health research must, at minimum, involve researchers who understand and respect Indigenous communities with whom they work, involve co-ownership of the research results to ensure community access to the data, account for participatory-action methods where community can give direction on the project, and must benefit both the community and researcher (see CIHR 2007). In this context, our relationships and methods are worth noting.

Members of this research team have worked together for over a decade, resulting in the development of trust, a well-documented component of successful research in Indigenous communities (Bartlett et al. 2007; Ball and Janyst 2008; Panelli 2008; Ford and Airhihenbuwa

2010). After codeveloping research methods with Carrier Sekani Family Services (CSFS), a "for First Nations, by First Nations" agency guided by First Nations elders that offers comprehensive social, health, and legal programs, the project was vetted through university and First Nations' ethics processes. Two Carrier Sekani youth partnered with an undergraduate medical student as research assistants with CSFS researchers and university academics. During fifty-eight interviews ranging in length from thirty minutes to two hours, which occurred in people's homes or public community spaces, participants answered open-ended questions designed to elicit reflection about their experiences accessing medical and health services, the majority of which are off-reserve. Resulting transcripts were thematically analyzed by CSFS and university partners using the qualitative analysis software HyperRESEARCH coding processes and close reading and narrative analysis practices. Results were validated with community members and participants and all resultant data and research are co-owned.

Results and Discussion

Three dominant themes about First Nations' experiences of accessing health care services arose on analysis of data. First, First Nations living on reserves in northern interior British Columbia are spatially and materially distant from services. Second, and more important, physical distances underpin a social distancing and a subsequent social construction and racialization of First Nations premised on imagined understandings of reserves as particular kinds of spaces. Finally, we identified what we call "uncommon ground," or divisions between First Nations worldviews and biomedical models of health care.

From the onset of a project initiated to assess levels of First Nations trust toward mainstream medicine, the role of geography in accessing health care was impossible to ignore. Many participants expressed profound connection with place, linking health and well-being to their land and the activities they undertook on their territories. Conversely, ill health, and not accessing health care services, was often linked to isolation, historically unceded ownership of lands, and social stigmatization because identity was tied to colonial geographies.

Spatial Distancing

Most participants voiced concerns about being physically far away from services. Distances in remote north central British Columbia are hazardous: "What happens here," noted one participant, "is horrendous. [One person in my reserve] travels 2,800 miles a month. This person is quite ill . . . the travel isn't doing her any good." Transportation is expensive: "I've got two daughters to feed . . . and I might be down a hundred bucks because I spent the day seeing a doctor." Hitchhiking, the most viable form of transportation to and from health care services for many people, has led in the last fifteen years to approximately eighteen unsolved murders or disappearances of First Nations women and girls along the primary highway[5] through Carrier Sekani Territory. Motor vehicle accidents account for 3.4 deaths per 10,000 in Status Indian populations in northern interior British Columbia, more than triple the number (0.9) in non-Indigenous populations across the rest of the province and a 36 percent greater number than Status Indians across British Columbia (2.5 per 10,000).[6]

Many (more than twenty) participants identified unsafe travel as a deterrent in accessing health services: "It's too complicated [leaving the reserve]. The things you have to go through, you just don't feel comfortable at all. So you put your health issues aside until it gets bad," stated one woman. Another woman, in need of regular and specialized care for a heart murmur, regularly did not access care: "If you're talking about our people [First Nations], they need safe transportation. Not having a reliable ride [means] missing a lot of appointments." Although doing so was dangerous, participants overcame physical distances. More challenging was the social distancing they faced when leaving places imagined and constructed as always peripheral by broader, off-reserve, non-Indigenous society.

Reserves and Racialization

First Nations in northern interior British Columbia are unequivocally remote and rural by provincial and national standards. Unlike non-Indigenous peoples who live remotely, being an on-reserve Indigenous person and leaving one's community means crossing the spatial and legal boundaries of the colonially constructed Indian reserve. Reserve spaces are literally and discursively cordoned off from the rest of the province. Living on an Indian reserve in northern interior British Columbia means being bound—materially and imaginarily—to an othered territory. Leaving that territory means being socially and territorially out of place. Participants sensed that their identities, anchored inextricably and legislatively to being a Status Indian from the uniquely demarcated space of an In-

dian reserve, are almost always judged negatively when interacting with the health care system:

> I just want to feel welcomed. When we go to the different places [emergency rooms and hospitals], we're kind of judged. I want to be taken seriously and not judged because we're Native. . . . Each Native person [from my reserve], when they go to the hospital in town, they bring this form [indicating being Status] and they have to get it stamped with a time and a date and everything. It looks like [the intake nurse] gets pretty annoyed with those forms. It's more work for her. She looks pretty cranky.

Being First Nations from a reserve, who, by necessity, steps out of place to access health care, requires overcoming significant physical distancing and remote geographies. This results in a social distancing whereby one's status—which easily translates into "race"—becomes immediately visible. Participants experienced active racialization based on being from, and leaving, the constructed spaces of Indian reserves:

> You have to go into town for emergencies and everything. With us out here, you either [have to] hitchhike or pay people to bring you into town. Lots of times, when you ask [a non-Native] for a ride, they don't want to give you a ride. They look [at you] and think "This Indian's a druggie . . . or a drunk." If you stick a camera in that medical building [in the city], you could see it. I spend my whole day getting into town to get treated like shit. . . . [Doctors should be] concerned no matter what color the person is.

Somewhat paradoxically, the majority (more than thirty-two) of participants expressed deep feelings of trust toward what was often referred to as Needo (Dakelh for Whiteman) medicine and simultaneously spoke about racism when they visited health care service sites. They believed and trusted that more abstract and placeless scientific research would positively impact their health. They were often uncomfortable, however, interacting with place-specific expressions of the health care system (e.g., hospitals, doctors' offices, clinics, and pharmacies), believing that professionals therein equated being from an Indian reserve with being a particular kind of person:

> I'd like people to know that just because we're Indians, we're not a lower life form. Which is really what you get [in the city, off-reserve]. It's a fact of life, we all know it. We try to avoid going into to town, to the hospital or anywhere because [we] can't stand being treated with such a condescending manner. It just makes me mad, so I just don't bother.

In describing her two children's negative experiences with the hospital in a city 400 km away from her reserve, one woman stated simply, "The medical system and the hospital are racist. My daughter once ended up there because of a seizure. They asked if she drank. She said 'Yup, I had one beer.' As soon as she said that, they discharged her. And my daughter is not a drinker." When asked what would make her more comfortable in that hospital, she replied "a new attitude toward Native people."

The conflation of living on reserve with stereotypes about being a certain kind of person (a drunk, poor, backward, hostile to medical advancements, and uneducated "Indian") was mentioned by more than one quarter of participants. Where a person lived, according to participants, immediately results in being racialized: "I took my cousin with chest pains into a doctor [off-reserve]. The nurse asked 'How many drinks have you had?' Automatically. My cousin doesn't even drink and she was automatically asked that. That's totally racist. Just because you're Native [from a reserve], they automatically think that you're drinking." As stories about racialized encounters travel, they impact peoples' perceptions about the health care system, independent of direct experience. One participant, who "wasn't sure [she'd personally] had bad experiences" with the health care system, nevertheless observed: "There's one Elder . . . he says to everybody that [the reason] Aboriginal people are dying off quickly is because of the hospital. Because we're not treated right. I really believe that." For some participants, the health care system is tantamount to some of the most egregious historic colonial practices: "It's another way of destroying our people. Same thing as residential school. . . ."

Uncommon Ground

Participants avoided homogenous definitions of Indigenous health practices. Many, however, suggested that integrating elements of "Indian medicine" into off-reserve *Needo* health systems could result in First Nations feeling more welcomed, despite being far away from their communities and thus, automatically, out of place. Language was central to these considerations, itself a form of "Indian medicine" and a literal utterance of culture and health. When an Indigenous language was not spoken, the situation was immediately unhealthy. A fifty-three-year-old Wet'suet'en woman who came "from a long line of traditional leaders" explained the challenges of accompanying an elder to the hospital:

He suffers from high blood pressure. The doctor was attempting to explain and I couldn't explain it to him in Wet'suet'en, which if someone would have interpreted it for him, he would have gotten. [Non-Native doctors] need to be more accepting of our ways. Start listening. I wish they could hire more people that speak our language.

Off-reserve health care spaces, for many participants, were expressions of colonialism designed to expand previous efforts of eradicating Indigenous systems of knowledge, including First Nations' medical knowledge. Not being able to discuss medicines and practices still actively used in their reserves added to some participants' beliefs that their ways of being were still actively under attack:

Go back to the beginning. Look at Canada and British Columbia. They came in here with their western law. We've become wards of Canada. They think they're going to protect us . . . they put down reserves . . . they say we're part of the western system. [Now] the whole hospital system is [that] system. We got no trust to follow it. Because our culture is different. To me, when I look at the medicine we make, it's a gift. But [that concept] is not valued.

In addition to integrating and valuing Indigenous languages and cultures within all aspects of health care provision, participants offered a number of other solutions. Increasing the number of Native doctors and health care professionals, who were seen by some participants as "understanding two worlds," and purposefully educating non-Indigenous health care professionals about Indigenous peoples—particularly their histories and the places where they lived—were all suggestions for overcoming the divisions experienced by First Nations from northern interior British Columbia.

Conclusions

Health and well-being are increasingly conceptualized as socially determined and as deeply affected by access, or lack of access, to systems of health care, which, in turn, reflect various hierarchies of social, cultural, political, and economic power. Social determinants of health and health care systems are geographical; they take place somewhere, shift and change according to their spatial and temporal contexts, and are altered by their geographic confines and geographies of power. If poor health is an outcome of poor social determinants and disconnect with health care systems, it is also an outcome of the physical and social geographies that shape and determine those social determinants and health care systems. Participants in this research made theses linkages clearly, suggesting that their health cannot be fully conceptualized, or improved, without understanding it as geographically contingent and, consequently, as an outcome of long-standing colonial work to spatially, socially, and corporeally dislocate and distance them from non-Indigenous British Columbians.

The 2009 British Columbia Provincial Health Officer's report offers a comprehensive picture of Indigenous peoples' health in the province (see also British Columbia Provincial Health Officer 2002). Since 2001, no substantive changes occurred in thirty-nine of fifty-seven indicators of health and well-being. British Columbia's Indigenous peoples experienced worsening trends in ten of fifty-seven indicators, including increased rates of HIV/AIDS, use of prescription drugs, poor housing, low birth weights, and the number of children who live as governmental wards (see also Hughes 2006; Blackstock 2007; de Leeuw, Greenwood, and Cameron 2010). Since 1992 Status Indian men maintained the lowest life expectancy of any group in the province, dying almost ten years earlier than non-Indigenous women, the people who live longest in the province. Status Indian women die over a year earlier than non-Indigenous men in the province and the gap between the two groups has widened by over a year in the last fifteen years.

These data capture health disparities between different peoples in different places, but neither physical nor social geography are explicitly theorized as determinants of health. For a significant number of First Nations in northern interior British Columbia, however, their Indian reserves and homeplaces, to which they were and continue to be confined, are inextricably linked to being, or not being, healthy. The health of First Nations living on reserves is determined by physical geographics (e.g., distance from services) but, even more difficult to overcome, by socio-imaginative geographies that racialize people precisely based on being from an Indian reserve. This place-based racialization determines whether First Nations access the health care system and how, if they do, they are treated once within the system. Too often, an experience of geographically linked racialization extends beyond interactions with the health care system; distance from an array of services and deep poverty in reserves demand that many First Nations leave their communities to access other services or find employment, both of which figure into peoples' overall health and well-being. Leaving the colonially produced spaces of Indian reserves also means leaving support structures that result from being in place, including understandings about

Indigenous languages and cultures, resulting in reduced resiliency against racialization. This, too, becomes a factor in being, or not being, healthy. Understanding disparities between Indigenous and non-Indigenous people's health means understanding those disparities as geographically expressed and determined. Accounting for racializations linked to colonial geographies is thus paramount in ameliorating the enduring and unequal health disparities lived by Indigenous peoples.

Notes

1. Canada's constitution recognizes unique relations with three distinct groups of peoples: First Nations, Inuit, and Métis. "Indian" is a problematic misnaming of these Indigenous peoples used herein for historical accuracy only.
2. In accordance with ethical standards of research with Indigenous people, communities had an opportunity to consider drafts of this research. Lee Edmond is not a pseudonym, fictitious character, or composite: It is the real name of a young man who insisted on his name appearing in this article.
3. According to the Provincial Health Officer's report published in 2009, deaths from "unintentional injuries" account for 8.5 deaths per 10,000 Status Indians in northern interior British Columbia, more than double the rate (4.2) of other residents in the region and over three times higher (2.6) than other residents across British Columbia. Mortality rates from unintentional injuries for Status Indians in northern interior British Columbia are also higher than the those of Status Indians across the province as a whole (7.3 per 10,000).
4. The government of Canada distinguishes between registered (or Status) and nonregistered (or non-Status) Indians. For further reading on these very complicated divisions, and resultant impacts, see de Leeuw and Greenwood (2011) and Aboriginal Affairs and Northern Development Canada's (2010) "You Wanted to Know: Federal Programs and Services for Registered Indians."
5. This highway, Highway 16, is known nationally as The Highway of Tears.
6. Data are age-standardized for mortality rates.

References

Aboriginal Affairs and Northern Development Canada. 2010. You wanted to know: Federal programs and services for registered Indians. http://ainc-inac.gc.ca/ai/pubs/ywtk/ywtk-eng.asp (last accessed 11 June 2011).

Adelson, N. 2005. The embodiment of inequalities: Health disparities in Aboriginal Canada. *Canadian Journal of Public Health* 96:45–61.

Ball, J., and P. Janyst. 2008. Enacting research ethics in partnerships with Indigenous communities in Canada: "Do it in a good way." *Journal of Empirical Research on Human Research Ethics* 3:33–51.

Bartlett, J. G., Y. Iwasaki, B. Gottlieb, D. Hall, and R. Mannell. 2007. Framework for Aboriginal-guided decolonizing research involving Métis and First Nations persons with diabetes. *Social Science and Medicine* 65:2371–82.

Baum, R., and L. Harris. 2006. Equity and the social determinants of health. *Health Promotion Journal of Australia* 17:163–65.

Blackstock, C. 2007. Residential schools: Did they close or just morph into child welfare. *Indigenous Law Journal 6* (1): 71–78.

Blomley, N. K. 1996. "Shut the province down": First Nations' blockades in British Columbia. *B.C. Studies* 3:5–35.

Brant-Castellano, M. 2004. Ethics of Aboriginal research. *Journal of Aboriginal Health* 1:98–114.

Brealey, K. 1995. Mapping them "out": Euro-Canadian cartography and the appropriation of the Nuxalk and Ts'ilhqot'in First Nations territories, 1793–1916. *Canadian Geographer* 39:140–68.

British Columbia Provincial Health Officer. 2002. *The health and well-being of Aboriginal people in British Columbia.* Provincial Health Officer's Annual Report 2001. Victoria, BC, Canada: Ministry of Health Planning.

———. 2009. *Pathways to health and healing: 2nd report on the health and well-being of Aboriginal people in British Columbia.* Provincial Health Officer's Annual Report 2007. Victoria, BC, Canada: Ministry of Healthy Living and Sport.

Buzzelli, M., and G. Veenstra. 2007. New approaches to researching environmental justice: Combining critical theory, population health and geographical information science (GIS). *Health and Place* 13:1–2.

Cameron, E., S. de Leeuw, and M. Greenwood. 2009. Indigeneity. In *International encyclopedia of human geography,* ed. N. Thrift and R. Kitchin, 352–57. London: Elsevier.

Canadian Institutes of Health Research (CIHR). 2007. *CIHR guidelines for health research involving Aboriginal people.* Ottawa, ON, Canada: Canadian Institutes of Health Research. http://www.cihr-irsc.gc.ca/e/documents/ethics_aboriginal_guidelines_e.pdf (last accessed 10 November 2010).

Carmalt, J. C., and T. Faubion. 2009. Normative approaches to critical health geography. *Progress in Human Geography* 34:292–308.

Castleden, H., V. Crooks, N. Hanlon, & N. Schuurman. 2010. Providers' perceptions of Aboriginal palliative care in British Columbia's rural interior. *Health and Social Care in the Community* 18 (5): 483–91.

Clayton, D. W. 2000. *Islands of truth: The imperial fashioning of Vancouver Island.* Vancouver, BC, Canada: UBC Press.

———. 2001–2002. Absence, memory, and geography. *B.C. Studies* 132:65–79.

Commission on the Social Determinants of Health. 2007. *Achieving health equity: From root causes to fair outcomes.* Geneva, Switzerland: World Health Organization.

Curtis, S., and M. Riva. 2010a. Health geographies I: Complexity theory and human health. *Progress in Human Geography* 34:215–23.

———. 2010b. Health geographies II: Complexity and health care systems and policy. *Progress in Human Geography* 34:513–20.

de Leeuw, S. 2007. Intimate colonialisms: The material and experienced places of British Columbia's residential schools. *Canadian Geographer* 51:339–59.

———. 2009. "If anything is to be done with the Indian, we must catch him very young": Colonial constructions of Aboriginal children and the geographies of Indian residential schooling in British Columbia, Canada. *Children's Geographies* 7:123–40.

de Leeuw, S., and M. Greenwood. 2011. Beyond borders and boundaries: Addressing Indigenous health inequities in Canada through theories and methods of intersectionality. In *Health inequalities in Canada: Intersectional frameworks and practices,* ed. O. Hankivshy, 53–70. Vancouver, BC, Canada: UBC Press.

de Leeuw, S., M. Greenwood, and E. Cameron. 2010. Deviant constructions: How governments preserve colonial narratives of addictions and poor mental health to intervene into the lives of Indigenous children and families in Canada. *International Journal of Mental Health and Addictions* 8:282–95.

Fiske, J.-A., and B. Patrick. 2000. *Cis Dideen Kat, When the plumes rise: The way of the Lake Babine Nation.* Vancouver, BC, Canada: UBC Press.

Ford, C. L., and C. O. Airhihenbuwa. 2010. The public health critical race methodology: Praxis for antiracism research. *Social Science & Medicine* 71:1390–98.

Giles, A. R., and H. Castleden. 2008. Community co-authorship in academic publishing: A commentary. *Canadian Journal of Native Education* 31: 208–16.

Gilmartin, M., and L. Berg. 2007. Locating postcolonialism. *Area* 39:120–24.

Harris, C. 1997. *The resettlement of British Columbia: Essays on colonialism and geographical change.* Vancouver, BC, Canada: UBC Press.

———. 2002. *Making native space: Colonialism, resistance, and reserves in British Columbia.* Vancouver, BC, Canada: UBC Press.

———. 2004. How did colonialism dispossess? Comments from an edge of empire. *Annals of the Association of American Geographers* 94:165–82.

Health Canada. 2008. *First Nations, Inuit and Aboriginal health.* Ottawa, ON, Canada: Health Canada. http://www.hc-sc.gc.ca/fniah-spnia/pubs/aborig-autoch/2009-stats-profil/index-eng.php (last accessed 10 November 2010).

Hughes, T. 2006. *BC children and youth review: An independent review of BC's child protection system.* Victoria, BC, Canada: Queen's Printer. http://www.mcf.gov.bc.ca/bc-childprotection/pdf/BC_Children_and_Youth_Review_Report_FINAL_April_4.pdf (last accessed 12 November 2010).

Irlbacher-Fox, S. 2009. *Finding Dahshaa: Self-government, social suffering, and Aboriginal policy in Canada.* Vancouver, BC, Canada: UBC Press.

Kelly, M. P., A. Morgan, J. Bonnefoy, J. Butt, and V. Bergman. 2007. *The social determinants of health: Developing an evidence base for political action.* London: The Measurement and Evidence Knowledge Network for the World Health Organization. http://www.who.int/social_determinants/resources/mekn_final_report_102007.pdf (last accessed 10 November 2010).

Kelm, M.-E. 1999. *Colonizing bodies: Aboriginal health and healing in British Columbia, 1900–50.* Vancouver, BC, Canada: UBC Press.

Kirmayer, L. J., G. M. Brass, and C. Tait. 2000. The mental health of Aboriginal peoples: Transformations of identity and community. *The Canadian Journal of Psychiatry* 45:607–16.

Kobayashi, A., and S. de Leeuw. 2010. Colonialism and the tensioned landscapes of Indigeneity. In *The handbook of social geography*, ed. S. Smith, R. Pain, S. Marston, and J. P. Jones III, 118–39. London: Sage.

Larson, A., M. Gillies, P. J. Howard, and J. Coffin. 2007. It's enough to make you sick: The impacts of racism on the health of Aboriginal Australians. *Australian and New Zealand Journal of Public Health* 31:322–29.

Lawrence, B. 2003. Gender, race, and the regulation of Native identity in Canada and the United States: An overview. *Hypatia* 18:3–31.

———. 2004. *"Real" Indians and others: Mixed-blood urban native peoples and Indigenous nationhood.* Vancouver, BC, Canada: UBC Press.

Legal Services Society of British Columbia. 2007. *Benefits, services, and resources for Aboriginal peoples.* Vancouver, BC, Canada: Legal Services Society.

Marmot, M. 2005. Social determinants of health inequalities. *The Lancet* 365:1099–1104.

Marmot, M., S. Friel, R. Bell, T. A. J. Howeling, and S. Taylor. 2008. Closing the gap in a generation: Health equity through action on the social determinants of health. *Lancet* 372:1661–69.

Nash, C. 2002. Genealogical identities. *Environment and Planning D: Society and Space* 20:27–52.

Nettleton, C., D. A. Napolitano, and C. Stephens, eds. 2007. An overview of current knowledge of the social determinants of Indigenous health. Working Paper commissioned by the Commission on Social Determinants of Health, World Health Organisation, for the Symposium on the Social Determinants of Indigenous Health, Adelaide, Australia.

Oliver, J. 2010. *Landscapes and social transformations on the Northwest Coast: Colonial Encounters in the Fraser Valley.* Vancouver, BC, Canada: UBC Press.

Panelli, R. 2008. Social geographies: Encounters with Indigenous and more-than-white/Anglo geographies. *Progress in Human Geography* 32:801–11.

Parr, H. 2004. Medical geography: Critical medical and health geography? *Progress in Human Geography* 28:246–57.

Pualani-Louis, R. 2007. Can you hear us now? Voices from the margin: Using Indigenous methodologies in geographic research. *Geographical Research* 45:130–39.

RHS National Team. 2007. *First Nations Regional Longitudinal Health Survey (RHS) 2002/2003: Results for adults, youth, and children living in First Nations communities.* Ottawa, ON, Canada: Assembly of First Nations/First Nations Information Governance Committee.

Richmond, C. A. M., and N. A. Ross. 2009. The determinants of First Nation and Inuit health: A critical population health approach. *Health and Place* 15:403–11.

Rossiter, D., and P. K. Wood. 2005. Fantastic topographies: Neo-liberal responses to Aboriginal land claims in British Columbia. *Canadian Geographer* 49:352–67.

Schnarch, B. 2004. Ownership, control, access, and possession (OCAP) or self-determination applied to research: A critical analysis of contemporary First Nations research and some options for First Nations communities. *Journal of Aboriginal Health* 1:80–95.

Schouls, T. 2003. *Shifting boundaries: Aboriginal identity, pluralistic theory and the politics of self-government.* Vancouver, BC, Canada: UBC Press.

Shaw, W. W., R. D. K. Herman, and G. R. Dobbs. 2006. Encountering Indigeneity: Re-imaging and decolonizing geography. *Geographiska Annaler* 88B:267–76.

Smith, L. T. 1999. *Decolonizing methodologies: Research and Indigenous peoples.* New York: Zed Books.

Smylie, J. 2009. The health of Aboriginal peoples. In *Social determinants of health: Canadian perspectives.* 2nd ed., ed. D. Raphael, 280–301. Toronto, ON, Canada: Canadian Scholars Press.

Sparke, M. 1998. A map that roared and an original atlas: Canada, cartography, and the narration of nation. *Annals of the Association of American Geographers* 88:463–95.

Stanger-Ross, J. 2008. Municipal colonialism in Vancouver: City planning and the conflict over Indian reserves. *The Canadian Historical Review* 89:541–80.

Sterritt, N. J., S. Marsden, R. Galois, P. Grant, and R. Overstall. 1998. *Tribal boundaries in the Nass watershed.* Vancouver, BC, Canada: UBC Press.

Tarantola, D. 2007. The interface of mental health and human rights in Indigenous peoples: Triple jeopardy and triple opportunity. *Australian Psychiatry* 15:10–17.

Valentine, G. 2007. Theorizing and researching intersectionality: A challenge for feminist geography. *The Professional Geographer* 59:10–21.

Waldram, J. B., A. Herring, and T. K. Young. 1995. *Aboriginal health in Canada: Historical, cultural, and epidemiological perspectives.* Toronto, ON, Canada: University of Toronto Press.

Watson, A., and O. H. Huntington. 2008. They're here—I can feel them: The epistemic spaces of Indigenous and Western knowledges. *Social and Cultural Geography* 9:257–81.

Williams-Braun, B. 1997. Buried epistemologies: The politics of nature in (post)colonial British Columbia. *Annals of the Association of American Geographers* 87:3–31.

Smoking, Ethnic Residential Segregation, and Ethnic Diversity: A Spatio-temporal Analysis

Graham Moon,* Jamie Pearce,† and Ross Barnett‡

*Geography and Environment, University of Southampton
†School of GeoSciences, University of Edinburgh
‡Department of Geography, University of Canterbury

Ethnic residential segregation is a profound social divide in many societies. In the health arena, U.S. work has been influential in demonstrating the impact of ethnic residential segregation on child health outcomes, showing how it can compound other forms of disadvantage. This article builds on and extends this research by examining the transferability of conclusions concerning the health impact of ethnic residential segregation to a non-U.S. context and to the field of health behavior. Using complete adult population data from the 1996 and 2006 New Zealand Censuses of Population geocoded to local and urban area levels, we examine smoking prevalence and cessation in relation to ethnic segregation and diversity. The article employs a repeated cross-sectional multilevel modeling strategy with smoking and cessation as outcome variables. The differential impact of segregation and diversity on smoking behavior by different ethnic groups is considered, taking into account the confounding effect of socioeconomic status and demographic variation. Conclusions suggest that Māori isolation has little overall effect but ethnic diversity has some relevance. Individual ethnic status and area-level deprivation are more important in understanding smoking behavior.

种族居住的隔离在许多社会里是一个深刻的社会鸿沟。在卫生领域，美国的研究工作在证明种族居住隔离对儿童的健康结果上是有影响力的，并表明了它如何能与其他形式的不利条件合成在一起。本文以这种研究为基础，并对该研究进行拓展，研究有关种族居住隔离对非美国环境和健康行为领域的影响结论的应用转让性。使用从 1996 年和 2006 年，完整的地理编码到本地和市区各级的新西兰成人人口普查数据，我们考察吸烟的流行和停止与民族分离和多样性的关系。本文采用重复的横断面多层次建模策略，把吸烟和戒烟作为结果变量。研究不同民族在吸烟行为上的隔离和多样性的差别影响，并考虑到社会经济地位和人口变化的干扰作用。结论表明，毛利人的隔离有小的整体效果，但种族多样性具有一定的相关性。个人的民族地位和区级的剥夺在了解吸烟行为上更为重要。
关键词：种族多样性，种族隔离，新西兰，吸烟行为。

La segregación residencial por razones étnicas constituye una profunda divisoria social en muchas sociedades. En la arena de la salud pública se puede apreciar que el trabajo realizado en EE.UU. ha sido influyente para demostrar el impacto de la segregación residencial étnica sobre la salud del niño, mostrando cómo esto puede desarrollar otras formas de desventaja. Este artículo se basa en aquellas investigaciones y las extiende examinando la transferabilidad de las conclusiones sobre el impacto que tiene sobre la salud la segregación residencial étnica a otros contextos fuera de los EE.UU. y al campo de conductas relacionadas con la salud. Utilizando datos completos de población adulta de los Censos de Población de 1996 y 2006 de Nueva Zelandia, georreferenciados a los niveles de áreas locales y urbanas, examinamos la prevalencia y cesación de fumar en relación con la segregación y diversidad étnicas. El artículo utiliza una estrategia de modelación repetida de nivel múltiple trans-seccional, usando el hábito de fumar y su cesación como variables resultantes. Se considera el impacto diferencial que tiene la segregación y la diversidad sobre el hábito de fumar por diferentes grupos étnicos, tomando en cuenta el efecto alterante del estatus socioeconómico y la variación demográfica. Las conclusiones sugieren que el efecto del aislamiento maorí es limitado pero en cambio la diversidad étnica tiene alguna relevancia. Más importantes para entender el comportamiento fumador son el estatus étnico individual y la privación a nivel de área.

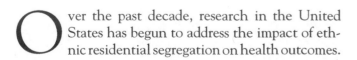

Over the past decade, research in the United States has begun to address the impact of ethnic residential segregation on health outcomes.

In contrast to an established research tradition that has considered ethnicity in terms of the proportion of the population drawn from particular ethnic groups, this recent work has focused on spatial segregation indexes, such as the well-known indexes of isolation and dissimilarity familiar to geographers interested in ethnic residential structure (Johnston, Poulsen, and Forrest 2007; Phillips 2007).

The United States has been the main laboratory for research on ethnic segregation and health outcomes. Deep social divides and income differentials persisting between ethnic groups result in well-established geographical health inequalities. In highly segregated communities, the evidence generally (although not exclusively) suggests that inequalities are deepened, impacting negatively on black communities. The purpose of this article is to extend research on the interplay of ethnic segregation and health to a non-U.S. setting. Our setting is New Zealand, an ethnically plural society with a high degree of mixing between ethnic groups: people of European origin, Māori (the indigenous people), Pacific peoples drawn from elsewhere in the South Pacific, and more recent immigrants from East Asia. Levels of ethnic segregation in New Zealand do not approach those in the United States. Our article thus also extends knowledge of the health impacts of ethnic segregation to a less segregated setting. For a health issue, we focus on smoking behavior. In New Zealand, inequalities in smoking behavior between ethnic groups are particularly marked and, a priori, we might expect ethnic segregation to play a part in these inequalities, as smoking is heavily influenced by local community norms and social practices. Choosing smoking behavior as an outcome extends the literature on ethnic segregation and health away from its established focal areas of maternal and child health.

Background

U.S. Studies of Segregation and Health

Our hypothesized association between smoking behavior and ethnic segregation indexes rests on a more general literature that has emerged in the last decade, largely in the United States (Acevedo-Garcia and Osypuk 2008; Pickett and Wilkinson 2008). Recent studies have considered birth outcomes (Bell et al. 2006; Grady 2006; Osypuk and Acevedo-Garcia 2008), obesity (Chang, Hillier, and Mehta 2009), drug use (Cooper et al. 2007, 2008), breast cancer (Dai 2010), acute myocardial infarction (Vaughan Sarrazin et al. 2009), and self-rated health (Subramanian, Acevedo-Garcia, and Osypuk 2005). Observed associations are often confounded by other variables or the spatial basis of analysis (Darden et al. 2010) but generally poorer

outcomes are associated with being black and living in an area of high black–white segregation. There is some evidence for segregation being protective for Hispanic groups, although this is contested, and for Asian Americans. Outcomes tend to be worse for women. A systematic review of work on these "consistent but complex" associations has called for longitudinal studies within a multilevel framework (Kramer and Hogue 2009).

There has been little U.S. work on ethnic segregation and smoking. What has been published is equivocal. Landrine and Klonoff (2000) suggested that greater segregation is associated with higher rates of smoking and the association is not significantly confounded by socioeconomic status. Ellen (2000) argued that smoking by black women during pregnancy is not related to the dissimilarity index of segregation. Bell et al. (2007) suggested that smoking in pregnancy follows a U-shaped relationship with segregation, being more common at the lowest and highest levels of segregation. Messer, Oakes, and Mason (2010) claimed that white women students are more likely to smoke than black women students if they attend schools with high levels of black–white segregation, whereas Shaw, Pickett, and Wilkinson (2010) reported that segregation reduces smoking in pregnancy for both black and Hispanic mothers.

Although almost entirely based on observational designs that cannot identify causality, these varying conclusions draw dominantly on psychosocial explanations for health inequality. On the one hand, these theories suggest that segregation, when accompanied by disadvantage, might cause community stress with consequent negative outcomes for health. Such negative outcomes can be compounded by racism, poorer health care provision, lower quality built environments, and toxic exposures (Landrine and Corral 2009). Alternatively, and equally psychosocial in origin, highly segregated groups might have high levels of internal solidarity, raised social capital, and aspirational norms that promote positive health outcomes. In this latter analysis, poorer health outcomes could be more appropriately anticipated in areas of high ethnic diversity. Recent work on the ethnic density hypothesis explores this contradiction in both the United States and the United Kingdom (Bécares, Nazroo, and Stafford 2009; Shaw, Pickett, and Wilkinson 2010).

The New Zealand Context

Health disparities between ethnic groups in New Zealand are strongly associated with deprivation and discrimination (Harris et al. 2006). Work on smoking has pointed to associations with income, employment, education, and housing tenure (Borman, Wilson, and Mailing 1999; Crampton et al. 2000; Tobias and Cheung 2001; Hill, Blakely, and Howden-Chapman 2003). Ethnic disadvantage is a particularly significant factor. Smoking prevalence among Māori is approximately double that for Europeans and polarization is increasing (Blakely et al. 2010). Historically, smoking played no part in Māori culture and its current significance results from a complex interplay of community norms, the historic impact of colonialism, and the trajectory of the smoking epidemic among Māori. It is notably high for Māori women, reflecting gendered disadvantage. Interactions with deprivation play a significant part in maintaining ethnic differences (Barnett, Moon, and Kearns 2004).

Urban ethnic segregation in New Zealand is less marked than in the United States (Johnston, Poulsen, and Forrest 2007). The major movement of Māori to urban areas came only in the 1950s and there were no laws restricting residence. Due to economic disadvantage, however, Māori were more likely to live in social housing. This was not initially concentrated spatially, although later developments were focused in outer suburbs. Consequently, such segregation as is present overlaps with socioeconomic disadvantage. Increases in immigration over the past twenty years mean that Pacific peoples are now the most segregated from the European majority, although this segregation is declining, as is that between Māori and Europeans (Grbic, Ishizawa, and Crothers 2010). Recent Asian immigrants to New Zealand are presently experiencing rising levels of segregation (Johnston, Poulsen, and Forrest 2008).

This article builds most directly on the one study, to date, that has used ethnic segregation indexes to examine smoking prevalence in New Zealand (Moon, Barnett, and Pearce 2010). This precursor study concluded that apparent segregation effects were largely artifacts of differential levels of deprivation. Four factors distinguish this study. First, it is now possible to examine data from 2006, providing an updated analysis rooted in more recent socioeconomic circumstances. Second, and more important, the earlier paper had access only to a restricted data set. This meant that analysis was limited to an assessment of the impact of ethnic segregation on age and sex groups. Data restrictions have been resolved in this article, enabling an examination of the possibility of a differential segregation impact on separate ethnic groups. Third, the earlier study, in common with the majority of U.S. studies, measured segregation at the level of the urban area. This article adds a consideration of intraurban ethnic diversity. Finally, to secure a more comprehensive assessment of segregation and smoking behavior, we move beyond current smoking prevalence and additionally consider ex-smokers.

Data and Methods

Research on smoking in New Zealand benefits immensely from the inclusion of smoking in the national census. This gives high-quality nationwide data. We used the 1996 and 2006 censuses, the most recent to cover smoking. Data were made available on smoking and quitting, cross-tabulated by age bands, sex, and ethnic group for all New Zealanders over sixteen years of age. Smoking was defined as a positive response to this question: "Do you smoke cigarettes regularly, that is one or more a day?" Quitters were people who indicated that they no longer smoked but had previously smoked at least one cigarette each day. Data on ethnic status used prioritized ethnicity as defined by Statistics New Zealand to distinguish Europeans, Māori, other ethnicities, and people of unknown ethnicity. Smoking and ethnicity status were self-assigned by census respondents. Data were extracted for the thirty-nine main and secondary urban areas of New Zealand.

To draw comparisons with the previous work by Moon and colleagues and in recognition of it being the measure of choice in most U.S. studies, a Māori isolation index was computed for each urban area. The isolation index is a measure of exposure (Massey and Denton 1988). It captures concentrated disadvantage (Subramanian, Acevedo-Garcia, and Osypuk 2005). We used a form of the standard index that controls for the local variations in the size of the Māori population (Johnston, Poulsen, and Forrest 2005):

$$I^* = \sum_{i=1}^{n} \left[\left[\frac{x_i}{X} \right] \left[\frac{x_i}{t_i} \right] \right] - \left[\frac{X}{T} \right] \qquad (1)$$

Here x denotes the Māori population in each i of n census area units (CAUs) within a particular urban area, t is the total CAU population, X is the Māori population, and T is the total population across the whole of the relevant urban area. The index varies between zero and one, with higher scores denoting greater segregation.

To capture the ethnic mix within urban areas, we employed a diversity index originally used to research plant species diversity but now widely used in

Table 1. Baseline smoking and ex-smoking prevalences (%), by age, sex, year, and ethnicity

Age	Year	European		Māori		Other	
		F	M	F	M	F	M
Smoking							
15–29	96	26.72	24.83	44.62	33.55	16.32	21.60
	06	24.76	24.95	42.45	34.71	14.85	20.15
30–44	96	24.54	26.41	46.70	40.62	14.04	28.23
	06	23.18	25.07	46.45	39.48	14.58	22.33
45–59	96	19.57	22.29	37.39	31.92	10.40	23.49
	06	19.21	20.63	40.39	33.73	11.58	17.13
60+	96	8.44	11.91	19.37	17.71	7.33	15.07
	06	7.91	9.65	21.03	15.79	4.49	8.20
Ex-smoking							
15–29	96	29.46	21.52	19.82	16.75	18.45	16.12
	06	33.59	25.82	22.28	16.62	33.98	24.01
30–44	96	46.78	42.96	29.13	31.76	26.61	28.48
	06	51.72	44.28	31.82	30.48	50.60	42.37
45–59	96	52.03	58.30	33.84	46.03	26.08	38.42
	06	59.06	60.22	40.72	48.03	59.71	60.91
60+	96	63.33	76.49	46.20	60.75	29.57	46.42
	06	77.71	83.35	58.53	72.21	78.43	81.59

population geography (Rees and Butt 2004):

$$D = 1 - \sum_{i=1}^{n} \left[\frac{x_i}{t_i} \right]^2 \qquad (2)$$

In this case x refers to an ethnic group but the notation is otherwise the same. The diversity index recognizes the mix of ethnicities in a CAU. It ranges from a low of zero when an area is ethnically homogeneous to a maximum determined by the number of ethnic categories and obtained when each ethnic group has an equal share of an area's population.

Our analysis employed multilevel modeling. We used MLwiN version 2.2 (Rasbash et al. 2009) and binomial proportional modeling (Subramanian, Duncan, and Jones 2001; Moon and Barnett 2003; Moon, Barnett, and Pearce 2010). Our design had three levels. At the individual level, status as a smoker or as an ex-smoker (quitter) was framed by age, sex, and ethnic status. To take into account socioeconomic confounders, we included a selection of deprivation indicators from the census at a second CAU level. The diversity index was included at this level, and the isolation index was a level-three, urban area variable. Deprivation confounders were grand mean centered and isolation and diversity were categorized to quintiles to enable the identification of nonlinearities. A staged approach to modeling was employed, sequentially bringing in iso-

lation and diversity, then deprivation, and then both together to assess the stability of effects.

Results and Discussion

Table 1 sets out the sociodemographic situation regarding smoking prevalence and quitting in New Zealand in 1996 and 2006. At both time points, levels of smoking are far higher among Māori, for both sexes and at all ages. People coded to other ethnicities tended to smoke least; this is likely to be an artifact of coding, with the "other" category concealing both high and low smoking ethnicities. The relationship between smoking and sex differs between the European and Māori populations. Across all ages, Māori women are more likely to smoke. This is in line with past research and reflects long-running trends and gendered disadvantage. Quitting behavior reveals some similar patterns. A markedly lower percentage of Māori self-report being ex-smokers. This is evident for both men and women. Ex-smoking rates have generally risen and tend to increase with age, although this is not the case for younger Māori men.

The mean level of Māori isolation in 1996 was 0.03, suggesting a low probability that Māori will meet only Māori within New Zealand urban areas. The highest isolation index was 0.08. By 2006, levels of Māori isolation had fallen even lower to a mean of 0.01, although the highest level of isolation increased marginally to

Table 2. Current smoking: Effects for Māori isolation and ethnic diversity, odds ratios [95% confidence interval]

	Isolation model		Ethnic diversity model	
	1996	2006	1996	2006
Ethnic group effects				
European	1	1	1	1
Māori	1.909	2.077	1.907	2.077
	[1.890, 1.928]	[2.057, 2.098]	[1.888, 1.926]	[2.057, 2.098]
Other	0.695	0.716	0.695	0.716
	[0.686, 0.703]	[0.710, 0.722]	[0.686, 0.703]	[0.710, 0.722]
Isolation and diversity effects				
Lowest quintile	1	1	1	1
Mid-low	1.017	1.061	1.021	1.034
	[0.927, 1.114]	[0.978, 1.150]	[1.019, 1.023]	[1.032, 1.037]
Mid	1.011	0.998	1.006	1.008
	[0.930, 1.099]	[0.917, 1.085]	[1.005, 1.008]	[1.006, 1.010]
Mid-high	0.990	1.016	1.047	1.063
	[0.921, 1.065]	[0.927, 1.113]	[1.037, 1.058]	[1.050, 1.076]
Highest quintile	0.918	1.094	1.009	0.998
	[0.843, 1.000]	[0.983, 1.218]	[0.997, 1.021]	[0.984, 1.012]
Deprivation effects				
No educational qualifications	1.020	1.035	1.003	1.007
	[1.018, 1.022]	[1.032, 1.038]	[1.000, 1.005]	[1.004, 1.011]
Renting tenure	1.007	1.008	1.062	1.069
	[1.005, 1.008]	[1.006, 1.009]	[1.023, 1.102]	[1.024, 1.115]
Receiving domestic purposes benefit[a]	1.052	1.063	1.101	1.092
	[1.041, 1.063]	[1.050, 1.076]	[1.056, 1.148]	[1.044, 1.142]
Unemployed	1.011	0.998	1.112	1.097
	[1.000, 1.023]	[0.985, 1.012]	[1.060, 1.167]	[1.046, 1.151]
Single-person household	1.003	1.009	1.091	1.057
	[1.001, 1.006]	[1.006, 1.012]	[1.023, 1.162]	[0.999, 1.118]

Note: Significant effects are shown in darker type.

[a]Domestic purposes benefit is state financial support primarily for single parents with dependent children or for people caring for someone who is not a spouse or partner.

0.09. The picture is one of low and relatively uniform segregation at the level of the urban area. At the more local scale of the CAU in 1996 the mean ethnic diversity was 0.35 and the highest diversity score was 0.72. In 2006 the corresponding figures were 0.47 and 0.67. These scores suggest that ethnic diversity at the CAU level was increasing overall but becoming more uniform in its distribution. Put simply, the probability of any two people in a CAU being from different ethnic groups generally rose between 1996 and 2006, indicating an increasingly diverse society but also, with reference to the falling isolation index, a society that was becoming less characterized by Māori isolation.

To elucidate how these changes affect smoking, once other relevant factors are taken into account, we turn now to our multilevel models. Our staged approach to modeling revealed consistent effects for age, sex, and deprivation at all stages. In contrast, isolation and di-

versity effects varied, suggesting that they are attenuated by deprivation and pointing to the need to include both deprivation and isolation and diversity. In the interests of space and clarity, we report only the independent effects for variables of interest from our final full models.

Smoking Prevalence

Table 2 shows the effects that ethnicity and isolation and diversity have on smoking prevalence, controlling for deprivation. Age and sex effects are not shown.

The isolation results build most immediately on the work of Moon, Barnett, and Pearce (2010), who found that a weak association with smoking in 1981 had disappeared in 1996. Here we confirm this disappearance and note that it remained absent in 2006. Moreover, we can now note that, although the area-based measure

Table 3. Ex-smoking: Effects for Māori isolation and ethnic diversity, odds ratios [95% confidence interval]

	Isolation model		Ethnic diversity model	
	1996	2006	1996	2006
Ethnic group effects				
European	1	1	1	1
Māori	0.582	0.561	0.583	0.562
	[0.574, 0.591]	[0.553, 0.569]	[0.575, 0.592]	[0.554, 0.569]
Other	0.452	0.900	0.452	0.900
	[0.444, 0.460]	[0.890, 0.910]	[0.444, 0.461]	[0.890, 0.910]
Isolation and diversity effects				
Lowest quintile	1	1	1	1
Mid-low	0.945	0.939	0.981	0.965
	[0.868, 1.028]	[0.879, 1.004]	[0.979, 0.982]	[0.963, 0.967]
Mid	0.954	0.972	0.996	0.994
	[0.884, 1.030]	[0.907, 1.041]	[0.995, 0.998]	[0.993, 0.996]
Mid-high	0.971	0.926	0.995	0.991
	[0.909, 1.036]	[0.859, 0.998]	[0.987, 1.002]	[0.981, 1.002]
Highest quintile	1.014	0.807	0.979	0.946
	[0.939, 1.095]	[0.739, 0.881]	[0.970, 0.988]	[0.935, 0.957]
Deprivation effects	0.980	0.964	0.994	0.997
No educational qualifications	[0.979, 0.982]	[0.962, 0.966]	[0.992, 0.996]	[0.995, 1.000]
	0.995	0.995	0.939	0.920
Renting tenure	[0.994, 0.996]	[0.993, 0.996]	[0.914, 0.965]	[0.890, 0.951]
	0.990	0.993	0.910	0.906
Receiving domestic purposes benefit[a]	[0.982, 0.998]	[0.983, 1.004]	[0.882, 0.938]	[0.874, 0.939]
	0.968	0.946	0.868	0.896
Unemployed	[0.960, 0.977]	[0.935, 0.957]	[0.838, 0.899]	[0.863, 0.931]
	0.994	0.996	0.813	0.928
Single-person household	[0.992, 0.997]	[0.993, 0.998]	[0.775, 0.853]	[0.887, 0.971]
	0.980	0.964	0.994	0.997

Note: Significant effects are shown in darker type.
[a]Domestic purposes benefit is state financial support primarily for single parents with dependent children or for people caring for someone who is not a spouse or partner.

of isolation is not a significant factor for smoking, the ethnic status of the individual is of considerable relevance. Māori smoke significantly more and the "other" ethnic category smoke significantly less. The deprivation indicators have the expected positive associations with one nonsignificant exception.

In the ethnic diversity models, deprivation effects are again almost all in the expected direction and the effect of individual ethnicity is virtually identical. Māori are more likely to smoke and smoking is significantly associated with area level deprivation. In contrast to isolation, the independent effect for ethnic diversity is, however, generally statistically significant. Within the limits of our research design, diversity might bring community stresses that raise smoking levels. Alternatively, it might reflect other unmeasured confounding factors or processes implicating community norms, practices, and neighborhood effects

(Pearce, Moon, and Barnett 2012). The diversity effect strengthened between 1996 and 2006, although not significantly. Deprivation has generally remained a more important influence and a suggested dose–response pattern to the association between smoking and diversity evident prior to the inclusion of deprivation in the model (not shown) disappears once deprivation is taken into account, suggesting that deprivation significantly attenuates the diversity effect in the most diverse CAUs.

Smoking Cessation

In Table 3 we examine the relationship between our target variables and ex-smoking. It is clear that both Māori and "other" ethnicities are significantly less likely than Europeans to be ex-smokers.

Table 4. Predicted probability of smoking, by year, age, sex, and ethnicity controlling for deprivation and Māori isolation and ethnic diversity

Sex	Age	Ethnicity	Segregation[a]	Māori isolation model		Ethnic diversity model	
				1996	2006	1996	2006
Woman	15–29	European	Lowest	0.255	0.239	0.250	0.228
Man				0.273	0.254	0.268	0.253
Woman	60 +			0.103	0.075	0.095	0.074
Man				0.112	0.081	0.104	0.084
Woman	15–29	Māori		0.395	0.378	0.380	0.371
Man				0.418	0.398	0.402	0.403
Woman	60 +			0.180	0.137	0.162	0.139
Man				0.194	0.147	0.176	0.156
Woman	15–29	European	Highest	0.241	0.219	0.243	0.221
Man				0.237	0.270	0.256	0.237
Woman	60 +			0.097	0.099	0.116	0.086
Man				0.095	0.126	0.123	0.093
Woman	15–29	Mäori		0.375	0.372	0.368	0.362
Man				0.370	0.439	0.385	0.382
Woman	60 +			0.168	0.188	0.192	0.158
Man				0.166	0.234	0.203	0.170

[a]Lowest/highest quintile on the Māori isolation index in the case of the isolation models; lowest/highest quintile on the ethnic diversity index for the diversity models.

Deprivation effects remain stable over time, but the effect of Māori isolation alters between 1996 and 2006. In 1996 the isolation effect was associated with reduced smoking at all but the highest levels of isolation, but no associations were statistically significant. In 2006 this pattern was generally reversed and higher levels of isolation emerged as significantly associated with a reduced probability of being an ex-smoker. This difference suggests increasing difficulties in achieving cessation in communities with high Māori isolation. Māori are widely held to be relatively organized and empowered in comparison to indigenous groups in other countries (Bramley et al. 2005) and it could be that such internal solidarity is higher in more isolated communities and results in smoking norms that reduce cessation.

The relationship between ex-smoking and ethnic diversity is generally suggestive of lower quit rates in ethnically diverse communities. It also appears that ex-smoking has become more deeply associated with greater ethnic diversity between 1996 and 2006. People in more diverse CAUs have become less likely to be ex-smokers. Nonetheless, it is individual ethnic status rather than ethnic diversity that is more strongly associated with reduced levels of ex-smoking. In 1996 people in the "other" ethnic category were most likely not to have stopped smoking; in 2006 Māori were the least likely to be ex-smokers. This changing pic-

ture for "other" ethnicities might reflect the changing patterns of immigration to New Zealand with a dominantly Pacific Islander "other" group in 1996 being confounded in 2006 by generally more affluent Asian groups with higher quitting propensity. In both 1996 and 2006 the associations with the deprivation variables were in the expected direction, indicating lower levels of ex-smoking from more deprived areas.

The Interaction of Ethnic Status and Spatial Ethnic Segregation

To extend knowledge of the relationship of segregation and smoking in New Zealand further, we explored what ethnic status and our two indexes jointly mean for smoking among different age–sex groups by examining their cross-level interaction, controlling for deprivation. We confine our attention to the highest and lowest quintiles of Māori isolation and ethnic diversity, Māori and European ethnicity, and the youngest and oldest age groups. We also limit consideration to smoking.

Table 4 shows little variation in the predicted probability of smoking between the highest and lowest levels of Māori isolation. Confidence intervals (not shown) all overlap. With this caveat of statistical nonsignificance in mind, we can note that smoking prevalence is generally predicted to be lower in CAUs with the highest levels of Māori isolation. This hint of a protective effect, possibly stemming from the positive side of a more collective culture, appears to have benefited both Māori and Europeans in 1996. In 2006 the situation changed; younger men and older people, both European and Māori, were more likely to smoke if they lived in a CAU with high Māori isolation. Generally there were small reductions in prevalence between 1996 and 2006 in areas with low Māori isolation and small rises in areas with high isolation. Again, however, these changes were not statistically significant.

There are also small but statistically nonsignificant differences in smoking prevalence by age, sex, and ethnicity in areas with high or low levels of ethnic diversity. In 1996, older people, both Māori and European, were marginally more likely to smoke if they lived in a more ethnically diverse area. This pattern persisted in 2006. Predicted smoking rates tended to reduce between 1996 and 2006 in areas of both high and low ethnic diversity, and this reduction was greatest, although still not statistically significant, for older Māori living in areas of greatest ethnic diversity.

Conclusion

We have extended and developed previous work, using more recent data, controlling for individual ethnic status, and employing smoking and cessation outcome variables and isolation and diversity predictors. In terms of ethnicity and smoking in New Zealand, it appears that what matters most, both for current and ex-smoking, is individual ethnic status. The Māori isolation index is generally unimportant, although there are suggestions that people living in areas with greater Māori isolation might be less likely to quit smoking. In contrast, the ethnic diversity index has a relatively clear association with smoking behavior and its effect on smoking prevalence appears to have strengthened between 1996 and 2006. There is, however, little patterning or gradient to this association. We were unable to identify significant differential impacts on age, sex, or ethnic groups by either Māori isolation or ethnic diversity. This suggests that diversity and isolation affect smoking by different demographic groups in similar but largely marginal ways.

We have worked with New Zealand census data. This gives the most comprehensive coverage of smoking behavior, ethnicity, and our other modeled variables. We must, however, acknowledge the limitations to our analysis. First, the recording of ethnicity in the New Zealand census is complicated by issues of self-assignment and mixed ethnicities that were automatically prioritized to one ethnicity. Consequently, our indexes of isolation and diversity as well our individual ethnicity indicators cannot be regarded as precise measures. Second, the census uses rounding to limit disclosure and some data were not provided where cell counts were too small. This issue is partly resolved through technical aspects of multilevel modeling such as its iterative approach to estimation and the idea of borrowing strength, whereby greater credence is given to cells with larger volumes of data. Third, we must acknowledge that the "other" ethnic category in our data is a chaotic conception; it groups together Pacific peoples and Asian ethnicities, as well as others. These groups have very different smoking behaviors and deprivation profiles. As we have focused on Māori–European distinctions and general ethnic diversity, this is not a major problem in this analysis, but it does mean that little can be drawn from our findings regarding the "other" ethnic group categorization. Fourth, we have been constrained to work with the census hierarchy of geographies; CAUs and urban areas might not be sensible spatial units for capturing ethnic difference and we must acknowledge the possibility that confounding processes, such as contagion between neighboring spatial units, might have affected our results.

The academic implication from this article is that the U.S. work highlighting the importance of ethnic isolation for health has limited applicability in the far less segregated society of New Zealand. The residential geography of New Zealand does not feature levels of Māori isolation that are sufficiently high for psychosocial community stresses to drive up levels of smoking. Indeed, ethnic intermarriage and processes of residential mobility have worked to ensure lower levels of segregation and, although racism, prejudice, and discrimination have certainly not been absent, there have been no parallels to the governmental, legal, and institutional roles played in segregation in the United States. Local ethnic diversity, on the other hand, is relevant to an understanding of smoking prevalence and cessation in New Zealand. The policy implication that follows is the need to recognize that more diverse communities might find cessation more difficult. Cessation campaigns need to recognize the importance of local ethnic diversity through social marketing to different groups and careful targeting that takes into account differences and the local mix of ethnic groups. Recognition of diversity should not, however, detract from policies that focus more directly on ethnic groups (rather than their mix) and on reducing the prevalence of smoking in deprived communities, for it is these factors that appear to be the main drivers of geographical inequalities in smoking behavior in New Zealand.

References

Acevedo-Garcia, D., and T. L. Osypuk. 2008. Invited commentary: Residential segregation and health—The complexity of modeling separate social contexts. *American Journal of Epidemiology* 168 (11): 1255–58.

Barnett, R., G. Moon, and R. Kearns. 2004. Social inequality and ethnic differences in smoking in New Zealand. *Social Science & Medicine* 59 (1): 129–43.

Bécares, L., J. Nazroo, and M. Stafford. 2009. The buffering effects of ethnic density on experienced racism and health. *Health & Place* 15 (3): 700–708.

Bell, J. F., F. J. Zimmerman, G. R. Almgren, J. D. Mayer, and C. E. Huebner. 2006. Birth outcomes among urban African-American women: A multilevel analysis of the role of racial residential segregation. *Social Science & Medicine* 63 (12): 3030–45.

Bell, J. F., F. J. Zimmerman, J. D. Mayer, G. R. Almgren, and C. E. Huebner. 2007. Associations between residential segregation and smoking during pregnancy among urban African-American women. *Journal of Urban Health* 84 (3): 372–88.

Blakely, T., K. Carter, N. Wilson, R. Edwards, A. Woodward, G. Thomson, and D. Sarfati. 2010. If nobody smoked tobacco in New Zealand from 2020 onwards, what effect would this have on ethnic inequalities in life expectancy? *New Zealand Medical Journal* 123 (1320): 6–8.

Borman, B., N. Wilson, and C. Mailing. 1999. Socio-demographic characteristics of New Zealand smokers: Results from the 1996 census. *New Zealand Medical Journal* 112 (1101): 460–63.

Bramley, D., P. Hebert, L. Tuzzio, and M. Chassin. 2005. Disparities in indigenous health: A cross-country comparison between New Zealand and the United States. *American Journal of Public Health* 95 (5): 844–50.

Chang, V. W., A. E. Hillier, and N. K. Mehta. 2009. Neighborhood racial isolation, disorder and obesity. *Social Forces* 87 (4): 2063–92.

Cooper, H. L. F., S. R. Friedman, B. Tempalski, and R. Friedman. 2007. Residential segregation and injection drug use prevalence among black adults in U.S. metropolitan areas. *American Journal of Public Health* 97 (2): 344.

———. 2008. Residential segregation and the prevalence of injection drug use among black adult residents of U.S. metropolitan areas. In *Geography and drug addiction*, ed. Y. Thomas, D. Richardson, and I. Cheung, 145–57. New York: Springer.

Crampton, P., C. Salmond, A. Woodward, and P. Reid. 2000. Socioeconomic deprivation and ethnicity are both important for anti-tobacco health promotion. *Health Education & Behavior* 27 (3): 317–27.

Dai, D. 2010. Black residential segregation, disparities in spatial access to health care facilities, and late-stage breast cancer diagnosis in metropolitan Detroit. *Health and Place* 16 (5): 1038–52.

Darden, J., M. Rahbar, L. Jezierski, M. Li, and E. Velie. 2010. The measurement of neighborhood socioeconomic characteristics and black and white residential segregation in metropolitan Detroit: Implications for the study of social disparities in health. *Annals of the Association of American Geographers* 100 (1): 137–58.

Ellen, I. G. 2000. Is segregation bad for your health? The case of low birth weight. *Brookings-Wharton Papers on Urban Affairs* 1:203–29.

Grady, S. C. 2006. Racial disparities in low birthweight and the contribution of residential segregation: A multilevel analysis. *Social Science & Medicine* 63 (12): 3013–29.

Grbic, D., H. Ishizawa, and C. Crothers. 2010. Ethnic residential segregation in New Zealand, 1991–2006. *Social Science Research* 39 (1): 25–38.

Harris, R., M. Tobias, M. Jeffreys, K. Waldegrave, S. Karlsen, and J. Nazroo. 2006. Effects of self-reported racial discrimination and deprivation on Maori health and inequalities in New Zealand: Cross-sectional study. *Lancet* 367 (9527): 2005–2009.

Hill, S., T. Blakely, and P. Howden-Chapman. 2003. *Smoking inequalities: Policies and patterns of tobacco use in New Zealand, 1981–1996.* Public Health Monograph Series. Wellington, New Zealand: University of Otago School of Medicine.

Johnston, R., M. Poulsen, and J. Forrest. 2005. On the measurement and meaning of residential segregation: A response to Simpson. *Urban Studies* 42 (7): 1221–27.

———. 2007. The geography of ethnic residential segregation: A comparative study of five countries. *Annals of the Association of American Geographers* 97 (4): 713–38.

———. 2008. Asians, Pacific Islanders, and ethnoburbs in Auckland, New Zealand. *Geographical Review* 98 (2): 214–41.

Kramer, M. R., and C. R. Hogue. 2009. Is segregation bad for your health? *Epidemiologic Reviews* 31 (1): 178–94.

Landrine, H., and I. Corral. 2009. Separate and unequal: Residential segregation and black health disparities. *Ethnicity and Disease* 19 (2): 179–84.

Landrine, H., and E. A. Klonoff. 2000. Racial segregation and cigarette smoking among blacks: Findings at the individual level. *Journal of Health Psychology* 5 (2): 211–19.

Massey, D. S., and N. A. Denton. 1988. The dimensions of residential segregation. *Social Forces* 67 (2): 281–315.

Messer, L. C., J. M. Oakes, and S. Mason. 2010. Effects of socioeconomic and racial residential segregation on preterm birth: A cautionary tale of structural confounding. *American Journal of Epidemiology* 171 (6): 664–73.

Moon, G., and R. Barnett. 2003. Spatial scale and the geography of tobacco smoking in New Zealand: A multilevel perspective. *New Zealand Geographer* 59:6–15.

Moon, G., R. Barnett, and J. Pearce. 2010. Ethnic spatial segregation and tobacco consumption: A multi-level repeated cross-sectional analysis of smoking prevalence in urban New Zealand, 1981–1996. *Environment and Planning A* 42:469–89.

Osypuk, T. L., and D. Acevedo-Garcia. 2008. Are racial disparities in preterm birth larger in hypersegregated areas? *American Journal of Epidemiology* 167 (11): 1295–1304.

Pearce, J., G. Moon, and R. Barnett. 2012. Socio-spatial inequalities in health-related behaviours: Pathways linking place and smoking. *Progress in Human Geography* 36:3–24.

Phillips, D. 2007. Ethnic and racial segregation: A critical perspective. *Geography Compass* 1 (5): 1138–59.

Pickett, K., and R. Wilkinson. 2008. People like us: Ethnic group density effects on health. *Ethnicity & Health* 13 (4): 321–34.

Rasbash, J., F. Steele, W. Browne, and H. Goldstein. 2009. *A user's guide to MLwiN version 2.2.* Bristol, UK: Centre for Multilevel Modelling, University of Bristol.

Rees, P., and F. Butt. 2004. Ethnic change and diversity in England, 1981–2001. *Area* 36 (2): 174–86.

Shaw, R. J., K. E. Pickett, and R. G. Wilkinson. 2010. Ethnic density effects on birth outcomes and maternal smoking during pregnancy in the U.S. linked birth and infant death data set. *American Journal of Public Health* 100 (4): 707–13.

Subramanian, S. V., D. Acevedo-Garcia, and T. L. Osypuk. 2005. Racial residential segregation and geographic heterogeneity in black/white disparity in poor self-rated health in the U.S.: A multilevel statistical analysis. *Social Science & Medicine* 60 (8): 1667.

Subramanian, S. V., C. Duncan, and K. Jones. 2001. Multilevel perspectives on modeling census data. *Environment and Planning A* 33 (3): 399–417.

Tobias, M., and J. Cheung. 2001. *Inhaling inequality: Tobacco's contribution to health inequality in New Zealand.* Wellington, New Zealand: Ministry of Health.

Vaughan Sarrazin, M. S., M. E. Campbell, K. K. Richardson, and G. E. Rosenthal. 2009. Racial segregation and disparities in health care delivery: Conceptual model and empirical assessment. *Health Services Research* 44 (4): 1424–44.

Spatial Methods to Study Local Racial Residential Segregation and Infant Health in Detroit, Michigan

Sue Grady and Joe Darden

Department of Geography, Michigan State University

Over the last several decades, blacks in the United States have experienced substantial health disadvantages compared to other racial and ethnic groups. These disadvantages have been observed for important types of morbidity and early mortality, which public health interventions have achieved limited progress in improving. A promising new direction in health geographic research investigates the relationships among racial residential segregation, neighborhood socioeconomic inequality, and racial health disparities in urban areas of the United States. Historical evidence shows that as class isolation increases in racially segregated neighborhoods, poverty is concentrated, resulting in reduced opportunities and available and accessible amenities and resources, important factors in the promotion and maintenance of population health and well-being. Contemporary evidence shows that the ability to modify the structural constraints that create and exacerbate these unhealthy "place" environments are limited by social and public health policies. This study therefore explores modifiable pathways by which to inform social and public health policy to improve the health of black residents living in concentrated poverty. The historical context of racial residential segregation and neighborhood socioeconomic inequality in the United States is reviewed. A contemporary case study of racial disparities in low birth weight in the Detroit, Michigan, metropolitan area is also presented to demonstrate the persistence of racial health disparities. To address racial health disparities it is recommended that future health policy be linked to housing policy as a way to provide social mobility options for residents living in racially segregated, concentrated poverty neighborhoods.

在过去的几十年中，美国的黑人，相比其他种族和族裔群体，有更多的健康缺陷。虽然已发现这些缺陷与重病发病率和早期死亡率相关，公共卫生干预措施在改善这些问题上只取得了有限的进展。卫生地理研究的一个有前途的新方向是探讨在美国城市地区，种族居住隔离，邻里社会经济不平等，和种族健康差距之间的关系。历史证据表明，随着在种族隔离街区的阶级隔离的增加，贫困更集中，导致机会以及可获得的和可利用的设施和资源的减少，它们是促进和维护居民的健康和福祉的重要因素。当代证据表明，改变结构局限性的能力被社会和公众健康政策所限制，这些结构局限性创建并加剧了这些不健康"地方"的环境。因此，本研究探讨可修改的途径，通过它来启发社会和公共卫生政策，以提高集中贫困地区的黑人居民的健康状况。对美国的种族居住隔离的历史背景和街区社会经济不平等进行审查。并给出密歇根州底特律大都市区低出生体重的种族差异的一个当代案例研究，以证明种族健康差距的持续性。为了解决种族的健康差距，我们建议把未来的卫生政策与住房政策联系起来，作为一种给生活在种族隔离区的集中的贫困社区的居民提供社会流动性选择的方式。*关键词：健康地理，低出生体重，种族的健康差距，种族居住隔离，社会经济地位，城市地理学。*

Durante las pasadas últimas décadas, los negros de los Estados Unidos han experimentado sustanciales desventajas en salud en comparación con otros grupos raciales y étnicos. Tales desventajas han sido observadas con referencia a importantes tipos de morbilidad y mortalidad precoz, sobre las cuales muy limitado es el progreso logrado para remediarlas mediante intervenciones de salud pública. Una nueva dirección prometedora en investigación geográfica de la salud examina las relaciones que existen entre la segregación racial por residencia, la desigualdad socioeconómica vecinal y las disparidades raciales por salud en áreas urbanas de los Estados Unidos. La evidencia histórica muestra que a medida que el aislamiento por clase aumenta en vecindarios segregados racialmente, la pobreza se concentra, resultando en una reducción de oportunidades y comodidades, y de recursos disponibles y accesibles; todos estos son factores importantes para la promoción y permanencia de la salud y bienestar de la población. La evidencia contemporánea indica que la capacidad para modificar los obstáculos estructurales que crean y exacerban estos entornos de "lugares" deletéreos, está limitada por las políticas sociales y de salud pública. Por tanto, este estudio explora rutas modificadas mediante las cuales informar la política social y de salud pública para mejorar la salud de los residentes negros que viven en pobreza concentrada. Se hace una revisión del contexto histórico de la segregación residencial por raza y la desigualdad socioeconómica vecinal en los Estados Unidos. También se presenta un estudio de caso contemporáneo de disparidades raciales en el área metropolitana de Detroit, Michigan, en lo que concierne al bajo peso al nacer, para demostrar la persistencia de disparidades raciales en salud. Para enfrentar este tipo de disparidades raciales, se recomienda que la futura política de salud pública vaya enlazada con las políticas de vivienda, como medio para proporcionar opciones de movilidad social a los residentes que viven en barriadas racialmente segregadas y de pobreza concentrada.

Racial disparities in low birth weight (infants born at less than 2,500 g) is a major public health problem in the United States. Historically, studies have focused on maternal-level risk factors and area-level poverty to explain the higher prevalence among black compared to white mothers. Targeted public health prevention and intervention programs have thus focused on changing maternal behaviors and reducing poverty to eliminate racial disparities. Although these activities have resulted in maternal health improvements, racial disparities in low birth weight persist. Over the last decade, health geographers and social epidemiologists have expanded on the research used to inform these public health initiatives to include an assessment of racial residential segregation and its impact on racial disparities in health in general and adverse birth outcomes in particular. Racial residential segregation refers to the physical separation of blacks from whites in neighborhoods across a metropolitan area. According to Massey and Denton (1988), segregation can be captured in five dimensions:

1. *Evenness:* Blacks can be distributed across neighborhoods so that they are overrepresented in some neighborhoods and underrepresented in others, leading to unevenness in the spatial distribution between blacks and whites.
2. *Isolation:* Blacks can be distributed so that they are racially isolated in that they rarely share neighborhoods with whites.
3. *Clustering:* Blacks can be distributed in a spatial pattern that constitutes a large area of contiguous neighborhoods.
4. *Concentration:* Blacks can be concentrated within a small area.
5. *Centralization:* Blacks can be spatially centralized within the urban core.

Recently, researchers (Reardon and O'Sullivan 2004) have argued that these five dimensions of segregation can be reduced to two spatial dimensions—spatial unevenness (or spatial clustering) and spatial exposure (or spatial isolation)—both of which consider the proximity of neighborhoods, in addition to the racial composition. Clustering is grouped with the unevenness measure because individuals who are redistributed at a larger scale (e.g., census tracts) can be clustered at a smaller scale (e.g., block groups). Similar distributions relating to scale might also occur with the concentration and centralization dimensions. Most social scientists use one or all of these dimensions of segregation when describing the geographic patterns of racial and ethnic groups in urban areas (Massey and Denton 1993).

Importantly, a high score on any one of these dimensions can have serious consequences by limiting black women's equal access to resources and opportunities, which could result in poor pregnancy outcomes (Acevedo-Garcia et al. 2008). In this article, racial residential segregation is viewed as a spatial manifestation of institutional racism that evolved out of the racialized social system (Bonilla-Silva 2001) established in the United States during the period of slavery (i.e., when population groups based on phenotype were placed within a social hierarchy, motivated by the ideology of white superiority and black inferiority; Blalock 1967). Over time, whites created and maintained social and economic advantages for whites through spatial structures that ensured racial inequality (Bonilla-Silva 2001). The dimensions of racial residential segregation are important examples of social–spatial structures designed by whites to foster social and economic disadvantages and maintain racial inequality. Over the last century, the restriction of life chances based on race has been most prominent in the housing market, as blacks have remained highly residentially segregated from whites—regardless of their class status (Darden and Kamel 2000; Iceland and Wilkes 2006). For poor blacks, racial residential segregation has resulted in black concentrated poverty (i.e., census tracts where 40 percent or more of the residents are poor; Jargowsky 2003). Thus, everyday life burdens of poor blacks are magnified compared to those of poor whites, who might also live in poverty but are less likely than blacks to live in concentrated poverty. An established body of research shows that the socioeconomic characteristics of neighborhoods are important determinants of health (Kawachi and Berkman 2003; Acevedo-Garcia and Osypuk 2008a). This research therefore conceptualizes residential segregation as a *distal* determinant of poor health, with the potential to induce *proximal* determinants defined as individual-level risk factors acquired from living in concentrated poverty; that is, unhealthy "place" environments. These distal and proximal determinants of health might operate independently or interactively to deteriorate the health status of all residents living in racially segregated concentrated poverty neighborhoods.

The purpose of this research is to present a contemporary case study of racial disparities in low birth weight (infants born at < 2,500 g) in the Detroit Metropolitan Area. In 2000, Metropolitan Detroit was one of the most highly segregated metropolitan areas in the United States, with most blacks highly clustered in the city of Detroit and most whites living in the suburbs. It is well known that pregnant black women living in highly segregated neighborhoods are at increased risk of having a low birth weight infant. The pathway(s) by which maternal exposure to racial residential segregation impacts low birth weight outcomes, however, is still under investigation. This study provides a novel contribution to the study of racial health disparities by (1) implementing a methodology by which to construct neighborhood zone designs that are specific to the study of low birth weight in Metropolitan Detroit; (2) applying the spatial segregation dimension of black clusters and the modified Darden–Kamel Composite Index (Darden et al. 2010) of socioeconomic position (SEP) to define these zones; and (3) estimating the effect(s) of black clusters and SEP level on two measures of low birth weight: intrauterine growth retardation and premature birth. Black clusters are referred to in the literature as *ghettoization* (Massey and Denton 1988) because of their extreme size of contiguous neighborhoods and long travel distances to encounter other racial and ethnic groups in the metropolitan area; they are therefore used as the segregation dimension in this study.

Background

In this research, the health of black mothers and their infants is studied because of their increased vulnerability to stressors associated with racism and neighborhood deprivation (Osypuk and Acevedo-Garcia 2008a; Kramer and Hogue 2009; Kramer et al. 2010). Infants born with low birth weight are at increased risk of several developmental and neurological disorders (McCormick 1985; Hack, Klein, and Taylor 1995) and infant mortality (MacDorman and Mathews 2011).

The distal pathway(s) by which racial residential segregation and low SEP might contribute to poor maternal health are complex (Williams and Collins 2001) and associated with untoward exposures such as stressors relating to lack of resources and amenities and familiar and social support (Williams and Jackson 2005). Mothers living in racially clustered ghettos might have to travel far distances to grocery stores or pharmacies, and inadequate access to nutritious foods and multivitamins (e.g., folic acid) can reduce their nutritional health, which is important in the prevention of low birth weight (Bell et al. 2006; Bukowski et al. 2009). Mothers might also have to find work longer distances from home, and transportation and time costs associated with this travel could make such work unfeasible (Acevedo-Garcia and Osypuk 2008a). Without employment and interpersonal connections, pregnant black mothers might also lack emotional or other means of support, leading to additional stressors on fetal growth and development (R. Alexander, Kogan, and Nabukera 2002). High unemployment among fathers might also strain the household, which could lead to abuse and violence, the breakup of the family, or both, resulting in single-parent households (Wilson 1987). Women living in homes and neighborhoods that lack social cohesion will have less social capital available to them and their children (Acevedo-Garcia and Osypuk 2008a). Furthermore, reduced social cohesion can lead to social disarray and increased crime and violence, resulting in further blight of the community (Massey and Denton 1993). Living in unsafe homes and neighborhoods could exacerbate stress and feelings of isolation and, over time, lead to stress-related morbidity and chronic disabilities (Williams and Jackson 2005), a phenomenon referred to in the literature as the *weathering hypothesis* (Geronimus 1996). It has been demonstrated that the chronic wear and tear on women's body organ systems (i.e., allostatic load) results in overactive hypothalamic–pituitary responses to everyday stressors and decreased immune-inflammatory responses as a result of increased circulating stress hormones (Kapcala, Chautard, and Eskay 1995). These unnatural responses increase susceptibility to developing hypertension and diabetes (Cutolo and Straub 2006) and acquiring infectious diseases (Kapcala, Chautard, and Eskay 1995).

During pregnancy, these stressors can contribute to poor fetal organ system development, reduced fetal growth, and premature birth. Although preconceptual and prenatal health care have the potential to offset some of the detrimental risks associated with living in racially segregated and poor neighborhoods, the provision of high-quality care requires appropriate staffing of health care facilities and the technology and resources to care for high-risk populations, which might also be limited in racially segregated and poor areas (Greene, Blustein, and Weitzman 2006).

Case Study: Metropolitan Detroit

The focus of this study is the three-county metropolitan area of Wayne, Oakland, and Macomb Counties (see Darden et al. 2010). This geographic area reveals a stark inequality between poor blacks in the central city of Detroit and the more affluent whites in the suburbs. Accompanying such extreme racial residential segregation is extreme class segregation (e.g., the city of Detroit had the highest poverty rate, at 33.8 percent, among large cities of 250,000 or more in 2007, when the national poverty rate was 9.8 percent; U.S. Bureau of the Census 2008). These geographic disparities in race and class make the Detroit Metropolitan Area an ideal location to improve our understanding of racial disparities in low birth weight.

Data

Mothers–Infants. Individual birth records were obtained from the Michigan Department of Community Health, Office of Vital Statistics for the years 1995 to 2006 ($N = 637,342$). Following the removal of pleural births and records with missing information ($n = 81,270$) the data set was reduced to 556,072 records. Of these, 33.4 percent ($n = 186,180$) of mothers were black and 66.5 percent ($n = 369,892$) of mothers were white. The dependent variables used in this analysis to examine low birth weight were intrauterine growth restriction (IUGR; $n = 12,722$) or infants born at less than the 10th percentile of weight for gestational age and preterm birth ($n = 20,367$), defined as infants born at less than thirty-seven weeks gestation and above the 10th percentile of weight for gestational age (National Reference for Fetal Growth Curve; G. R. Alexander et al. 1996). Other maternal and infant characteristics that were studied included mother's race (1 = black, 0 = white), age in years (continuous), educational level in years (continuous), and smoking history during pregnancy (1 = yes, 0 = no). Age is an important risk factor for low birth weight if black mothers are experiencing early health deterioration. Increased education should protect maternal health through knowledge and increased opportunities. Smoking is the most important predictor of poor fetal growth and low birth weight (MacDorman and Mathews 2011).

Neighborhoods. In the analysis, census tracts are surrogates for neighborhoods. The three-county metropolitan area of Detroit consisted of 1,165 total census tracts. Census tracts were excluded from the current analyses if they had fewer than 100 people ($n = 10$ census tracts), had only juvenile institutions ($n = 3$ census tracts), and had Arab or Hispanic populations that exceeded 10 percent of the census tract ($n = 61$ tracts). A total of 1,091 census tracts were thus eligible for analysis. The census tracts ($n = 1,091$) of mother's residence at the time of infant birth was the geographic unit from which new neighborhood zones were constructed (methodology follows). Birth records were geocoded to the residential street address of the mother and the "points" were assigned a census tract identifier in ArcGIS 10.0 (Environmental Systems Research Institute [ESRI] 2010). Data to calculate the segregation and SEP indexes were collected from the 2000 U.S. Bureau of the Census (2002).

Methodology

Automated Zone Matching Methodology. Constructing a new set of neighborhood zones for the study of low birth weight in metropolitan Detroit was necessary because of the small number of low birth weight births in many census tracts and the mix and match of maternal-level risk factors in very high and very low SEP and racially segregated and nonsegregated neighborhoods, a phenomenon referred to as *structural confounding* (Messer, Oakes, and Mason 2010). The new neighborhood zone design was derived using the AZTool software (Martin and Cockings 2011). AZTool operates by recombining census tracts into a different set of output zones using an iterative process by which one tract is randomly selected and attribute constraint parameter(s) are evaluated. If the constraint parameters are not met, automated zone matching (AZM) will search contiguous tracts until they are achieved, thereafter aggregating and dissolving internal boundaries to create a new zone. In this study, a minimum threshold of thirty low birth weight births was used to ensure an adequate sample size (Hox 1995; Kreft 1996) for subsequent multilevel regression analysis. The other constraint parameters were the SEP level of census tracts and maximum shape compactness. If census tracts with more than one SEP level were included in a new AZM zone, the SEP level with a majority of births was assigned to that zone.

Segregation Index. The spatial clustering dimension of segregation was computed by aggregating the number of blacks and total population at the census tract level to the AZM zone. The local G-statistic (Getis and Ord 1992) in ArcGIS 10.0 (ESRI 2010) was calculated using the percentage black population by AZM zone. The G-statistic estimates how correlated (clustered) contiguous zones of percentage black population are with one another. The output is a standardized score made up of a z score and p value. In this study, a z score > +1.96 (two standard deviations from the mean p value 0.05) was used to detect high black clusters. Those zones with a z score > -1.96 and $\leq$ +1.96 were considered racially integrated or "mixed," and a z score $\leq$ -1.96 was considered low black segregation (referent). For a complete description of how the spatial cluster index is calculated, please see Lee and Wong (2001).

Socioeconomic Position Index. Socioeconomic inequality is measured using the modified Darden–Kamel Composite Index, which measures SEP and assigns an SEP score to census tracts. A high SEP score reflects better socioeconomic quality of tract characteristics, and a low SEP score reflects poor-quality tract characteristics. The modified Darden–Kamel index incorporates nine variables, including the percentage of residents with university degrees, median household income, percentage of managerial and professional positions, median value of housing dwellings, median gross rent of dwellings, percentage of homeownership, percentage of households with a vehicle, unemployment rate, and the percentage of the population living below poverty. The SEP index was divided into quintiles (i.e., very low [SEP = 0], low [SEP = 1], medium [SEP = 2], high [SEP = 3], and very high [SEP = 4]). In this analysis, very low SEP was the primary exposure of interest with medium, high, and very high SEP as the referent. For a complete description of the Darden–Kamel Composite Index, see Darden et al. (2010).

Statistical Analysis. To estimate the effect of racial clustering and SEP levels on low birth weight, two-level hierarchical generalized linear models with a logit link function were used to predict low birth weight

as a function of maternal and neighborhood-level characteristics (Raudenbush and Bryk 2002). To examine the odds of low birth weight via IUGR as compared to preterm birth, a multinomial logistic hierarchical generalized linear model was used to assess whether an infant was born normal weight (referent), low birth weight due to IUGR, or preterm as competing risks. The analyses were conducted using HLM 6.0 (Scientific Software International 2011). All variables input into the models were grand centered. For a more detailed description of these multilevel models, please see Raudenbush and Bryk (2002).

Results

Table 1 shows the incidence of IUGR and preterm births for black and white infants by SEP level in high and low black segregated and racially mixed AZM zones. Overall the incidence of IUGR and preterm birth was higher for black versus white infants, with incidence decreasing in both groups with increasing SEP across all levels of segregation. Figure 1 shows a map of SEP and segregation levels in metropolitan Detroit. Zones with very low SEP are primarily located in highly segregated neighborhoods in Detroit. There are also many highly segregated and racially integrated or mixed zones of high and very high SEP. Table 2 shows the results from the multilevel models. The first model (Model 1), containing only individual-level risk factors, showed an increased odds of IUGR for black compared to white infants, odds ration (OR) = 2.323 (95 percent confidence interval [CI], 2.227, 2.425) controlling for mother's age, educational level, and smoking status. When high black segregation and very low SEP were added to the model (Model 2), the odds of IUGR for black compared to white infants was slightly attenuated, OR = 1.973 (95 percent CI, 1.852, 2.103), with very low SEP (not segregation) significantly reducing the racial disparity in IUGR incidence. Model 3 included the interaction terms black race * high black segregation and very low SEP. In this model, the odds of IUGR for black compared to white infants decreased substantially, OR = 1.737 (95 percent CI, 1.567, 1.926), with very low SEP (not segregation) significantly explaining this reduction. Living in very low SEP zones is therefore an important risk factor for IUGR independent of high black segregation.

In Model 1 the odds of preterm birth for black infants was OR = 2.664 (95 percent CI, 2.570, 2.762) controlling for maternal age, education, and smoking.

Table 1. Singleton low birth weight births (intrauterine growth restriction and preterm birth) stratified by mother's race and segregation and socioeconomic position levels in automated zone matching zones: Detroit Metropolitan Area, 1995–2006

Race * SEP	Births No.	IUGR No.	Rate[a]	Preterm No.	Rate[a]
High black segregation (n = 203)					
Black					
SEP = 0	88,235	3,744	4.2	5,781	6.6
SEP = 1	27,544	952	3.5	1,730	6.3
SEP = 2	3,278	87	2.7	174	5.3
SEP = 3	789	21	2.7	55	7.0
SEP = 4	1,811	45	2.5	93	5.1
White					
SEP = 0	4,442	144	3.2	180	4.1
SEP = 1	1,570	39	2.5	69	4.4
SEP = 2	195	—	—	—	—
SEP = 3	223	—	—	—	—
SEP = 4	1,625	21	1.3	39	2.4
Integrated or racially mixed neighborhoods (n − 276)					
Black					
SEP = 0	16,373	643	3.9	1,072	6.5
SEP = 1	18,436	613	3.3	1,084	5.9
SEP = 2	8,761	270	3.1	480	5.5
SEP = 3	3,318	88	2.7	186	5.6
SEP = 4	2,626	73	2.8	134	5.1
White					
SEP = 0	12,056	347	2.9	398	3.3
SEP = 1	26,438	479	1.8	727	2.7
SEP = 2	57,317	881	1.5	1,302	2.3
SEP = 3	29,539	486	1.6	726	2.5
SEP = 4	32,698	413	1.3	650	2.0
Low black segregation (n = 201)					
Black					
SEP = 0	7,609	223	2.9	429	5.6
SEP = 1	1,983	59	3.0	107	5.4
SEP = 2	1,992	61	3.1	123	6.2
SEP = 3	1,937	60	3.1	103	5.3
SEP = 4	1,488	36	2.4	84	5.6
White					
SEP = 0	—	—	—	—	—
SEP = 1	195	—	—	—	—
SEP = 2	51,949	813	1.6	1,322	2.5
SEP = 3	91,107	1,348	1.5	2,106	2.3
SEP = 4	60,538	768	1.3	1,186	2.0
Overall	556,072	12,722	2.3	20,367	3.7

Note: Dashed line represents < 20 births and infers an unstable rate. IUGR = intrauterine growth restriction; SEP = socioeconomic position.
[a]Rate per 100 live births.

After adding high black segregation and very low SEP to the model (Model 2) the odds of preterm birth for black compared to white infants was slightly attenuated, OR = 2.459 (95 percent CI, 2.339, 2.586). Model 3 included the interaction term black race * high black segregation and very low SEP. In this model the odds of preterm birth for black compared to white infants decreased substantially, OR = 2.124 (95 percent CI, 1.941, 2.326), with both high black segregation and very low SEP significantly explaining this reduction.

In summary, the findings from this study showed that the level of SEP in one's neighborhood was more important in predicting IUGR than the level of racial residential segregation (i.e., the gap in IUGR between black and white infants decreased with increasing SEP). Mothers living in high black segregated and very low

41

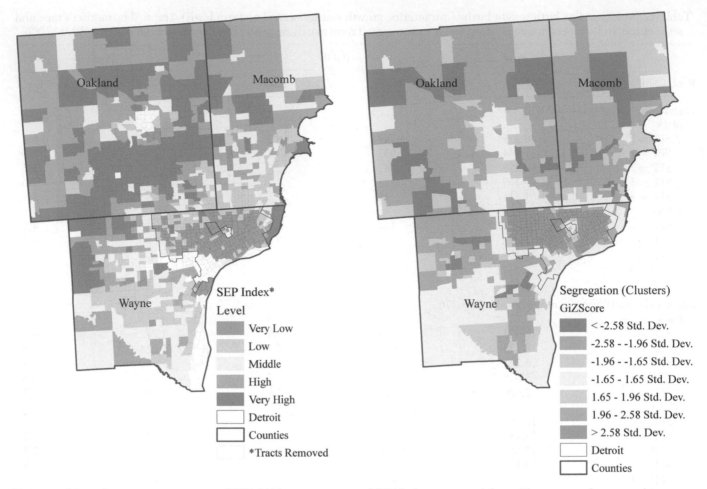

Figure 1. Maps of socioeconomic position (SEP) (A) by census tracts and (B) black segregation (clusters) by automated zone matching zones: Metropolitan Detroit, Michigan, 2000. (Color figure available online.)

SEP neighborhoods, however, were more likely to have a preterm birth than similar mothers living in racially integrated, mixed, or low segregated and higher SEP neighborhoods (i.e., the gap in preterm birth between black and white infants decreased with decreased segregation and increased SEP). Racially segregated and concentrated poverty neighborhoods thus appear to be strong distal determinants of poor health. The proximal determinant of smoking was an important risk factors for IUGR and preterm birth; increasing maternal age was an important risk factor for preterm birth but not IUGR; and increasing education was significantly protective of IUGR and preterm birth for all mothers, regardless of the neighborhood in which they lived.

Discussion

We have argued that the effects of racial residential segregation and black concentration in very low socioeconomic status (SES) neighborhoods have con-

tributed to the gap in IUGR and preterm births between black and white mothers. The outcome appears to be the result of institutional racism; that is, unfair policies and practices in the housing industry and other institutions that have produced racially inequitable outcomes for black mothers and advantages for white mothers (Keleher 2000; Bonilla-Silva 2001). Preterm births and other types of black–white disparities have existed historically, and our data and analyses demonstrate that such disparities still persist today in the United States (for other types of disparities, see Centers for Disease Control and Prevention 2011).

Acevedo-Garcia and Osypuk (2008b, 1255) reported that racial disparities are "so pronounced that ignoring segregation may lead to misestimating the effect of individual-level risk factors on health." Our case study highlighted this point by showing a significant reduction in the racial disparities of IUGR and preterm birth in metropolitan Detroit when high black segregation and very low SEP were not ignored. Our case study also

Table 2. Odds ratios for hierarchical multinomial logistic regression (fixed effect models) stratified by race estimating competing risk of low birth weight due to intrauterine growth restriction and preterm birth compared to normal birth weight in the Detroit Metropolitan Area, 1995–2006

Characteristics	Odds ratio	95% CI	p value	Odds ratio	95% CI	p value	Odds ratio	95% CI	p value
IUGR									
Black	**2.323**	2.227, 2.425	0.000	**1.973**	1.852, 2.103	0.000	**1.737**	1.567, 1.926	0.000
High black segregation[a]							0.914	0.751, 1.114	0.375
SEP very low[b]							0.812	0.706, 0.936	0.004
Age	1.002	0.999, 1.006	0.187	1.002	1.000, 1.006	0.087	1.002	1.000, 1.006	0.085
Education	0.947	0.940, 0.955	0.000	0.949	0.941, 0.957	0.000	0.949	0.942, 0.957	0.000
Smoking	2.322	2.220, 2.429	0.000	2.288	2.187, 2.394	0.000	2.285	2.185, 2.391	0.000
High black segregation				1.020	0.951, 1.094	0.572	1.109	0.968, 1.237	0.138
SEP very low				1.285	1.196, 1.382	0.000	1.368	1.257, 1.491	0.000
Preterm									
Black	**2.664**	2.570, 2.762	0.000	**2.459**	2.339, 2.586	0.000	**2.124**	1.941, 2.326	0.000
High black segregation							0.840	0.710, 0.996	0.045
SEP very low							0.880	0.780, 0.995	0.040
Age	1.014	1.012, 1.018	0.000	1.015	1.012, 1.018	0.000	1.015	1.012, 1.018	0.000
Education	0.965	0.959, 0.973	0.000	0.965	0.959, 0.973	0.000	0.966	0.960, 0.973	0.000
Smoking	1.714	1.647, 1.785	0.000	1.700	1.633, 1.771	0.000	1.698	1.632, 1.768	0.000
High black segregation				0.993	0.927, 1.063	0.858	1.126	1.002, 1.267	0.046
SEP very low				1.177	1.099, 1.261	0.000	1.232	1.141, 1.332	0.000

Note: Bold is used to highlight the reduction in racial disparities in IUGR and preterm birth and the contribution of high black segregation and very low SEP. IUGR = intrauterine growth restriction; SEP = socioeconomic position.
[a]Referent group = racially mixed and low black segregated automated zone matching zones.
[b]Referent group = mixed, high, and very high SEP (excludes low SEP).

demonstrated the construction and implementation of neighborhood zones that were ideal for the study of low birth weight in metropolitan Detroit. AZM zones were derived to address the "complexity of modeling separate social contexts" (Acevedo-Garcia and Osypuk 2008b, 1255) in regard to sample size constraints in segregation and health studies. Although focused on the three-county area of metropolitan Detroit, these methods and findings have implications for other metropolitan areas in the United States and in other industrialized countries (Darden et al. 2010). Improving our understanding of neighborhood design characteristics and the conditions under which all residents receive optimal health protection is an important area of future health geographic research with implications for public health policy.

nomic neighborhood inequality, and health outcomes has been established, any future health policy to address racial disparities should be linked to housing policy as a way to provide spatial mobility options for residents who live in racially segregated concentrated poverty neighborhoods (Acevedo-Garcia, Osypuk, and Werbel 2004; Osypuk and Acevedo-Garcia 2010). Moreover, with assistance from nonprofit organizations such as fair housing centers, there is some evidence that moving to greater opportunity in the suburbs away from low-SES neighborhoods is one realistic alternative for many residents of segregated areas in concentrated poverty in Detroit (Darden and Thomas forthcoming). Such housing policy will enable black mothers to close the gap between their rate of preterm births and the rate for white mothers.

Conclusions

We have demonstrated the historical evolution of the effects of racial residential segregation and socioeconomic neighborhood inequality on the population health of black women and their disproportionately low birth weight infants compared to white women. Because the link among racial residential segregation, socioeco-

References

Acevedo-Garcia, D., and T. L. Osypuk. 2008a. Impacts of housing and neighborhoods on health: Pathways, racial/ethnic disparities and policy directions. In *Segregation: The rising costs for America*, ed. J. Carr and N. Kutty, 197–235. London and New York: Routledge.

Acevedo-Garcia, D., and T. L. Osypuk. 2008b. Invited commentary: Residential segregation and health—The complexity of modeling separate social contexts. *American Journal of Epidemiology* 168 (11): 1255–58.

Acevedo-Garcia, D., T. L. Osypuk, N. McArdle, and D. R. Williams. 2008. Toward a policy-relevant analysis of geographic and racial/ethnic disparities in child health. *Health Affairs* 27 (2): 321–33.

Acevedo-Garcia, D., T. L. Osypuk, and R. E. Werbel. 2004. Does housing mobility policy improve health? *Housing Policy Debate* 15 (1): 49–97.

Alexander, G. R., J. H. Himes, R. B. Kaufman, J. Mor, and M. Kogan. 1996. A United States national reference for fetal growth. *Obstetrics and Gynecology* 87 (2): 163–68.

Alexander, R., M. Kogan, and S. Nabukera. 2002. Racial differences in prenatal care use in the United States: Are disparities decreasing? *American Journal of Public Health* 92 (12): 1970–74.

Bell, J. F., F. J. Zimmerman, G. R. Almgren, J. D. Mayer, and C. E. Huebner. 2006. Birth outcomes among urban African American women: A multilevel analysis of the role of racial residential segregation. *Social Science and Medicine* 63:3030–45.

Blalock, H. 1967. *Towards a theory of minority group relations.* New York: Capricorn Books.

Bonilla-Silva, E. 2001. *White supremacy and racism in the post-civil rights era.* Boulder, CO: Lynne Rienner.

Bukowski, R., F. D. Malone, T. Porter, D. A. Nyberg, C. H. Comstock, G. D. V. Hankins, K. Eddleman, et al. 2009. Preconceptual folate supplementation and the risk of spontaneous preterm birth: A cohort study. *PLoS Medicine* 6 (5): e10000361.

Centers for Disease Control and Prevention. 2011. Health disparities. http://www.cdc.gov/omhd/Topic/HealthDisparities.html (last accessed 10 November 2011).

Cutolo, M., and R. H. Straub. 2006. Stress as a risk factor in the pathogenesis of rheumatoid arthritis. *Neuroimmunomodulation* 13 (5–6): 277–82.

Darden, J. T., and S. Kamel. 2000. Black residential segregation in the city and suburbs of Detroit: Does socioeconomic status matter? *Journal of Urban Affairs* 22 (1): 1–13.

Darden, J., M. Rahbar, L. Jezierski, M. Li., and E. Velie. 2010. The measurement of neighborhood socioeconomic characteristics and black and white residential segregation in metropolitan Detroit: Implications for the study of social disparities in health. *Annals of the Association of American Geographers* 100 (1): 1–22.

Darden, J. T., and R. Thomas. *Forthcoming. Detroit: Race riots, racial conflicts and efforts to bridge the racial divide.* East Lansing: Michigan State University Press.

Environmental Systems Research Institute (ESRI). 2010. ArcGIS 10.0. Redlands, CA: ESRI. http://www.esri.com (last accessed 10 November 2011).

Geronimus, A. T. 1996. Black/white differences in the relationship of maternal age to birthweight: A population-based test of the weathering hypothesis. *Social Science and Medicine* 42 (4): 589–97.

Getis, A., and J. K. Ord. 1992. The analysis of spatial association by use of distance statistics. *Geographical Analysis* 24 (3): 189–207.

Greene, J., J. Blustein, and B. C. Weitzman. 2006. Race, segregation, and physicians' participation in Medicaid. *Milbank Quarterly* 84 (2): 239–72.

Hack, M., N. Klein, and G. Taylor. 1995. Long-term development outcomes of low birth weight infants. *The Future of Children* 5 (1): 1–21.

Hox, J. 1995. *Applied multilevel analysis.* Amsterdam, The Netherlands: TT-Publikaties.

Iceland, J., and R. Wilkes. 2006. Does socioeconomic status matter? Race, class and residential segregation. *Social Problems* 53 (2): 248–73.

Jargowsky, P. 2003. *Stunning progress, hidden problems: The dramatic decline of concentrated poverty in the 1990s.* Washington, DC: Brookings Institution.

Kapcala, L. P., T. Chautard, and R. L. Eskay. 1995. The protective role of the hypothalamic–pituitary–adrenal axis against lethality produced by immune, infectious, and inflammatory stress. *Annals of the New York Academy of Science* 771:419–37.

Kawachi, I., and L. F. Berkman. 2003. *Neighborhoods and health.* Oxford, UK: Oxford University Press.

Keleher, T. 2000. *Racial justice leadership.* Los Angeles: Applied Research Center.

Kramer, M. R., H. L. Cooper, C. D. Drews-Botsch, L. A. Waller, and C. R. Hogue. 2010. Metropolitan isolation segregation and black–white disparities in very preterm birth: A test of mediating pathways and variance explained. *Social Science and Medicine* 71:2108–16.

Kramer, M. R., and C. R. Hogue. 2009. Is segregation bad for your health? *Epidemiological Reviews* 31: 178–94.

Kreft, I. G. G. 1996. *Are multilevel techniques necessary? An overview including simulation studies.* Los Angeles: California State University Press.

Lee, J., and D. W. S. Wong. 2001. *Statistical analysis with ArcView GIS.* New York: Wiley.

MacDorman, M. F., and T. J. Mathews. 2011. Understanding racial and ethnic disparities in US infant mortality rates. National Center for Health Statistics (NCHS) Data Brief No. 74, National Center for Health Statistics, Hyattsville, MD.

Martin, D., and S. Cockings. 2011. AZTool. Southampton, UK: Southampton University. http://www.geodata.soton.ac.uk/software/AZTool/ (last accessed 26 October 2011).

Massey, D., and N. Denton. 1988. The dimensions of residential segregation. *Social Forces* 67:281–315.

———. 1993. *American apartheid: Segregation and the making of the underclass.* Cambridge, MA: Harvard University Press.

McCormick, M. C. 1985. The contribution of low birthweight to infant mortality and childhood morbidity. *The New England Journal of Medicine* 312:82–90.

Messer, L. C., J. M. Oakes, and S. Mason. 2010. Effects of socioeconomic and racial residential segregation on preterm birth: A cautionary tale of structural confounding. *American Journal of Epidemiology* 171 (6): 664–73.

Osypuk, T. L., and D. Acevedo-Garcia. 2008. Are racial disparities in preterm birth larger in hypersegregated areas? *American Journal of Epidemiology* 167:1295–1304.

———. 2010. Beyond individual neighborhoods: A geography of opportunity perspective for understanding racial/ethnic health disparities. *Health and Place* 16 (6): 1113–23.

Raudenbush, S. W., and A. S. Bryk. 2002. *Hierarchical linear models*. London: Sage.

Reardon, S., and D. O'Sullivan. 2004. Measures of spatial segregation. *Sociological Methodology* 34:121–62.

Scientific Software International (SSI). 2011. *HLM 6.0*. Lincolnwood, IL: SSI. http://www.ssicentral.com (last accessed 19 October 2011).

U.S. Bureau of the Census. 2002. *2000 census of population and housing (SF3): Michigan*. Washington, DC: U.S. Bureau of the Census, Data User Services Division.

———. 2008. *2007 American community survey*. Washington, DC: U.S. Government Printing Office.

Williams, D. R., and C. Collins. 2001. Racial residential segregation: A fundamental cause of racial disparities in health. *Public Health Reports* 116:404–16.

Williams, D. R., and P. B. Jackson. 2005. Social sources of racial disparities in health. *Health Affairs* 24 (2): 325–34.

Wilson, J. 1987. *The truly disadvantaged: The inner city, the underclass and public policy*. Chicago: University of Chicago Press.

Connecting the Dots Between Health, Poverty, and Place in Accra, Ghana

John R. Weeks,* Arthur Getis,* Douglas A. Stow,* Allan G. Hill,[†] David Rain,[‡] Ryan Engstrom,[‡]
Justin Stoler,* Christopher Lippitt,* Marta Jankowska,* Anna Carla Lopez-Carr,* Lloyd Coulter,*
and Caetlin Ofiesh[‡]

*Department of Geography, San Diego State University
[†]Department of Global Health and Population, Harvard School of Public Health
[‡]Department of Geography, George Washington University

West Africa has a rapidly growing population, an increasing fraction of which lives in urban informal settlements characterized by inadequate infrastructure and relatively high health risks. Little is known, however, about the spatial or health characteristics of cities in this region or about the spatial inequalities in health within them. In this article we show how we have been creating a data-rich field laboratory in Accra, Ghana, to connect the dots between health, poverty, and place in a large city in West Africa. Our overarching goal is to test the hypothesis that satellite imagery, in combination with census and limited survey data, such as that found in demographic and health surveys (DHSs), can provide clues to the spatial distribution of health inequalities in cities where fewer data exist than those we have collected for Accra. To this end, we have created the first digital boundary file of the city, obtained high spatial resolution satellite imagery for two dates, collected data from a longitudinal panel of 3,200 women spatially distributed throughout Accra, and obtained microlevel data from the census. We have also acquired water, sewerage, and elevation layers and then coupled all of these data with extensive field research on the neighborhood structure of Accra. We show that the proportional abundance of vegetation in a neighborhood serves as a key indicator of local levels of health and well-being and that local perceptions of health risk are not always consistent with objective measures.

西非人口迅速增长，居住在城市非正式定居点的人口比例越来越多，这些定居点以基础设施不足和相对高的健康风险为特点。然而，我们既对这一地区的城市空间或健康的特点，也对它们中的空间健康不平等所知甚少。在这篇文章中，我们表明如何在加纳的阿克拉创建数据丰富的野外实验室，在一个西非的大型城市里，把卫生，贫困，和地方之间的节点连接起来。我们的首要目标是测试一个假设，即卫星图像，结合普查和有限的调查数据，如在人口与健康调查 (DHSs) 中所发现的数据，来提供一些城市里健康不平等的空间分布，这些城市比起阿克拉有较少的数据存在。为此，我们已经创建了第一个城市的数字边界文件，获得了两个时次的高空间分辨率卫星图像，从纵向面收集了在空间上分布于整个阿克拉的3200位妇女的数据，并从人口普查获得了微观层面的数据。我们还收购了供水，污水和高度数据层，然后把所有这些数据与阿克拉的邻里结构上的广泛的实地研究结合。我们的研究表明，邻里的植被比例丰度可作为地方各级卫生和福祉的关键指标，并且那些健康风险的地方看法并不总是与目标措施一致。*关键词: 非洲，加纳，健康，邻里，遥感。*

África Occidental cuenta con una población en rápido crecimiento, una creciente fracción de la cual vive en asentamientos urbanos informales caracterizados por su infraestructura inadecuada y riesgos contra la salud relativamente altos. Sin embargo, muy poco se sabe de las características espaciales o sanitarias de las ciudades de esta región, ni de las desigualdades espaciales en salud existentes en las mismas. En este artículo mostramos lo que estamos haciendo en Accra, Ghana, buscando crear un laboratorio de campo bien surtido de datos, para tratar de unir los puntos entre salud, pobreza y lugar en una ciudad grande del África Occidental. Nuestro principal objetivo es poner a prueba la hipótesis de que las imágenes satelitales, combinadas con datos censales y los limitados datos de campo, como los que se encuentran en los estudios demográficos y de salubridad (DHSs), pueden generar indicios sobre la distribución espacial de desigualdades de la salud en ciudades donde existen menos datos de los que nosotros logramos conseguir para Accra. Con tal propósito, creamos el primer archivo de límites digitales de la ciudad, obtuvimos un conjunto de imágenes satelitales de alta resolución espacial de dos fechas de observación, recogimos datos entre un panel longitudinal de 3.200 mujeres distribuidas espacialmente a través de Accra, y obtuvimos datos del censo a nivel micro. Obtuvimos también datos sobre agua, alcantarillado y elevaciones, para luego cotejar toda esta información con un amplio estudio de campo sobre la estructura barrial de Accra. Mostramos que la abundancia proporcional de vegetación en un vecindario sirve de indicador clave de los niveles locales de salud y bienestar y que las percepciones locales sobre riesgos de la salud no son siempre consistentes con las mediciones objetivas.

Sustainable development in Africa, as elsewhere in the world, requires that future population growth be absorbed by cities, because only in or near cities can we anticipate the kind of economic and employment growth needed for people to rise above the poverty level. At the same time, sustainable development requires a healthy population to generate rising levels of economic productivity (Bloom, Canning, and Sevilla 2001; López-Casasnovas, Rivera, and Currais 2005). The health of the urban population thus takes on political and economic significance, yet the literature on health within cities of developing nations is very limited. Our goal is to provide answers to the question of how to move the urban health transition forward in sub-Saharan Africa through an analysis of data for Accra, the capital city of Ghana. Because surveying health patterns within a city is a very expensive process, we test the idea that satellite imagery can be used to help identify places with the worst and best health. If so, we might be able to efficiently model the spatial inequality in health in any city in a developing country. Our work in Accra aims to connect these dots and develop a model that can be used in other cities.

The theoretical framework that underpins our work is that urban health and well-being are part of a complex social–ecological–biomedical system, consistent with the view that "the health and well-being of the whole population may be best conceptualized as a 'systems' problem, occurring on a continuum over the human lifespan as well as across a variety of levels of analysis, ranging from the cellular and molecular to individual and interpersonal behaviors, to the community and society and to macro-socioeconomic and global levels" (Mabry et al. 2008, S215). The value of this approach, as Diez Roux (2007, 572) noted, is that "a systems approach focuses on understanding the system functioning so that changes in response to an intervention can be predicted." The health of an urban resident cannot be understood, nor effective interventions implemented, without taking into account the sociocultural and ecologic structure of the city and the economic and health infrastructure in which residents are embedded. Individuals are not simply autonomous biological entities. Rather, the biological characteristics are constantly interacting with the broader environment in a system of feedback loops. Our specific focus within this systems approach is to emphasize that neighborhoods, broadly defined, are important when it comes to the health and well-being of people living in those places (Kawachi and Berkman 2003; Entwisle 2007). Taking it one step further, we hypothesize that when neighborhoods become healthier as a system, not just as a set of individuals, they become environments in which economic productivity and income are more likely to increase, which in turn will accelerate the overall improvement in the health of the people living and working in those environments. As Dunn and Cummins (2007) have noted:

> One of the tantalizing features of research on context and health is that it may lead to more effective interventions to improve health and reduce health disparities. Traditional behaviouralist perspectives, which saw health behaviours and other determinants of health as simple individual attributes, have now been eclipsed by a perspective that emphasizes human behaviour and activity as significantly influenced by contextual factors. It follows from this perspective that changes in context may produce changes in the risk profile for whole populations, rather than just for the people who receive and are successful with individually-oriented interventions. (1822)

The local context in a city influences health indirectly by promoting or constraining the knowledge of disease transmission and the ability to access health providers. It also directly affects health through water quality, sewage and waste disposal methods, crowded and dirty homes and yards, indoor air pollution from unhealthy cooking fuels, or lack of refrigeration. Both the direct and indirect contextual influences are related closely to socioeconomic status (SES) and housing quality. Because people of similar SES and similar housing quality tend to live in proximity to one another (although there are exceptions to this pattern), identifying regions of the city by these characteristics provides an important indicator of the population's health and well-being. Our work builds on the idea that neighborhood-level patterns of morbidity and mortality ("health") are importantly influenced by the interaction and feedback between poverty and place and that these patterns can be discerned from the classification of high-resolution satellite imagery.

Connecting People to Places

The 2000 Census of Population and Housing in Ghana was the first geographically detailed census in the nation's history, although none of the maps created for the census had yet been digitized when we began our research. The Ghana Statistical Service (GSS) provided us with a set of paper maps, along with a formal set of boundary definitions (based on physical points of interest in a neighborhood), and we used this information in combination with high-resolution satellite imagery to create a digital boundary file of the 1,731 enumeration areas (EAs) in Accra. We then connected boundaries to the census data, allowing us to characterize each EA on the basis of census-derived data.

EAs in Accra are designed to encompass approximately 1,000 persons and are generally too small in area, population, or both to be considered neighborhoods. In recognition of this fact, GSS created a set of eighty-eight neighborhood boundaries that agglomerated EAs into areas that we call "vernacular" neighborhoods, referring to "neighborhood boundaries that are broadly recognized and agreed to by residents of a given city—in this case Accra, Ghana—even if they may have no premeditated and formal definition. These are the place names, for example, that would be provided to a taxi driver, especially since there is no comprehensive street address system in Accra" (Weeks et al. 2010, 563). These boundaries are not unlike those generated on the basis of local knowledge without access to GSS census data (Songsore et al. 2005; Agyei-Mensah and Owusu 2010) and are similar to what one would find in printed tourist maps of Accra.

One of the goals of our research has been to conduct fieldwork to validate and reconcile differing neighborhood boundaries. The result of this effort has been a modification of the original GSS vernacular neighborhoods to reflect the perceptions of residents of the local boundaries. We call these the field modified vernacular (FMV) neighborhoods. Most of the difference between the original and FMV neighborhood definitions is that the latter provide a more nuanced and finer gradation, dividing the city into 108 neighborhoods (Rain et al. 2011), as shown in Figure 1. For purposes of this article,

Figure 1. Housing quality index quintiles by field modified vernacular neighborhoods in Accra, Ghana, based on 2000 census data. (Color figure available online.)

we are treating the FMV neighborhoods as the environment in which people live and thus the context in which their health can be evaluated.

Connecting People and Places to the Environment

We first measure the environmental context with an index that combines aspects of the built and social environment, created by running the 2000 census data through a principal components analysis to produce a "housing quality" (HQ) index. The variables included type of floor, whether or not the house has electricity, the source of water, type of toilet, type of bathing facility, methods of waste disposal, cooking fuel, kitchen facility, and number of persons per sleeping room. We used the first component from the rotated matrix as our index of HQ. It had an eigenvalue of 4.5, which combined characteristics of toilet, liquid waste, source of water, cooking fuel, and type of kitchen facility. Because factor scores are centered on a mean of zero, with a minimum score in this case of −2.22, we added a constant to each score so that the range was from zero, for a house with the lowest quality of housing, to a score slightly above 5 for the highest quality. These scores were then averaged for any given neighborhood definition.

Figure 1 shows the spatial variability in Accra of HQ by FMV neighborhoods, with the color ramp representing the range from the lowest quintile of housing quality (darkest red) to the highest quintile of housing quality (darkest green). The ten neighborhoods with the lowest housing quality (and potentially at higher risk of poor health outcomes) are labeled in Figure 1, as are the neighborhoods with the highest housing quality (potentially at lower risk of poor health outcomes). For example, the five neighborhoods with the lowest HQ are, in order of HQ, Sodom & Gomorrah, Jamestown, Nima, Teshie Old Town, and Mamobi. The five neighborhoods with the highest HQ are, in order, East Ridge, Achimota Forest, Ringway Estate, North Ridge, and Awudome Estate. Of particular note is that some of the neighborhoods with very low HQ are spatially proximate to neighborhoods with high HQ.

Our second measure of the environmental context combines aspects of the natural and built environments. From the imagery we are able to map land cover features such as landscape vegetation and the size and geometry of residential dwellings that serve as proxies for SES and, potentially, health status. We measured land cover and land use change in Accra with georeferenced Quick-Bird satellite multispectral image data for April 2002 and January 2010. We refined the co-registration of the images and used an empirical line approach to normalize the two dates of imagery radiometrically (Coulter et al. 2011).

The proportion of vegetation cover was estimated for EAs of greater Accra that were imaged in both 2002 and 2010 by summing the number of pixels classified as green vegetation and dividing by the total number of pixels contained within each EA boundary. Difference of proportions for each EA were calculated and used to generate maps of absolute change in green vegetation proportions and percentage change relative to proportions in 2002. A sample of 1,000 visually interpreted pixels for each image date was utilized as reference data for assessing the accuracy of the 2002 and 2010 vegetation maps at the pixel level. Vegetation and nonvegetation pixels were classified with overall, user's, and producer's accuracies in the 93 to 94 percent range. For the entire map extent, the estimate of vegetation change from the reference data is a decrease of 9.3 percent and the semiautomated image change analysis estimates a reduction of 5.7 percent. Due to the difficulty of finding cloud-free imagery at a time near the 2000 census, the high-resolution imagery does not quite cover the entire study site for the year 2002. For the remaining EAs, we calculated vegetation fractions from 15 m spatial resolution ASTER imagery, after determining that there was a very good fit between results from the very high and moderate resolution data (Stoler et al. 2012). Figure 2 shows the results of the vegetation calculations, with each neighborhood in Accra categorized according to the quintile of its percentage of vegetation cover. The patterns were very similar in both years ($r = .90$), despite the substantial vegetation loss throughout the city between 2002 and 2010 due to increased population pressure on the land.

The most revealing difference between Accra residential neighborhoods of varying housing quality and SES is the relative abundance of vegetation cover and the size and density of residential structures (not shown; Weeks et al. 2007; Stow, Lippitt, and Weeks 2010). High-housing-quality areas tend to have a high proportion of landscape vegetation, whereas low-housing-quality areas have little vegetation ($r = .73$). Overall, 49 percent of the neighborhoods were in the same quintile with respect to housing quality and percentage vegetation, and the spatial cooccurrence of these two ways of differentiating environments helps us define the extremes of neighborhood context in Accra. Thus, there are eleven neighborhoods in the first (lowest) housing

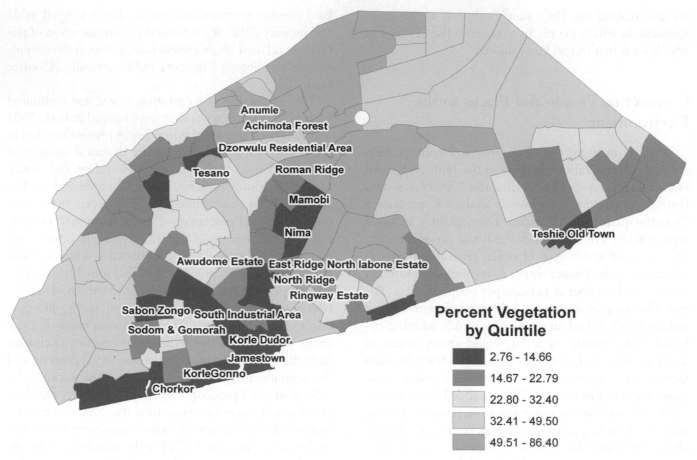

Figure 2. Percentage vegetation quintiles (based on classification of 2010 Quickbird imagery) by field modified vernacular neighborhoods, Accra, Ghana. (Color figure available online.)

quality quintile and also in the first (lowest) vegetation quintile in both 2003 and 2010 (see Figure 3). We hypothesize that these will be the neighborhoods with the poorest health outcomes. There are eleven neighborhoods in the highest quintile with respect to both housing quality and vegetation (both years) and we expect these to have the best health outcomes. It can be seen in Figure 3 that there is clear spatial clustering of the extremes.

Connecting People, Place, Environment, and Health

Do patterns of health follow the expectations raised by the cooccurrence of housing quality and vegetation? To answer that question we turn to the results of in-depth surveys that we conducted in a longitudinal panel of nearly 3,200 women spatially distributed around Accra—the Women's Health Survey of Accra (WHSA). The first round (WHSA–I) was conducted in

2003. A sample of 200 EAs was selected with probabilities proportional to population size. Enumerators then visited those EAs and eligible women (residents of Accra aged eighteen and older) were listed by name and address. Respondents were then selected with probabilities fixed according to the SES of the EA and the age group of the women (older women were progressively oversampled). From the cohort of 3,183 women who completed the household survey in 2003, the first 1,328 women were also asked to participate in a comprehensive medical and laboratory examination. The women interviewed in 2003 consented to be revisited in the future, and this panel of women was reinterviewed between October 2008 and March 2009 (WHSA–II). In WHSA–II there were neither blood tests nor medical examinations, but height, weight, blood pressure, and visual acuity were measured in the home at the end of the household interview.

Tracking women in 2008 and 2009 who had been interviewed in 2003 led to several problems of identification, and different approaches were adopted

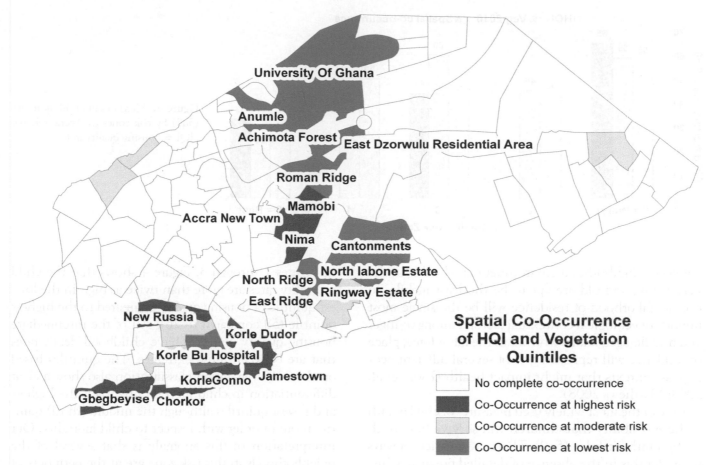

Figure 3. Spatial cooccurrence of housing quality index (HQI) and vegetation quintiles for both 2002 and 2010 by field modified vernacular neighborhood, Accra, Ghana. (Color figure available online.)

depending on the circumstances. An important lesson learned is that tracking individuals is very challenging in a community with a poorly developed system of street addresses and a tendency to use a variety of personal names, depending on the context. Nonetheless, almost two thirds of the originally surveyed women were identified and successfully reinterviewed five to six years after the initial contact. This was aided substantially by a street map we created from a purchased set of CAD data files (tiles) from the Ghana Department of Lands and Surveys that included local detail obtained from digitizing aerial photographs, street names, and elevation detail. These data were combined with our digital boundaries and satellite imagery into a geodatabase from which details could be printed out for sections of the city, thus allowing interviewers to navigate each neighborhood of the city successfully.

Follow-up was also aided immeasurably by the remarkable penetration of mobile phone use. We found that 90 percent of households in the survey own a mobile telephone—a much higher rate than expected,

even though Africa has been experiencing the most rapid increase in use of this technology (Tryhorn 2009). For women who were found to have moved within the Accra metropolitan area (AMA), the team made every attempt to locate them in their new residence and interview them as part of the study. For women who were found to have moved outside the AMA, replacements matched by age and EA of residence were identified and asked to join the study. For women who were found to have died, a later study was carried out, in which a verbal autopsy was conducted to ascertain probable cause of death. The number of completed interviews in WHSA–II was 2,814, of which 1,819 were reinterviews of women from WHSA–I and 995 were replacements for women lost to follow-up.

In analyzing neighborhoods as health contexts, building on our systems approach, we recognize that in a city with relatively high mortality such as Accra, the biggest differences in health will be found especially at the youngest and oldest ages. These are the ages for which we can anticipate the strongest interaction

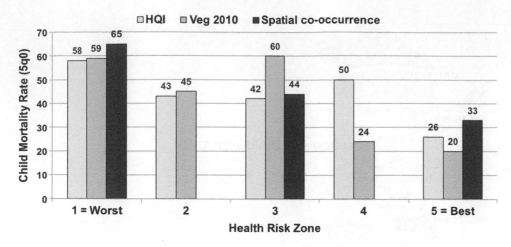

Figure 4. Measures of child mortality (5q0) by risk zones in Accra, Ghana. HQI = housing quality index.

between neighborhood environment and health, as the very young and old are apt to be the least mobile, so the neighborhood of residence will be the single most important ecological contextual factor. Among younger and middle-aged adults who are in the labor force, place of residence will represent one of several different ecological contexts that might impact health (Kwan et al. 2008; Matthews 2008).

One of the most widely used measures of child health is the probability that children will survive from birth to their fifth birthday (5q0). There are a variety of ways to model this in the absence of detailed pregnancy histories or complete vital statistics, drawing on two key questions asked of women in many surveys: (1) number of children ever born and (2) number of those who have died. It turns out that the ratio of children dead to children born at different ages of motherhood provides a close approximation to 5q0 (Rajaratnam et al. 2010). For women who have ever had a live birth, we calculate whether or not any of their children have died to date. We do this calculation by age group of women and then calculate age-standardized proportions of women who have lost a child, using the Accra 2000 census population as the standard. We calibrated this measure at the regional level in Ghana, using data from the 2003 Demographic and Health Survey. The adjusted R^2 between 5q0 and the age-adjusted proportion of women who have lost at least one child was 0.88. We employed the regression equation from these data to convert proportions who had a lost a child into the predicted child mortality rates (5q0; formulas are available from the authors on request).

We then used the data from the pregnancy histories of women interviewed in WHSA–II to calculate the proportion of women with a live birth who had lost at least one child in each of the health-risk zones shown

in Figures 1 through 3. Figure 4 shows that the child mortality rates are more than twice as high in the lowest quintile of housing quality compared to the highest quintile (58 compared to 26). All of the intermediate housing quality quintiles have childhood death rates that are between those extremes. The quintiles based on the 2010 vegetation classification also show a clear differentiation in child mortality between the highest and lowest quintiles although the middle (third) quintile is out of order with respect to child mortality. Our interpretation of this anomaly is that several of the neighborhoods in this risk zone are at the periphery of the city and are occupied by migrants to the city who have above-average mortality but are living in an area that has been settled recently enough that it has not yet been completely devegetated. Finally, it can be seen that the zones created by the spatial cooccurrence of quintiles based on housing quality in 2000 and vegetation in 2003 and 2010 had a clear spread, in which child mortality was almost exactly twice the level (65) in the highest risk area compared to the lowest risk area (33), with the moderate risk area falling between those extremes (44) in terms of child mortality.

We now turn to the other end of the age structure and ask whether there is a relationship between the health of older women (age fifty-five and older) and the risk zones in which they live. Keep in mind that we have, by definition, interviewed survivors, so we can anticipate differences less dramatic than for childhood mortality. Some parts of the city might have higher survival rates from given diseases and conditions, but we have no data to measure this. Rather, we focus on the relative conditions of the women interviewed. For the sake of brevity we have focused on one example, and Figure 5 shows the percentage of women reporting that their health is only fair or poor by risk zone quintiles.

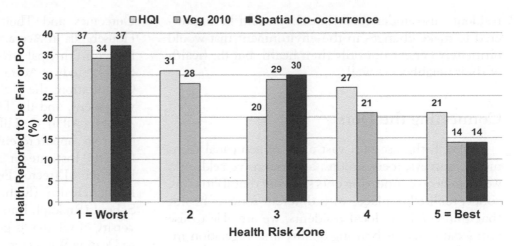

Figure 5. Percentage of women 55+ who report that their health is fair or poor by risk zones in Accra, Ghana. HQI = housing quality index.

Consistent with our expectations, the percentages are clearly highest in the higher risk neighborhoods and lowest in the lower risk neighborhoods, regardless of whether we measure risk in terms of housing quality, vegetation, or the cooccurrence of the two.

Connecting with the People About Their Health

During the summer of 2010, the research team conducted a series of focus groups in an SES-stratified selection of ten neighborhoods within the AMA. To arrange each focus group, our facilitators at the University of Ghana contacted the assemblyperson or local chief for each neighborhood and then worked through community groups to organize participants. Focus groups of eight to fifteen persons were conducted at a variety of venues (including a church, a bar, and a private residence), with discussions lasting between one and two and one half hours.

We know from the WHSA–II that people living in high-risk zones were least likely to understand the exact cause of malaria, although the self-report data suggest that they were not more likely to have ever had malaria. Across all focus groups, however, participants perceived malaria as the number one health issue in their neighborhoods. As one Nima neighborhood resident stated, "The problem here is malaria." Nima is one of the poorest neighborhoods in Accra, but residents in every focus group regardless of neighborhood SES expressed the same perception that malaria was their neighborhood's primary health concern. Participants in focus groups seemed to blame most fevers on malaria from the mosquitoes that breed in trash-choked gutters near their homes. Even if this is not the type of place in which the *Anopheles gambiae* mosquito that carries the malaria parasite typically breeds, the residents understood a connection between their environment and health when it comes to malaria. Other diseases like typhoid, diarrhea, respiratory issues, and even chronic diseases like stress and diabetes, however, were not so readily connected to the environment. Participants in Nima were also aware that people are more likely to get sick during the rainy season when the gutters are overflowing from all of the trash being thrown in them, but there were no specific diseases attributed to this problem.

Focus groups also revealed a contrast between neighborhoods where residents felt a strong sense of personal responsibility for their own health and others where people expressed that they felt more susceptible to their environment. For instance, in the relatively wealthy Cantonments neighborhood, where the majority of residents are government and embassy workers, focus group participants expressed that "it is your individual house that is the cause, not the environment," and that environmental elements did not have a significant impact on residents' health. One participant stated, "The air is clean, there are trees all around." In contrast, in the relatively poorer Sabon Zongo and Jamestown neighborhoods, focus group participants mentioned the collection of environmental factors that they felt impacted their health, such as low water quality, reliance on public latrines, and poor solid waste management. Participants in Sabon Zongo noted that nearby neighborhoods where people are neater, cleaner, and better educated about health and hygiene were also healthier. These comments are consistent with our view of health as a system. People saw themselves as being at risk of disease because of the combination of personal habits and environmental context, but some of the neighborhood residents also understood that they could be empowered to make changes in the environment that would ultimately benefit not only their health but the health of their neighbors as well.

Connecting the Dots

Through the use of a vast array of geospatial techniques, surveys, focus groups, and extensive fieldwork, we have been connecting dots showing that if cities are conceptualized as places with different zones of health that transcend the local residents, we are able to use proxy data such as housing quality and vegetation indexes from satellite imagery to identify places within the city that are likely to have the worst and the best health outcomes among people living there. More than one in five (22 percent) of Accra's population live in the highest risk neighborhoods (the first quintile on both the HQ index and vegetation), whereas only 3 percent live in the lowest risk neighborhoods (the fifth quintile on both HQ index and vegetation). The burden of health is thus heavily and disproportionately borne by lower income neighborhoods. Some of them have active nongovernmental organizations working to empower local residents to improve their lives, including their health, but most neighborhoods do not. Harpham (2009) has suggested that something akin to street-level advocacy might be appropriate in the Global South to create the kind of social movement for health in that part of the world that has successfully brought health improvement to the policy table in developed countries. In Accra and other sub-Saharan African cities, the systems approach suggests that the most efficient way to improve health is to improve the environment, at the same time teaching people more effectively how to protect themselves from disease and other health hazards.

Acknowledgments

This research was funded in part by grant number R01 HD054906 from the Eunice Kennedy Shriver National Institute of Child Health and Human Development ("Health, Poverty and Place in Accra, Ghana," John R. Weeks, Project Director/Principal Investigator). The content is solely the responsibility of the authors and does not necessarily represent the official views of the National Institute of Child Health and Human Development or the National Institutes of Health. Additional funding was provided by Hewlett/PRB ("Reproductive and Overall Health Outcomes and Their Economic Consequences for Households in Accra, Ghana," Allan G. Hill, Project Director/Principal Investigator). The 2003 Women's Health Study of Accra was funded by the World Health Organization, the U.S. Agency for International Development, and the Fulbright New Century Scholars Award (Allan G. Hill, Principal Investigator). The generous support received during all phases of this study from the Institute for Statistical, Social and Economic Research (Director Ernest Aryeetey); the School of Public Health (Richard Adanu); and the Medical School (Rudolph Darko and Richard Biritwum), University of Ghana, is gratefully acknowledged. Thanks to Deanna Weeks for expert manuscript editing.

References

Agyei-Mensah, S., and G. Owusu. 2010. Segregated by neighborhoods? A portrait of ethnic diversity in the neighborhoods of the Accra metropolitan area, Ghana. *Population, Space and Place* 16:499–516.

Bloom, D. E., D. Canning, and J. Sevilla. 2001. The effect of health on economic growth: Theory and evidence. NBER Working Paper No. 8587, National Bureau of Economic Research, Cambridge, MA.

Coulter, L., A. Hope, D. Stow, C. Lippitt, and S. Lathrop. 2011. Time-space radiometric normalization of TM/ETM+ images for land-cover change detection. *International Journal of Remote Sensing* 32 (2): 7539–56.

Diez Roux, A. V. 2007. Integrating social and biologic factors in health research: A systems view. *Annals of Epidemiology* 17 (7): 269–74.

Dunn, J. R., and S. Cummins. 2007. Placing health in context. *Social Science and Medicine* 65:1821–24.

Entwisle, B. 2007. Putting people into place. *Demography* 44 (4): 687–703.

Harpham, T. 2009. Urban health in developing countries: What do we know and where do we go? *Health & Place* 15 (1): 107–16.

Kawachi, I., and L. F. Berkman, eds. 2003. *Neighborhoods and health.* New York: Oxford University Press.

Kwan, M.-P., R. D. Peterson, C. R. Browning, L. A. Burrington, C. A. Calder, and L. J. Krivo. 2008. Reconceptualizing sociogeographic context for the study of drug use, abuse, and addiction. In *Geography and drug addiction*, ed. Y. F. Thomas, D. Richardson, and I. Cheung, 437–46. Berlin: Springer.

López-Casasnovas, G., B. Rivera, and L. Currais, eds. 2005. *Health and economic growth: Findings and implications.* Cambridge, MA: MIT Press.

Mabry, P. L., D. H. Olster, G. D. Morgan, and D. B. Abrams. 2008. Interdisciplinarity and systems science to improve population health: A view from the NIH Office of Behavioral and Social Sciences Research. *American Journal of Preventive Medicine* 35 (2 Suppl.): S211–S224.

Matthews, S. A. 2008. The salience of neighborhood: Some lessons from sociology. *American Journal of Preventive Medicine* 34 (3): 257–59.

Rain, D., R. Engstrom, C. Ludlow, and S. Antos. 2011. Accra, Ghana: A city vulnerable to flooding and drought-induced migration. Case study prepared for cities and climate change 2011, global report on human settlements 2011. http://www.unhabitat.org/downloads/docs/GRHS2011CaseStudyChapter04Accra.pdf (last accessed 15 March 2012).

Rajaratnam, J. K., L. N. Tran, A. D. Lopez, and C. J. L. Murray. 2010. Measuring under-five mortality: Validation of new low-cost methods. *PLoS Medicine* 7 (4): e1000253.

Songsore, J., J. S. Nabila, Y. Yanyuoru, E. Amuah, E. K. Bosque-Hamilton, K. K. Etsibah, J.-E. Gustafsson, and G. Jacks. 2005. *State of environmental health report of the greater Accra metropolitan area 2001*. Accra, Ghana: Ghana Universities Press.

Stoler, J., D. Daniels, J. R. Weeks, D. Stow, L. Coulter, and B. K. Finch. 2012. Assessing the utility of satellite imagery with differing spatial resolutions for deriving proxy measures of slum presence in Accra, Ghana. *GIScience & Remote Sensing* 49 (1): 31–52.

Stow, D., C. Lippitt, and J. R. Weeks. 2010. Delineation of neighborhoods of Accra, Ghana based on Quickbird satellite data. *Photogrammetric Engineering and Remote Sensing* 76:907–14.

Tryhorn, C. 2009. Nice talking to you . . . Mobile phone use passes milestone. 3 March. http://www.guardian.co.uk/technology/2009/mar/03/mobile-phones1?INTCMP-SRCH (last accessed 15 March 2012).

Weeks, J. R., A. Getis, A. G. Hill, S. Agyei-Mensah, and D. Rain. 2010. Neighborhoods and fertility in Accra, Ghana: An AMOEBA-based approach. *Annals of the Association of American Geographers* 100 (3): 558–78.

Weeks, J. R., A. G. Hill, D. Stow, A. Getis, and D. Fugate. 2007. Can you spot a neighborhood from the air? Defining neighborhood structure in Accra, Ghana. *GeoJournal* 69:9–22.

Geospatial Methods for Reducing Uncertainties in Environmental Health Risk Assessment: Challenges and Opportunities

Nina Siu-Ngan Lam

Department of Environmental Sciences, Louisiana State University

As in many studies involving disparate data and methodologies, environmental health risk assessment research is prone to errors and uncertainties. Unfortunately, uncertainties associated with the research findings can seldom be eliminated, and policy decisions often need to be made based on the uncertain findings. Hence methods that can help reduce uncertainties associated with the research findings would be most useful and need to be developed. This article highlights, through a case study, four major sources of uncertainties, which include uncertainties from data, methods of analysis, interpretations of the findings, and reactions to the findings. Five groups of geospatial methods that can help reduce the uncertainties as a result of combinations of the four sources are identified, including methods for visualization and measurement, cluster detection, exposure modeling, scale analysis, and decision support. The article concludes that future effort should focus more on the development of decision tools to enable decision making based on uncertain findings.

与涉及许多不同的数据和方法的研究一样，环境健康风险评估研究容易出现错误和不确定性。不幸的是，研究结果的不确定性，很少能够被去除，决策往往需要在不确定的结果的基础上作出。因此，可以帮助减少研究结果中的不确定性的方法将是很有用的和所需的。本文通过案例研究强调不确定性的四个主要来源，其中包括从数据，分析方法，研究结果的诠释，和对结果的反应而来的不确定性。五组地理空间方法，包括可视化和测量，群集检测，曝光造型，尺度分析和决策支持的方法组合，可以帮助减少由以上四个来源综合产生的结果的不确定性。文章得出结论认为，今后的努力应着眼于发展决策工具，使决策基于不确定的结果。*关键词：环境健康风险评估，地理空间分析，不确定性*。

Como ocurre en muchos estudios que incluyen datos y metodologías dispares, la investigación relacionada con evaluación del riesgo de origen ambiental contra la salud es propensa a errores e incertidumbres. Infortunadamente, las dudas asociadas con los hallazgos de una investigación rara vez pueden eliminarse, por lo que a menudo las decisiones sobre políticas deben tomarse con base en hallazgos inciertos. De ahí que los métodos que puedan reducir las incertidumbres asociadas con los hallazgos de una investigación serían de la mayor utilidad, y por eso deben ser desarrollados. Por medio de un estudio de caso, este artículo destaca cuatro fuentes principales de incertidumbre, que incluyen aquellas derivadas de los datos, los métodos de análisis, interpretaciones de los hallazgos y reacciones ante los hallazgos. Se identificaron cinco grupos de métodos geoespaciales que pueden ayudar a disminuir las incertidumbres, como resultado de la combinación de las cuatro fuentes, incluyendo los métodos de visualización y medición, detección de aglomeraciones, modelización de exposición, análisis de escala y respaldo a la decisión. El artículo concluye en que el esfuerzo futuro debe enfocarse más en el desarrollo de herramientas de decisión para habilitar la toma de decisiones basadas en hallazgos inciertos.

As in many studies involving disparate data and methodologies, environmental health risk assessment research is prone to errors and uncertainties. This article focuses on the uncertainty issues in the research area that studies the spatial relationships between environmental factors (exposures) and the risk of disease. A common set of questions in this research area is as follows: Do certain hazardous facilities or industrial plants pose an adverse health effect on

people living nearby? If so, what are the effects? How far is the zone of influence? Although this set of questions has been asked and studied repeatedly through many case studies, conclusive statements have seldom been reached. Conflicting results can be generated using different definitions of data, different spatial and time scales, and different spatial analysis methods using varying parameter values. The consequences of uncertain findings in health risk assessment can be very serious, as uncertain findings make it difficult for policymakers to design a course of action. At the same time, opponents of public health and environmental regulations often "manufacture" uncertainty of the scientific results to avoid regulations (Michaels and Monforton 2005). Hence, in addition to the problem of the science itself, politics plays a large part in managing uncertainty.

Because uncertainties of the research findings exist and can seldom be eliminated, it is important to identify the sources of uncertainties and develop methods to characterize and possibly reduce them (May 2001). There is a vast literature on uncertainty in the spatial analysis of health and environment (Cox and Ricci 1992; Pickle et al. 2006; Beale et al. 2008). Moreover, there are many aspects and meanings of uncertainty, and a variety of approaches and methods for dealing with uncertainties have been proposed (Foody and Atkinson 2002; Shi 2010). It would be useful for this article to provide a brief overview of what the common uncertainty issues are, what geospatial methods have been proposed to manage them, what progress has been made recently, and what the challenges and opportunities for future research are. This article limits the scope by focusing on four issues of uncertainties that are more distinctive to health risk assessment involving spatial data and spatial analysis. It is by no means comprehensive; rather, its purpose is to increase awareness of the issue, which is central to the study of health and geography.

Sources of Uncertainties

In environmental health research, uncertainty issues stem from at least four areas: data, methods, interpretation of the findings, and reactions to the findings (Lam 2001). These similar issues have been further elaborated in recent literature by various groups (e.g., Beale et al. 2008), and they are summarized next.

Data

Uncertainty issues surrounding data are multifaceted. There are generic uncertainty issues regarding spatial and health data, such as location and attribute uncertainty, sampling variability, and other spatial data quality issues. These types of uncertainties are very fundamental, and models have been developed to understand and estimate them (Shi 2010). This article addresses a problem that will occur even if there are no data quality issues. Most environmental health studies require both health and demographic data to derive incidence or mortality rates. Disease occurrence data of-

ten come in the form of points (e.g., patient's residential address), whereas demographic data are available mostly in a spatially aggregated form, such as by census block or ZIP code area. For health data, issues such as data availability, confidentiality, changes in disease definition, and the time and spatial scales used for reporting are all critical issues contributing to uncertainty in analysis results (Pickle et al. 2006). For demographic data, a simple but essential question arises: How should space be partitioned for reporting the population and disease occurrence statistics? Figure 1 illustrates how a spatial configuration can obscure the detection of a clustered pattern of disease occurrence. When the study area is delineated into nine squares of equal population, a uniform incidence rate results (three cases per zone), masking the underlying disease cluster. (Strictly speaking, the number of cases in each square should be greater than twenty before one calculates the rate reliably. Figure 1 is a simplification for illustration purposes.) These are classic modifiable areal unit problems that persist in spatial analysis. In addition, population changes due to migration into and out of a region would make it even more difficult to detect the effects of environment on health, especially for those diseases that have long incubation periods (Lam 1986; Jarup 2004). Furthermore, administrative boundary changes and areal data incompatibility (e.g., census block vs. ZIP code boundaries) pose additional uncertainties for the study. Analysis using these disparate spatial data sets will need methods

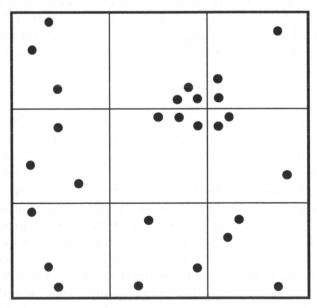

Figure 1. A simplified example showing how a spatial configuration (nine squares) could obscure an underlying clustered pattern of disease occurrence (see explanation in text).

such as areal interpolation to transform data from one areal unit delineation to another before further analysis can be conducted (Lam 1983). Conflicting results are easily generated using different definitions of data (e.g., standardized population, disease incidence vs. mortality), different spatial and time scales of data, and different methods for interpolating data into a uniform database.

Methods

Geospatial methods applied to analyzing health data can further add uncertainty to the final results. Different methods or the same method with different parameters often lead to varying results, such as the weighting scheme used in calculating the spatial autocorrelation statistics. Moreover, a spatial model might not be valid everywhere. Obviously, the analytical methods used are confined by the quality of the data, and the results generated from the methods can only be as good and reliable as the data warrant. In the cancer literature, the famous controversy on whether a significant increase in leukemia incidence is related to nearby nuclear plants in England during the late 1980s has generated numerous studies and disparate results (e.g., Gardner 1989). The techniques used in these studies varied: Some used cancer incidence and others used mortality; some used rates derived from observed with expected numbers and others used Poisson probabilities; some used existing administrative boundaries and others used arbitrary circles drawn around a presumed source (e.g., Figure 1). Disease clusters could be made to appear or disappear by simple statistical manipulation (e.g., by using different statistical confidence intervals; Openshaw et al. 1988).

Interpretations

Interpretation issues that add to the uncertainties in the study of health and geography are especially difficult to deal with. First, it is commonly known that correlations found by the analysis do not necessarily imply causation. Model specification, or rather misspecification, plays a big role in the correct interpretation of model results (Cox and Ricci 1992). Uncertainties would likely increase when the analysis is exploratory in nature and is not guided by theory (Rushton and Lolonis 1996). Second, confounding effects often exist in this type of study. For example, high cancer mortality found near a suspected source (i.e., focused cluster detection) might not be due to contamination near the source; rather, it could be a result of other socio-economic factors such as inadequate health care and prevention. For this reason, incidence rates are often preferred over mortality rates for environmental health analysis. Third, for disease cluster detection without a suspected source (i.e., general cluster detection), once a cluster is detected it is difficult to attribute to a source without more comprehensive investigations. Both false-positive and false-negative results could occur. Therefore, interpretations of the findings should be made in accordance with the degree of uncertainty, which unfortunately is difficult to quantify.

Reactions

Because environmental health studies often involve sensitive topics that could have serious political, economic, and legal implications, reactions to the uncertain findings can be very intense. If no correlation is found between a suspected pollution source and health outcomes, the nearby community would find it difficult to accept. Likewise, if correlation (not causation) is found, it could easily be exaggerated and interpreted as the cause. Furthermore, studies that found no correlation are less frequently reported, leading to a bias in the literature (National Research Council 1997). For example, in the late 1980s, a study was conducted to examine whether there was an increase in miscarriage cases in St. Gabriel, Louisiana, a small town located on the east bank of the Mississippi River in the industrial area known as "Cancer Alley." The study found no statistically significant relation between miscarriage rates and proximity to industrial sites and concluded that further studies of the rates were not warranted. This generated much criticism from the community and intense media coverage. One point is especially relevant here: It was noted that the study did find some correlation, but its confidence level did not quite reach 95 percent; hence the conclusion of no correlation. "In science, a conclusion has to be 95% certain before it is acceptable" (Lewis, Keating, and Russell 1992, 34). This example is not unique. Similar problems occurred in the 1980s when the AIDS epidemic terrified the world. Among the various problems at that time, one heated debate in the United States was on whether blood supply should be regulated to curb infection when scientific evidence was still uncertain (Shilts 1987). Given the uncertainties involved in the findings, the challenge to today's researchers and decision makers is this: When and how should environmental or health policies be enacted to mitigate adverse health effects?

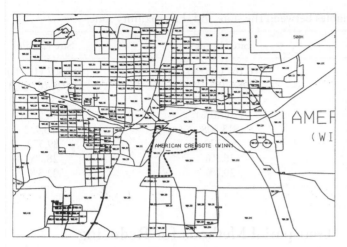 

Figure 2. An example of health risk assessment surrounding a Superfund hazardous waste site called American Creosote Works, Inc., in Winn Parish, Louisiana. Left: The boundary of the site is shown overlaid with the census block boundaries. Right: A Google screenshot showing the same study area (last accessed 30 November 2011). (Color figure available online.)

An Environmental Health Risk Assessment Example

The following summarizes a study conducted in the 1990s to illustrate the uncertainty issues commonly encountered in this type of study and provides a background on the need for better geospatial methods to reduce uncertainties. The findings of the study could not be published at that time, precisely because of the four issues mentioned previously (Lam 2001). It is now reported in this article after further study was conducted a few years later (Agency for Toxic Substances and Disease Registry [ATSDR] 2004). The purpose of the study was to examine whether there were adverse health effects for people living near a Superfund hazardous waste site. The site, called American Creosote Works, Inc., is located near the town of Winnfield in central Louisiana, which is a primarily rural, low-income town with a 1990 population of about 6,000. The 34.21-acre site was a wood treating facility that was abandoned in 1985. Hazardous substances, including polycyclic aromatic hydrocarbons (PAHs), pentachlorophenol (PCP), benzene, and dioxins, were found in the soils and liquids in the site and in nearby drainages, which posed a public health hazard. The "standard" geographic information system (GIS) buffering method was employed to retrieve the number of cancer incidence cases (for 1988–1992) and the 1990 population counts of the census block groups intersecting the buffer boundaries (Figure 2).

The standardized incidence ratios (SIRs) were used for the main analysis. The SIRs were calculated by di-

viding the observed number of cases in the proximity zone by the expected number of cases in a comparison region (an indirect standardization method). The expected number was derived by multiplying a comparison region's age- and sex-specific incidence rates and the buffer's age- and sex-specific population data. Both state and state subregion's average annual incidence rates for the same time period were used to derive the expected number of cases. If the observed number of cases equals the expected number of cases, the SIR will equal 1.0. When the SIR is less than 1.0, fewer cases are observed than expected and vice versa. The 95 percent confidence interval was used to evaluate the statistical significance of the SIRs. If the confidence interval includes 1.0, then the SIR is not considered to be significantly different from 1.0 (Kahn and Sempos 1989).

The 1-mile and 2-mile radius criteria, which have been traditionally used in this type of study, were applied to retrieve the census block groups intersecting the buffers. Table 1 compares some key socioeconomic characteristics between the proximity zones and the state as a whole, which shows a pattern of higher concentration of African American population and poverty within the 1-mile proximity zone. For brevity, the SIR ratios for only two cancer types are listed in Table 2. For the 1-mile proximity zone, incidence rates for males (all cancers) and for male non-Hodgkin's lymphoma were found to be statistically elevated compared with the state rates. For the 2-mile proximity zone, none of the SIRs were statistically elevated; instead, female (all cancers) was found to be statistically low. The statistically elevated non-Hodgkin's lymphoma

Table 1. Comparisons of key socioeconomic variables between the state and the proximity zones

	State	Site (1 mile)	Site (2 mile)
All persons	4,219,973 (100%)	6,767 (100%)	10,302 (100%)
White	2,839,138 (67.3%)	3,605 (53.3%)	6,924 (67.2%)
Black	1,299,281 (30.8%)	3,090 (45.7%)	3,273 (31.8%)
Other	81,554 (1.9%)	72 (1.0%)	105 (1.0%)
Median household income	$21,949	$13,508	$16,922
Per capita income	$10,663	$8,379	$9,572
% persons below poverty	23.6%	36.4%	28.0%
% families below poverty	19.4%	29.7%	22.3%

incidence rates found in this study were a real concern, as the chemicals found in the site had been suggested to be a potential cause of such cancer, but because of the inconsistent results generated mainly from using different proximity zones, further studies were suggested. A follow-up study was conducted a few years later to include more years of cancer incidence data (from 1988–1997). Unfortunately, the data contained many post office box or rural route addresses, and only about 44.2 percent of the cases could be geocoded. It was determined that such a low geocoding rate would likely yield inaccurate results; hence, a cancer review could not be completed (ATSDR 2004).

This example of conflicting results generated from the uses of 1-mile versus 2-mile proximity zone is only one of the many examples in the literature. It highlights the uncertainty issues surrounding the four sources discussed previously. Unfortunately for this example, the matter has not been resolved. Nevertheless, the poor socioeconomic condition surrounding the site (Table 1) epitomizes the issue of environmental justice, which is often intertwined with the issue of health disparity.

Geospatial Methods for Reducing Uncertainties: Recent Developments, Challenges, and Opportunities

Because uncertainties in analysis results can seldom be eliminated but possibly reduced, geospatial methods designed to explicitly identify, visualize, and quantify uncertainties should be developed as the first step toward managing and reducing uncertainties (Cox and Ricci 1992; May 2001). Indeed, there has been significant progress in tool and software development in the past decade that has helped in managing uncertainties (Beale et al. 2010). This article categorizes these geospatial methods into five groups: measurement and visualization, cluster detection, focused exposure modeling, scale analysis, and decision support (Table 3). These methods could be integrated into a GIS framework as a first step toward reducing uncertainties that might come from combinations of the sources discussed earlier. This article is not an evaluation of software, and some of these tools might not directly address uncertainty, but they can be modified to include functions to

Table 2. The standardized incidence ratios (SIRs) for selected cancer types, 1988–1992

Cancer type	Sex	SIR (1 mile)	95% CI	SIR (2 mile)	95% CI
All cancers[a]	Male	1.31*	1.08, 1.54	0.98	0.82, 1.14
	Female	0.94	0.74, 1.17	0.79**	0.64, 0.97
Lung/bronchus	Male	1.36	0.91, 1.96	0.98	0.67, 1.38
	Female	0.89	0.42, 1.63	0.62	0.30, 1.14
Non-Hodgkin's lymphoma	Male	3.21*	1.47, 6.10	2.09	0.96, 3.97
	Female	1.03	0.21, 3.02	0.98	0.27, 2.50

Note: The expected cancer incidence was calculated using the state as a comparison region. CI = confidence interval.

[a]The total cancer incidence cases within the 2-mile zone were 140 for males and 92 for females. The numbers of cases for other cancer types are not listed due to small numbers.

*Statistically elevated at the $p < 0.05$ level.

**Statistically low at the $p < 0.05$ level.

Table 3. Geospatial methods and recent software that could be used to examine uncertainties in environmental health risk assessment

Module	Functions/analytical tools	Example software
Measurement and visualization	Spatial trends and statistics	GeoDa, RIF
	Disease mapping, Bayesian smoothing	
	Standardized ratios	
Cluster detection	General cluster detection	SaTScan, ClusterSeer
	Focus cluster detection	
Exposure modeling	Spatial interpolation	Spherekit
Scale analysis	Monte Carlo simulation	WinBUGS
	Multilevel modeling	
Decision support	Decision analytic methods	SADA, PrecisionTree,
	Predictive modeling	DTREG

Note: GeoDa (Anselin et al. 2006); RIF (Rapid Inquiry Facility; Beale et al. 2010); SaTScan, ClusterSeer (Robertson and Nelson 2010); SADA (Spatial Analysis and Decision Assistance; Institute of Environmental Modeling, University of Tennessee 2010); Spherekit (http://www.ncgia.ucsb.edu/pubs/spherekit/ [last accessed 15 January 2011]); DTREG (Software for predictive model and forecasting; http://www.dtreg.com/ [last accessed 15 January 2011]).

reveal and characterize uncertainties. Two of the latest software developments, the Rapid Inquiry Facility (RIF; Beale et al. 2010) and the Spatial Analysis and Decision Assistance (Institute of Environmental Modeling, University of Tennessee 2010), are especially promising. Due to space limitations, the following can only briefly describe the major functions. Readers are directed to the Web sites of the software packages for detailed descriptions.

Measurement and Visualization

Calculating and mapping the spatial and temporal trends of disease occurrence rates is the first step toward understanding the patterns. Useful hypotheses guided by theory could be generated from visualizing the maps. Hence, better methods for mapping disease risks will help in generating the "right" hypotheses. For example, a choropleth mapping routine that generates class intervals based on statistical significance is better than other nonstatistical methods (e.g., natural breaks, equal intervals) in pinpointing areas of concern. The 1979 cancer mortality maps of China, which were displayed in an atlas using different choropleth mapping methods, provide a classic example of how mapping methods affect the resultant maps and people's perception (Lam 1986). In addition, software routines that can compute Bayesian smoothing might be used to add more certainty and credibility to the analysis. The RIF software, recently developed in the United Kingdom as an extension to ESRI/ArcGIS (ESRI 2011), provides a suite of tools for disease mapping and risk analysis. Beale et al. (2010) applied the software to map the risks of esophageal cancer incidence in Norfolk, UK, using its various options of Bayesian smoothing, the results of

which helped in deriving a more confident conclusion that the study area did not have higher relative risks of esophageal cancer. Other mapping methods, such as the spatial adaptive filtering mapping method developed by Rushton, might also be used (Beyer and Rushton 2009). For the health risk assessment example discussed earlier, mapping the cancer distribution for a larger study area would have helped in generating a better understanding of the spatial relationship between the Superfund site and cancer incidence.

Spatial measurement tools, such as spatial autocorrelation indexes, variograms, and correlograms computed on each health and environmental variable, provide a basic characterization of how the variable changes in space, which should help in understanding the potential effects of scale and subsequent exposure modeling. Spatial autocorrelation is also the basis for cluster detection, and adjustment for spatial autocorrelation is needed for accurate health risk studies (Beale et al. 2010). Variograms and correlograms have been applied to demonstrate how the HIV/AIDS virus spread through time and space in the United States (Lam, Fan, and Liu 1996). Some of these spatial measurement methods are available from the GeoDa software (Anselin et al. 2006). These spatial measurement methods, together with the mapping tools, should provide researchers with a good overall picture of how different variables are related spatially to increase confidence in analysis and interpretation.

Cluster Detection

Cluster detection includes both general cluster detection (without a presumed source) and focused cluster detection (with a presumed source), and both might be needed in the same study. For example, if disease clusters are found only in the areas surrounding a hazardous waste site after the general cluster detection process, then we can focus on those areas and examine in more detail the factors leading to high disease risks using focused cluster detection as shown in the preceding case study. On the contrary, if disease clusters also exist in other areas that have no hazardous waste sites, then it could imply that proximity to those sites is neither a factor nor the only factor in increasing disease risks. In the latter case, more studies will be needed to examine the various possibilities. This phenomenon occurred in Gardner's (1989) study, in which he found that the immediate areas surrounding a nuclear installation in England had higher rates of leukemia but they were not unique; other areas without nuclear installations were also found to have higher rates.

Many methods have been developed to detect spatial and temporal clusters. Cluster detection is probably the most intensively researched topic related to the use of GIS in health analysis (e.g., Marshall 1991). Currently, there are more than 100 cluster detection methods available (Kingsley, Schmeichel, and Rubin 2007). In a recent review of space–time cluster detection software, Robertson and Nelson (2010) found that SaTScan (Kulldorff and Information Management Services 2010) is the most developed and robust software package, and ClusterSeer (Jacquez et al. 2002) is more suited for data exploration. These tools need to be integrated with other tools in a single GIS or cyberinfrastructure for researchers to explore and examine uncertainties in cluster investigations.

Focused Exposure Modeling

Environmental health risk assessment needs better modeling of the environmental data to reduce uncertainties, such as how pollutants travel through air, water, and soil. In focused exposure modeling, uncertainty often arises due to the use of an arbitrarily delineated proximity zone (see the earlier case study). A guided approach would be to delineate the proximity zone according to a dispersion model that utilizes spatial interpolation. Consider a hazardous waste site that has three known contaminants—x, y, and z—that are of potential threat to human health (e.g., PAH). Spatial interpolation methods such as kriging can be used to create a surface of grids showing the concentration of each contaminant (Lam 1983). An exposure index E can be derived for each grid such that it is a func-

tion of the three contaminants: $E = f(x, y, z)$. One can use the E values to delineate a threshold boundary, which should be more meaningful than the arbitrarily defined proximity zone. Another approach is to aggregate the grid values into small areas and then use a regression model to examine if high E is associated with a health outcome. Although these are quite common methods in GIS (e.g., interpolation, grid-to-polygon conversions), they have not been integrated effectively with other tools in a single platform for health risk assessment.

Scale Analysis

Monte Carlo simulation routines that can simulate the scale effects for sensitivity analysis will help reduce uncertainties. Schneider et al. (1993) demonstrated that analysis of cancer incidence data and cancer clusters at multiple geographic scales can help provide confidence in the results, alleviate the fears of the general public, and prevent costly and unwarranted epidemiologic studies. Similarly, Ozdenerol and Lam (2004) performed cluster detection of high rates of low birth weight in East Baton Rouge Parish, Louisiana, and found that the same cluster persisted at all three geographical scales, thus adding confidence in the results. Real data at multiple scales are seldom available, however, and simulation is needed instead to examine the uncertainties of model results at different spatial scales. Many related studies have already used Monte Carlo simulation, but specialized software targeted for this application is not commonly available (Luo, McLafferty, and Wang 2010). Future development should include such a routine to support uncertainty analysis. In addition, software that enables multilevel modeling should be integrated to help evaluate the changing effects of variables on a predictive model at different scales (Webster et al. 2008).

Decision Support

This is the area that is least developed in environmental health risk assessment, although it is the most critical linkage to policy development. Decision making under uncertainty is a very challenging issue in many fields, and the trade-offs between regulation and health impacts need to be addressed. In decision sciences, there are elaborated decision software tools such as Decision Tree or PrecisionTree that could be adapted to the analysis of health and environment (Clemen and Reilly 2001). In a nutshell, four main approaches can be used to model uncertainties for decision making. The first is to translate uncertainty into probability numbers so that they can then be included in a decision tree. The second approach is to represent uncertainty with a standard mathematical model and then derive probabilities on the basis of the mathematical model. Use of Bayes's theorem is common in deriving the probabilities. The third approach uses Monte Carlo simulation to simulate the outcomes of a model. By tracking the outcomes, the decision maker can obtain a fair idea of probabilities associated with various outcomes. The fourth approach is to develop different models for the same problem and then use the model averaging technique to evaluate the uncertainties (Roberts and Martin 2010). Environmental health risk assessment needs these tools to help make the best decisions. Integrating these decision tools with the geospatial methods discussed earlier is a critical task that deserves further research and development.

Conclusions

This article highlights through a case study four uncertainty issues commonly found in the study of health and geography. It identifies five groups of geospatial methods that could help in reducing uncertainties that arise from combinations of the four sources. Indeed, there has been significant progress in the development of geospatial methods and associated software in the past decade, such as disease mapping using Bayesian smoothing, cluster detection, and Monte Carlo simulation. A major gap in geospatial method development is the lack of decision-making tools to enable decision making, which is critical to advancing environmental health studies. Future research should focus on the development of decision tools to enable decision making under uncertainty. These decision-making tools, together with the existing geospatial software tools, should be integrated into a common GIS platform, or rather a cyberinfrastructure, to support research and decision making. Environmental health risk assessment, health disparity, and environmental justice are increasingly part of the dialogue in environmental health and policy analysis, wherein issues of error and uncertainty must be addressed. A cyberinfrastructure for health and environment that contains data and methods to educate the public of the uncertainty issue as well as to respond to their concerns of the health and environment would seem to be inevitable.

Acknowledgments

I thank the three anonymous reviewers and Mei-Po Kwan for their constructive comments. I acknowledge funding from the Louisiana Office of Public Health and the National Science Foundation (Award BCS-0729259). The statements and findings are those of the author and do not necessarily reflect the views of the funding agencies.

References

Agency for Toxic Substances and Disease Registry (ATSDR). 2004. Health consultation: Assessment of cancer incidence from the Louisiana Tumor Registry from 1988–1997 for American Creosote Works, Incorporated, Winnfield, Winn Parish, Louisiana. http://www.atsdr.cdc.gov/HAC/pha/PHA.asp?docid=710&pg=0 (last accessed 15 January 2011).

Anselin, L., I. Syabri, and Y. Kho. 2006. GeoDa: An introduction to spatial data analysis. *Geographical Analysis* 38:5–22.

Beale, L., J. J. Abellan, S. Hodgson, and L. Jarup. 2008. Methodologic issues and approaches to spatial epidemiology. *Environmental Health Perspectives* 116 (8): 1105–10.

Beale, L., S. Hodgson, J. J. Abellan, S. LeFevre, and J. Jarup. 2010. Evaluation of spatial relationships between health and environment: The rapid inquiry facility. *Environmental Health Perspectives* 118 (9): 1306–12.

Beyer, K. M. M., and G. Rushton. 2009. Mapping cancer for community engagement. *Preventing Chronic Disease* 6 (1): A03. http://www.cdc.gov/pcd/issues/2009/jan/08_0029.htm (last accessed 28 October 2011).

Clemen, R. T., and T. Reilly. 2001. *Making hard decisions with DecisionTools.* Pacific Grove, CA: Duxbury Thompson Learning.

Cox, L. A., Jr., and P. F. Ricci. 1992. Dealing with uncertainty: From health risk assessment to environmental decision making. *Journal of Energy Engineering* 118 (2): 77–94.

ESRI. 2011. ArcGIS desktop: Release 10. Redlands, CA: Environmental Systems Research Institute.

Foody, G. M., and P. M. Atkinson. 2002. Current status of uncertainty issues in remote sensing and GIS. In *Uncertainty in remote sensing and GIS*, ed. G. M. Foody and P. M. Atkinson, 287–302. West Sussex, UK: Wiley.

Gardner, M. J. 1989. Review of reported increases of childhood cancer rates in the vicinity of nuclear installations in the UK. *Journal of Royal Statistical Society A* 152:307–25.

Institute of Environmental Modeling, University of Tennessee. 2010. Spatial Analysis and Decision Assistance (SADA) home page. http://www.tiem.utk.edu/~sada/index.shtml (last accessed 15 January 2010).

Jacquez, G. M., D. A. Greiling, H. Durbeck, L. Estberg, E. Do, A. Long, and B. Rommel. 2002. *ClusterSeer User Guide*

2: Software for identifying disease clusters. Ann Arbor, MI: TerraSeer Press.

Jarup, L. 2004. Health and environment information systems for exposure and disease mapping, and risk assessment. *Environmental Health Perspectives* 112 (9): 995–97.

Kahn, H. A., and C. T. Sempos. 1989. *Statistical methods in epidemiology.* New York: Oxford University Press.

Kingsley, B. S., K. L. Schmeichel, and C. H. Rubin. 2007. An update of cancer cluster activities at the Centers for Disease Control and Prevention. *Environmental Health Perspectives* 115 (1): 165–71.

Kulldorff, M., and Information Management Services. 2010. SaTScan v.9.1.0: Software for the spatial, temporal, and space-time scan statistics. http://www.satscan.org/ (last accessed 15 January 2011).

Lam, N. S. N. 1983. Spatial interpolation methods: A review. *The American Cartographer* 10 (2): 129–49.

———. 1986. Geographical patterns of cancer mortality in China. *Social Science and Medicine* 23 (3): 241–47.

———. 2001. Spatial analysis for reducing uncertainties in human health risk assessment. *Acta Geographica Sinica* 56 (2): 239–47.

Lam, N. S. N., M. Fan, and K. B. Liu. 1996. Spatial-temporal spread of the AIDS epidemic: A correlogram analysis of four regions of the United States. *Geographical Analysis* 28 (2): 93–107.

Lewis, S., B. Keating, and D. Russell. 1992. Inconclusive by design: Waste, fraud and abused in federal environmental health research. Unpublished report. Larkspur, CA: Environmental Health Network.

Luo, L., S. McLafferty, and F. Wang. 2010. Analyzing spatial aggregation error in statistical models of late-stage cancer risk: A Monte Carlo simulation approach. *International Journal of Health Geographics* 9:51.

Marshall, R. J. 1991. A review of methods for the statistical analysis of spatial patterns of disease. *Journal of Royal Statistical Society* A 154 (3): 421–41.

May, R. 2001. Risk and uncertainty. *Nature* 411:891.

Michaels, D., and C. Monforton. 2005. Manufacturing uncertainty: Contested science and the protection of the public's health and environment. *American Journal of Public Health* 95 (S1): S39–S48.

National Research Council. 1997. *Environmental epidemiology:Vol. 2. Use of the gray literature and other data in environmental epidemiology.* Washington, DC: National Academy Press.

Openshaw, S., A. W. Craft, M. G. Charlton, and J. M. Birch. 1988. Investigation of leukemia clusters by use of a geographical analysis machine. *Lancet* 6:272–73.

Ozdenerol, E., and N. S. N. Lam. 2004. Detecting spatial clusters of cancer mortality in East Baton Rouge Parish, Louisiana. In *WorldMinds: Geographical perspectives on 100 problems,* ed. D. G. Janelle, B. Warf, and K. Hansen K, 75–80. Dordrecht, The Netherlands: Kluwer Academic.

Pickle, L. W., M. Szczur, D. R. Lewis, and D. G. Stinchcomb. 2006. The crossroads of GIS and health information: A workshop on developing a research agenda to improve cancer control. *International Journal of Health Geographics* 5:51.

Roberts, S., and M. A. Martin. 2010. Bootstrap-after-bootstrap model averaging for reducing model uncertainty in model selection for air pollution mortality studies. *Environmental Health Perspectives* 118 (1): 131–36.

Robertson, C., and T. A. Nelson. 2010. Review of software for space-time disease surveillance. *International Journal of Health Geographics* 9:16.

Rushton, G., and P. Lolonis. 1996. Exploratory spatial analysis on birth defect rates in an urban population. *Statistics in Medicine* 15:717–26.

Schneider, D., M. R. Greenberg, M. H. Donaldson, and D. Choi. 1993. Cancer clusters: The importance of monitoring multiple geographic scales. *Social Science and Medicine* 37 (6): 753–59.

Shi, W. 2010. *Principles of modeling uncertainties in spatial data and spatial analysis.* Boca Raton, FL: CRC Press.

Shilts, R. 1987. *And the band played on: Politics, people, and the AIDS epidemic.* New York: St. Martin's.

Webster, T. F., K. Hoffman, J. Weinberg, V. Vieira, and A. Aschengrau. 2008. Community- and individual-level socioeconomic status and breast cancer risk: Multilevel modeling on Cape Cod, Massachusetts. *Environmental Health Perspectives* 116 (8): 1125–29.

Opening Up the Black Box of the Body in Geographical Obesity Research: Toward a Critical Political Ecology of Fat

Division of Social Sciences, University of California, Santa Cruz

Geographic treatments of the etiology of obesity tend to turn on the obesogenic environment thesis and investigate the relationship between urban form and obesity. With their emphasis on environmental features that mediate eating and exercise activities, these explorations fundamentally rest on behavioral models of obesogenesis. As such, they tend to black-box the biological body as the site where excess calories are putatively metabolized into fat and made unhealthy. Drawing on critical political ecology, this article discusses the limitations of this dominant approach. First it provides some anomalies not well explained by the energy balance model. Then it reports on emerging biomedical research regarding the role of the endocrine system and endocrine-disrupting chemicals in transforming body ecologies to make them more susceptible to adiposity, regardless of caloric intake. This research also points to the active role of adipose tissue in regulating fat. In light of this evidence, the article argues for a rethinking of current geographical approaches to obesity and health more generally, with due attention to the ecologies of bodies as well as the interpretation of science.

对肥胖症成因的地理治疗通常注重有关肥胖基因的环境论文，和注重于调查城市形态和肥胖之间的关系。因其强调能调和饮食和锻炼活动的环境特征，这些探索从根本上依赖肥胖成因的行为模式。正因为如此，他们往往把生物体黑箱化，把它作为一个多余的让人不健康的场所，在此场所中，热量近似地代谢转化为脂肪。本文借鉴重要的政治生态，讨论了这种主导方法的局限性。首先，它提出了能量平衡模型不能很好解释的一些异常现象。然后，它报告新兴生物医学研究在有关内分泌系统和干扰内分泌的化学物质在转化身体生态环境过程中的作用，这些化学物质使身体生态环境无论摄入多少热量，都更容易变肥胖。这项研究还指出，脂肪组织在调节身体脂肪中的积极作用。文章主张反思当前有关肥胖和更普遍的健康方面的地理方法，对身体生态以及科学的解释给了应有的重视。*关键词：环境毒素，肥胖，肥胖基因环境，政治生态，科学研究。*

Los tratamientos geográficos de la etiología de la obesidad tienden a poner en funcionamiento la tesis del entorno obesogénico e investigar la relación entre forma urbana y obesidad. Con su énfasis en los rasgos ambientales que median las actividades de comer y ejercitarse, estas exploraciones descansas fundamentalmente en modelos conductistas de obesogénesis. Como tales, estos tienden a ver el cuerpo biológico como el sitio cerrado donde las calorías en exceso son putativamente metabolizadas en forma de grasa, haciéndolas perjudiciales para la salud. Con base en la ecología más política, el artículo discute las limitaciones de este enfoque dominante. En primer término, se muestran algunas anomalías que no se explican bien por el modelo del balance de energía. Luego, se informa sobre la nueva investigación biomédica del papel que tienen el sistema endocrino y los químicos que lo afectan en la transformación de las ecologías corporales para hacerlas más susceptibles a la adiposidad, sin consideración del consumo calórico. Este artículo también se ocupa del papel activo del tejido adiposo en la regulación de la grasa. A la luz de esta evidencia, el artículo pretende repensar los actuales enfoques geográficos sobre la obesidad y la salud, de manera más general, con debida atención a las ecologías de los cuerpos lo mismo que a la interpretación de la ciencia.

Although not without skeptics, the so-called obesity epidemic has become one of the most dominant public health concerns of the day. Accordingly, scholars have generated a good deal of research to understand obesity's etiology and thereby contribute to appropriate interventions. Within geography, the dominant approach to the etiology of obesity turns on what has come to be known as the *obesogenic environment thesis*. (A discussion of research within geography that is more discursive and phenomenological is beyond the scope of this article.) This thesis, based on the term first coined by Swinburn, Eggar, and Raza (1999), posits that the increased prevalence of obesity owes to the ubiquity of cheap, fast, nutritionally inferior food, as well as the dearth of physical activity opportunities in the built environment (Hill and Peters 1998). Studies that test this thesis tend to use geographic information systems and spatial analysis to map and test associations between spatial features and obesity prevalence (e.g., Stafford et al. 2007; Morland and Evenson 2009, among many), with varying degrees of recognition that nonecological socioeconomic factors could also play a role (e.g., Cummins and Macintyre [2006], for an overview of studies).

In some respects this approach is salutary, as it shifts focus from the putative moral failings of fat people to the structural or environmental causes of obesity. This puts it in keeping with political ecology perspectives that investigate the broader political economic conditions that shape potentially ecologically deleterious behaviors (Robbins 2004). Nevertheless, the environmental features that are objects of study still point to behavioral causes of obesity. For example, studies in this vein tend to employ variables such as distances to grocery stores, drive-to-work times, or access to parks, such that problem environments induce too much eating or not enough physical activity. As it turns out, studies that convincingly demonstrate the effects of the built environment on obesity have been limited, and it is increasingly acknowledged that the relationship between urban form and obesity is not well understood (Cummins and Macintyre 2006; White 2007; Townshend and Lake 2009). Part of the problem is that the approach elides the question of how human subjects actually negotiate these environments, absences that qualitative health geographers have critiqued (Kearns and Moon 2002; Parr 2002).

An additional problem with this approach is that it treats the human body mechanistically. Indeed, virtually all research in this vein absolutely presumes the energy balance model of obesogenesis, such that obesity is a straightforward consequence of an excess of caloric intake relative to expenditure. (Researchers who take this approach also tend to assume that the problem of obesity in relation to poor health is clear-cut and noncontroversial.) So paradigmatic is the energy balance model that to suggest otherwise inevitably raises eyebrows. Yet, there is emerging biomedical research that shows environmentally induced pathways to obesity that are not directly related to calories—and hence to behaviors. Evidence I discuss in this article, especially regarding the role of the endocrine system and endocrine-disrupting chemicals in obesity, is therefore potentially paradigm-shifting.

The purpose of this article is thus both methodological and substantive. Methodologically, I want to contribute to a deepening of political ecology perspectives in health geography and contribute to what might be termed a political ecology of the body. Specifically, I want to encourage opening up the black box of the human body and paying attention to its biophysical workings, albeit with cognizance of how scientific understandings of the body, as well as the environment, are socially filtered. Tools from critical political ecology, which draws from science and technology studies, are particularly helpful for such an analysis. Substantively, I want to use this emerging research to show how such an approach reveals ecological causes of obesity not reducible to individual behaviors and ecological roles of fat that are not pathological. Because what I present challenges most geographical approaches to obesity, it pays to look first at how explanations that rely on the energy balance model fall short.

Tipping the Energy Balance

The basic question that all causal explanations of obesity have to address is what accounts for the significant and abrupt rise in size since 1980, as well as variation among different subpopulations. Obesity is generally measured through body mass index (BMI), a ratio of weight to height squared. For the United States, there was little change in age-adjusted mean BMI between 1960 and 1980, hovering around 25 for both adult men and women. By 2002, it had increased to 27.9 and 28.2 for men and women, respectively, although since then its increase has slowed or leveled off for all population groups (Kuczmarski et al. 1994; Flegal et al. 2002; Ogden et al. 2006). Translated, a change in women's BMI from 25 to 28.2 is equivalent to about twenty pounds for a woman who is five foot, five inches tall.

Because genetic variation is supposedly evolutionary in its temporality, genetic arguments are generally discounted in explanations for the sudden rise in obesity (Crossley 2004). Most assume that it lies in changes in eating and physical activity related to modern lifestyles. The fact that on average whites, Latinos, and African Americans of lower socioeconomic status are bigger than those with high income and educational attainment (Zhang and Wang 2004; Chang and Christakis 2005; Wang and Beydoun 2007) is usually attributed to their eating cheap, processed, high-caloric food, as well as bigger portions (Critser 2003). The fact that there is significant geographic variation in mean BMI is generally attributed to urban form, in particular those properties enshrined in the obesogenic environment thesis (see Papas et al. [2007], for an overview of typical measures).

Yet, it has not been convincingly demonstrated that people in industrialized countries eat more and exercise less than they did a generation or two ago. In an exhaustive review of the literature on caloric intake and expenditure, Gard and Wright (2005) found no definitive proof that food intake has risen and activity levels have declined, particularly in the years since 1980, although they noted plenty of presumptions made based on "common sense." Nor is subpopulation variation in food intake well supported by existing data. A report called *What We Eat in America*, based on national health surveys of daily caloric intake in 2007–2008, showed remarkable similarity across racial and income categories, challenging assumptions that poor and low-income people eat more (U.S. Department of Agriculture 2010). Although self-reporting is notoriously unreliable, one cannot assume that some groups fudge more than others.

Two other trends are even more puzzling. One is the significant increase in very big adults (or so-called morbid obesity) since 1980. When BMIs are plotted on a bell curve, not only has the curve shifted to the right but the skew of the curve is such that the right tail is further above the x axis than the left (Freedman et al. 2002; Flegal et al. 2010). Parallel evidence exists that very big children have gotten bigger, whereas other increases in size among children have flattened (Ogden et al. 2010). The other is the significant increase in obese newborns (0–6 months). Although few studies actually measure infants as a stand-alone category, a major study of well-care visits at a Massachusetts health maintenance organization found a 73.5 percent increase in prevalence of overweight infants (from 3.4 to 5.9 percent) between 1980 and 2001 (Kim et al. 2006). "This epidemic of obese 6-month-olds," as pediatric endocrinologist Robert Lustig calls it, "poses a problem for conventional explanations of the fattening of America . . . since they're eating only formula or breast milk, and never exactly got a lot of exercise the obvious explanations for obesity don't work for babies" (Begley 2009). It appears, then, that the energy balance model is not sufficient to explain the post-1980 rise in size.

A Critical Political Ecology Approach

To the extent that current geographical research on obesity takes the energy balance model for granted, it is akin to what Forsyth refers to as an *environmental orthodoxy*. Forsyth takes issue with environmental research that builds taken-for-granted assumptions into both explanations and solutions and privileges urgency to do something over careful scientific examination (Forsyth 2003, 37–38). Forsyth offers up "critical political ecology" as an approach that challenges these orthodoxies by "highlight[ing] as far as possible the implicit social and political models built into statements of supposedly neutral explanation" (20). For Forsyth, a critical political ecology goes beyond the epistemology offered by political economy (7) and hence "traditional" political ecology, which "combines the concerns of ecology and a broadly defined political economy" (Blaikie and Brookfield 1987, 17). Although Forsyth does not negate the role that political economic factors might play in environment degradation, his concern is that traditional political ecology overlooks biophysical factors that exist independently of political economic conflicts or outside of human experience.

Clearly most traditional political ecology has been concerned with the ecology of nature "out there" and not in the human body—and so is Forsyth. Yet, insofar as dominant geographical explanations of obesity are rooted in, or at least share certain elements with, traditional political ecology, I take Forsyth's call as applicable to what some have called the *political ecology of health*. For even those who claim that field tend to focus on the broader political and social contexts of disease etiology as opposed to the materiality of disease itself (e.g., Mayer 1996; King 2010). In contrast, recent work in environmental history has attempted to integrate political and economic contexts of disease causation with accounts of biological pathways, particularly in relation to environmentally induced illness (e.g., Langston 2010). That is the path on which I seek to continue. But in the vein of critical political ecology, such investigation also requires attention to the politics of disease construction, interpretation, and representation (Craddock 2000; Harper 2004).

The science studies approach on which Forsyth draws does not negate the real properties of the scientific object in question. Rather, it acknowledges that scientific discovery and reporting about biophysical processes take place in social contexts, with socially derived instruments and metrics, such that knowledge of health and environment problems necessarily reflects the manifold social relations that affect science: from grantor funding priorities to the state of technology, as well as economic interests and personal relationships and values (Hess 1997; Demeritt 1998; Proctor 1998). One of the critical ways in which this "social construction" manifests is in the framing of the scientific problem. Framing delimits the universe of further scientific inquiry and, hence, political discourse and possible policy options (Jasanoff and Wynne 1998, 5). A critical political ecology approach thus pays attention to how existing framings shape scientific understandings of environmental problems and how those framings foreclose other explanations (Forsyth 2003). In the case of obesity etiology, it is arguable that the paradigmatic status of the energy balance model has foreclosed looking elsewhere for explanation. To be sure, one of the scientists with whom I spoke about his discoveries of environmental obesogens (discussed later) had his research proposals rejected several times because reviewers found his hypotheses unthinkable.

Two additional concepts are particularly relevant for the environmental etiology of obesity. One is problem closure, which occurs when a specific definition of a problem is used to frame subsequent study of the problem's causes and consequences (Hajer 1995, 22). Problem closure can entail defining the purpose of inquiry to, for example, determining causes of a problem rather than mitigating consequences or even defining the problem in relation to socially acceptable solutions. In the case of climate change policy, for example, problem closure is evident in the focus on reducing atmospheric gases rather than reducing vulnerability to climate change (Forsyth 2003, 79). In the case of obesity, problem closure appears evident in efforts to reshape urban environments to have more outlets for fresh fruits and vegetables and more bike trails, because these are expected to alter caloric intake and expenditure. This is not to disparage fresh fruits and vegetables or bike trails. It is to note that the solutions in some sense wag the dog of the problem statement, for it assumes that obesity is caused by a lack of either in the built environment.

The second concept is black boxing, taken from Latour's (1987) *Science in Action*. A scientific concept or term is black-boxed when its internal nature is taken to be objectively established, immutable, or beyond the possibility of human action to reshape it (Forsyth 2003). Therefore, it seems to bear no scrutiny. As I have already suggested, current geographic perspectives on obesity black-box the human body and treat it as a machine that processes calories in a predictable manner. Although obesity scientists certainly allow for population variation in size, they attribute that variation entirely to genetic predisposition. Gard and Wright (2005) have thus concluded that genetic arguments have become a red herring of sorts, employed to hold the energy balance model together. What new worlds of explanation become possible when we open up that black box and the problem statement more generally? Many, as the following brief overview of this emerging research on environmental toxins and endocrinal regulation of obesity shows.

Environmental Toxins and Obesity: Emerging Research

The first person to make the case for the role of environmental toxins in obesity was Baillie-Hamilton (2002). In an article in the *Journal of Alternative and Complementary Medicine*, she pointed the finger at the proliferation of synthetic organic and inorganic chemicals since 1940, in the form of pesticides, dies, perfumes, cosmetics, medicines, food additives, plastics, fire retardants, solvents, surfactants, and so forth. In making her case, she cited a number of studies that had shown that lab animal exposure to various pesticides had produced significant weight gain, in one case when caloric availability was cut in half. As she noted, these results had been ignored, explained away, or even missed by other scientists because weight gain was a nonhypothesized consequence and thus had not been included in scientific abstracts. Only through independent discoveries of their own have a handful of scientists noted that she was on to something (Heindel 2003; Grun and Blumberg 2009).

The substances of specific concern are external agents (xenobiotics) that behave like, interact with, or alter the function of hormones produced by the body, either mimicking, enhancing, or inhibiting them, to alter developmental pathways (Krimsky 2000, 116). Synthetic xenobiotics are more typically called *endocrine-disrupting chemicals* (EDCs). Thus far, the direct evidence for the role of EDCs has come from animal experiments. For example, one experiment gave both low and high doses of diethylstilbestrol (DES) to mice during gestation and immediately following birth. Although the estrogen seemed to have no or a negative effect on size at birth and during infancy, in adulthood these mice had significantly higher body weight than control groups (Newbold et al. 2008). Similar results have come from mice whose mothers were exposed to bisphenol A (BPA), tributlyltin (TBT), and perfluorooctanoic acid (PFOA; Grun et al. 2006; Hines et al. 2009; Rubin and Soto 2009). In many of these studies, the changes induced by prenatal and early postnatal exposures did not become visible until adulthood (Grun and Blumberg 2009). This lag time between exposure and manifestation could thus help explain how chemicals that began to proliferate in the 1950s and 1960s would result in increases in adult BMI after 1980.

A related scientific discovery that bears on the etiology of obesity is the existence of what are called *epigenetic mechanisms*: changes in gene function that are not a result of changed DNA sequences but are nonetheless heritable. Because postpartum changes in the endocrine system can control the expression of inherited genes, exposure to EDCs can permanently transform bodily form and function, and these changes can be passed on to offspring (Crews and McLachlan 2006, S4). Demonstrating such an epigenetic effect, one of the preceding studies found that the genes that direct fat distribution had been permanently altered by DES exposure, meaning that the tendency toward high-fat tissue would be heritable were the mice to reproduce (Newbold et al. 2008). DES, incidentally, was widely prescribed to pregnant women in the 1950s and early 1960s to prevent miscarriages; it was later banned because many of the progeny of these women developed reproductive cancers. Epigenetic effects thus help account for generational lags, and exposure to EDCs analogous to DES could account for the rise in infant obesity. The epidemiological data, although thus far limited by the number of studies, do provide important associative evidence (see Hatch et al. [2010], for an overview). For example, scientists in North Carolina found that children exposed to higher levels of polychlorinated biphenyls (PCBs) and dichlorodiphenyldichloroethylene (DDE) before birth were fatter than those exposed to lower levels (Gladen, Bagan, and Rogan 2000).

Perhaps the most paradigm-threatening finding from the laboratory studies is that some of these substances appear to provoke the division of fat cells into more fat cells or stimulate cells having no particular destination to become fat cells (Masuno et al. 2002). This means that they can induce fat creation regardless of caloric intake. Insofar as existing fat cells are not only very hard to get rid of but can produce even more fat cells, there is potential for large gains of adiposity from nonexceptional exposures to EDCs. This mechanism might explain the skew of the bell curve, or why big people are getting even bigger, a phenomenon that intuitively does not square with the energy balance model, because bigness puts high energetic demands on the body.

To the extent that fat plays a role in producing more fat, as appears to be the case in extreme obesity, the role of fat does appear pathological. By the same token, within this body of research is the hypothesis that fat can play a protective role. This hypothesis is based in part on the 1990s discovery that adipose tissue secretes hormones that are active in regulating appetite and metabolism, meaning that adipose tissue not only stores fat but actually participates in the body's signaling function. Adiponectin, for example, plays a protective role against insulin resistance and inflammation and also appears to discourage further fat cell production (Kershaw and Flier 2004). More generally, adiposity can be protective because it stores lipids away from places it could be toxic to bodily function, delaying, for example, the onset of metabolic syndrome (Unger and Scherer 2010). As Grun and Blumberg (2009, 1131) stated, "the physiological process of diverting excess calories toward lipid storage may be benign or pathological depending on the mechanism and site." If this is the case, it is at least plausible that increases in adiposity are adaptive responses to environmental toxicity, such that bodies are transforming to deal with environmental changes. In short, this emerging research, although far from conclusive, potentially addresses significant gaps in existing explanations and, more generally, provides different ways of thinking about the role and function of fat.

Rethinking Geographic Perspectives

Although the research just all too briefly discussed is still in its early stages, it strongly suggests that environmental toxins are at least partially responsible for the rise in size and specifically points to noncaloric pathways to obesity. As such, this research not only challenges existing assumptions about obesity's etiology in excesses of calories ingested, but it throws up serious challenges to geographical approaches to health more generally that tend to treat the body as an expression of individual behaviors, even if those behaviors are understood to be influenced by the external environment. It also shows that ecological influences do not stop at the skin, mouth, or other entry point into the body. In this case, the active role of the endocrine system and fat itself in regulating fat creation and disposition serves as an important reminder that bodies are ecological, too, not separate from but ultimately a part of the socionatural environment. Even more subtly, it shows that environmental influences (and toxic exposures) can be temporal as well as spatial, which certainly complicates methodological approaches that privilege space over time. Together, it suggests the imperative to open up the black box of the body and explore it as an ecological, geographical, and historical object.

Still, the focus on bodily ecologies cannot come at the expense of attention to the discursive aspects of knowledge. The kind of analysis I have offered here only becomes possible with an approach that encourages examination beyond common sense, the kind of approach afforded by critical political ecology. By engaging with the science interpretively, it opens up lines of inquiry that in the case of obesity escape the broad condemnation that it receives in more paradigmatic perspectives. For, among other things, this research not only acknowledges the adaptive role of fat; it might also serve as a caution that there are chemically induced bodily changes that might be quite toxic that do not happen to manifest as fat. Like all environmental transformation, that is, the most visible change might not be the worst problem, and the worst problem might not be the most visible.

References

Baillie-Hamilton, P. 2002. Chemical toxins: A hypothesis to explain the global obesity epidemic. *Journal of Alternative and Complementary Medicine* 8 (2): 185–92.

Begley, S. 2009. Born to be big. *Newsweek* 21 September: 56–58, 62.

Blaikie, P., and H. Brookfield. 1987. *Land degradation and society.* London: Methuen.

Chang, V. W., and N. A. Christakis. 2005. Income inequality and weight status in U.S. metropolitan areas. *Social Science & Medicine* 61 (1): 83–96.

Craddock, S. 2000. Disease, social identity, and risk: Rethinking the geography of AIDS. *Transactions of the Institute of British Geographers* NS 25:153–68.

Crews, D., and J. A. McLachlan. 2006. Epigenetics, evolution, endocrine disruption, health, and disease. *Endocrinology* 147 (6): s4–s10.

Critser, G. 2003. *Fat land: How Americans became the fattest people in the world.* Boston: Houghton Mifflin.

Crossley, N. 2004. Fat is a sociological issue: Obesity rates in late modern, "body-conscious" societies. *Social Theory & Health* 2 (3): 222–53.

Cummins, S., and S. Macintyre. 2006. Food environments and obesity—Neighbourhood or nation? *International Journal of Epidemiology* 35 (1): 100–104.

Demeritt, D. 1998. Science, social constructivism, & nature. In *Remaking reality: Nature at the millennium,* ed. B. Braun and N. Castree, 173–93. London and New York: Routledge.

Flegal, K. M., M. D. Carroll, C. L. Ogden, and L. R. Curtin. 2010. Prevalence and trends in obesity among U.S. adults, 1999–2008. *JAMA* 303 (3): 235–41.

Flegal, K. M., M. D. Carroll, C. L. Ogden, and C. L. Johnson. 2002. Prevalence and trends in obesity among U.S. adults, 1999–2000. *JAMA* 288 (14): 1723–27.

Forsyth, T. 2003. *Critical political ecology: The politics of environmental science.* London and New York: Routledge.

Freedman, D. S., L. K. Khan, M. K. Serdula, D. A. Galuska, and W. H. Dietz. 2002. Trends and correlates of class 3 obesity in the United States from 1990 through 2000. *JAMA* 288 (14): 1758–61.

Gard, M., and J. Wright. 2005. *The obesity epidemic: Science, morality, and ideology.* London and New York: Routledge.

Gladen, B. C., N. B. Bagan, and W. J. Rogan. 2000. Pubertal growth and development and prenatal and lactational exposure to polychlorinated biphenyls and dichlorodiphenyl dichloroethene. *The Journal of Pediatrics* 136 (4): 490–96.

Grun, F., and B. Blumberg. 2009. Minireview: The case for obesogens. *Molecular Endocrinology* 23 (8): 1127–34.

Grun, F., H. Watanabe, Z. Zamanian, L. Maeda, K. Arima, R. Cubacha, D. M. Gardiner, J. Kanno, T. Iguchi, and B. Blumberg. 2006. Endocrine-disrupting organotin compounds are potent inducers of adipogenesis in vertebrates. *Molecular Endocrinology* 20 (9): 2141–55.

Hajer, M. A. 1995. *The politics of environmental discourse: Ecological modernization and the policy process.* New York: Oxford University Press.

Harper, J. 2004. Breathless in Houston: A political ecology of health approach to understanding environmental health concerns. *Medical Anthropology: Cross-Cultural Studies in Health and Illness* 23 (4): 295–326.

Hatch, E. E., J. W. Nelson, R. W. Stahlhut, and T. F. Webster. 2010. Association of endocrine disruptors and obesity: Perspectives from epidemiological studies. *International Journal of Andrology* 33 (2): 324–32.

Heindel, J. J. 2003. Endocrine disruptors and the obesity epidemic. *Toxicological Sciences* 76 (2): 247–49.

Hess, D. J. 1997. *Science studies: An advanced introduction.* New York: New York University Press.

Hill, J. O., and J. C. Peters. 1998. Environmental contributions to the obesity epidemic. *Science* 280 (5368): 1371–74.

Hines, E. P., S. S. White, J. P. Stanko, E. A. Gibbs-Flournoy, C. Lau, and S. E. Fenton. 2009. Phenotypic dichotomy following developmental exposure to perfluorooctanoic acid (pfoa) in female cd-1 mice: Low doses induce elevated serum leptin and insulin, and overweight in mid-life. *Molecular and Cellular Endocrinology* 304 (1–2): 97–105.

Jasanoff, S., and B. Wynne. 1998. Scientific and decision-making. In *Human choice and climate change,* ed. S. Rayner and E. Malone, 1–87. Columbus, OH: Battelle Press.

Kearns, R., and G. Moon. 2002. From medical to health geography: Novelty, place and theory after a decade of change. *Progress in Human Geography* 26 (5): 605–25.

Kershaw, E. E., and J. S. Flier. 2004. Adipose tissue as an endocrine organ. *Journal of Clinical Endocrinology and Metabolism* 89 (6): 2548–56.

Kim, J., K. E. Peterson, K. S. Scanlon, G. M. Fitzmaurice, A. Must, E. Oken, S. L. Rifas-Shiman, J. W. Rich-Edwards, and M. W. Gillman. 2006. Trends in overweight from

1980 through 2001 among preschool-aged children enrolled in a health maintenance organization. *Obesity* 14 (7): 1107–12.

King, B. 2010. Political ecologies of health. *Progress in Human Geography* 34 (1): 38–55.

Krimsky, S. 2000. *Hormonal chaos: The scientific and social origins of the environmental endocrine hypothesis.* Baltimore: Johns Hopkins University Press.

Kuczmarski, R. J., K. M. Flegal, S. M. Campbell, and C. L. Johnson. 1994. Increasing prevalence of overweight among U.S. adults: The national health and nutrition examination surveys, 1960 to 1991. *JAMA* 272 (3): 205–11.

Langston, N. 2010. *Toxic bodies: Hormone disrupters and the legacy of DES.* New Haven, CT: Yale University Press.

Latour, B. 1987. *Science in action.* Cambridge, MA: Harvard University Press.

Masuno, H., T. Kidani, K. Sekiya, K. Sakayama, T. Shiosaka, H. Yamamoto, and K. Honda. 2002. Bisphenol A in combination with insulin can accelerate the conversion of 3t3–11 fibroblasts to adipocytes. *Journal of Lipid Research* 43 (5): 676–84.

Mayer, J. D. 1996. The political ecology of disease as one new focus for medical geography. *Progress in Human Geography* 20 (4): 441–56.

Morland, K. B., and K. R. Evenson. 2009. Obesity prevalence and the local food environment. *Health and Place* 15 (2): 491–95.

Newbold, R., E. Padilla-Banks, W. Jefferson, and J. Heindel. 2008. Effects of endocrine disruptors on obesity. *International Journal of Andrology* 31:201–208.

Ogden, C. L., M. D. Carroll, L. R. Curtin, M. M. Lamb, and K. M. Flegal. 2010. Prevalence of high body mass index in us children and adolescents, 2007–2008. *JAMA* 303 (3): 242–49.

Ogden, C. L., M. D. Carroll, L. R. Curtin, M. A. McDowell, C. J. Tabak, and K. M. Flegal. 2006. Prevalence of overweight and obesity in the United States, 1999–2004. *JAMA* 295 (13): 1549–55.

Papas, M. A., A. J. Alberg, R. Ewing, K. J. Helzlsouer, T. L. Gary, and A. C. Klassen. 2007. The built environment and obesity. *Epidemiologic Reviews* 29 (1): 129–43.

Parr, H. 2002. Medical geography: Diagnosing the body in medical and health geography, 1999–2000. *Progress in Human Geography* 26 (2): 240–51.

Proctor, J. D. 1998. The social construction of nature: Relativist accusations, pragmatist and critical realist responses. *Annals of the Association of American Geographers* 88 (3): 352–76.

Robbins, P. 2004. *Political ecology.* Oxford, UK: Blackwell.

Rubin, B. S., and A. M. Soto. 2009. Bisphenol A: Perinatal exposure and body weight. *Molecular and Cellular Endocrinology* 304 (1–2): 55–62.

Stafford, M., S. Cummins, A. Ellaway, A. Sacker, R. D. Wiggins, and S. Macintyre. 2007. Pathways to obesity: Identifying local, modifiable determinants of physical activity and diet. *Social Science & Medicine* 65 (9): 1882–97.

Swinburn, B., G. Egger, and F. Raza. 1999. Dissecting obesogenic environments: The development and application of a framework for identifying and prioritizing environmental interventions for obesity. *Preventative Medicine* 29 (6): 563–70.

Townshend, T., and A. A. Lake. 2009. Obesogenic urban form: Theory, policy and practice. *Health and Place* 15 (4): 909–16.

Unger, R. H., and P. E. Scherer. 2010. Gluttony, sloth and the metabolic syndrome: A roadmap to lipotoxicity. *Trends in Endocrinology & Metabolism* 21 (6): 345–52.

U.S. Department of Agriculture. 2010. *What we eat in America 2010.* http://www.ars.usda.gov/Services/docs.htm?docid=18349# (last accessed 27 October 2010).

Wang, Y., and M. A. Beydoun. 2007. The obesity epidemic in the United States—Gender, age, socioeconomic, racial/ethnic, and geographic characteristics: A systematic review and meta-regression analysis. *Epidemiological Review* 29 (1): 6–28.

White, M. 2007. Food access and obesity. *Obesity Reviews* 8:99–107.

Zhang, Q., and Y. Wang. 2004. Socioeconomic inequality of obesity in the United States: Do gender, age, and ethnicity matter? *Social Science & Medicine* 58 (6): 1171–80.

The Uncertain Geographic Context Problem

Mei-Po Kwan

Department of Geography, University of California, Berkeley

Any study that examines the effects of area-based attributes on individual behaviors or outcomes faces another fundamental methodological problem besides the modifiable areal unit problem (MAUP). It is the problem that results about these effects can be affected by how contextual units or neighborhoods are geographically delineated and the extent to which these areal units deviate from the true geographic context. The problem arises because of the spatial uncertainty in the actual areas that exert the contextual influences under study and the temporal uncertainty in the timing and duration in which individuals experienced these contextual influences. Using neighborhood effects and environmental health research as a point of departure, this article clarifies the nature and sources of this problem, which is referred to as the *uncertain geographic context problem* (UGCoP). It highlights some of the inferential errors that the UGCoP might cause and discusses some means for mitigating the problem. It reviews recent studies to show that both contextual variables and research findings are sensitive to different delineations of contextual units. The article argues that the UGCoP is a problem as fundamental as the MAUP but is a different kind of problem. Future research needs to pay explicit attention to its potential confounding effects on research results and to methods for mitigating the problem.

任何探讨以区域为基础的属性对个体的行为或结果之影响的研究，除了可变面元问题（MAUP），还都面临着另一个基本方法的问题。该问题影响了这些研究结果，使它们受到背景单位或社区是如何被地理划定的影响，并受到这些面元偏离真实的地理环境的程度的影响。问题的产生是因为在这些实际区域的空间不确定性的情景，以及个人在经历这些情景影响时，在时间和持续时间上的不确定性的影响。使用邻里效应和环境健康研究作为出发点，本文阐明了这个被称为不确定的地理环境问题（UGCoP）的性质和来源。它突出了一些UGCoP可能导致的推理错误，并讨论了减轻这个问题的一些手段。它回顾了最近的研究，揭示了情景变量和研究成果对情景单位的不同划定是敏感的。文章认为，UGCoP是与MAUP类似的根本性问题，但又是不同类的问题。未来的研究需要明确注意其潜在的对研究结果的复杂影响，并关注减轻这一问题的方法。*关键词：情景的不确定性，环境卫生，邻里效应，地理环境不确定性问题，UGCoP。*

En cualquier estudio que examine los efectos que tienen atributos basados en área sobre conductas o logros individuales se tiene que enfrentar un problema metodológico fundamental aparte del problema de la unidad espacial modificable (PUEM). El inconveniente es que los resultados acerca de estos efectos pueden afectarse por la manera como las unidades contextuales o vecindarios sean delineados geográficamente y el grado en que dichas unidades espaciales se apartan del contexto geográfico verdadero. El problema surge debido a la incertidumbre espacial en las áreas reales que ejercen las influencias contextuales bajo estudio y la incertidumbre temporal presente en el cronograma y duración con que los individuos experimentaron estas influencias contextuales. Utilizando los efectos de vecindad y la investigación sobre salubridad ambiental como punto de partida, este artículo hace claridad sobre la naturaleza y orígenes de este problema, que se denomina *problema de incertidumbre del contexto geográfico* (PICoG). Se destacan algunos de los errores de inferencia que el PICoG podría ocasionar y se discuten algunos medios para mitigar el problema. En el artículo se revisan también estudios recientes para mostrar que tanto las variables contextuales como los hallazgos de la investigación son sensibles a las diferentes demarcaciones de las unidades contextuales. En el artículo se arguye que el PICoG es un problema tan fundamental como el PUEM, aunque se trata de un tipo diferente de problema. La futura investigación sobre el particular deberá poner atención explícita a los potenciales efectos desorientadores de los resultados que se logren y a los métodos para mitigar el problema.

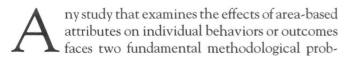

Any study that examines the effects of area-based attributes on individual behaviors or outcomes faces two fundamental methodological problems. One is that results about these effects can be affected by the zoning scheme or geographic scale of the areal units used (Openshaw 1984; Fotheringham and

Wong 1991). This is the well-known modifiable areal unit problem (MAUP), which has received much attention to date. The other problem, however, has received much less attention. This is the problem that findings about the effects of area-based attributes could be affected by how contextual units or neighborhoods are geographically delineated and the extent to which these areal units deviate from the "true causally relevant" geographic context (the precise spatial configuration of which is unknown in most studies to date; Diez-Roux and Mair 2010, 134). This problem is referred to as the *uncertain geographic context problem* (UGCoP) in this article. It arises because of the spatial uncertainty in the actual areas that exert contextual influences on the individuals being studied and the temporal uncertainty in the timing and duration in which individuals experienced these contextual influences. The UGCoP is a significant methodological problem because it means that analytical results can be different for different delineations of contextual units even if everything else is the same. It is perhaps a major reason why research findings concerning the effects of social and physical environments on health behaviors and outcomes are often inconsistent, given that past studies on the same issue (e.g., obesity) often used different contextual units (e.g., Inagami, Cohen, and Finch 2007; Black and Macinko 2008; Wilks et al. 2010).

This article argues that the UGCoP is a problem as fundamental as the MAUP for any study that uses area-based contextual attributes, but it is a different kind of problem because it is not due to the use of different zonal schemes or spatial scales for area-based variables. Because methods for addressing the MAUP do not automatically solve the UGCoP, future research needs to pay explicit attention to its potential confounding effects on research results and to methods for mitigating the problem. The article clarifies the nature and sources of the UGCoP and highlights the inferential errors it might cause. It reviews recent studies to show that both contextual variables and research findings are sensitive to different delineations of contextual units. It suggests some means for addressing the problem, such as delineating more appropriate contextual units based on people's actual or potential activity spaces. The article uses neighborhood effects and environmental health research as a point of departure to highlight the methodological challenges the UGCoP poses to health research in particular and to geographic research in general.[1] It addresses a fundamental methodological problem for any study that uses area-based attributes as explanatory variables.

Contextual Uncertainty: The Spatial Dimension

Studies that examine the effect of contextual or environmental influences (e.g., neighborhood physical and social features) on health using ecological designs often begin by constructing a conceptual model that specifies the causal pathways among the contextual attributes (or variables) and health outcomes (Diez-Roux and Mair 2010). Based on the conceptual model, contextual units or geographic areas for evaluating individual exposure to these contextual influences are then identified. After these units are defined, values of the relevant contextual variables (e.g., neighborhood deprivation) for these contextual units are derived and used as indicators of the exposure individuals in particular contextual units experienced. Effects of the contextual variables on health are finally evaluated using appropriate statistical models (e.g., multilevel models). In this process, two important sources of contextual uncertainty contribute to the UGCoP. One is the uncertainty in the spatial configuration of the appropriate contextual units for assessing the influence of environmental variables on health outcomes, and the other is the uncertainty about the timing and duration to which individuals are actually exposed to these contextual influences (the temporal dimension of contextual uncertainty is discussed in the next section under the rubric of dynamics of geographic context).

Health researchers normally have little or no prior knowledge about the precise spatial configuration and boundary of the geographic area that, through its physical or social characteristics, has significant influence on an individual's health. The "true causally relevant" geographic context is thus unknown in most studies to date (Diez-Roux and Mair 2010, 134). A common practice in the past has been to use residential neighborhoods—operationalized as static administrative areas such as census tracts or postal code areas or buffer areas around individuals' home addresses or centroids of their home census tracts—as contextual units. Leal and Chaix (2011), for instance, observed that 90 percent of studies on environmental influences on cardiometabolic risk factors used the residential neighborhood as the contextual unit. These units not only are convenient but often are the only viable option because available data of censuses and surveys that can be used to derive contextual measures are tied to them.

But residential neighborhoods might not accurately represent the actual areas that exert contextual influences on the health outcome under study (Diez-Roux 1998; Cummins 2007; Matthews 2008; Chaix 2009;

Kwan 2009). For instance, adolescent risk behavior like substance use might be influenced not only by socioeconomic deprivation in the residential neighborhood but also by interactions with friends and peers in various nonresidential contexts (e.g., schools and places for various leisure activities). For working adults, opportunities for physical activities and the quality of food near the workplace could also have important effects on their health. The boundaries of these multiple contexts are often difficult to clearly delineate; even when it is possible, some of them might not be continuous in geographic space (i.e., one contextual unit might consist of several discrete geographic areas) and thus cannot be represented or analyzed in any simple manner even using advanced geographic information systems (GIS) methods (Wiehe et al. 2008).

Further, social contexts such as families, friends, or peers are not in themselves geographically defined and thus cannot be easily delineated as geographic areas with precise boundaries (Diez-Roux 2001; Macintyre, Ellaway, and Cummins 2002). Delineations of contextual units in these cases need to take into account how social networks constituted through people's routine activities and social interactions express themselves in geographic space (Grannis 2009). In other cases, neighborhoods defined on the basis of people's perceptions might be more relevant. The perceived neighborhood for different individuals might not coincide with or might even deviate significantly from the administratively defined home neighborhood or people's activity space, however. For instance, Basta, Richmond, and Wiebe (2010) found that participants' perceived neighborhoods did not correspond to the boundaries of the home census tracts, and time they spent in close proximity to alcohol outlets during their daily activities was not correlated with the prevalence of alcohol outlets in the census tract of their residence. Vallée et al. (2010) found that over 80 percent of the participants have an activity space larger than their perceived neighborhood. This means that even when the boundaries of perceived neighborhoods are identified using appropriate procedures, a considerable portion of people's normal daily activities could still fall outside these boundaries. In these cases, perceived neighborhoods might not correspond well with, and thus are not good proxies for, true geographic contexts.

Besides administrative areas and perceived neighborhoods, other geographic areas have been used as proxies to the true geographic context. For instance, studies on the effect of neighborhood features such as land-use mix and residential density on people's physical activity or body weight have defined neighborhood around each participant's home as a 1-km or 3-km circular zone (Berke et al. 2007), as a 1-km road network buffer (Frank et al. 2005), as a 0.5-mile radius or a ten-minute walk from the respondent's home for some variables, and as a 10-mile radius or a twenty-minute drive from the respondent's home for several other variables (Brownson et al. 2004). It is far from clear, however, which of these areal units appropriately represents the areal extent and spatial configuration of the true geographic context. The mixed results of past studies on neighborhood effects (e.g., neighborhood income inequality and racial composition) on health (e.g., obesity) could thus be partly due to the different neighborhood delineations used (Black and Macinko 2008).

Finally, the relevant contextual unit might vary depending on the population groups under study and according to different factors and processes hypothesized to influence the health outcome in question (Subramanian, Jones, and Duncan 2003). For instance, results in Oliver, Schuurman, and Hall (2007) indicated that larger or smaller contextual units might be more appropriate for different types of built environments and population groups (e.g., smaller for elderly people because of their lower out-of-home mobility). Further, some contextual influences might operate in the block on which a person resides, some might operate in a larger area around the block, some might operate near the person's workplace or school, and still others might exert their influence in the area in which specific types of stores or institutions are located (Diez-Roux 2001; Macintyre and Ellaway 2003). The multilevel and multiscale nature of contextual influences greatly complicates the task of accurately delineating the appropriate contextual units, which could be nested or overlapped in a complex manner. Part of the uncertainty in the spatial configuration and boundaries of contextual units arises from the dynamic characteristics of individuals and contextual influences (Gatrell 2011).

Dynamics of Geographic Context

People move around to undertake their daily activities. They often traverse the boundaries of multiple neighborhoods during the course of a day and come under the influence of many different neighborhood contexts besides their residential neighborhoods (Sampson, Morenoff, and Gannon-Rowley 2002; Chaix 2009). The majority of physical and social resources they use (which affect their health and well-being) might be located outside of or far from their home neighborhoods. Geographers have long observed the spatial and temporal variability of people's daily activities (including those performed in evening hours), and it was also noted that individuals of different social groups tend to have distinctive activity patterns in space–time (e.g., Hanson and Hanson 1981; Kwan 2000; Lee and Kwan 2011).

Recent studies that collected detailed data about people's out-of-home activities and travel routes over the course of one to many days using Global Positioning System (GPS) or other location-aware devices provide further evidence about where and when people spent time in their daily life. Elgethun et al. (2003), for instance, found that participants (children two to eight years old) on average spent most of their time inside schools on weekdays, while spending most of their time in establishments like restaurants and cinemas on weekend days. Basta, Richmond, and Wiebe (2010) observed that half of the sampled participants (fifteen to nineteen years old) spent 92 percent of their time outside of their neighborhood. Wiehe et al. (2008) found that participants (female adolescents) spent one third of their time in locations more than 1 km from home, which is the distance used in many previous studies for defining neighborhood. This means that the participants spent a considerable amount of time in their daily lives outside of what has conventionally been defined as geographic context or neighborhood. The study also found considerable day-to-day variability in participants' activity locations besides their variability by time of day (Wiehe et al. 2008). The daily and day-to-day variability in human activity locations not only raises concerns about using conventional static contextual units in health research but also calls into question the appropriateness of the notion of daytime population, which does not take temporal variability over the course of a day into account.

As Gatrell (2011) cogently argued, exposure to health risks, spread of diseases, and use of health care facilities are inextricably connected to human movements at various spatial and temporal scales. The spatial and temporal variability of human activities thus has significant implications for any study that examines the effect of contextual influences on health. It means that people's activities (and thus exposures) do not take place at one time point and wholly within any conventionally defined neighborhood. Their use of different physical resources and their social interactions with friends, peers, and others might take place at different times of the day and in disparate geographic areas outside of their home neighborhoods (Kwan 2009). The

neighborhood of residence is only one of the places people spend their time, and it might not adequately capture people's exposure to relevant contextual influences. Further, besides moving around to undertake their daily activities, people also move around over time. They could change their residence in the same city (residential mobility) or move to another (migration). As a result of moving to different neighborhoods, people's exposure to environmental influences might also change over time. A study on people's exposure to carcinogenic risk factors, for instance, needs to consider their residential history (in addition to individual factors including family predisposition), as knowing where and for how long a person has lived in the past might help more accurately estimate his or her cumulative or lifetime exposure to radioactive substances (Löytönen 1998).

Contextual influences can vary over space and time in a highly complex manner. They might vary with different temporal patterns or time frames. As people move through the changing pollution field over time during the day, for instance, their exposure to traffic-related air pollution also changes (Gulliver and Briggs 2005). Some environmental influences could change over the twenty-four-hour period of a day (e.g., pollutants from truck traffic), and some might change over the seasons. The physical and social characteristics of neighborhoods can also change over time (Entwisle 2007). Population composition and local social ties might change as a result of residential mobility and migration. Government and people's actions could change the physical features and health facilities in a neighborhood over time. When environmental or neighborhood influences have considerable spatial and temporal variability, their health impact often cannot be adequately assessed using data for just one time point (Setton et al. 2010). It might also be difficult to identify which portion of them is causally relevant to a particular individual in relation to the person's daily movement in the study area.

Further, most studies to date assume that the effects of contextual influences on health are most appropriately assessed using data collected at or around the same time point (Entwisle 2007). There might be a variety of response lags that mediate the causal pathways between contextual factors and health outcomes, however. The outcome variable at time point t, for instance, might be determined by the value of a contextual variable at a particular time point or period of time before t (Wheaton and Clarke 2003). For instance, there is some evidence that variation in people's health behaviors and outcomes is related to their exposure to neighborhood characteristics during childhood and adolescence (e.g., Monden, Van Lenthe, and Mackenbach 2006). In addition, the outcome variable might not be determined by the value of the contextual variables at a particular time point before t but by their cumulative effect over a period of time before t. As a result of these and other cause–effect lags, there could be considerable uncertainty regarding the best time point or time period for deriving the values of the contextual attributes. The lack of significant association between the contextual variables and the outcome variable in any study might be due to a failure to account for this aspect of contextual dynamics (e.g., using the wrong time lag or using a particular time point instead of considering the cumulative effect of a contextual factor).

Inferential Challenges Posed by the UGCoP

The spatial uncertainty and dynamics of geographic context associated with the UGCoP greatly complicate any examination of the effect of contextual influences on health. The error of misspecifying the true geographic context might lead to inconsistent results and inferential errors. Consider a case in which the outcome variable Y (e.g., body mass index) is hypothesized to be determined by n contextual variables X_i (e.g., street network density, land-use mix, and social disadvantage) after taking into account all relevant individual- and household-level factors. Now suppose that significant association (either positive or negative) is found between one or more contextual variables and the outcome variable, assuming that there is no "misspecification of the model at the individual level" (Diez-Roux 1998, 219). This result is normally interpreted in a straightforward manner with few qualifications on the possible confounding effects of the UGCoP. For instance, a study might conclude that neighborhood physical features that encourage physical activity were associated with decreased body mass index (e.g., Berke et al. 2007). The best possibility for this result is that it is true, which not only means that the contextual variables worked as hypothesized in the causal model; it also means that the areal extent and spatial configuration of the true geographic context were correctly identified and used to derive the contextual variables in the study. Because in most cases the true geographic context is not known, the researcher cannot be certain that it had been used in the study. This uncertainty implies that the significant relationships observed might be false

Table 1. Inferential errors due to the uncertain geographic context problem

True state of contextual effect	Observed state of contextual effect	
	Has effect	No effect
Has effect	Contextual units correct	Contextual units incorrect
	Correct inference	False negatives (obscured contextual effect)
No effect	Contextual units incorrect	Contextual units correct
	False positives (spurious association)	Correct inference

positives: There was actually no association between the contextual variables and the outcome variable as hypothesized, but significant association between them was still observed.

Table 1 shows how true or incorrect contextual units could lead to different inferential errors. Inferential errors in the form of false positives can occur when the contextual variable actually has no effect on the outcome variable but significant association between them is observed. This situation is called *spurious association* in the parlance of statistical inference and is similar to a Type I error in hypothesis testing (the null hypothesis is rejected although it is true). It can arise for two different reasons. First, it can occur purely by chance even when the causal pathway hypothesized between the contextual variable and the outcome variable is illusionary. Second, the erroneously defined contextual units might have introduced variations in the contextual variable such that positive association between the contextual variable and the outcome variable is observed, even when there is actually no association between them. This type of inferential error could considerably confound research results.

Now suppose that no significant association is observed between the contextual variable and the outcome variable. This result is often interpreted in a straightforward manner with few qualifications, and the best possibility is that it is true: The contextual variable did not work as hypothesized in the causal model. Because the true geographic context is often not known, however, and the researcher cannot be sure that it has been used in the study, failure to observe a significant relationship between the contextual and outcome variables might be due to other reasons. There might be significant association between the contextual variable and the outcome variable as hypothesized, but such association was not observed. In the parlance of statistical inference, this type of error is similar to a Type II error in hypothesis testing (failure to reject the null hypothesis when it is false). One reason why this failure (or false negative) occurs is that the contextual variable might have been misspecified and thus does not correctly capture the true contextual effect. Another reason for the error is that the spatial extent and configuration of the true geographic context were not correctly identified and used in the study. In addition, the contextual influence might be characterized with wrong temporal attributes (e.g., incorrect time point, time lag, or duration). As a result, the effect of the contextual variable on the outcome variable was obscured by the erroneously defined contextual units or inappropriate temporal characterization of the contextual influence.

These two types of error arising from the misspecification of contextual units or inappropriate temporal characterization of the contextual influence might significantly confound research results. For instance, Spielman and Yoo (2009) used simulation experiments to show that linear models tend to underestimate the effects of contextual influences on health outcomes when the size of the true geographic context is underestimated (and vice versa). The study also showed that variation in the characteristics of the population group being studied and the study area can pose a significant problem for inference about neighborhood effects. Kwan et al. (2009) found significant differences in the size and shape of three different delineations of geographic context: two delineations of activity space (the standard deviational ellipse and the kernel density surface) and the home census tract. The study observed that for certain gender and racial groups, neighborhood effects based on people's home census tracts tend to overestimate their actual exposure to social disadvantage (because characteristics of the nonresidential neighborhoods people visit might mitigate the disadvantage they experience in their residential neighborhood). Further, Troped et al. (2010) examined associations among five physical features within 1-km road network buffers of participants' homes and workplaces and the amount of moderate to vigorous physical activity. The study found that three features around the participants' homes were associated with their physical activity near their homes, and two features around their workplaces were associated with their physical activity around their workplaces. None of the five features, however, showed associations with participants' total physical activity. The study not only shows that people's physical activity might vary according to where they are but also suggests that a study that uses only participants' home neighborhoods as the contextual unit might not find any association between its physical features and participants' body mass indexes, because body mass index depends on total physical activity, not just activity around one's home or workplace.

Addressing the UGCoP

As argued in this article, the UGCoP might introduce inferential errors and confound research results in studies that examine the effects of area-based attributes on individual behaviors or outcomes. The problem arises because of our limited knowledge of the precise spatial and temporal characteristics of the true geographic context. The main difficulty it poses is that we cannot tell whether our results are true or confounded and, if confounded, which type of error is involved and to what extent it has obfuscated the results. The UGCoP is a problem as fundamental as the MAUP, but it is a different kind of problem because it is not due to the use of different zonal schemes or spatial scales for deriving area-based variables. These two problems are not necessarily related to each other and can both be present in a particular study. Methods for addressing one of them might not automatically solve the other. For instance, using the best zoning scheme or spatial scale does not help us identify the true geographic context or characterize the temporal attributes of contextual influences, but using delineations of geographic context that capture people's movement in space–time seems to mitigate both the UGCoP and the MAUP (Kwan and Weber 2008).

Given the potential confounding effect of the UGCoP on research results, it is important that future research takes the problem seriously and considers steps to mitigate its impact when using area-based contextual variables. An important initial step is to develop an adequate theoretical model for taking spatial and temporal contextual uncertainties into account (Macintyre, Ellaway, and Cummins 2002). After constructing such a dynamic conceptualization of contextual influences, an explicit statement about what contextual units will be used, how well they approximate the true geographic context, and what temporal attributes of individuals and contextual influences will be taken into account by these contextual units should be given. For example, to evaluate the health impact of traffic-related air pollution, contextual units should be conceived so that they can take the spatiotemporal variations of both air pollution and people's daily movement into account (Hoek et al. 2008).

Further, it is important to recognize that observing significant association between a contextual variable and an outcome variable does not in itself validate the contextual units—as we cannot tell whether our results are produced by erroneous contextual units that have introduced variations in the contextual variable such that association between the contextual variable and the outcome variable is observed, even when there is actually no association between them. It is also important to recognize that, as Spielman and Yoo (2009) have shown using simulation experiments, the model with the best fit is not necessarily the one that uses the true geographic context. This means that appropriateness of contextual units cannot be justified using model fit as a criterion.

In view of these difficulties, it would be particularly helpful (when resources allow) to perform sensitivity analysis to assess how different delineations of contextual units might affect contextual variables and study results (Shi 2010). There is some evidence to date that both are sensitive to the choice of contextual units (e.g., Kwan et al. 2009; Troped et al. 2010; Zenk et al. 2011). Kwan et al (2011), for instance, observed significant difference between the composite deprivation index (as a contextual variable) derived from circular buffers around participants' home addresses and those derived from half-mile road network buffers around participants' GPS tracks. The deprivation index derived using the minimum convex polygon is significantly different from those derived using participants' home census tracts. With respect to research findings, Oliver, Schuurman, and Hall (2007) found that the use of different kinds of buffers around participants' homes (based on centroids of their home postal codes) as contextual units has a considerable influence on the results: Land-use characteristics tend to show greater associations with walking using line-based road network buffers than circular buffers; circular and polygon buffers tend to underestimate the effects of land-use characteristics on walking because they might include large areas that are irrelevant to walking (e.g., industrial land) or inaccessible. These studies indicate that both contextual variables and study results are sensitive to the choice of contextual units. It is thus important to undertake sensitivity analysis to determine their stability and the extent to which they will be affected.

In recent years, geographers and health researchers have explored various methods to address the UGCoP. A promising direction is the use of individual activity space to approximate the true geographic context. An activity space is the area containing all locations that an individual visits as a result of his or her daily activities and travel (Golledge and Stimson 1997). Because humans tend to exhibit a high degree of habitual behavior on most days and circulate on a relatively small island in space–time, their actual or potential activity spaces could provide better proxies to true geographic contexts than conventional administrative areas (cf. Kwan 1999, 2000; González, Hidalgo, and Barabási 2008). Another advantage of this approach is that exposure to contextual influences is evaluated based on personalized contextual units that allow exposure level to vary even for individuals within the same neighborhood or household (Kwan 2009). It also helps transcend the traditional division of health-determining factors into either neighborhood or individual characteristics. Because personalized contextual units are constructed based on people's daily activities and travel as well as their interactions with various places, values of the contextual variables based on these units reflects both individual and place characteristics at the same time.

Some studies construct individual activity spaces using GIS and activity survey data (e.g., Arcury et al. 2005; Sherman et al. 2005; Kwan et al. 2009; Vallée et al. 2010). In these studies the standard deviational ellipse, the kernel density surface, the road network buffer, and the minimum convex polygon are common methods for deriving activity spaces. Although activity surveys provide useful data for delineating people's activity spaces, information about the location and timing of these activities is often very limited. To overcome this limitation, researchers have begun to explore the use of GPS or other location-aware devices in collecting detailed space–time data of people's activities and routes (e.g., Wiehe et al. 2008; Maddison et al. 2010; Troped et al. 2010; Zenk et al. 2011). For example, Kwan et al. (2011) used an integrated GPS–activity diary approach to examine the effect of exposure to protobacco advertisement and socioeconomic deprivation on the use of smokeless tobacco in the Appalachian region of Ohio. The study found that those who traveled in areas with lower socioeconomic status are more likely to use smokeless tobacco heavily.

Because GPS data can record where and how much time people spend as they undertake their daily activities with very high spatial and temporal resolutions, these data allow us to assess people's environmental exposures much more accurately.[2] Detailed GPS data also allow us to perform time-geographic 3D visualizations of people's space–time paths, which will be particularly valuable for studying the health risk of individuals without a stable home or those who live in multiple places such as the homeless (Hägerstrand 1970; Kwan 2004; Kwan and Ding 2008; Lee and Kwan 2011). Using a person's GPS tracks collected over many days (e.g., a week), we can estimate the probability distribution of his or her activities and routes over space and time and more accurately approximate the true geographic context. This will help us move beyond deterministic approaches to the delineation of contextual units and facilitate the development of new stochastic approaches for doing so. Further, using GPS data also helps overcome the conventional dichotomy between daytime and nighttime populations because these data capture people's continuous space–time trajectories, and analysis does not require dividing a day into two distinct segments of time.

Using GPS data to delineate activity spaces and approximate true geographic contexts represents a significant step forward in addressing the UGCoP. Some health behaviors, however, are affected by *situational contingencies* (e.g., interactions with others in real time) that cannot be captured by GPS data alone. For instance, a particular situation might promote or inhibit an adolescent's substance use depending on who is with him or her (e.g., friends or parents) and what they are doing together even for the same context (e.g., movie theater). To take into account the full spectrum of contextual influences on certain health behaviors or outcomes, it might be important to also consider the characteristics of relevant real-time contexts and social interactions. An emerging and promising approach along this line is to integrate GPS methodologies with *ecological momentary assessment* (EMA) and *social network analysis* (SNA). EMA has been used in a wide range of health studies to collect data on people's real-time situations (Shiffman 2009). It involves using wireless devices (e.g., mobile phones) to prompt and collect information from participants about their moods, perceptions, behaviors, and features of the environment as they occur in real time. On the other hand, data about people's social networks—such as attributes of their peers and friends and attributes and structure of the relationships among them—will help shed light on how a person's interactions with others in particular spaces and times could affect their health behaviors (Mennis and Mason 2011). An integrated GPS–EMA–SNA approach seems particularly promising for understanding the transmission of infectious diseases—as information about who has contact with whom and what they are doing together at what times will help shed light on the sociogeographic processes involved.

Another important component in attempts to address the UGCoP involves measuring the spatiotemporal variation of contextual or environmental influences (e.g., airborne pollutants), identifying when individuals are affected by them, and assessing the cumulative exposure of each individual with respect to his or her movement in space–time. This is a highly challenging research area because, for instance, the spatiotemporal dynamics of contextual influences and detailed space–time trajectories of individuals need to be integrated into a suitable analytical framework to accurately assess people's exposures. Some recent studies indicate how this could be accomplished (e.g., Setton et al. 2010). For instance, Gulliver and Briggs (2005) collected twenty-four-hour activity diary data from participants and developed a space–time exposure modeling method to evaluate their journey-time exposure to traffic-related pollution. The method integrated four submodels in a GIS: a traffic model, an air pollution dispersion model, a background pollution model, and a time-activity-based exposure model. Research in this area is sorely needed to fully address the UGCoP.

Promising as these developments might sound, there are still many challenges and limitations. First, collecting GPS, EMA or social network data is costly and time consuming. These methods are thus not suitable for obtaining data for large populations in a short period of time. Second, GPS have their own limitations (e.g., cannot collect reliable indoor data) and are sometimes error prone. Third, collecting space–time data with GPS greatly increases the volume of data, and methods for analyzing these data in health research are still limited to date. This could increase our analytical burden and undermine our ability to identify the true geographic context or to accurately assess people's exposure to contextual or environmental influences (Kwan 2004).

To conclude, the UGCoP is a problem as fundamental as the MAUP, but it is a different kind of problem that calls for new research on its confounding effects and mitigation. Recent studies have shown that both contextual variables and research findings are sensitive to different delineations of contextual units, and model fit by itself is not a reliable criterion for deciding the most appropriate contextual units. It is time to go beyond the static concepts and methods of conventional notions of geographic context and exposure measures. People move around to undertake their daily activities and come under the influence of many different neighborhood contexts besides their residential neighborhood. Their movement, their routes, the places they visit, and the time they spend there are no less important than their residential neighborhood in determining their exposure to contextual influences. Studies on the effects of contextual influences on health thus need to consider where and how much time people spend while engaged in their daily activities in relation to the spatiotemporal patterns of relevant contextual influences. This dynamic conceptualization of geographic context is very much in line with the "new mobilities paradigm" that emerged in social science in the last decade or so (Sheller and Urry 2006). The mobilities turn asserts the ontological significance of people's movement and expands our attention to what people experience while traveling. For health research it helps turn our focus from location to movement, from place to mobility, and from space to space–time. In the final analysis, humans are active agents who construct their own geographic contexts and tie together different spatial scales through their daily activities, movements and social interactions. The interconnections among individuals and places are vastly complex and vibrantly dynamic, and they should be conceptualized and examined as such.

Acknowledgments

I thank Antoinette WinklerPrins for handling the blind review process and making editorial decisions for this article (including the abstract). Her helpful suggestions and the thoughtful comments of three anonymous reviewers have helped improve the article considerably. This article was written while I was supported by the following grants: NSF BCS-0729466, NCI R21 CA129907, and NIDA R01DA025415-01.

Notes

1. Two important qualifications of the article's focus on health studies are in order. First, arguments in this article mainly apply to health studies that are based on ecological designs because other research designs are not primarily concerned with identifying contextual influences from geographic factors and processes (and thus do not need to explicitly delineate contextual areas). Second, discussion in this article is relevant mainly to studies in which area-based contextual variables (e.g., neighborhood poverty) are used to explain or predict individual health behaviors or outcomes. An important goal of many health studies, however, is to identify at-risk populations or areas where the health outcomes are significantly worse than expected. Given their analytical focus on the relationship between area-based contextual variables and area-based outcome variables (e.g., low birthweight rates of census tracts), us-

ing conventional administrative units like census tracts in this kind of study is needed and is often the only viable option.

2. A major concern with collecting GPS data in health research is participants' privacy and data confidentiality, because it could be possible to identify a person's identity through reverse geocoding if data are not handled carefully. In countries with strict human subject protection regulations (such as the United States), all persons involved in collecting and analyzing the data are required to go through rigorous human subject protection training and be certified before any involvement in research activities. They are obliged legally and ethically to protect participants' privacy and data confidentiality, and all research procedures (including recruitment, informed consent, data analysis, and dissemination of results) require prior approval from and are closely monitored through continuing review by their institutional or ethical review boards. For instance, in the Appalachian smokeless tobacco usage study, all data were deidentified before being incorporated into the database and no one handling those data or seeing them by chance will be able to identify any participant. Further, no maps or displays of the home or activity sites visited by participants or their daily paths can be printed or disseminated. The GPS data cannot be used in any form other than for deriving activity spaces and related measures or generating aggregate statistical results. In countries without strict regulations and in situations where people provide information without knowing or agreeing to its use for research purposes (e.g., social networking Web sites such as Facebook Places, Twitter, and Foursquare), there might be no informed consent and human subject protection protocol and issues of privacy violation can be a serious concern. It is not clear how the use of location data can be justified with respect to the norms of human subject protection in these contexts.

References

Arcury, T. A., W. M. Gesler, J. S. Preisser, J. E. Sherman, J. Spencer, and J. Perin. 2005. The effects of geography and spatial behavior on health care utilization among the residents of rural region. *Health Services Research* 40 (1): 135–55.

Basta, L. A., T. S. Richmond, and D. J. Wiebe. 2010. Neighborhoods, daily activities, and measuring health risks experienced in urban environments. *Social Science & Medicine* 71 (11): 1943–50.

Berke, E. M., T. D. Koepsell, A. V. Moudon, R. E. Hoskins, and E. B. Larson. 2007. Association of the built environment with physical activity and obesity in older persons. *American Journal of Public Health* 97 (3): 486–92.

Black, L. J., and J. Macinko. 2008. Neighborhoods and obesity. *Nutrition Review* 66 (1): 2–20.

Brownson, R. C., J. J. Chang, A. A. Eyler, B. E. Ainsworth, K. A. Kirtland, B. E. Saelens, and J. F. Sallis. 2004. Measuring the environment for friendliness toward physical activity: A comparison of the reliability of 3 questionnaires. *American Journal of Public Health* 94 (3): 473–83.

Chaix, B. 2009. Geographic life environments and coronary heart disease: A literature review. Theoretical contributions, methodological updates, and a research agenda. *Annual Review of Public Health* 30:81–105.

Cummins, S. 2007. Investigating neighborhood effects on health—Avoiding the "local trap." *International Journal of Epidemiology* 36:355–57.

Diez-Roux, A. V. 1998. Bringing context back into epidemiology: Variables and fallacies in multilevel analysis. *American Journal of Public Health* 88 (2): 216–22.

———. 2001. Investigating neighborhood and area effects on health. *American Journal of Public Health* 91 (11): 1783–89.

Diez-Roux, A. V., and C. Mair. 2010. Neighborhood and health. *Annals of the New York Academy of Sciences* 1186:125–45.

Elgethun, K., R. A. Fenske, M. G. Yost, and G. J. Palcisko. 2003. Time-location analysis for exposure assessment studies of children using a novel global positioning system instrument. *Environmental Health Perspectives* 111 (1): 115–22.

Entwisle, B. 2007. Putting people into place. *Demography* 44 (4): 687–703.

Fotheringham, A. S., and D. W. S. Wong. 1991. The modifiable areal unit problem in multivariate statistical analysis. *Environment and Planning A* 23 (7): 1025–44.

Frank, L. D., T. L. Schmid, J. F. Sallis, J. Chapman, and B. E. Saelens. 2005. Linking objectively measured physical activity with objectively measured urban form—Findings from SMARTRAQ. *American Journal of Preventive Medicine* 28:117–25.

Gatrell, A. C. 2011. *Mobilities and health.* Aldershot, UK: Ashgate.

Golledge, R. G., and R. J. Stimson. 1997. *Spatial behavior: A geographic perspective.* New York: Guilford.

González, M. C., C. A. Hidalgo, and A.-L. Barabási. 2008. Understanding individual human mobility patterns. *Nature* 453:779–82.

Grannis, R. 2009. *From the ground up: Translating geography into community through neighbor networks.* Princeton, NJ: Princeton University Press.

Gulliver, J., and D. J. Briggs. 2005. Time–space modeling of journey-time exposure to traffic-related air pollution using GIS. *Environmental Research* 97:10–25.

Hägerstrand, T. 1970. What about people in regional science? *Papers of Regional Science Association* 24:7–21.

Hanson, S., and P. Hanson. 1981. The travel-activity patterns of urban residents: Dimensions and relationships to sociodemographic characteristics. *Economic Geography* 57:332–47.

Hoek, G., R. Beelen, K. de Hoogh, D. Vienneau, J. Gulliver, P. Fischer, and D. Briggs. 2008. A review of land-use regression models to assess spatial variation of outdoor air pollution. *Atmospheric Environment* 42:7561–78.

Inagami, S., D. A. Cohen, and B. K. Finch. 2007. Nonresidential neighborhood exposures suppress neighborhood effects on self-rated health. *Social Science & Medicine* 65:1779–91.

Kwan, M.-P. 1999. Gender and individual access to urban opportunities: A study using space–time measures. *The Professional Geographer* 51 (2): 210–27.

———. 2000. Interactive geovisualization of activity-travel patterns using three-dimensional geographical informa-

tion systems: A methodological exploration with a large data set. *Transportation Research C* 8:185–203.

———. 2004. GIS methods in time-geographic research: Geocomputation and geovisualization of human activity patterns. *Geografiska Annaler B* 86 (4): 267–80.

———. 2009. From place-based to people-based exposure measures. *Social Science & Medicine* 69 (9): 1311–13.

Kwan, M.-P., and G. Ding. 2008. Geo-narrative: Extending geographic information systems for narrative analysis in qualitative and mixed-method research. *The Professional Geographer* 60 (4): 443–65.

Kwan, M.-P., T. Hawthorne, C. Calder, W. Darneider, A. Jackson, and L. J. Krivo. 2009. Activity-space measures for studying spatial crime and social isolation. Paper presented at the annual meeting of the Association of American Geographers, Las Vegas, NV.

Kwan, M.-P., L. L. Kenda, M. E. Wewers, A. K. Ferketich, and E. G. Klein. 2011. Sociogeographic context, protobacco advertising, and smokeless tobacco usage in the Appalachian Region of Ohio (USA). Paper presented at the 2011 International Medical Geography Symposium, Durham, UK.

Kwan, M.-P., and J. Weber. 2008. Scale and accessibility: Implications for the analysis of land use–travel interaction. *Applied Geography* 28:110–23.

Leal, C., and B. Chaix. 2011. The influence of geographic life environments on cardiometabolic risk factors: A systematic review, a methodological assessment and a research agenda. *Obesity Reviews* 12 (3): 217–30.

Lee, J. Y., and M.-P. Kwan. 2011. Visualization of sociospatial isolation based on human activity patterns and social networks in space-time. *Tijdschrift voor Economische en Sociale Geografie* 102 (4): 468–85.

Löytönen, M. 1998. GIS, Time geography and health. In *GIS and health,* ed. A. C. Gatrell and M. Löytönen, 97–110. London and New York: Taylor and Francis.

Macintyre, S., and A. Ellaway. 2003. Neighborhoods and health: An overview. In *Neighborhoods and health,* ed. I. Kawachi and L. F. Berkman, 20–42. Oxford, UK: Oxford University Press.

Macintyre, S., A. Ellaway, and S. Cummins. 2002. Place effects on health: How can we conceptualise, operationalise and measure them? *Social Science & Medicine* 55:125–39.

Maddison, R., Y. Jiang, S. Vander Hoorn, D. Exeter, C. N. Mhurchu, and E. Dorey. 2010. Describing patterns of physical activity in adolescents using global positioning systems and accelerometry. *Pediatric Exercise Science* 22 (3): 392–407.

Matthews, S.A. 2008. The salience of neighborhood: Some lessons from sociology. *American Journal of Preventive Medicine* 34 (3): 257–59.

Mennis, J., and M. J. Mason. 2011. People, places, and adolescent substance use: Integrating activity space and social network data for analyzing health behavior. *Annals of the Association of American Geographers* 102 (2): 279–91.

Monden, C. W. S., F. J. Van Lenthe, and J. P. Mackenbach. 2006. A simultaneous analysis of neighborhood and childhood socio-economic environment with self-assessed health and health-related behaviors. *Health & Place* 12:394–403.

Oliver, L. N., N. Schuurman, and A. W. Hall. 2007. Comparing circular and network buffers to examine the influence of land use on walking for leisure and errands. *International Journal of Health Geographics* 6:41.

Openshaw, S. 1984. *The modifiable areal unit problem.* Norwich, UK: Geo Books.

Sampson, R. J., J. D. Morenoff, and T. Gannon-Rowley. 2002. Assessing "neighborhood effects": Social processes and new directions in research. *The Annual Review of Sociology* 28:443–78.

Setton, E., C. P. Keller, D. Cloutier-Fisher, and P. W. Hystad. 2010. Gender differences in chronic exposure to traffic-related air pollution—A simulation study of working females and males. *The Professional Geographer* 62 (1): 66–83.

Sheller, M., and J. Urry. 2006. The new mobilities paradigm. *Environment and Planning A* 38:207–26.

Sherman, J. E., J. Spencer, J. S. Preisser, W. M. Gesler, and T. A. Arcury. 2005. A suite of methods for representing activity space in a healthcare accessibility study. *International Journal of Health Geographics* 4:24.

Shi, W. 2010. *Principles of modeling uncertainties in spatial data and spatial analyses.* Boca Raton, FL: CRC Press.

Shiffman, S. 2009. Ecological momentary assessment (EMA) in studies of substance use. *Psychological Assessment* 21 (4): 486–97.

Spielman, S. E., and E.-H. Yoo. 2009. The spatial dimensions of neighborhood effects. *Social Science & Medicine* 68 (6): 1098–1105.

Subramanian, S. V., K. Jones, and C. Duncan. 2003. Multilevel methods for public health research. In *Neighborhoods and health,* ed. I. Kawchi and L. F. Berkman, 65–111. Oxford, UK: Oxford University Press.

Troped, P. J., J. S. Wilson, C. E. Matthews, E. K. Cromley, and S. J. Melly. 2010. The built environment and location-based physical activity. *American Journal of Preventive Medicine* 38 (4): 429–38.

Vallée, J., E. Cadot, F. Grillo, I. Parizot, and P. Chauvin. 2010. The combined effects of activity space and neighborhood of residence on participation in preventive health-care activities: The case of cervical screening in the Paris metropolitan area (France). *Health & Place* 16:838–52.

Wheaton, B., and P. Clarke. 2003. Space meets time: Integrating temporal and contextual influences on mental health in early adulthood. *American Sociological Review* 68:680–706.

Wiehe, S. E., S. C. Hoch, G. C. Liu, A. E. Carroll, J. S. Wilson, and J. D. Fortenberry. 2008. Adolescent travel patterns: Pilot data indicating distance from home varies by time of day and day of week. *Journal of Adolescent Health* 42:418–20.

Wilks, D., H. Besson, A. Lindroos, and U. Ekelund. 2010. Objectively measured physical activity and obesity prevention in children, adolescents and adults: A systematic review of prospective studies. *Obesity Reviews* 12 (5): 119–29.

Zenk, S. N., A. J. Schulz, S. A. Matthews, A. Odoms-Young, J. Wilbur, L. Wegrzyn, K. Gibbs, C. Braunschweig, and C. Stokes. 2011. Activity space environment and dietary and physical activity behaviors: A pilot study. *Health & Place* 17 (5): 1150–61.

Environmental Health as Biosecurity: "Seafood Choices," Risk, and the Pregnant Woman as Threshold

Becky Mansfield

Department of Geography, The Ohio State University

Environmental pollutants are now widespread not only in the environment but in human bodies. Seafood is one of the main sources of human exposure to many of these chemicals, with concern mainly about effects on fetal development. This article examines toxicological, epidemiological, and other public health scholarship on contaminated seafood to understand how these environmental health concerns are constituted as an object of knowledge and management. Analyzing current approaches to contaminated seafood through the lens of biopolitics, the article shows that the dominant approach is one in which risk is used to secure the population by calculating the net benefits and hazards of women's seafood consumption, with the goal being to influence women's "seafood choices." This marks a new arena of biopolitical concern about how the contaminated environment is changing the nature of humans by altering cognitive development and intelligence. It also marks a new, highly gendered regime of biosecurity in which women are positioned as protectors of the population against the threat of contamination. By calculating proper seafood choices, risk spatializes reproductive women as the bodily threshold between the contaminated environment and the population. Risk thus places responsibility for environmental health on women, who must make proper choices to manage movement of pollutants through their own bodies. In making this argument, the article opens up a new field of inquiry into health as a nature–society issue and contributes new insight about the constitutive role of both gender and environment in biopolitics.

环境污染物现在不仅在环境中也在人体内很普遍。海鲜是人体暴露于这些化学物质的主要来源之一，其主要的关注是对胎儿发育的影响。本文探讨毒理学，流行病学和其他有关受污染的海鲜方面的公共卫生课题，以了解这些环境健康的关注是如何被构建成一个知识和管理的对象的。通过生命政治的镜头，分析目前对污染海鲜的研究方法，本文显示，在主导力法中，通过计算的净福利和妇女的海鲜消费危害，以确保妇女群体对"海鲜的选择。"这标志着生命政治所关注的一个新舞台，即污染的环境是如何通过改变认知的发展和智力来改变人类的本质的。它也标志着一个新的，高度性别化的生物安全制度，在该制度中妇女被定位为受保护人群，以反对污染的威胁。通过计算适当的海鲜选择，风险把育龄妇女空间化为污染的环境和人口之间的身体门槛值。因此，风险对妇女的健康环保带来责任，她们必须对管理污染物在其自身移动作出正确的选择。本文作出的这一结论，开辟了将健康作为一个自然—社会问题的新的调查领域，并在有关性别和环境在生命政治领域的构成作用问题上，贡献了新的见解。*关键问: 生命政治，环境健康，性别，繁殖，风险。*

Los contaminantes ambientales ahora son ubicuos, encontrándose no solo en el medio ambiente sino dentro de los cuerpos humanos. Los productos marinos son una de las principales fuentes de exposición a muchos de estos químicos, con especial preocupación por sus efectos sobre el desarrollo fetal. Este artículo examina las fuentes eruditas disponibles sobre cuestiones toxicológicas, epidemiológicas y temas relacionados de salubridad pública sobre la comida de mar contaminada, para entender cómo han llegado a constituirse todas estas preocupaciones de salud ambiental en objeto de conocimiento y manejo. Analizando con la lente de la biopolítica los actuales enfoques sobre los alimentos marinos contaminados, el artículo muestra que el enfoque dominante es aquel en que el riesgo es utilizado para asegurar la población, calculando los beneficios netos y los peligros del consumo de comida de mar por las mujeres, tomando como meta influir en la "elección de comida de mar" que ellas puedan hacer. Esto pone de presente un nuevo escenario de preocupación biopolítica sobre cómo el medio ambiente contaminado está cambiando la naturaleza de los humanos al alterar el desarrollo cognoscitivo y la inteligencia. También marca un nuevo régimen de bioseguridad altamente sesgado por género en el que las mujeres se posicionan como defensoras de la población contra las amenazas de la contaminación. Calculando las escogencias apropiadas de comida marina, el riesgo espacializa a las mujeres reproductivas como el umbral corpóreo entre el entorno contaminado y la población. Así el riesgo asume responsabilidad sobre la salubridad ambiental para las mujeres, que deben hacer elecciones apropiadas para manejar el desplazamiento de contaminantes a través de sus propios cuerpos. Al formular este argumento, el artículo inaugura un campo nuevo de indagación sobre la salud como un asunto de la relación naturaleza-sociedad y contribuye nuevas aproximaciones inteligentes acerca del papel constitutivo del género y el medio ambiente en la biopolítica.

Measurement of "body burdens" of environmental pollutants such as dioxins and mercury suggests that these chemicals are now widespread not only in the environment but also in human bodies (Centers for Disease Control and Prevention 2009). Seafood is one of the main sources of human exposure to many of these chemicals, particularly those that "bioaccumulate" along the food chain, with concern centering on effects of such exposure on fetal neurodevelopment (Institute of Medicine [IOM] 2007). This article examines recent toxicological, epidemiological, and other public health scholarship on contaminated seafood to understand how these environmental health concerns are constituted as an object of knowledge and management, particularly in the United States. I analyze current approaches to contaminated seafood through the lens of biopolitics, defined as procedures for regulating and securing the "population" as a biological and statistical materialization of humans as a species (Foucault 2003, 2007, 2008). Whereas a public health focus on environmental contaminants seems to challenge ongoing industry denials and neoliberal regulatory failures regarding chemicals (Michaels 2005; Langston 2010), I argue that current fears and attempts to protect the population from these contaminants mark a new arena of biopolitical concern about how the contaminated environment is changing the nature of humans by altering cognitive development and intelligence. Further, this biopolitics of environmental health is also a biopolitics of reproduction, in which women's "reproductivity" (Deutscher 2010) serves as a spatial threshold for securing the future of the population.

The dominant approach to contaminated seafood is one in which risk is used to secure the population by calculating the net benefits and hazards of women's seafood consumption, with the goal being to influence women's "seafood choices." Understanding the role of risk analysis requires distinguishing analytically between uncertainty and risk. Uncertainty regarding contaminants in seafood is ubiquitous and commonly noted (IOM 2007), arising due to the need to understand all of the links in the pathway between a chemical source and particular health outcome; diversity of chemicals, types of seafood, and human practices; and as-yet unsettled methodological questions about how to measure chemical exposure and health outcomes (Spurgeon 2006; Domingo et al. 2007; Dorea 2008). Risk is not equivalent to uncertainty but is a biopolitical technology for managing uncertainty through particular forms of calculation, and as such risk is a technique of government. Risk is "a social technology by means of which the uncertain future . . . is rendered knowable and actionable" (Aradau, Lobo-Guerrero, and Van Munster 2008, 150; also Weir 2006). Risk makes the future governable by taking hold of life as a contingent yet calculable emergent potential (Dillon 2008). Thus, risk is a biopolitical technology for governing the population in the name of security. Evident in the biopolitics of contaminated seafood is that risk governs the population both in the aggregate and at the level of the self-regulating individual through whom population security is achieved.

This article argues that risk analysis of contaminated seafood marks a new, highly gendered regime of biosecurity in which women are positioned as protectors of the population against the threat of environmental contamination. In so doing, the article opens up inquiry into health as a nature–society issue, encompassing simultaneously the physical relationship between bodies and the environment and how health, bodies, and environments are understood and managed (Mansfield 2008); it also contributes new insight about the constitutive role of both gender (see Stoler 1995; Miller 2007; Deutscher 2008) and environment in biopolitics. By calculating proper seafood choices, risk spatializes reproductive women as the physical, bodily threshold between the outside contaminated environment and the population, represented by the fetus. But the threshold is less one of biophysical porosity and more about the choices that women are supposed to make. Women are to ensure population security by carefully managing movement of chemicals through their own bodies. This gendered spatial imaginary of environmental biosecurity contrasts with alternatives that suggest not thresholds but fluid relations between bodies and the wider contaminated environment. Attention to gender does more than show that "the population" is not gender neutral; rather, the reproductive woman is a key figure in contemporary biopolitics, securing the human–environment divide to protect the future against environmental threats we ourselves have created.

Environmental Contamination, Fetal Life, and "the Pregnant Woman"

Contaminated seafood offers a valuable lens on the biopolitics of contamination because it crystalizes current anxieties about food, fetuses, and maternal responsibility. Although food in general has attracted heightened attention (Guthman 2011), seafood captures current anxieties more than many foods because of the ways in which it is simultaneously healthful and hazardous. Both the main source of exposure to many contaminants and a low-fat protein loaded with healthful omega-3 fatty acids, seafood is touted as a wonder food and a danger. The public health response to this dilemma is risk assessment to calculate the proper "balance" of benefits and harms and to issue "seafood consumption advisories" that tell different segments of the population which fish they should eat and how often to get benefits without harms (on these advisories, see Mansfield forthcoming). The seafood dilemma then becomes emblematic of neoliberal public health, which aims to manage "lifestyle" choices of individual consumers (Peterson and Lupton 1996).

Seafood also captures growing fears about the effects of various industrial chemicals and pharmaceuticals on the developing fetus (Steingraber 2001; Langston 2010). Research on such fetal effects has led to a marked, yet incomplete, upheaval in toxicology. Traditionally focused almost exclusively on doses of chemicals that cause harm, the new focus is timing: For some chemicals, exposure to minute doses in utero can have measurable, even profound, effects (Vogel 2008). For contaminated seafood, concern is about both dose and timing, and the most prominent worries swirl around fetal neurodevelopment, as research has focused on the long-term effects of fetal exposures on outcomes such as intelligence and behavior. Thus, the fear is that by affecting fetuses, the toxic environment is changing seemingly innate population characteristics.

This combination of public health efforts to manage lifestyle and current fears about fetal life gives rise to a new locus of environmental health control: the pregnant woman (Mansfield forthcoming). Public health seafood advice is targeted mainly to "women who are or who could become pregnant"—women who are "vulnerable" by virtue of potential effects on their "unborn children." Importantly, this target population need not be pregnant but only potentially pregnant at some time in the future, and hence it includes all premenopausal females. Throughout the article, the term *pregnant woman* indicates this figure of the always already pregnant woman, who is urged to freely choose to make good seafood choices.

The public health problematic spatializes the pregnant woman as the threshold—the narrow margin—between the contaminated environment and the fetus, which figures not just as an individual but as the future of the population; as a threshold, women's bodies are biopolitical space. Since the nineteenth century, "women's reproductivity [has been] the critical threshold" (Deutscher 2010, 219) between the vitality of the individual and the population, thus "turn[ing] the woman-citizen into both an inhabitant of biopolitical space and also the setting in which this space would be carved out" (Miller 2007, 39). This has authorized ongoing intervention into women's bodies in the name of collective "life" (see also Stormer 2000; Weir 2006). Central twentieth-century developments to which feminist scholars have given extensive attention include the medicalization of pregnancy and invention of prenatal care, the associated rise of "fetal personhood," and enduring ideals of motherhood (see Mansfield forthcoming for discussion). Risk analysis of environmental contaminants expands this into a gendered spatial imaginary of human–environment relations. In this imaginary, the population can be secured against environmental threats through women's self-control of their bodies as thresholds. These thresholds are not just between the individual and population but also between the environment and population (such that women are part of both).

The pregnant woman as biopolitical-environmental threshold also intersects with the toxicological threshold defined as a dose below which a toxin is safe. Nash (2006, 2008) has argued that toxicological thresholds are a means to narrowly define environmental threats and separate the body from the environment. Although this notion of threshold as a dose is undermined by evidence of effects from fetal exposure to very low doses of some chemicals, the idea of a threshold persists in the form of women's bodies. This is not a return to the notion that the body "naturally" forms a barrier, as in now discredited ideas about the placenta as barrier. Instead, it is specifically women's bodily choices that serve as the protecting threshold. Although prenatal choices are presented as freely made by good mothers, the biopolitically proper choices are increasingly obligatory and serve "to limit the field of possible conduct in response to pregnancy" (Tremain 2006, 36; Samerski 2009). This gendered geography of responsibility suggests that the promise of better living through health is not just about individualized responsibility and outcomes—Rose's (2007) individualized optimization. Rather, better living through health is also about individualized responsibility for biosecurity, in which it is the job of the always already pregnant woman to maintain the nature of humans as a species.

Risk and Advice about Women's Seafood Choices

Although research into the health effects of contaminated seafood is carried out by a range of scholars—including toxicologists, epidemiologists, risk specialists, and other public health policy researchers—a notable feature of the collective research agenda is the extent to which it is framed in terms of risk analysis. The dominant focus is "balancing" benefits and harms, and the collective research endeavor comes together around risk calculation to generate advice about what seafood individual women should choose to eat to protect their (present or future) fetuses. This section first demonstrates the dominance of this problematic by showing convergence across the scientific literature. It then describes alternatives that challenge the dominant approach by focusing on harm rather than balance. My focus is how these concerns are problematized in the United States, and I discuss information from multiple locations when it figures in U.S. public health knowledge.

Balancing Benefits and Harms

Public health risk assessment studies offer a useful starting point because they explicitly entail calculating risk. These studies provide quantitative assessment using existing data to understand exposures and dose–response relationships and then model overall health effects. These meta-analyses are explicit that their research is about "balancing" benefits and harms, often proclaiming this in their title. Two prominent examples are *Seafood Choices: Balancing Benefits and Risks*, the comprehensive, 700-page report from the U.S. IOM (2007), and "Fish: Balancing Health Risks and Benefits," the lead-off commentary for a series of

articles on this topic from the Harvard Center for Risk Analysis (Willett 2005). Most such studies focus on methylmercury; the IOM study additionally assesses organic contaminants such as dioxins. These studies also model wide-ranging population-level health effects on both fetuses (measured as milestones in childhood development) and adults (e.g., cardiovascular health, cancers).

These risk assessments largely suggest that eating fish is still more healthful than not eating it: In all but a few cases, the benefits outweigh the harms. Thus, the IOM report's first recommendation is that "dietary advice to the general population from federal agencies should emphasize that seafood is a component of a healthy diet" (IOM 2007, 7). It then divides "the general population" into four subgroups and recommends restrictive consumption advice (regarding species that should not be eaten or eaten less frequently) to females who are or may become pregnant or are breastfeeding (a group that includes adolescent females) and children under twelve. This disaggregation of the general population—echoed in many other studies and government advisories—reflects mounting evidence that seafood is beneficial to all, but exposure to methylmercury through seafood is on balance harmful to fetal neurodevelopment (IOM 2007, 124–33).

Already evident is that the question of "balancing" benefits and harms becomes inseparable from the question of how to guide individuals' consumption choices. This "lifestyle" approach is explicit in the IOM (2007) report, the specific aim of which is to "examine relationships between benefits and risks associated with seafood consumption to help consumers make informed choices" (2). This is also the explicit goal of the Harvard studies; these five quantitative assessments of the balance of benefits and harms were designed to understand "the possible health effects from policies to alter fish consumption" and their findings support "current guidelines that focus on changes in the type of fish eaten by women in the reproductive age" (Willett 2005, 320). Risk approaches that balance benefits and harms are thus explicitly tied to individual consumption advice, particularly to pregnant women.

This assemblage of risk, consumption advice, and the always already pregnant woman is dominant not only in risk assessment itself but in scholarship that deals with other stages of the research endeavor. Unsurprisingly, scholarship on risk communication, which schematically comes "after" that on risk assessment, adopts the risk-advice approach, focusing on how to maintain seafood consumption while guiding pregnant women to make disciplined choices (e.g., Scherer et al. 2008; Ginsberg and Toal 2009; Zhang, Nakai, and Masunaga 2009). But research on schematically "prior" stages is also framed in terms of risk, balancing, and advice to pregnant women. Thus, research on the health effects of toxic exposure through seafood is not only on effects of maternal consumption on fetal neurodevelopment but is largely framed in terms of "women making dietary choices" (Oken et al. 2005, 1376) and the need to "refine recommendations to optimize outcomes for mothers and children" (Oken and Bellinger 2008, 178; see also Hibbeln et al. 2007). Even articles on the health benefits of seafood not only mention the issue of contaminants but cast this in terms of consumption advice (Kris-Etherton, Harris, and Appel 2002; He 2009).

More removed from direct questions of risk is research on levels of toxins in different types of seafood, yet much of this research, too, is framed as "balancing" the health benefits and harms of individual seafood consumption. Although the actual research involves chemically measuring contaminants in samples of fish, the titles of such articles include "Benefits and Risks of Fish Consumption" (Domingo et al. 2007) and "Mercury in Commercial Fish: Optimizing Individual Choices to Reduce Risk" (Burger, Stern, and Gochfeld 2005). These articles start with a rehearsal of the healthful–unhealthful dilemma and conclude with a discussion about implications for dietary advice. The most high profile of these articles—"Global Assessment of Contaminants in Farmed Salmon" by Hites et al. (2004)—never makes explicit claims about dietary advice, but it does emphasize that less contaminated sources of salmon exist and concludes with the need for additional labeling of seafood, thus implying that what is needed is change in consumers' individual choices. Further, other efforts by this same group involve risk analysis to generate consumer advice regarding farmed salmon (Foran, Carpenter, et al. 2005; Foran, Good, et al. 2005; Huang et al. 2006). Strikingly, despite the fact that their quantitative analyses do not focus on pregnant women, these articles emphasize this in their conclusions, even stating that "girls and young women" are "exceptionally vulnerable" because of adverse effects on their future fetuses due to the long half-life of dioxins and other dioxin-like chemicals (Foran, Carpenter, et al. 2005, 555).

This consistency across the research shows that, at least in the first decade of this century, it is not the case that finding toxins leads to a study of health effects, which leads to risk assessment and policy interventions. Rather, the prior fact of risk assessment to generate a particular policy intervention (seafood consumption advice) reverberates through the research required to generate the evidence to make such risk assessments necessary and policy interventions meaningful. Risk is not a mere stage of research but is a technology that problematizes contaminated seafood as an issue of balancing benefits and harms through women's consumption choices.

From Balance to Harm

Although uncommon, there exist other ways of problematizing contaminated seafood. Crucially, the question "What is the balance of benefits and hazards?" is not equivalent to "Do contaminants in seafood cause harm?" The first, a product of risk analysis, asks how benefits might offset harms and thus whether people should continue to eat contaminated seafood. The second asks about the harms associated with toxins. Certainly risk calculations can conclude that on balance harms outweigh benefits, and such findings underlie efforts to define pregnant women as a vulnerable subpopulation. Net harm is also the finding of newer studies examining organic pollutants and disaggregating species and populations of fish from the category of seafood (Foran, Carpenter, et al. 2005; Foran, Good, et al. 2005; Huang et al. 2006; Domingo et al. 2007). Net harm, however, is not the same as harm, which can exist even with a finding of net benefit. That is, benefits could be reduced in such a way as to cause population effects, even though overall the benefits outweigh the harms. Identifying harm thus requires disaggregating benefits and harms, not balancing them. The policy implication is that dietary advice becomes irrelevant and other actions are required to prevent harm.

One hint of this problematic of harm is a study reanalyzing data from an influential epidemiological study on mercury and neurodevelopment in the Faroe Islands. The reanalysis treats the relationship between the benefits and hazards of seafood not as a balance but as a problem of "confounding," in which both positive effects of nutrients and negative effects of contaminants are underestimated unless their countervailing effects are disaggregated (Budtz-Jorgensen, Grandjean, and Weihe 2007). The reanalysis finds that methylmercury is more harmful than their earlier analyses estimated, thus contradicting critics who argued that previous results overestimated effects.

A stronger alternative emerges in the Mohawk territory of Akwesasne, an indigenous territory that straddles the U.S.–Canada border (Arquette et al. 2002).

Fish in this territory are contaminated with PCBs and other toxic substances, and fish advisories alert that they should not be eaten at all by reproductive-aged women or children and only occasionally by men and older women. Once residents stopped eating these contaminated fish, their body burdens dropped significantly, which was taken by experts as a sign of success. Rather than seeing this as a triumph of risk analysis, however, Mohawk people have argued that these advisories force them to curtail traditional practices of eating fish, which they trade for an unhealthful, high-fat diet. For them, the substitution of unhealthful meat for traditional fish is another form of harm caused by contamination. This contrasts with standard risk approaches in which such dietary shifts do need to be taken into account when calculating the balance of benefits and harms, but they count on the "benefit" side for fish (e.g., IOM 2007). That is, the substitution of unhealthful meat is an argument for continuing to eat contaminated fish—it pushes the "balance" toward net benefit rather than net harm. For the Mohawk, in contrast, these substitutions cannot be calculated as part of a balance, but rather are additional harms. This is even as Mohawk women have changed their consumption patterns and reduced their body burdens. The necessity of women's self-discipline and resulting dietary shifts is not the solution but rather is part of the problem, and Mohawk people demand action from polluters and regulators to address the contaminated environment.

Such attention to changing the environment rather than changing women's behaviors is largely missing in the toxicological and public health literature. One rare exception is a review that concludes that there are "proven health advantages" of seafood, and the harms caused by exposure "can be minimized by societal action" (Dorea 2008, 105). These societal actions do not include consumption advisories—to pregnant women or anyone else—but instead "regulation of fishing in contaminated waters and environmental industrial practices to improve environmental health standards" (Dorea 2008, 105). Thus, asking about harm broadens what can be examined and shifts attention away from individual women's actions (and hence advice) and to the contaminated environment itself.

The Pregnant Woman as Threshold

These two different problematics of contaminated seafood—as an object of risk, balance, and advice or as an object of multiple harms—each spatializes the pregnant woman in very different ways. This is so even as the effects of exposure on fetal development are a central concern in both approaches.

Through the singular focus on guiding the seafood consumption choices of the always already pregnant woman, risk spatializes the pregnant woman as the narrow, bodily threshold between the environment and the fetal body, which are treated as ontologically separate. The translation from fetal harm to "women who are or may become pregnant" as a unique and vulnerable population occurs because evidence of fetal harm is always produced and interpreted in the context of consumption advice: It is through individual lifestyle choices that the balance of benefit and harm occurs. The problem is not in the wider environment but is in how individual action affects dose and timing of exposure. The porosity of woman and fetus is treated as the separability of fetal life, which needs protection from its "environment"—the pregnant woman. The pregnant woman as threshold serves as a border between the environment (positioned as the outside) and the population (the inside), and movement across this border must be policed to ensure population security. In this liberal biopolitics, policing the border comes from within, as freely chosen self-discipline. Pregnant women are to treat themselves as porous thresholds by closely managing movements through their own bodies while the outside environment remains beyond control.

In contrast, focusing on multiple and compounding harms—including to fetuses—and on the wider contaminated environment spatializes the pregnant woman quite differently. Rather than treating the fetus as a separate entity in need of protection from its mother, evidence of fetal harm and the porosity of pregnant women's and fetal bodies are taken as a lesson in the potential openness of all bodies to all environments, with recognition of how different people are imbricated differently in this open environment (e.g., Gohlke and Portier 2007). This recalls nineteenth-century materializations of bodies and environments, which focused on porosity rather than separation (Nash 2004, 2006, 2008; see also Mitman, Murphy, and Sellers 2004). This presents an ontology of bio-insecurity, in which there are no borders to be managed, no identifiable thresholds to be crossed, and no safe space inside to be guarded. In this problematic, the lines between environment and fetus or environment and species not only cannot be drawn through women and their bodily choices but cannot be drawn at all.

That both of these very different problematics are tied to concerns about how chemicals affect fetal development suggests that a biopolitics of risk that spatializes woman-as-threshold is not the "natural" consequence of evidence of fetal harm. Demanding that women change their consumption habits is not the only possible response to evidence of fetal effects. That both problematics center on fears regarding reproduction and intelligence, however, suggests that we are increasingly concerned about how humans are destroying not only external nature but also internal human nature, which reveals biopolitical and normalizing concerns about what it means to be "truly" and naturally human. Although this deserves further elaboration, it does seem to challenge claims that the contemporary biopolitical order is one in which individualized optimization replaces normalization (see also Braun 2007; Roberts 2009). In this optimistic view, biology is no longer destiny but is something in which we can intervene, such that it is now "a field of choice" that can be optimized (Rose 2007, 40). The effects of contaminants do suggest that biology is mutable, but risk in environmental health emerges as a technology of biosecurity that places responsibility for population health on the pregnant woman and her (obligatory) choices. Further, our broader fears about how changes to fetal development affect the population and its future should give pause to any sense that optimization replaces normalization. Environmental health technologies today do not offer a postpopulation, postnormalization biopolitics of individualized optimization but instead offer a gendered biopolitics of responsibility for guarding norms of what it means to be human.

Acknowledgments

I would like to thank Mat Coleman, Julie Guthman, Jill Harrison, Aaron Bobrow-Strain, members of the University of California Studies of Food and the Body Multicampus Research Program, and the anonymous reviewers for their excellent feedback on earlier versions of this article.

References

Aradau, C., L. Lobo-Guerrero, and R. Van Munster. 2008. Security, technologies of risk, and the political: Guest editors' introduction. *Security Dialogue* 39 (2–3): 147–54.

Arquette, M., M. Cole, K. Cook, B. LaFrance, M. Peters, J. Ransom, E. Sargent, V. Smoke, and A. Stairs. 2002. Holistic risk-based environmental decision making: A native perspective. *Environmental Health Perspectives* 110 (S2): 259–64.

Braun, B. 2007. Biopolitics and the molecularization of life. *Cultural Geographies* 14:6–28.

Budtz-Jorgensen, E., P. Grandjean, and P. Weihe. 2007. Separation of risks and benefits of seafood intake. *Environmental Health Perspectives* 115 (3): 323–27.

Burger, J., A. Stern, and M. Gochfeld. 2005. Mercury in commercial fish: Optimizing individual choices to reduce risk. *Environmental Health Perspectives* 113 (3): 266–71.

Centers for Disease Control and Prevention. 2009. *Fourth national report on human exposure to environmental chemicals*. Atlanta, GA: Centers for Disease Control and Prevention.

Deutscher, P. 2008. The inversion of exceptionality: Foucault, Agamben, and "reproductive rights." *South Atlantic Quarterly* 107 (1): 55–70.

———. 2010. Reproductive politics, biopolitics, and auto-immunity: From Foucault to Esposito. *Bioethical Inquiry* 7:217–26.

Dillon, M. 2008. Underwriting security. *Security Dialogue* 39 (2–3): 309–32.

Domingo, J. L., A. Bocio, G. Falco, and J. M. Llobet. 2007. Benefits and risks of fish consumption: Part I. A quantitative analysis of the intake of omega-3 fatty acids and chemical contaminants. *Toxicology* 230:219–26.

Dorea, J. 2008. Persistent, bioaccumulative and toxic substances in fish: Human health considerations. *Science of the Total Environment* 400:93–114.

Foran, J. A., D. O. Carpenter, M. C. Hamilton, B. A. Knuth, and S. J. Schwager. 2005. Risk-based consumption advice for farmed Atlantic and wild Pacific salmon contaminated with dioxins and dioxin-like compounds. *Environmental Health Perspectives* 113 (5): 552–55.

Foran, J. A., D. H. Good, D. O. Carpenter, M. C. Hamilton, B. A. Knuth, and S. J. Schwager. 2005. Quantitative analysis of the benefits and risks of consuming farmed and wild salmon. *The Journal of Nutrition* 135:2639–43.

Foucault, M. 2003. *Society must be defended: Lectures at the College de France 1975–1976*, trans. D. Macey, ed. M. Bertani and A. Fontana. New York: Picador.

———. 2007. *Security, territory, population: Lectures at the College de France 1977–1978*, trans. G. Burchell, ed. M. Senellart. Basingstoke, UK: Palgrave Macmillan.

———. 2008. *The birth of biopolitics: Lectures at the College de France 1978–1979*, trans. G. Burchell, ed. M. Senellart. Basingstoke, UK: Palgrave Macmillan.

Ginsberg, G., and B. Toal. 2009. Quantitative approach for incorporating methylmercury risks and omega-3 fatty acid benefits in developing species-specific fish consumption advice. *Environmental Health Perspectives* 117:267–75.

Gohlke, J. M., and C. J. Portier. 2007. The forest for the trees: A systems approach to human health research. *Environmental Health Perspectives* 115 (9): 1261–63.

Guthman, J. 2011. *Weighing in: Obesity, food justice, and the limits of capitalism*. Berkeley: University of California Press.

He, K. 2009. Fish, long-chain omega-3 polyunsaturated fatty acids and prevention of cardiovascular disease—Eat fish or take fish oil supplement? *Progress in Cardiovascular Diseases* 52:95–114.

Hibbeln, J., J. Davis, C. Steer, P. Emmett, I. Rogers, C. Williams, and J. Golding. 2007. Maternal seafood consumption in pregnancy and neurodevelopmental outcomes in childhood (ALSPAC study): An observational cohort study. *The Lancet* 369:578–85.

Hites, R. A., J. A. Foran, D. O. Carpenter, M. C. Hamilton, B. A. Knuth, and S. J. Schwager. 2004. Global assessment of organic contaminants in farmed salmon. *Science* 303:226–29.

Huang, X., R. A. Hites, J. A. Foran, M. C. Hamilton, B. A. Knuth, S. J. Schwager, and D. O. Carpenter. 2006. Consumption advisories for salmon based on risk of cancer and noncancer health effects. *Environmental Research* 101:263–74.

Institute of Medicine. 2007. *Seafood choices: Balancing benefits and risks.* Washington, DC: Institute of Medicine, National Academy of Sciences.

Kris-Etherton, P., W. Harris, and L. Appel, for the Nutrition Committee of the American Heart Association. 2002. Fish consumption, fish oil, omega-3 fatty acids, and cardiovascular disease. *Circulation* 106:2747–57.

Langston, N. 2010. *Toxic bodies: Hormone disruptors and the legacy of DES.* New Haven, CT: Yale University Press.

Mansfield, B. 2008. Health as a nature–society question. *Environment and Planning A* 40:1015–19.

———. Forthcoming. Gendered biopolitics of public health: Regulation and discipline in seafood consumption advisories. *Environment and Planning D: Society and Space.*

Michaels, D. 2005. Scientific evidence and public policy. *American Journal of Public Health* 95 (S1): S5–S7.

Miller, R. A. 2007. *The limits of bodily integrity.* Aldershot, UK: Ashgate.

Mitman, G., M. Murphy, and C. Sellers. 2004. Landscapes of exposure: Knowledge and illness in modern environments: Introduction: A cloud over history. *Osiris* 19:1–20.

Nash, L. 2004. The fruits of ill-health: Pesticides and workers' bodies in post-World War II California. *Osiris* 19:203–19.

———. 2006. *Inescapable ecologies: A history of environment, disease, and knowledge.* Berkeley: University of California Press.

———. 2008. Purity and danger: Historical reflections on the regulation of environmental pollutants. *Environmental History* 13:651–58.

Oken, E., and D. C. Bellinger. 2008. Fish consumption, methylmercury and child neurodevelopment. *Current Opinion in Pediatrics* 20:178–83.

Oken, E., R. Wright, K. P. Kleinman, D. Bellinger, C. Amarasiriwardena, H. Hu, J. W. Rich-Edwards, and M. W. Gillman. 2005. Maternal fish consumption, hair mercury, and infant cognition in a U.S. cohort. *Environmental Health Perspectives* 113 (10): 1376–80.

Peterson, A., and D. Lupton. 1996. *The new public health: Health and self in the age of risk.* London: Sage.

Roberts, D. E. 2009. Race, gender, and genetic technologies: A new reproductive dystopia. *Signs: Journal of Women in Culture and Society* 34 (4): 783–804.

Rose, N. 2007. *The politics of life itself: Biomedicine, power, and subjectivity in the twenty-first century.* Princeton, NJ: Princeton University Press.

Samerski, S. 2009. Genetic counseling and the fiction of choice. *Signs: Journal of Women in Culture and Society* 34 (4): 735–61.

Scherer, A., A. Tsuchiya, L. Younglove, T. Burbacher, and E. Faustman. 2008. Comparative analysis of fish consumption advisories targeting sensitive populations. *Environmental Health Perspectives* 116 (12): 1598–1605.

Spurgeon, A. 2006. Prenatal methylmercury exposure and developmental outcomes: Review of the evidence and discussion of future directions. *Environmental Health Perspectives* 114 (2): 307–12.

Steingraber, S. 2001. *Having faith: An ecologists' journey to motherhood.* New York: Berkley Books.

Stoler, A. 1995. *Race and the education of desire.* Durham, NC: Duke University Press.

Stormer, N. 2000. Prenatal space. *Signs: Journal of Women in Culture and Society* 26 (1): 109–44.

Tremain, S. 2006. Reproductive freedom, self-regulation, and the government of impairment in utero. *Hypatia* 21 (1): 35–53.

Vogel, S. A. 2008. From "the dose makes the poison" to "the timing makes the poison": Conceptualizing risk in the synthetic age. *Environmental History* 13:667–73.

Weir, L. 2006. *Pregnancy, risk, and biopolitics: On the threshold of the living subject.* London and New York: Routledge.

Willett, W. C. 2005. Fish: Balancing health risks and benefits. *American Journal of Preventive Medicine* 29 (4): 320–21.

Zhang, Y., S. Nakai, and S. Masunaga. 2009. Simulated impact of a change in fish consumption on intake of n-3 polyunsaturated fatty acids. *Journal of Food Composition and Analysis* 22:657–62.

The Mutual Conditioning of Humans and Pathogens: Implications for Integrative Geographical Scholarship

Christopher A. Scott,*,† Paul F. Robbins,* and Andrew C. Comrie*

*School of Geography & Development, University of Arizona
†Udall Center for Studies in Public Policy, University of Arizona

We highlight an emerging mode of human–environment enquiry that is executed by cross-disciplinary teams, spurs innovation of hybrid methods, and leads to nonintuitive findings relevant beyond disciplinary framings or specific cases. The extension of this approach in health geography is particularly instructive. By focusing on material objects like soils, insects, or sewage, researchers from diverse epistemologies are compelled to translate conceptual models of disease causation, risk, and vulnerability. Humans and pathogens mutually condition one another, a result of continuously changing exposures (settlement and development patterns that modify pathogen and vector ecology) and institutional processes (legal, economic, and organizational contexts in which environments are modified and agents respond to risk). The dynamic interactions of pathogen ecologies and human institutions produce a type of coevolution, as evidenced by three cases we consider: bacteriological and helminth infections from urban wastewater irrigation, West Nile virus and its mosquito vector in the built environment, and Valley Fever and fungal distribution under changing climate and land disturbance. Place-based, contextual exposure pathways are shown to provide only a partial explanation of disease transmission and must be complemented by insights into individual and organizational agents' motivations, logics, and responses. The object in its context holds the key to understanding the intersection between physical and environmental, and human and governance geographies. Interactively identifying and pursuing theoretical and applied challenges in this manner allows researchers to move beyond entrenched subdisciplinary understandings to frame new supradisciplinary questions.

我们强调人类与环境调查的一个新兴模式，该模式出跨学科的团队执行，促使混合方法的创新，导致了超越纪律框架或具体案件的不直观的相关调查结果。这种方法在健康地理上的扩展是尤其具有指导意义的。通过以实物，如土壤，昆虫，或污水为重点，具不同的认识论的研究人员正在被迫转化疾病的因果关系，风险和脆弱性的概念模型。人类和病原体互为条件，是不断变化的风险（修改病原体和载体生态的安置和发展模式）和体制进程的结果（环境在法律，经济和组织的情况下被修改，和代理人对风险的应对）。病原体的生态环境和人类机构的动态相互作用产生出一种共同的进化类型，以我们考虑的三种情况为证：来自城市污水灌溉的细菌和寄生虫感染，西尼罗河病毒和它内置环境中的蚊子媒介，以及依气候变化和土地变更的裂谷热和真菌的分布。基于地点，情景的暴露途径只能给疾病的传播提供部分的解释，同时必须由对个人和组织的代理人的动机，逻辑，反应的洞察所补充。在这种情况下的对象掌握了理解物理和环境，人权和治理地域之间的交叉点的关键。以这种方式交互识别和追求理论和应用的挑战，使研究人员能够超越根深蒂固的子学科的理解，以构建新的超学科的问题。*关键词：人类病原体相互作用，机构，蚊子，裂谷热，废水。*

Destacamos un emergente modo de investigación humano-ambiental que se ejecuta por equipos interdisciplinarios, alienta la innovación de métodos híbridos y conduce a descubrimientos no intuitivos relevantes, más allá de marcos disciplinarios o de casos específicos. La aplicación de este enfoque en geografía de la salud es particularmente instructivo. Al concentrarse en objetos materiales como suelos, insectos, o alcantarillado, los investigadores provenientes de diversas epistemologías son inducidos a traducir los modelos conceptuales de la causación de la enfermedad, riesgo y vulnerabilidad. Los humanos y los patógenos se condicionan mutuamente, resultado que proviene de cambiar continuamente las exposiciones (patrones de poblamiento y desarrollo que modifican la ecología del patógeno y el vector) y de procesos institucionales (contextos legales, económicos y organizacionales en que los entornos ambientales son modificados y los agentes responden al riesgo). Las dinámicas interacciones de ecologías patogénicas e instituciones humanas producen un tipo de co-evolución, como se evidencia en tres casos que nosotros consideramos: infecciones bacteriológicas y de helmintos derivadas de irrigación con aguas residuales urbanas, virus del occidente del Nilo y su mosquito vector en el entorno ambiental construido, y la Fiebre del Valle y la distribución micótica bajo condiciones de cambio climático y perturbaciones de las tierras. Se muestran las rutas de exposiciones contextuales, basadas en lugar, para suministrar solo una explicación parcial de la transmisión de la enfermedad, la cual debe complementarse con el estudio profundo de las motivaciones, lógica y respuesta individual y de los agentes organizacionales. El objeto en su contexto tiene la clave para comprender la intersección entre las geografías física y ambiental, humana y de la gobernaza. De esta manera, identificando interactivamente y persiguiendo retos teóricos y aplicados, los investigadores pueden ir más allá de entendimientos subdisciplinarios arraigados para formular nuevas cuestiones supradisciplinarias.

T heory and practice in natural and social sciences have tended to develop along classical, disciplinary lines. Focusing on system subcomponents permitted rapid advances in scientific understanding as well as often-dramatic results derived from applied research. In geography, this entailed specialization in topical and conceptual development, innovation with methods, but also, inexorably, narrowing domains of enquiry (Turner 2002). Splintered communities of enquiry remained interconnected through scholarly exchange and publication, now boosted by the Internet. Narrow focus, however, tended to proliferate research communities within subdisciplinary specialist silos (Bierly and Gatrell 2004; Skole 2004). This trend culminated in periodic contestation as disciplinary entrenchment, official calls for academic transparency, and heightened competition for dwindling resources pitched many scholarly communities into the "science wars," where critics focused attention on divisions across epistemology and methods (Smith 1992; Schuurman 2009). This contestation, however, also led to the reinterpretation of research agendas and new modes of enquiry (Kwan 2004; Skole 2004; Jones 2005; Menand 2010).

Here we aim to expand on a particular form of scholarship that emerged from such reconciliation—an object-centered and coevolutionary model of human–environment interaction—with emphasis on its particular manifestation in health-geography research (schematized in Figure 1). First, we stress enquiry based on cross-disciplinarity of individual researchers combined with interdisciplinarity among teams that enrich both research and practice. This is made possible by focusing on specific objects as influenced by human and natural processes. In the process we emphasize object–context mutual conditioning. Second, with application to health geography, we consider human–pathogen mutual influences and coevolution, not in biological terms, although this does occur over longer time frames, but as the dynamic, coadaptive interaction between pathogen ecologies and human institutions. After situating our conceptual approach, we explore new understandings of how human–pathogen interactions change over time by considering cases from our ongoing research. In the conclusion we draw out implications of this emergent mode of enquiry for geographical scholarship more broadly.

Conceptual Development: The Role of Institutions in Human–Pathogen Coevolution

The opportunities for specializations to converge are born of two simultaneous acknowledgments across the human–environment and health sciences. On the one hand, there is increasing recognition that knowledge systems and practices, however scientifically and medically founded, are deeply influenced by institutional context (Bowker and Star 1999). At the same time, scholars historically stressing the social and historical construction of environmental problems increasingly account for material and environmental factors (Whatmore 2002; Braun 2004; Robbins and Marks 2009). Research into lab practice and professional training has come to stress the differences in categorization, conceptualization, and assignment of causation rooted in cultures of practice. Despite the interdisciplinary nature of environmental and public health management problems, some organizations have come to be dominated by the outlook and methods of specific disciplinary groups. The resulting explanatory habits, systems of classification, and preference of data to confirm or deny claims differ widely (Becher 1989; Braxton and Hargens 1996) among both scientists (Latour 1999) and practitioners (Bowker and Star 1999). Concomitantly, research has come to stress how the material, grounded conditions of investigation and management can impinge

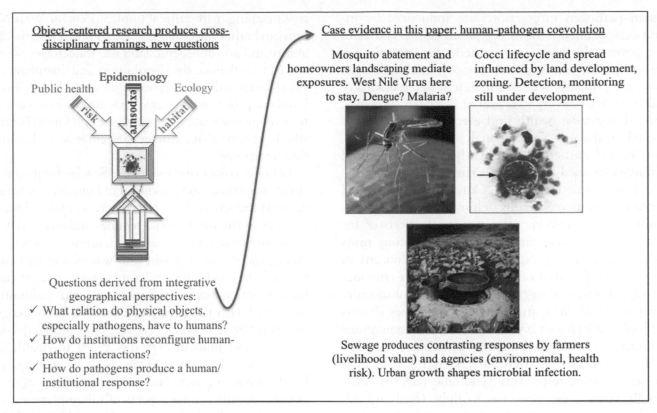

Figure 1. Object-centered conceptual framing of human–pathogen interaction. (Color figure available online.)

on the way these explanatory habits and categories emerge. We assert, therefore, that the ways in which objects become linked to people—here, sewage to farmers, mosquitoes to homeowners, and soils to range managers—continue to change over time as a result of ongoing iterations between social habits of knowledge and material conditions of research and practice. This view takes seriously the qualities, activities, and material specificities of nonhumans within what are otherwise considered strictly human processes (Mitchell 2002).

We are interested in material objects, mediated institutionally, as fundamental determinants of health and quality of life. The supradisciplinary challenge of health geography as we understand it is that although pathogens lack intention, they clearly produce a set of behaviors by humans, thus producing a human–environment dynamic that changes over time. The risk of human exposure is both a product of networks linking technologies, disease vectors, and pathogens as a result of the system of diagnosis and causal modeling. Understanding the emergence and trajectory of health risks and outcomes is therefore enhanced by the networking of pathogens, environments, and knowledge systems.

For example, spatial epidemiological models, which attempt to capture some of these complexities, assume that the degree and intensity of exposure respond to intrinsic dynamics of disease agents (DeGroote et al. 2008). Human interaction is understood to occur reflexively as mitigation and abatement and only in a secondary fashion through habitat modification. Model chains functionally underpin this understanding: ecological models for genetics and evolution, systems models for interactions, and population models for exposure. Disease ecology models that link pathogens and environment to epidemiology are overwhelmingly statistical (Kolivras and Comrie 2004; Comrie 2007). Some dynamic models exist, for example, with mosquito-borne diseases (Morin and Comrie 2010). Epidemiological models generally do not deal explicitly with ecological complexities and focus instead on pathogens and disease transmission to humans, and they are usually mathematical or compartmental (e.g., susceptible, exposed, infectious, recovered). Human–pathogen coevolution is sufficiently complex that this developing field will require additional applied cross-disciplinary work as outlined here.

Largely absent from spatial epidemiology and public health more broadly are the mechanisms by which human–pathogen interactions are influenced by institutions, defined here as organized norms and rules that govern the built environment; for example, urban sewage and zoning policy. Instead, institutions and policy are primarily seen as functional determinants of health service delivery (Mays et al. 2006) or outcomes. Methodologically, health and medical geography has focused on the unit area or small area problem (Riva, Gauvin, and Barnett 2007), in which variable spatial delimitations used for data reporting apparently influence health and well-being. Outcome measures and therefore specific reporting delimitations are institutional at root; however, the structural effects of institutions on disease exposures across reporting units remain inadequately explored. A second concern in health geography that only partially speaks to the mediating influence of institutions is the social inequity of access to services and, thereby, outcomes (Curtis and Jones 2008; Kwan 2009). Here, too, interpretations of mutual conditioning among humans, pathogens, and habitat require improved understanding. Epidemics like severe acute respiratory syndrome (SARS) occasionally bring these concerns to light (Keil and Ali 2007).

These problems in current understandings lead to our conceptual proposition that treats institutions as (1) cognitively and materially shaping human exposure to pathogens, (2) creating pathogen niche ecosystems in the built environment, and (3) presenting both enabling and limiting conditions for human–pathogen coevolution in spatial habitat and institutional terms.

Of Sewage, Mosquitoes, and Spores

We discuss evidence from three interactions, less as formal case studies than as an integrated set of observations highlighting principles and insights derived from cross-disciplinary, object-focused, coevolutionary framings. First, food crop irrigation using sewage is shown to be institutionally structured by agency perceptions and mandates, individual and collective agents' decision priorities, the cognitive and spatial terrain of sanitation, and the physical interactions between coliforms and helminthes on the one hand and producers and consumers of food on the other. The specific nature of these interactions strongly influences health and nutrition outcomes. Second, mosquito vectors' reoccupation of human settlements and the coevolution of their microhabitats and built environments is shown to respond to layered and often contradictory institutional cogni-

tive mapping, with critical implications for West Nile virus and other infectious diseases. Third, climatic variability and anthropogenic land use trajectories are observed to influence the distribution and disturbance of *Coccidioides* soil fungus spores that cause Valley Fever. Land use policy and practice plus human behaviors interactively influence spore exposure and fungal habitat, which in turn shape disease surveillance and institutional response.

Let us consider case evidence. Sewage farming continues a centuries-long tradition of human excreta used in food production. Two relatively recent (decades-old) changes to the institutional regimes and material flows of urban wastewater—urban agriculture as subject to management and oversight and wastewater treatment technology and infrastructure—have reconfigured human interaction with coliforms and helminthes associated with this practice. There has been a cognitive shift from resource recovery and productive use of water and nutrients (Asano 1988) to health and environmental hazard (Toze 2006), accompanied by the multiplication and specialization of agencies mandated with various aspects of urban planning and management (Scott, Faruqui, and Raschid-Sally 2004). Illustrative of institutional complexity, Hyderabad, India, has an urban development board, a water and wastewater utility, pollution control and environmental protection agencies, parks authorities, separate agriculture and irrigation departments, farmers' cooperatives, coalitions of the landless, formal and informal produce marketing bodies, and a public health department—all with multiple, often competing, objectives. Institutional (dis-)articulation has meant that the health risks to the general public from bacterial contamination of wastewater-irrigated food are identified with priority, notwithstanding research demonstrating that market wash water and other contamination immediately prior to the sale of produce poses greater risk of coliform transmission to consumers (Drechsel et al. 2010). Rarely have sewage farmers' own acute skin-contact helminth transmission, organo-toxins and heavy-metal exposures, or gastrointestinal disease burden been prioritized (Ensink et al. 2010). To the contrary, agency personnel see disease as retribution for urban farmers who transgress regulation, progress, and modernity. The second intervention—primary treatment (or secondary where affordable)—has modified not just wastewater quality but also access and ownership of water and nutrients. The Hyderabad utility produces its own fodder using effluent, although not as productively as neighboring farmers who also grow vegetables, rice, and nonedible crops. Under conditions of scarcity, the capture of water and nutrients by treatment plants and their subsequent commercialized distribution (in Hyderabad for landscaping and parks) have implications for pathogen exposure. Yet similar to the context of many developing countries with chronic background exposure levels and often-compromised immune response, the outcomes for disease burden attributable to advanced treatment in Hyderabad are questionable, to say nothing of the costs per disability-adjusted life years saved, as demonstrated for the Middle East by Fattal, Lampert, and Shuval (2004). Under these conditions, there is no fully prophylactic course of action for gastrointestinal disease. It is impossible to ensure full coverage of more failsafe therapies—oral rehydration for diarrhea and anthelminthic drugs for *Ascaris*, *Trichuris*, and other protozoa. In effect, humans and pathogens must coexist and continue to mutually influence spatial distribution and behaviors.

Integrative health geography research on wastewater and health, urban and periurban microhabitat spatial dynamics, and individual and collective agents' institutional positioning on risks and gains must account for stakeholder perspectives and the material nature of wastewater flows and pathogen exposures. New habitats for human–pathogen interactions result from urban growth—land conversion and settlement, sanitation practices and services, food provisioning, and health care. They also respond to specific dynamics that are place based but generalizable across the urban periurban gradient as evidenced in Hyderabad (Buechler and Mekala 2005). The urban core is particularly susceptible to progress and modernity conceptions, in terms of both land use and sewerage coverage; farmers and exposures are pushed toward the periphery—downstream along the Musi River—in search of productive assets (land, nutrients, and water) and decreased regulatory oversight of land use and environmental health concerns with food safety. One implication of the focus on food safety is that in Hyderabad, like other locations, wastewater farmers are constrained by the very interests—urban sanitation and public health authorities—who seek to continue dumping waste without remediation. The specific food production niches and practices of urban farming differ from conventional, rural fieldcropping (Freidberg 2001). Safeguarding vegetables on the vine from theft, frequent watering requirements of soils with poor water-holding capacity, and proximity to market outlets often entail seasonal cohabitation by producers in field tents, with markedly different levels

and durations of skin contact exposure to helminth ova than would be the case in other production contexts.

The focus on sewage, then, integrates perspectives of water resources geography, microbiology, public health, sociology, and agronomy, among others. Researching the coevolution of human–pathogen interactions in periurban sewage irrigation thus hybridizes spatial and institutional research with practitioners' emphases on measures taken by the utility, public health, and agricultural and irrigation agencies, as well as farmers' production and marketing arrangements. This integrated perspective raises several questions. How does sewage as resource and hazard alter the food security and public health outcomes of urban sanitation? How does urban agriculture's internalization of pathogens influence farmer, consumer, and manager behaviors and how do these mediate gastrointestinal and helminth infections?

Although operating in a very different economic and ecological context, the threat of West Nile virus in the U.S. Southwest, and the analysis of its drivers and risk management, shares several key elements with the previous case. Although southern Arizona and Sonora have long been the center of mosquito-borne insect problems and epidemics (Teeples 1929; Fink 1998), the problem is novel, in historical terms, insofar as the interactions of contemporary system elements driving the central vector—the mosquito—are largely recent. The decline of wetlands and surface-water flow combined with the widespread use of DDT and other chemical controls in the twentieth century to significantly reduce mosquito populations. By 2005, the region had seen the expansion of vector-carrying species of mosquito, including *Culex quinquefasciatus*, many of which are capable of overwintering in the region. These have been found in both wetland and residential areas—occupying built environment niches similar to the wastewater case—and continue to feed extensively on humans and birds, presenting a serious West Nile risk (Zinser, Ramberg, and Willott 2004).

The return of the insect vector is a response, in part, to the institutional logics and patterns of urban development in the region. Urban land development has been especially influential, including increases in backyard habitats like lush landscaping, neglected swimming pools, unmaintained fountains, and abandoned tires. For *quinquefasciatus*, a rise in breeding habitat has further followed the deliberate restoration for amenity and ecosystem services of previously drained wetlands, as well as the construction of wetlands for wastewater treatment (Willott 2004).

The development of the vector threat has simultaneously resulted in the promulgation of control techniques, each of which further reflects divergent assumptions about mosquito behaviors and causes. Adulticiding efforts utilizing methodical aerial spraying in some locations contrast with source-control strategies and larviciding in others (Shaw, Robbins, and Jones 2010). Each such response, and its differential impact on breeding, risk reduction, and disease distribution and evolution, is itself enmeshed in urban development history. Whereas some abatement offices and health departments seek reduction in public complaints, others respond to trap counts and geographic information systems (GIS) data. Landscapers focus on enhancing aesthetics and property values, whereas drainage infrastructure managers seek to minimize flooding. Planning for West Nile thus interacts with divergent incentives and institutional contexts to produce heterogeneous contexts for the disease and its vectors to evolve and adapt. Spatially heterogeneous built environments, dynamic hydrological systems, and shifting control philosophies have enabled the development of the vector in some areas and retarded it in others, with no sign that final resolution to the "mosquito wars" is at hand (Spielman and D'Antonio 2001). In a sense, then, through specific modalities of coexistence in the city, humans and mosquitoes have come to mutually influence one another's behaviors.

Although conventional spatial analysis can be used to determine important predictive variables (DeGroote et al. 2008), this case shows that emphasis on the mosquito as well as modeling and mapping of disease trajectories depend on the integration of multiple forms of data and characterizing of diverse system elements. Landscape-scale analysis of municipal morphology must be advanced utilizing mosquito-relevant land cover classes. These require integration into vector-population models sensitive to dynamic microscale temperature and precipitation impacts on differing life stages of the insect. These must be integrated with the jurisdictional behaviors of managers and neighborhood-scale practices of diverse publics—akin to institutional disarticulation for wastewater—that in turn are sensitive to the degree and distributions of disease outbreaks and differing institutional and community knowledges and priorities. Starting from the mosquito as the object of integration, the coadaptation (and coevolution) of vector distributions and mitigation strategies, mediated by a built environment that answers to exogenous political and economic drivers, becomes the central focus of meaningful epidemiological research, which is by its nature interdisciplinary. This raises the need for integrative enquiry on how institutional forms reconfigure human–mosquito interaction. Further research must address this question: Will preferences for specific built environments and managerial mandates perpetuate West Nile virus and produce new mosquito vector-based exposures, especially dengue and malaria?

The third case from our research, *Coccidioides* spp. soil fungi that cause Valley Fever (frequently abbreviated as cocci), are New World in origin and have evolved into two species (Fisher et al. 2001). Their spread from arid North America to Central and South America is thought to have accompanied the migrations of the first humans and their animals. Today, human–cocci mutual conditioning follows an exposure logic: Modification of the desert environment for human habitation disturbs the pathogen's soil habitat and heightens host contact with fungal spores. Addressing the primary interactions requires cross-disciplinary approaches to spores, soil, habitats, and inhabitants. Detection of cocci spores in the environment is very poor and therefore so is understanding of key social factors, necessitating a high level of research collaboration across complementary perspectives including medicine, vaccines, cellular and molecular biology, epidemiology, fungal genetics, fungal ecology, soil science, both micro- and landscape-scale ecology and biogeography of specific habitats, remote sensing and GIS, and climatology.

Physical contextual factors such as climate and weather variables (temperature, precipitation, humidity, wind, etc.) and ecology variables (vegetation type, phenology, hosts, vectors, and pathogens in the environment), when combined, are observable only at the landscape scale. Habitat is thus a scale proxy for pathogens (or vectors) that cannot be resolved individually. Observing hosts and their behaviors requires spatial data compromises to work around data unavailability at the pathogen scale; for example, mapping cocci-infected individuals by home address, knowing that the spore exposure might not have occurred at that specific location because of host movement or because of broader scale exposure via wind-blown transmission of the spores from elsewhere, causing exposure at that location. Overt collaborations among the range of researchers previously mentioned, specifically focused on elucidating fungal ecology, have partially overcome the data challenges to explain fungus–environment interactions, as in establishing and confirming key hypotheses regarding the role of climate in interannual disease seasonality and outbreaks (Tamerius and Comrie 2011).

Social and institutional contextual factors are similarly challenging. As with other diseases (Kwan 2009), uneven and changing case reporting for coccidioidomycosis exposure detection leads to significant public health surveillance uncertainties. Physicians seeing patients might not test for cocci or might not report it if identified. Central laboratories perform tests that can change with time, but their clinical data collection priorities do not necessarily align with epidemiological data priorities for public health, let alone environmental science data needs. Coccidioidomycosis incidence is increasing, but given the different institutional perspectives it is not surprising that its cause is contested, whether it be climate, urban expansion and soil disturbance, better reporting, or even whether the trend is real (Tamerius and Comrie 2011). Valley Fever has no cure per se, although antifungal drugs are effective in many cases and there is active work on a vaccine. Adaptive response is therefore the principal public health intervention, but it is difficult for two reasons. First, and most important, cocci spores are effectively ubiquitous; it is impossible to fully treat the environment or protect hosts across multiple environments in which they are active. Second, even if specific cocci source locations were identifiable, it is unknown what sort of soil remediation would be effective, practical, and safe. Thus, humans and pathogens must coexist in mutually conditioned habitats, just as we observed for the wastewater and mosquito cases earlier.

There are both lessons and challenges for health geography scholarship on cocci. Environmental and biological scientists are trying to detect an organism that is effectively invisible on the landscape—this remains the essential challenge. Finding answers to where, when, and how the spores are distributed presently requires an elaborate and expensive process of soil data collection and testing via mice to ascertain whether a sample is positive, with very low detection rates. Pathogen-focused cross-disciplinary collaboration has enabled advances in understanding geographic dimensions of cocci by working through the complex association with coccidioidomycosis. New genetic and molecular techniques offer the opportunity to make further advances by counting cocci directly, an approach currently outside the realm of conventional geographic tools but ripe for new collaborations between geographers and molecular and cellular bioscientists. Approaching the human–cocci challenge from social and institutional perspectives will entail investigation of behaviors that heighten exposure. Seemingly inexorable urban expansion and the institutions that produce it can clearly contract and expand cocci habitat and subsequent exposure, like periurban agriculture and mosquito vector habitats. Assuming that pathogen habitat was well understood, and acknowledging progress with institutionally produced data, the public health dimension will need to evolve from a medical response to unknown environmental and social strategies for mitigation. Cocci exposure, then, is produced via mutual conditioning of humans and pathogens in unique microhabitats and societal contexts, raising questions for continued research: How will improved cocci detection and monitoring inform epidemiological research? Will this translate to diminished Valley Fever incidence? Will economic drivers of open-space development account for the influence of land disturbance on cocci distribution?

Conclusion

The three cases of human–pathogen interaction presented have identified challenges and methods derived from a focus on institutionally mediated physical exposure: sewage-based transmission of gastrointestinal and helminth infection, mosquitoes in produced habitats for West Nile virus, and open-space development coupled with climate variations for Valley Fever. Despite providing diagnostic insights, there are limitations to understanding human–pathogen interaction as primarily place-based exposure. The effect of institutions in framing problems and risks and defining the scope of decision sets to address health challenges needs to be underscored. This requires new theoretical approaches—hybrid at times, clean-slate innovative at others—and new ways of seeing.

Cognitive maps resulting from focusing on physical objects contribute to new understandings of human–pathogen interaction and to coevolutionary effects of objects on researchers and managers. In this article, we have conceptualized this hybrid approach as shown schematically in Figure 1.

New hybrid framings stemming from this approach raise several propositions that are relevant for broader geography scholarship, most saliently:

- The mutual conditioning of objects and humans expands outcome possibilities; for health geography, human–pathogen interactions are bidirectionally causal.
- Object-oriented enquiry bridges multiple epistemologies; geographical research practice is thereby enriched, as presented here for health geography,

but also for water, energy, or other objects of concern occupying the *Annals*.

- Cross-disciplinary framing of questions and methods leads to fundamentally new understandings; geographers must continue at the forefront of this mode of dialogic enquiry.

In sum, it has long been acknowledged that health geography necessitates interdisciplinary and intersubdisciplinary perspectives (Turner 2002; Kwan 2004). We suggest that such integration is best achieved through an approach that starts from concrete objects, stresses the entanglement of material landscape conditions with knowledge and assumptions, and pursues the coevolutionary development of hazards and the institutions that mediate human interactions with them.

Acknowledgments

This material is based on work supported by the International Water Management Institute (Agreement 439230), the National Science Foundation (Grant No. 0948334), and the U.S. Environmental Protection Agency's Science to Achieve Results Program (Grant R8327540).

References

Asano, T. 1988. *Wastewater reclamation and reuse*. Boca Raton, FL: CRC Press.

Becher, T. 1989. *Academic tribes and territories: Intellectual enquiry and the cultures of the disciplines*. Bristol, PA: Open University Press.

Bierly, G. D., and J. D. Gatrell. 2004. Structural and compositional change in geography graduate programs in the United States: 1991–2001. *The Professional Geographer* 56 (3): 337–44.

Bowker, G., and S. Star. 1999. *Sorting things out: Classification and its consequences*. Cambridge, MA: MIT Press.

Braun, B. 2004. Modalities of posthumanism. *Environment and Planning A* 36:1352–55.

Braxton, J. M., and L. L. Hargens. 1996. Variation among academic disciplines: Analytic frameworks and research. In *Higher education: Handbook of theory and research*. Vol. 11, ed. J. C. Smart, 1–46. New York: Agathon.

Buechler, S. J., and G. D. Mekala. 2005. Local responses to water resource degradation in India: Groundwater farmer innovations and the reversal of knowledge flows. *Journal of Environment and Development* 14 (4): 410–38.

Comrie, A. C. 2007. Climate change and human health. *Geography Compass* 1:325–39.

Curtis, S., and I. R. Jones. 2008. Is there a place for geography in the analysis of health inequality? *Sociology of Health & Illness* 20 (5): 645–72.

DeGroote, J. P., R. Sugumaran, S. M. Brend, B. J. Tucker, and L. C. Bartholomay. 2008. Landscape, demographic,

entomological, and climatic associations with human disease incidence of West Nile virus in the state of Iowa, USA. *International Journal of Health Geographics* 7 (1): 19.

Drechsel, P., C. A. Scott, L. Raschid, M. Redwood, and A. Bahri, eds. 2010. *Wastewater irrigation and health: Assessing and mitigating risks in low-income countries*. London: Earthscan.

Ensink, J. H. J., C. A. Scott, S. Brooker, and S. Cairncross. 2010. Sewage disposal in the Musi River, India: Water quality remediation through irrigation infrastructure. *Irrigation and Drainage Systems* 24:65–77.

Fattal, B., Y. Lampert, and H. Shuval. 2004. A fresh look at microbial guidelines for wastewater irrigation in agriculture: A risk-assessment and cost-effectiveness approach. In *Wastewater use in irrigated agriculture: Confronting the livelihood and environmental realities*, ed. C. A. Scott, N. I. Faruqui, and L. Raschid-Sally, 59–68. Wallingford, UK: CAB International.

Fink, T. M. 1998. John Spring's account of "malarial fever" at Camp Wallen, A.T., 1866–1869. *Journal of Arizona History* 39:67–84.

Fisher, M. C., G. L. Koenig, T. J. White, G. San-Blas, R. Negroni, B. Wanke, G. Alvarez, and J. W. Taylor. 2001. Biogeographic range expansion into South America by *Coccidioides immitis* mirrors New World patterns of human migration. *Proceedings of the National Academy of Science* 98 (8): 4558–62.

Freidberg, S. E. 2001. Gardening on the edge: The social conditions of unsustainability on an African urban periphery. *Annals of the Association of American Geographers* 91 (2): 349–69.

Jones, J. P., III. 2005. Deleuze and Bush settle science wars. *AAG Newsletter* 40 (8): 16, 19.

Keil, R., and H. Ali. 2007. Governing the sick city: Urban governance in the age of emerging infectious disease. *Antipode* 39:846–73.

Kolivras, K. N., and A. C. Comrie. 2004. Climate and infectious disease in the southwestern United States. *Progress in Physical Geography* 28:387–98.

Kwan, M. P. 2004. Beyond difference: From canonical geography to hybrid geographies. *Annals of the Association of American Geographers* 94 (4): 756–63.

———. 2009. From place-based to people-based exposure measures. *Social Science and Medicine* 69 (9): 1311–13.

Latour, B. 1999. *Pandora's hope: Essays on the reality of science studies*. Cambridge, MA: Harvard University Press.

Mays, G. P., M. C. McHugh, K. Shim, N. Perry, D. Lenaway, P. K. Halverson, and R. Moonesinghe. 2006. Institutional and economic determinants of public health system performance. *American Journal of Public Health* 96:523–31.

Menand, L. 2010. *The marketplace of ideas: Reform and resistance in the American university*. New York: Norton.

Mitchell, T. 2002. *Rule of experts: Egypt, techno-politics, modernity*. Berkeley: University of California Press.

Morin, C., and A. C. Comrie. 2010. Modeled response of the West Nile virus vector *Culex quinquefasciatus* to changing climate using the dynamic mosquito simulation model. *International Journal of Biometeorology* 54:517–29.

Riva, M., L. Gauvin, and T. A. Barnett. 2007. Toward the next generation of research into small area effects on health: A synthesis of multilevel investigations published since July 1998. *Journal of Epidemiology and Community Health* 61:853–61.

Robbins, P., and B. Marks. 2009. Assemblage geographies. In *The Sage handbook of social geographies*, ed. S. Smith, R. Pain, S. A. Marston, and J. P. Jones III, 176–94. Beverly Hills, CA: Sage.

Schuurman, N. 2009. Critical GIScience in Canada in the new millennium. *The Canadian Geographer* 53 (2): 139–44.

Scott, C. A., N. I. Faruqui, and L. Raschid-Sally, eds. 2004. *Wastewater use in irrigated agriculture: Confronting the livelihood and environmental realities.* Wallingford, UK: CAB International.

Shaw, I. G. R., P. F. Robbins, and J. P. Jones III. 2010. A bug's life and the spatial ontologies of mosquito management. *Annals of the Association of American Geographers* 100 (2): 373–92.

Skole, D. L. 2004. Geography as a great intellectual melting pot and the preeminent interdisciplinary environmental discipline. *Annals of the Association of American Geographers* 94 (4): 739–43.

Smith, N. 1992. History and philosophy of geography: Real wars, theory wars. *Progress in Human Geography* 16:257–71.

Spielman, A., and M. D'Antonio. 2001. *Mosquito: A natural history of our most persistent and deadly foe.* New York: Hyperion.

Tamerius, J. D., and A. C. Comrie. 2011. Coccidioidomycosis incidence in Arizona predicted by seasonal precipitation. *PLoS ONE* 6 (6): e21009.

Teeples, C. A. 1929. The first pioneers of the Gila Valley. *Arizona Historical Review* 1 (4): 74–78.

Toze, S. 2006. Water reuse and health risks—Real vs. perceived. *Desalination* 187:41–51.

Turner, B. L. 2002. Contested identities: Human–environment geography and disciplinary implications in a restructuring academy. *Annals of the Association of American Geographers* 92 (1): 52–74.

Whatmore, S. 2002. *Hybrid geographies: Natures, cultures, spaces.* London: Sage.

Willott, E. 2004. Restoring nature, without mosquitoes? *Restoration Ecology* 12 (12): 147–53.

Zinser, M., F. Ramberg, and E. Willott. 2004. *Culex quinquefasciatus* (Diptera: Culicidae) as a potential West Nile virus vector in Tucson, Arizona: Blood meal analysis indicates feeding on both humans and birds. *Journal of Insect Science* 4 (20).

Moving Neighborhoods and Health Research Forward: Using Geographic Methods to Examine the Role of Spatial Scale in Neighborhood Effects on Health

Elisabeth Dowling Root

Department of Geography and Institute of Behavioral Sciences, University of Colorado at Boulder

A rich history of research documents the effects of neighborhood-level socioeconomic status (SES) conditions on health outcomes. Recent criticism of the neighborhoods and health literature, however, has stressed several conceptual and methodological challenges not adequately addressed in previous research. Critics suggest that early work on neighborhoods and health gave little thought to the spatial scale at which SES factors influence a specific health outcome. This article discusses the concept of neighborhoods and health, reviews recent criticisms of existing work, and provides a case study that exemplifies how geographic methods can address one such criticism. Using data on birth defects in North Carolina, the case study examines the relation of SES to orofacial clefts (cleft lip and cleft palate) at different spatial scales. The Brown–Forsythe test is used to select optimal neighborhood size. Results are evaluated using logistic regression models to examine the relationship between SES measures and orofacial clefts, controlling for individual-level risk factors. Results indicate modest associations between neighborhood-level measures of poverty and cleft palate but no associations with cleft lip with or without cleft palate.

丰富的研究历史记载了街区一级的社会经济地位条件 (SES) 对健康结局的影响。然而，最近对街区和健康文学的批评，强调了在以前的研究中没有得到充分解决的一些概念和方法上的挑战。评论家认为，早期的对居民区和健康的研究很少考虑到影响特定健康结果的 SES 因素的空间尺度。本文讨论了街区和健康的概念，回顾最近对现有研究的批评，并提供了一个案例研究，充分体现了地理的方法可以针对这样一个批评。本案例使用北卡罗莱纳州的出生缺陷数据，研究考察了 SES 与唇面裂（唇裂和颚裂）在不同空间尺度的关系。用布朗科西测试来选择最优邻域大小。评估结果采用逻辑回归模型评估检查 SES 指标和唇面裂之间的关系，它控制个体级别的风险因素。结果表明，在街区级的贫困指标和腭裂之间有适度相关，但是这些指标和唇裂之间没有关联，无论唇裂与腭裂同时发生与否。关键词：出生缺陷，地理信息系统，逻辑回归，街区和健康，空间尺度。

Los efectos que tienen las condiciones del estatus socioeconómico (SES) a nivel de barrio sobre el estado de la salubridad pública se hallan documentados en una rica historia de investigación. Una reciente crítica a la literatura de vecindarios y salud, no obstante, pone de presente varios retos conceptuales y metodológicos que no han sido tocados en la investigación precedente. Los críticos sugieren que en el trabajo inicial sobre barrios y salud se concedió muy poca atención a la escala espacial con la que los factores SES influyen un estado específico de salubridad pública. Este artículo discute el concepto de vecindarios y salud, revisa las críticas recientes sobre el trabajo que se ha hecho y suministra un estudio de caso a título de ejemplo sobre cómo pueden los métodos geográficos abocar una de esas críticas. Utilizando datos sobre defectos de nacimiento en Carolina del Norte, este estudio de caso examina la relación del SES con hendiduras orofaciales (labio leporino y paladar hendido), a diferentes escalas espaciales. Se utilizó el test Brown-Forsythe para seleccionar el tamaño óptimo del vecindario. Los resultados se evaluaron mediante modelos de regresión logística, para examinar la relación entre las mediciones SES y las hendiduras orofaciales, con un control de factores de riesgo a nivel individual. Los resultados muestran modestas asociaciones entre las mediciones de pobreza y paladar hendido a nivel de barrio, pero ninguna asociación con labio leporino, con o sin paladar hendido.

The medical geographic literature of research exploring differences in health outcomes among people residing in different geographic areas is extensive. Widespread documentation exists of small-area variations in morbidity, mortality, and health-related behaviors over the past 150 years (Macintyre and Ellaway 2003), and geography has often been used to pinpoint specific causes of disease. Recently, researchers have suggested that these observable differences in health between places could be due to neighborhood effects, or "the independent causal effect of a neighborhood ... on a number of health and/or social outcomes" (Oakes 2004, 1938). Observable differences in health outcomes between places might be due to differences in the kinds of people who live in these places (composition) or differences in the physical or social environment (contextual; Oakes 2004). In addition, variation in health outcomes could be influenced by complex interrelationships between characteristics of people and the places in which they live. The study of neighborhood influences on health is uniquely geographic, yet much of the early research in this area was centered in major schools of public health, and research teams did not routinely include geographers. Many of the recent criticisms of neighborhoods and health research are related to the "nongeographic" lens through which public health researchers developed neighborhood studies and statistical analyses. Although tools such as geographic information systems (GIS) have been employed, analysis is often not particularly sophisticated, and huge bodies of geographic theory on the modifiable areal unit problem (MAUP) and the relevance of scale have all but been ignored.

In 2001, Ana Diez Roux, a pioneer in neighborhoods and health research, published an article on the conceptual and methodological challenges to neighborhoods and health studies (Diez Roux 2001). One specific challenge presented was how to define "neighborhoods" or relevant geographic areas. She suggested that the size and definition of the geographic area relevant to studying a specific health outcome might vary according to the processes through which the area effect is hypothesized to operate and the outcome being studied. That is to say, there is a causal pathway by which area characteristics (e.g., social or economic climate) impact health, but researchers need to understand at what scale this causal pathway operates. This requires the development and testing of hypotheses regarding the geographic area that is relevant for health outcomes.

Despite the obvious importance of scale in neighborhood and health studies, very few early studies attempted to understand the limitations of using a single geographic unit for analysis. Most health research used (and continues to use) geopolitical boundaries such as counties, census tracts, or electoral wards to examine the relationship between area-level socioeconomic and environmental variables and health outcomes. Although use of these proxies is a practical alternative because data are readily available, they are limited in that they do not necessarily correspond to the theoretically relevant geographic neighborhood. Only in the past few years have geographers actively engaged in the debate, once again revisiting geographic theories such as the MAUP and applying them to health research (Haynes et al. 2007; Spielman and Yoo 2009) and examining the effect of spatial scale on neighborhood studies (Weiss et al. 2007; Flowerdew, Manley, and Sabel 2008). In addition, advances in real-time tracking technologies have allowed researchers to define neighborhoods based on an individual's use of and movement across space and time. Such methods challenge the assumption that neighborhood effects operate only through connections that exist among those residing in the same area and mitigate the possibility of incorrectly attributing exposure to neighborhood environments (Kwan 2009).

Overall, results from neighborhood studies tell us is that the definition of neighborhood does matter in statistical analyses of neighborhood effects. Given the importance of choosing the appropriate spatial scale for neighborhood studies, the literature is sparse and most studies choose neighborhood areas without much thought as to how or why these areas have been chosen. This article explores how geographic and statistical methods can assist in the definition, construction, and selection of neighborhoods using a case study of orofacial clefts in North Carolina.

Case Study: Orofacial Clefts in North Carolina

Orofacial clefts (OFCs) are among the most common types of birth defects in the United States. They occur when structures of the mouth fail to form properly and are divided into two distinct groups: cleft lip with or without cleft palate (CL $\pm$ P) and isolated cleft palate (CP). Nationwide, the three-year (2004–2006) estimated prevalence of CL $\pm$ P was 10.6 per 10,000 live births, and the prevalence of isolated CP was 6.4 per 10,000 live births (Parker et al. 2010). Orofacial clefts can impair the development of speech, teeth, and feeding capabilities and affected individuals require medical care from birth until adulthood. The defects therefore pose a substantial burden to individuals and their families and require significant health-related expenditures.

There is no agreement as to the effect, if any, of exposure to environmental toxicants on CP and CL ± P. What evidence we do have comes from conflicting results of studies examining presumed exposure (due to occupation or proximity) to sites that use or release specific chemical compounds (e.g., hazardous waste and toxic release sites or landfill; Dolk et al. 1998; Brender et al. 2006). Prior analysis of the data set used in this article found no association with proximity to a variety of industrial and hazardous waste sites and CP or CL ± P (Root unpublished research). Maternal smoking has also been linked to cleft lip and palate (Little, Cardy, and Munger 2004) and there is evidence of a gene–environment interaction between maternal or infant gene variations and smoking (Lammer et al. 2004). Neighborhood socioeconomic status (SES) context might also influence birth outcomes through a complex set of psychosocial, behavioral, and biological factors. Results from the few studies that have looked at the effect of both individual- and area-level characteristics on orofacial clefts are inconclusive. Some studies have found a positive association between neighborhood measures of low SES and orofacial clefts (Clark et al. 2003) and others have found no association at all (Vrijheid et al. 2000; Carmichael et al. 2003).

Given the uncertainties that exist about the etiology of orofacial clefts, examining the influence of neighborhood-level SES measures might assist in the generation of hypotheses about the underlying causal factors associated with socioeconomic risk factors. This study examines the relationship between CL ± P and isolated CP and three area-based measures of SES after adjusting for known individual-level risk factors. Three study questions guided this research: (1) To what extent are neighborhood-level SES variables related to the risk of an orofacial cleft? (2) Does this relationship differ when different spatial scales are used to define neighborhoods? (3) What do the results tell us about the intricacies of choosing a spatial scale at which to study health outcomes?

Methods

Birth Defects Data

The main source of birth defects data for this study was the North Carolina Birth Defects Monitoring Program (NCBDMP) run by the North Carolina State Center for Health Statistics (NC SCHS 2006). This study used a retrospective case-control design of North Carolina resident live births with an orofacial cleft (CL ± P or CP) between 1 January 1999 and 31 December 2004. To identify infants with an orofacial cleft, the NCBDMP database was searched using the *International Classification of Diseases* (ICD-9) code for orofacial clefts (codes 749.000–749.290). All live-born singleton infants with a birth certificate, without any birth defect, and born between 1 January 1999 and 31 December 2004 to North Carolina resident mothers contained in the NC Composite Linked Birth File were eligible to be controls. The data include residential address at birth, which was used to geocode cases and controls. A majority of the geocoding was completed by the NC SCHS, using geographic data technology (GDT) and parcel data from the North Carolina Department of Transportation. Records not matched by the NC SCHS were geocoded using a multistage geocoding method and different Web-based geocoding services (Lovasi et al. 2007). Using this process, 92.2 percent of cases and 93.3 percent of controls were matched.

The analysis was limited to urban areas because the appropriate scale for a spatial neighborhood might differ depending on rural or urban residence. Using a GIS, each urban area was buffered at a distance of 5,500 m (the largest neighborhood size tested). Urban areas were buffered so that complete neighborhoods for individuals living near the border of an urban area could be constructed. Only cases and controls within the urban area were included in the statistical analysis, but areas contained in the buffered area were used to construct the neighborhood measures and conduct the Brown–Forsythe test (refer to Figure 1 and "Construction of Neighborhood Variables" later). The final urban sample included 319 cases of CL ± P, 206 cases of isolated CP, and 7,663 controls.

Socioeconomic Data

Socioeconomic variables for census tracts, block groups, and blocks were obtained from the U.S. Census Bureau for the 2000 Census. Following the approach of several previous studies that examine area-level effects on various birth outcomes (Krieger et al. 2003; Messer et al. 2008), three census variables were used to estimate neighborhood-level SES characteristics: percentage of the population living below 100 percent of federal poverty level, percentage of the population with less than a high school education, and percentage of the population unemployed. These measures quantify

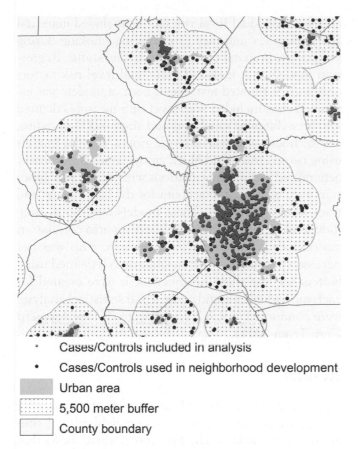

* · Cases/Controls included in analysis
* • Cases/Controls used in neighborhood development
* ▓ Urban area
* ⬚ 5,500 meter buffer
* ☐ County boundary

Figure 1. Example of geographic information system used to select urban births for analysis. (Color figure available online.)

several socioeconomic domains that effect health: poverty, employment, and education. Whereas some researchers have advocated the use of indexes to measure the cumulative effects of several measures of SES (Messer et al. 2008), others have found that estimates of effects detected using a single variable measure of poverty were similar to those based on indexes (Krieger et al. 2003). The effect of a composite measure of SES similar to the Carstairs Index on OFCs was found to be statistically similar to that of the single variable measure of poverty. Therefore, this article presents single variable measures to examine separate effects of each SES domain on orofacial clefts. Each census variable was classified into quartiles based on the distribution among controls and each census tract, block group, and block given a corresponding quartile score (1 = lowest, 4 = highest).

Construction of Neighborhood Variables

Cases and controls were assigned to year 2000 census block groups and tracts. A set of "neighborhoods" was developed by creating circular buffers of various

sizes around each study subject. Based on the size of the study area and the distribution of the population, the minimum size was set to a 1,000-m radius neighborhood and increased stepwise by 500 m until 5,500 m size was reached. This resulted in ten different neighborhood sizes from which to select the scale at which to conduct this analysis. The neighborhood-level SES variables were estimated by aggregating census block group data by each of the circular windows. In cases where the circular window contained only a portion of a census block group, variables were weighted by the proportion of the population from that block group that was encompassed by the window.

Selection of Neighborhood Scale

The neighborhood size used to conduct this analysis was chosen using the Brown–Forsythe (F_{BF}) test. This method assesses whether the variance of a variable of interest (e.g., disease incidence or risk factors) is equal between two different-sized neighborhood groups. The underlying assumption of this method is that there is greater variance in disease rates among smaller neighborhoods and lower variance among larger neighborhoods. A high variance value means that the data are local or individualistic, whereas a low variance means that they capture a more global process. The optimal neighborhood ensures that the aggregate disease incidence data is somewhere in between. Use of this method is explained in detail elsewhere (Root, Meyer, and Emch 2011). The test statistic, F_{BF}, was calculated as

$$F_{BF} = \frac{\frac{\sum_{i=1}^{t} n_i (\bar{D}_i - \bar{D})^2}{(t-1)}}{\frac{\sum_{i=1}^{t} \sum_{j=1}^{n_1} (D_{ij} - \bar{D}_1)^2}{(N-t)}},$$

where n_i is the number of samples in the ith neighborhood group (e.g., the number of individual neighborhoods constructed that are the same size), N is the total number of samples for all neighborhood groups, t is the number of neighborhood groups (e.g., two different-sized neighborhood groups are compared), y_{ij} is the disease rate of the jth sample from the ith neighborhood group, $\bar{y}_i$ is the median of disease rate from the ith neighborhood group, $D_{ij} = |y_{ij} - \bar{y}_i|$ is the absolute deviation of the jth observation from the ith neighborhood group median, $\bar{D}_i$ is the mean of D_{ij} for neighborhood group i, and $\bar{D}$ is the mean of all D_{ij} (e.g., both neighborhood groups combined). The test assumes that the variances of two different neighborhood groups are equal under the null hypothesis. The critical value was calculated using an F distribution with $(t - 1, N - t)$ degrees of freedom and $\alpha = 0.05$ was used to test for significance.

The F_{BF} test involves two steps. First, the variance between each neighborhood group (from 1,000 to 5,000 m) and the largest neighborhood group (5,500 m) is compared (for a total of nine separate calculations of F_{BF}, shown in the F_{BF1} column of Table 1). Second, the variance between each neighborhood group (from 1,500–5,500 m) and the smallest neighborhood group (1,000 m) is compared (for a total of nine separate calculations of the F_{BF}, shown in the F_{BF2} column). A significant value of F_{BF1} indicates that the neighborhood does not reveal the global structure of data; in essence each neighborhood is so small that it only captures disease dynamics for a small group of individuals. In contrast, a significant value in F_{BF2} implies that the neighborhood data are not individualistic; they are so large that local-level disease dynamics are "washed out" and undetectable. The neighborhoods between the lower and the upper limits identify a spatial scale at which local-level variation is still detectable but captures larger "neighborhood-level" disease dynamics.

Logistic Regression

To estimate the variation in risk of CL ± P and CP affected pregnancy associated with differences in neighborhood SES, maximum likelihood estimates of odds ratios (ORs) and 95 percent confidence intervals (CIs) were calculated from logistic regression models. For all analyses, models were estimated separately for CL ± P and CP, because initial models indicated that they are influenced by different individual-level risk factors. Individual-level risk factors included maternal age, infant sex and race or ethnicity, smoking during pregnancy, prenatal care, and Medicaid status. Regression models that included individual-level risk factors only were estimated first. Next, a set of models was estimated that included area-level SES measures defined for three different neighborhood sizes: 4,000-m radius, census tracts, and census block groups. For the models using census tracts and block groups as neighborhoods, generalized estimating equations with a logit link function were estimated to account for the block group and tract-level correlation. These models were built using independent and exchangeable within-area correlation matrices to control for the correlation. This was not necessary for models with neighborhoods defined using buffers because these neighborhoods were created for each individual case and control. All statistical analyses were conducted in R v.2.9.2 software (R Development Core Team 2011).

Results

The F_{BF} test results for homogeneity of variance for orofacial cleft rates under various neighborhood sizes are listed in Table 1. The F_{BF1} test statistic shows that a neighborhood size of approximately 3,500 m is optimal and the F_{BF2} test statistic shows that a neighborhood size of approximately 4,500 m is optimal. Below 3,500 m, neighborhood data might only capture the characteristics of each individual, whereas above 4,500 m the neighborhoods are so large that they do not capture the influence of an individual's proximal environment. Given these results, a neighborhood size of approximately 4,000 m was chosen for modeling. A

Table 1. Descriptive statistics and results for Brown–Forsythe test

	Population				Incidence rate/100,000 births					Brown–Forsythe					
r	Min	Max	M	Mdn	Min	Max	M	Mdn	Variance	F_{BF1}	df_1	CV	F_{BF2}	df_2	CV
1,000	1	33	6.1	5	0	66,667	1003.4	0	20,571,812	3.99	9	1.88	—	—	—
1,500	1	64	11.0	9	0	50,000	1056.0	0	14,896,288	5.53	8	1.94	0.63	1	3.84
2,000	1	90	17.3	14	0	50,000	1081.4	0	10,762,540	6.59	7	2.01	0.84	2	3.00
2,500	1	126	25.0	20	0	33,333	1046.4	0	7,735,246	5.53	6	2.10	0.64	3	2.61
3,000	1	175	33.9	28	0	33,333	1016.9	0	5,797,722	4.45	5	2.21	0.66	4	2.37
3,500	1	221	44.0	36	0	25,000	989.8	0	4,489,869	**3.36**	4	2.37	0.93	5	2.21
4,000	1	285	55.2	45	0	20,000	963.9	0	3,619,704	2.31	3	2.61	1.42	6	2.10
4,500	1	345	67.5	56	0	14,286	941.5	0	2,935,198	1.48	2	3.00	**2.09**	7	2.01
5,000	2	388	80.7	67	0	12,903	919.3	0	2,443,491	0.73	1	3.84	2.95	8	1.94
5,500	2	427	94.8	80	0	11,111	899.2	0	2,081,109	—	—	—	3.99	9	1.88

Note: Bold figures in the F_{BF1} and F_{BF2} columns are the upper and lower limit of neighborhood scale, and the bold figure in the r column is the neighborhood size chosen for this analysis. r = size of the neighborhood radius in meters; df = degrees of freedom; CV_1 and CV_2 = critical values at 95 percent confidence level for F_{BF1} and F_{BF2}.

neighborhood of approximately 4,000 m might seem large but equates to an area of approximately 5 miles across. In many U.S. urban areas, people routinely travel 5 to 10 miles to carry out many daily activities, such as shopping, taking children to school, or exercising.

Table 2 shows the prevalence of covariates among cases and controls. Risk factors were significantly different between cases and controls and between infants born with isolated CP and those born with CL ± P. Table 3 presents results from the logistic regression of individual-level variables only. Male infants showed an increased risk of CL ± P (OR = 1.61), whereas female infants showed an increased risk of isolated CP (OR = 0.66). Black infants showed a significantly decreased risk of both CL ± P (OR = 0.51) and isolated CP (OR = 0.71) compared to white infants. Finally, Medicaid status and late prenatal care (considered proxy indicators of individual-level SES) only showed an association with CL ± P.

Figure 2 shows the adjusted ORs for two of the neighborhood SES measures in relation to CL ± P and CP. Only neighborhood measures of unemployment and percentage of the population below 100 percent of federal poverty level are shown because high school education showed no significant effect at any neighborhood scale. Overall, measures of neighborhood SES showed no significant association with CL ± P at any neighborhood scale. Marginal effects were present for infants born with isolated CP. Residence in a neighborhood in the highest quartile of poverty compared to the lowest quartile was associated with an increased risk for CP at all neighborhood scales. The neighborhood defined by census tracts showed the strongest association (OR = 2.06).

Residence in a neighborhood in the highest quartile of unemployment compared to the lowest quartile was also associated with an increased risk for CP but only for the neighborhoods defined by the census tract and 4,000-m buffer. Women residing in a neighborhood in the highest quartile of unemployment were 1.5 times more likely to give birth to an infant with CP than women residing in the lowest quartile (4,000 m: OR = 1.56; tract: OR = 1.57). The census block group neighborhood unemployment rate showed almost no association and CIs overlapped the mean in both cases.

Discussion

This study indicates a weak association between residence in a lower SES neighborhood, as measured by poverty and unemployment, and the risk of having an infant with isolated CP, after adjusting for individual-level risk factors. The odds of a CL ± P–affected pregnancy did not appear to be affected by neighborhood-level SES, regardless of the scale at which SES was measured. Given these findings and the results of prior research, can theories about plausible links between specific socioeconomic features of the neighborhood and CP be developed? There are no data to suggest specific causal relationships, but hypotheses that merit further epidemiological research can be developed. For example, prior research on birth outcomes suggests that women who experience high levels of psychosocial stress are at greater risk for preterm and low-birth-weight births (see Hobel, Goldstein, and Barrett 2008). Although the results of this case study might be interesting by themselves, more important, they demonstrate several points about the challenge of choosing the geographic scale at which to test the effects of neighborhood-level SES on health outcomes.

At each SES quartile, the CIs for ORs measured by the 4,000-m buffer, census tract, and census block group overlapped, indicating no significant difference between ORs estimated at different scales. This suggests that too much emphasis on point estimates might be misleading, especially with cases where the point estimate at one scale (e.g., block group) is not significantly associated with the outcome, whereas the point estimate at another scale (e.g., 4,000-m buffer) is significantly associated. In this study, this was the case with fourth quartile neighborhood unemployment rates, where the buffer and tract-level OR estimates were statistically significant whereas the block group was not. The CIs for all three neighborhood scales overlapped, however, indicating that the point estimates across neighborhood scales were statistically similar. This has implications for all neighborhood studies, especially in the field of public health where so much emphasis is given to regression coefficients and ORs. In addition, it lends credibility to the ongoing debate about the usefulness of p values and significance tests in statistical inference (Cohen 1994; Stang, Poole, and Kuss 2010). In the social and medical sciences, the use of significance tests is ubiquitous, but placing too much emphasis on a p value of 0.05 might cause researchers to incorrectly reject or accept a null hypothesis. Disregarding neighborhood-level effects because CIs encompass 1.0 might impede important discoveries, especially for rare health conditions because p values and CIs are so dependent on sample size.

The neighborhood scale that is chosen for analysis is important, especially from a theoretical

Table 2. Descriptive characteristics of cases and controls included in analysis

	Cleft lip ± palate			Cleft palate		
	Cases[a] n (%)	Controls[b] n (%)	p	Cases[c] n (%)	Controls[d] n (%)	p
Age						
<25 Years	126 (39.5)	2,914 (38.0)	0.596	63 (30.6)	2,914 (38.0)	0.030
25–29 Years	81 (25.4)	2,009 (26.2)	0.743	62 (30.1)	2,009 (26.2)	0.212
30–34 Years	70 (21.9)	1,793 (23.4)	0.547	56 (27.2)	1,793 (23.4)	0.206
≥35 Years	42 (13.2)	947 (12.4)	0.668	25 (12.1)	947 (12.4)	0.924
Sex						
Female	116 (36.5)	3,697 (48.2)		120 (58.3)	3,697 (48.3)	
Male	202 (63.5)	3,965 (51.8)	<0.0001	86 (41.8)	3,965 (51.8)	0.005
Race						
White	202 (63.3)	4,212 (55.0)	0.003	134 (65.1)	4,212 (55.0)	0.004
Black	56 (17.5)	2,129 (27.8)	<0.0001	45 (21.8)	2,129 (27.8)	0.060
Hispanic	49 (15.4)	1,016 (13.3)	0.279	24 (11.7)	1,016 (13.3)	0.501
Other	12 (3.8)	306 (4.0)	0.836	3 (1.5)	306 (4.0)	0.064
Insurance						
Private insurer	188 (58.9)	4,775 (62.3)		131 (63.6)	4,775 (62.3)	
Medicaid	131 (41.1)	2,888 (37.7)	0.223	75 (36.4)	2,888 (37.7)	0.708
Prenatal care						
1st/2nd trimester	314 (98.4)	7,401 (96.6)		197 (95.6)	7,401 (96.6)	
3rd trimester	5 (1.6)	262 (3.4)	0.072	9 (4.4)	262 (3.4)	0.461

[a]n = 319. [b]n = 7,663. [c]n = 206. [d]n = 7,663.

Table 3. Adjusted odds ratio (95 percent confidence interval) for cleft lip ± palate and cleft palate

	Cleft lip ± palate			Cleft palate		
	AOR	95% CI	p	AOR	95% CI	p
Age						
<25 Years	—	—	—	0.73	0.51–1.05	0.0863
25–29 Years	—	—	—	Ref		
30–34 Years	—	—	—	0.98	0.68–1.41	0.8994
≥35 Years	—	—	—	0.82	0.51–1.32	0.4126
Sex						
Female		Ref			Ref	
Male	1.61	1.28–2.03	<0.0001	0.66	0.50–0.87	0.0038
Race						
White		Ref			Ref	
Black	0.51	0.37–0.69	<0.0001	0.70	0.49–0.98	0.0436
Hispanic	0.95	0.68–1.32	0.7737	0.77	0.49–1.21	0.2560
Other	0.84	0.47–1.53	0.5738	0.30	0.10–0.95	0.0407
Insurance						
Private insurer		Ref		—	—	—
Medicaid	1.36	1.06–1.73	0.0137	—	—	—
Prenatal care						
1st/2nd trimester		Ref		—	—	—
3rd trimester	0.45	0.18–1.10	0.0821	—	—	—

Note: AOR = adjusted odds ratio; CI = confidence interval.

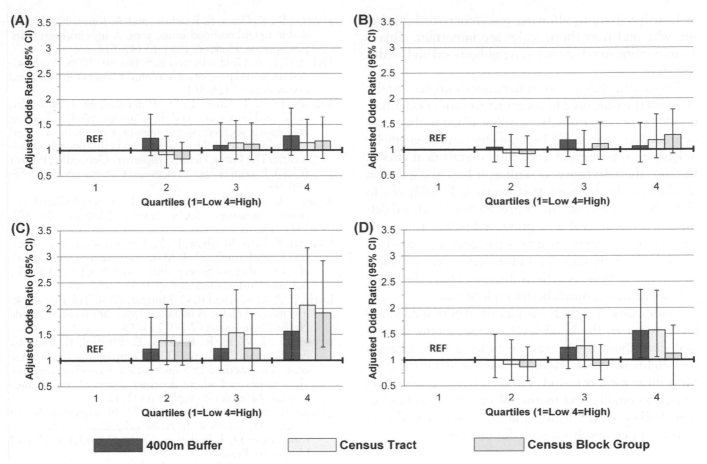

Figure 2. Adjusted odds ratios (95 percent confidence interval [CI]) for neighborhood risk quartiles for (A) cleft lip ± palate (CL ± P), 100 percent of federal poverty level; (B) CL ± P, unemployment rate; (C) cleft palate (CP), 100 percent of federal poverty level; and (D) CP, unemployment rate. CL ± P adjusted for sex, race, insurance, and prenatal care; CP adjusted for age, sex, and race. (Color figure available online.)

perspective. Researchers should use strong theory to choose a neighborhood scale that they believe represents the geographic scale at which causal pathways act between SES and the health outcome. Or, if there is uncertainty about the scale at which SES context influences a health outcome, several geographic scales could be explored and hypotheses drawn about potential causal mechanisms. In this study, for example, unemployment and poverty appear to exert some influence on CP but at different scales (poverty showed a stronger effect at smaller scales and unemployment at larger scales). It makes some sense that unemployment rate might influence CP at a higher level of geography because high unemployment tends to depress the economy in much larger areas (e.g., a whole city), whereas poverty is often localized, existing in "pockets" around a city.

Geographers have been struggling with the issue of scale for decades, discussing the MAUP in its many forms. The debate is particularly relevant to studies using GIS. The idea that aggregating data to different scales of geography might yield different results is no surprise but was given very little thought in public health, especially when neighborhoods and health studies began. Geographers have given a great deal of thought to the problems associated with scale and can contribute much to the debate of "appropriate" neighborhood scale in health studies. There are, of course, many different ways to define a neighborhood using real-time tracking, measures of distance, or even social connections or historical boundaries. The technique employed in this study (F_{BF} test) is another method of exploring and choosing neighborhood scale. It is not, of course, the only method and might not be appropriate for choosing neighborhoods constructed using methods other than buffering, but it adds to the toolbox of methods used to understand and adapt to the MAUP. It can uncover a set of geographic scales that are relevant for assessment of health variation, allowing researchers to hypothesize why and how those scales are important. This is an important step forward in neighborhood and health research.

Finally, the skill necessary to correctly implement the GIS and the time used to construct the buffer neighborhoods might far outweigh the value added to the statistical analysis, especially because the census geography neighborhoods showed very similar statistical effects. Because the results are so similar, it begs the question of whether the extra effort is worth it. Possibly not in all health studies, but a more precise neighborhood definition could (1) validate previous findings because it reflects the geographic reality of the disease process better or (2) be more important in studies of certain health outcomes, such as infectious diseases where the mode of transmission is much better understood.

In summary, this study has identified both individual-level and neighborhood-level socioeconomic factors that contribute to the risk of isolated CP, although no neighborhood relationship was found for CL $\pm$ P. Most important, this study has shown how geographers can conceptually and methodologically contribute to neighborhoods and health studies, especially to the selection of neighborhood scale for analysis.

References

Brender, J. D., F. B. Zhan, L. Suarez, P. Langlois, and K. Moody. 2006. Maternal residential proximity to waste sites and industrial facilities and oral clefts in offspring. *Journal of Occupational and Environmental Medicine* 48 (6): 565.

Carmichael, S. L., V. Nelson, G. M. Shaw, C. R. Wasserman, and L. A. Croen. 2003. Socio-economic status and risk of orofacial clefts. *Paediatric and Perinatal Epidemiology* 17 (3): 264.

Clark, J. D., P. A. Mossey, L. Sharp, and J. Little. 2003. Socioeconomic status and orofacial clefts in Scotland, 1989 to 1998. *Cleft Palate Craniofacial Journal* 40 (5): 481.

Cohen, J. 1994. The Earth is round ($p < .05$). *American Psychologist* 49:997–1003.

Diez Roux, A. V. 2001. Investigating neighborhood and area effects on health. *American Journal of Public Health* 91 (11): 1783.

Dolk, H., M. Vrijheid, B. Armstrong, L. Abramsky, F. Bianchi, E. Garne, V. Nelen, et al. 1998. Risk of congenital anomalies near hazardous-waste landfill sites in Europe: The EUROHAZCON study. *Lancet* 352 (9126): 423.

Flowerdew, R., D. Manley, and C. Sabel. 2008. Neighbourhood effects on health: Does it matter where you draw the boundaries? *Social Science Medicine* 66 (6): 1241–55.

Haynes, R., K. Daras, R. Reading, and A. Jones. 2007. Modifiable neighbourhood units, zone design and residents' perceptions. *Health & Place* 13 (4): 812–25.

Hobel, C. J., A. Goldstein, and E. S. Barrett. 2008. Psychosocial stress and pregnancy outcome. *Clinical Obstetrics and Gynecology* 51 (2): 333.

Krieger, N., J. T. Chen, P. D. Waterman, M.-J. Soobader, S. V. Subramanian, and R. Carson. 2003. Choosing area based socioeconomic measures to monitor social inequalities in low birth weight and childhood lead poisoning: The Public Health Disparities Geocoding Project (U.S.). *Journal of Epidemiology and Community Health* 57: 186–99.

Kwan, M. 2009. From place-based to people-based exposure measures. *Social Science Medicine* 69 (9): 1311–13.

Lammer, E. J., G. M. Shaw, D. M. Lovannisci, J. Van Waes, and R. H. Finnell. 2004. Maternal smoking and the risk of orofacial clefts: Susceptibility with NAT1 and NAT2 polymorphisms. *Epidemiology* 15 (2): 150–56.

Little, J., A. Cardy, and R. G. Munger. 2004. Tobacco smoking and oral clefts: A meta-analysis. *Bulletin of the World Health Organization* 82 (3): 213–18.

Lovasi, G., J. Weiss, R. Hoskins, E. Whitsel, K. Rice, C. Erickson, and B. Psaty. 2007. Comparing a single-stage geocoding method to a multi-stage geocoding method: How much and where do they disagree? *International Journal of Health Geographics* 6 (1): 12.

Macintyre, S., and A. Ellaway. 2003. Neighborhoods and health: An overview. In *Neighborhoods and health*, ed. I. Kawachi and L. F. Berkman, 20–42. Oxford, UK: Oxford University Press.

Messer, L. C., L. C. Vinikoor, B. A. Laraia, J. S. Kaufman, J. Eyster, C. Holzman, J. Culhane, I. Elo, J. G. Burke, and P. O'Campo. 2008. Socioeconomic domains and associations with preterm birth. *Social Science Medicine* 67 (8): 1247.

North Carolina State Center for Health Statistics (NC SCHS). 2006. *North Carolina Birth Defects Monitoring Program.* Raleigh: North Carolina State Center for Health Statistics. http://www.schs.state.nc.us/SCHS/bdmp (last accessed 1 November 2010).

Oakes, J. M. 2004. The (mis)estimation of neighborhood effects: Causal inference for a practicable social epidemiology. *Social Science Medicine* 58 (10): 1929–52.

Parker, S., C. Mai, M. Canfield, R. Rickard, Y. Wang, R. Meyer, P. Anderson, et al. 2010. Updated national birth prevalence estimates for selected birth defects in the United States, 2004–2006. *Birth Defects Research A: Clinical and Molelcular Teratology* 88 (12): 1008–16.

R Development Core Team. 2011. *R: A language and environment for statistical computing (v.9.2.9).* Vienna, Austria: R Foundation for Statistical Computing. http://www.R-project.org (last accessed 15 February 2012).

Root, E. D., R. Meyer, and M. Emch. 2011. Socioeconomic context and gastroschisis: Exploring associations at various geographic scales. *Social Science Medicine* 72 (4): 625–33.

Spielman, S. E., and E. H. Yoo. 2009. The spatial dimensions of neighborhood effects. *Social Science Medicine* 68 (6): 1098–1105.

Stang, A., C. Poole, and O. Kuss. 2010. The ongoing tyranny of statistical significance testing in biomedical research. *European Journal of Epidemiology* 25 (4): 225–30.

Vrijheid, M., H. Dolk, D. Stone, L. Abramsky, E. Alberman, and J. E. Scott. 2000. Socioeconomic inequalities in risk of congenital anomaly. *Archives of Disease in Childhood* 82 (5): 349.

Weiss, L., D. Ompad, S. Galea, and D. Vlahov. 2007. Defining neighborhood boundaries for urban health research. *American Journal of Preventive Medicine* 32 (6 Suppl.): S154–59.

Is a Green Residential Environment Better for Health? If So, Why?

Peter P. Groenewegen,* Agnes E. van den Berg,[†] Jolanda Maas,[‡] Robert A. Verheij,[§] and Sjerp de Vries[#]

*NIVEL—Netherlands Institute for Health Services Research, and Department of Human Geography and Department of Sociology, Utrecht University, Utrecht, The Netherlands
[†]Alterra Wageningen UR, Wageningen, The Netherlands
[‡]EMGO+ Institute, Department of Public and Occupational Health, VU University Medical Center, Amsterdam, The Netherlands
[§]NIVEL—Netherlands Institute for Health Services Research, Utrecht, The Netherlands
[#]Wageningen UR Landscape Centre, Wageningen, The Netherlands

Over the past years our group has been working on a coherent research program on the relationships between greenspace and health. The main aims of this "Vitamin G" program (where G stands for green) were to empirically verify relationships between greenspace in residential areas and health and to gain insight into mechanisms explaining these relationships. In this article, we bring together key results of our program regarding the relevance of three possible mechanisms: stress reduction, physical activity, and social cohesion. The program consisted of three projects in which relationships between greenspace and health were studied at national, urban, and local scales. We used a mixed-method approach, including secondary analysis, survey data, observations, and an experiment. The results confirmed that quantity as well as quality of greenspace in residential areas were positively related to health. These relationships could be (partly) explained by the fact that residents of greener areas experienced less stress and more social cohesion. In general, residents of greener areas did not engage in more physical activity. The article concludes with a discussion of the practical implications of these findings and identification of areas that need more in-depth research.

我们的研究小组在过去的几年一直致力于一个有关绿地和健康之间关系的连贯的研究计划。这种"维生素 G"方案（其中 G 代表绿色）的主要目的是，实证地验证居民区的绿色空间和健康之间的关系，并获得解释这些关系机制的洞察。在这篇文章中，我们将这个项目的三种可能机制的相关性，即压力减少，体力活动和社会凝聚力的关键成果放在一起。该项目包括三个课题，在国家，城市和地方的尺度，对绿地和健康之间的关系进行了研究。我们采用混合分析方法，包括二次分析，调查数据，观测和实验。结果证实，居民区绿地数量和质量与健康呈正相关。这些关系可能（部分）由更绿化地区的居民经历了较少的压力和更多的社会凝聚力的事实来解释。在一般情况下，更绿化地区的居民并没有参与更多的体力活动。文章结尾讨论这些调查结果的实际意义的和确认那些需要更深入的研究地区。关键词：绿地，健康，体力活动，社会凝聚力，压力。

Nuestro grupo ha estado trabajando los años pasados en un programa coherente de investigación sobre las relaciones entre espacio-verde y salud. Las metas principales de este programa de "Vitamina G" (donde la G se identifica con "green", verde, en inglés) eran verificar empíricamente las relaciones en áreas residenciales entre espacio-verde y la salud, y ganar mayor compenetración con los mecanismos que explican estas relaciones. En este artículo juntamos los resultados claves de nuestro programa en lo que concierne a la relevancia de tres posibles mecanismos explicativos: reducción del estrés, actividad física y cohesión social. El programa consistió de tres proyectos en los que las relaciones entre el espacio-verde y la salud fueron estudiadas a escala nacional, urbana y local. Utilizamos un enfoque de método mixto, que incluye análisis secundario, datos del estudio, observaciones y un experimento. Los resultados confirman que la cantidad lo mismo que la calidad del espacio-vede en áreas residenciales estuvieron positivamente relacionadas con la salud. Estas relaciones podrían ser (parcialmente) explicadas por el hecho de que los residentes de las áreas más verdes experimentaban menos estrés y mayor cohesión social. En general, los residentes de las áreas más verdes no estuvieron comprometidos con mayor actividad física. El artículo concluye con una discusión de las implicaciones prácticas de estos hallazgos y con la identificación de áreas que demandan más investigación de profundidad.

The idea that greenspace in the residential environment can promote health has a long history in environmental planning (Hartig et al. 2011). A notable example is Howard's ([1902] 1946) *Garden Cities of Tomorrow*. The author pleaded that the advantages of town and countryside should be integrated in garden cities with optimal opportunities for health and well-being. In practice, however, it has been difficult to realize this vision. Due to large-scale demands for housing and other services, urban greenspace has increasingly come under threat. What are the consequences of these developments for public health? Does a green residential environment indeed promote health? If so, why? The relationship between greenspace and health has only recently become a theme for empirical research (Hartig et al. 2011; De Vries et al. 2011). Prior to the Vitamin G project, only two epidemiological studies had investigated the direct relationship between greenspace and health (Takano, Nakamura, and Watanabe 2002; De Vries et al. 2003). These studies suggested a positive link between the amount of greenspace in the residential environment and health but also raised questions about the strength and robustness of this relation and the mechanisms behind it.

Possible Mechanisms Behind the Greenspace–Health Relationship

Three mechanisms—stress reduction, physical activity, and social cohesion—are frequently mentioned in the literature. They are well-known predictors of health (U.S. Department of Health and Human Services 1996) and might also be related to the availability of greenspace.

Stress Reduction

Many people consider contact with nature one of the most powerful ways to obtain relief from stress and mental fatigue (Grahn and Stigsdotter 2003). This mechanism can be regarded as a remnant of evolution in natural environments, during which humans have developed a partly genetic readiness to respond positively to unthreatening natural settings that were favorable to well-being and survival (Ulrich 1993). Natural scenes seem to have a unique attention-drawing quality (soft fascination; Berto, Massaccesi, and Pasini 2008), which repletes directed attention and replaces negative emotions (S. Kaplan 1995). Evidence for these restorative responses has been collected through field studies in such diverse settings as wilderness areas (Hartig, Mang, and Evans 1991) and gardens (Ottosson and Grahn 2005). Even viewing nature from the window of one's home can provide "micro-restorative" opportunities that can build resilience against stress (R. Kaplan 2001). Viewing slides or videos of natural environments leads to a faster and more complete stress recovery than viewing built environments (Ulrich et al. 1991; Van den Berg, Koole, and Van der Wulp 2003). In sum, there is substantial evidence that restoration from stress and mental fatigue is a potentially important mechanism in the relationship between greenspace and health.

Physical Exercise

People more easily undertake physical activities, such as cycling and walking, in aesthetically appealing environments (Pikora et al. 2003; Owen et al. 2004), perhaps even for longer periods of time (Pretty et al. 2007). The evidence for a positive influence of greenspace on physical activities, however, is as yet mixed and inconclusive (positive: Ellaway, Macintyre, and Bonnefoy 2005; McGinn et al. 2007; no relation: Hoehner et al. 2005; Hillsdon et al. 2006; McGinn et al. 2007; negative: Duncan and Mummery 2005). Apart from differences in the geographic context of studies, this mixed evidence might be related to the intrinsic character of the motivation to exercise. People who want to be active will find themselves a way to do so, even if it requires taking the car to travel to a faraway green area. Moreover, some greenspaces invite passive forms of recreation rather than active forms. Thus, although intuitively plausible, the literature suggests that physical activity is not a strong candidate for explaining the relationship between greenspace and health.

Social Cohesion

Meeting opportunities are important for social ties with neighbors, leading to more cohesive communities (Völker, Flap, and Lindenberg 2007). The presence of green in common spaces might attract residents to outdoor spaces, leading to more frequent contacts (Coley, Kuo, and Sullivan 1997). Studies in an underprivileged area of Chicago provide indications for a positive relation between green public facilities and social ties (Kuo, Sullivan, and Wiley 1998; Kweon, Sullivan, and Wiley 1998). Besides offering meeting opportunities, greenspace can promote a general sense of community, which might decrease feelings of loneliness and increase social support (Prezza et al. 2001). Overall, there are several indications for social cohesion as a possible mechanism underlying the relation between greenspace and health.

Against this background, the Vitamin G program aimed to empirically verify a positive relationship between greenspace in the residential environment and health and to evaluate the importance of the three mechanisms mentioned. In the remainder of this article, we first briefly describe the study method, followed by an overview of the main findings. In this overview, we bring together the findings regarding the relevance of the three explanatory mechanisms (stress reduction, physical activity, and social cohesion).

Study Design, Data, and Methods

The Vitamin G research program consisted of three projects, each with a different scope (Groenewegen et al. 2006). The first project had a national scope and used data from The Netherlands as a whole. The second project focused on eighty neighborhoods in four Dutch cities (urban scale). The third project studied allotment gardens in and around twelve Dutch cities (local scale).

In each project, greenspace indicators were linked to health indicators. All analyses were carried out at the level of individuals, and multilevel analysis was used to control for the nesting of individuals in neighborhoods wherever necessary. The analyses were controlled for characteristics known to be associated with health (e.g., age, gender, education, income, urbanity). Given the egalitarian health care system and the high density of health service provision in The Netherlands, there was no need to add additional variables regarding access to health care.

In the national study, greenspace around people's homes was related to subjective and objective health. Land use data were derived from the National Land Cover Classification database and aggregated to the percentage of greenspace in circles with a 1- and 3-km radius around the six-digit postal code of the respondents' addresses. Health data were derived from the second Dutch National Survey of General Practice (Westert et al. 2005). This was a two-stage sample of 104 family practices, a one-page mailed questionnaire among the practice population of approximately 400,000 people (in The Netherlands, virtually the whole population is registered with a practice), 300,000 respondents (76%), and health interviews among a random sample

of the practice population, resulting in 12,700 respondents (64%). Objective health data were derived from electronic medical records of the practice populations.

The urban neighborhood study related quantity and quality of greenspace to self-reported health. Data on health and explanatory variables (related to stress, physical activity, and social cohesion) were collected through mailed questionnaires (1,641 respondents) sent to residents of twenty neighborhoods (average size 2,200 inhabitants) in four cities. To measure the quantity of larger green areas (parks, etc.), we used land use data from Statistics Netherlands. Trained observers assessed the quality of three larger green areas as well as the amount and quality of streetscape greenery (not captured in the land use data) at four preselected points within each neighborhood. Quality indicators encompassed accessibility, maintenance, variation, naturalness, colorfulness, clear arrangement, shelter, absence of litter, safety, and general impression.

The allotment garden study involved a comparison of gardeners and their neighbors with no allotment. The study consisted of a survey among members of twelve allotment sites (121 respondents) and a control group of their neighbors without an allotment (63 respondents). The survey included self-reported health and measures of stress, physical activity, and social cohesion. A distinction was made between respondents younger or older than sixty-two. In addition to the survey, a field experiment among thirty allotment gardeners was conducted. The gardeners first performed a stressful task and were then randomly assigned to thirty minutes of gardening or indoor reading. Salivary cortisol levels and self-reported mood were repeatedly measured before and after the stressful task and during and after the restorative activity.

Results

Greenspace and Health

The national study showed that residents of greener areas felt healthier than residents of less green areas (Maas et al. 2006; see Figure 1). Furthermore, for fifteen out of twenty-four physician-assessed morbidity clusters, the incidence rates were negatively related to the amount of greenspace within a 1-km radius around people's homes (Maas, Verheij, et al. 2009). The relationship was strongest for anxiety disorders and depression; for depression the chances were 1.33 times higher in areas with little greenspace than in areas with very much greenspace. Across different studies, relationships

Relation between green space and health

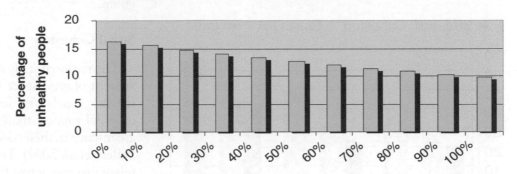

Figure 1. Estimated relationship between amount of greenspace (in a 3-km radius) and self-perceived health (percentage stating that their health is less than good) based on a logistic multilevel model controlling for age, gender, income, education, and urbanity (taken from Maas et al. 2006). (Color figure available online.)

between greenspace and health tended to be stronger for individuals of a lower socioeconomic status (SES), younger age groups, and the elderly.

In the urban neighborhood study, both the quantity and quality of nearby green areas and streetscape greenery were positively related to self-rated health, number of acute complaints, and mental health status (Van Dillen et al. 2011).

In the allotment garden study, gardeners sixty-two years and older scored significantly better than their neighbors on physical disabilities, health complaints, and family physician consultations and marginally better on perceived general health and chronic illness than their neighbors in the same age category (Van den Berg, Van Winsum-Westra, et al. 2010).

Stress Reduction

The national study investigated the extent to which greenspace can buffer adverse health impacts of stressful life events (Van den Berg, Maas, et al. 2010). Detrimental impacts of stressful life events on perceived general and acute health complaints were less strong for respondents with much greenspace in a 3-km radius around their home than for respondents with little greenspace in a 3-km radius (Figure 2). The same pattern was observed for perceived mental health, although it was marginally significant.

In the urban neighborhood study, stress turned out to be a full mediator of the relationship between the quantity of streetscape greenery and mental health and a partial mediator of the relationships with self-rated health and number of acute health complaints. Stress explained one fifth to over two fifths of the relationships of quantity of streetscape greenery with these health indicators (Figure 3). In the relationships of

health indicators with quality of streetscape greenery, stress was always a partial mediator (De Vries et al. forthcoming).

In the allotment gardening survey, 86 percent of the gardeners reported feeling less stressed after a visit to their allotment garden; 56 percent rated stress relief as a very important reason for gardening (Van den Berg, Van Winsum-Westra, et al. 2010). Older allotment gardeners had experienced less stress recently and seemed better able to cope with stress. The field experiment showed that gardening led to stronger decreases in the stress hormone cortisol than reading (Van den Berg and Custers 2011).

Physical Activity

At the national scale no relationships were found between the amount of greenspace in the area and time spent on physical activity in general, sports, and walking for commuting purposes (Maas et al. 2008). For walking

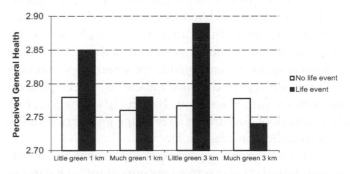

Figure 2. Estimated marginal means of perceived general health (1 = excellent; 5 = poor) as a function of stressful life events in past three months (white bars: no life events; black bars: life event(s)) and amount of greenspace in a 1-km and 3-km radius, controlling for age, gender, income, education, and of urbanity (taken from Van den Berg, Maas et al. 2010).

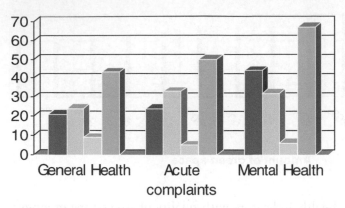

Mediated part of effect for quantity (%)

Figure 3. Indirect effect on health indicators as percentage of total effect of quantity of streetscape greenery for the three mediators separately and combined, controlling for gender, age, education, income, life events, having children living at home, smoking, and excessive drinking (taken from De Vries et al. forthcoming). PA = physical activity. (Color figure available online.)

and cycling in leisure time, residents of areas with much greenspace spent even less time on these activities than residents of areas with little greenspace. Greenspace and cycling for commuting purposes were positively related, but this could not explain the relationship between greenspace and health.

The urban neighborhood study also indicated that total physical activity could not explain the relationship between greenspace and health, as it was not related to greenspace (De Vries et al. forthcoming). Only "green" physical activities (activities that could be undertaken in the public space of the neighborhood, such as walking and cycling) were related to greenspace but to a lesser extent than stress. In all analyses, green physical activity was at best a partial mediator of the relationships between quantity and quality of streetscape greenery and the self-reported health indicators. Green physical activity explained less than 10 percent of the direct relationships of quantity of streetscape greenery with the health indicators (Figure 3).

In the allotment gardening study, 50 percent of gardeners found staying active very important for having an allotment garden (Van den Berg, Van Winsum-Westra, et al. 2010). Younger and older allotment gardeners had better scores on physical activity than the control group. In summer, 84 percent of the allotment gardeners met the Dutch public health recommendations for physi-

cal activity compared to only 62 percent of the control group.

Social Cohesion

The national study showed that residents with a high amount of greenspace in their residential environment felt less lonely and less often experienced a shortage of social support than residents with a low amount of greenspace in their residential environment (Maas, Van Dillen, et al. 2009). The amount of greenspace was not related to the actual frequency of contact with neighbors or the number of supportive interactions. All relationships between greenspace and self-reported health, number of health complaints, and mental health were at least partially mediated by loneliness and shortage of social support. The relationship between greenspace and mental health could even be fully explained by the fact that residents of greener areas less often experienced a shortage of social support.

In the urban neighborhood study, the finding that residents of neigborhoods with higher amounts of streetscape greenery reported less acute health complaints and better mental health could be fully explained by the stronger social cohesion in greener neighborhoods. Social cohesion also partly explained the positive relationship between quantity of streetscape scenery and self-rated general health. Overall, social cohesion explained one fifth to one third of the relationships between quantity of streetscape greenery and the health indicators (see Figure 3). Furthermore, social cohesion was always a partial mediator in the analyses of the relationships of health indicators with the quality of streetscape greenery (De Vries et al. forthcoming).

Only 17 percent of allotment gardeners mentioned social contacts as a very important motive for gardening (Van den Berg, Van Winsum-Westra, et al. 2010). Nevertheless, older gardeners felt significantly less lonely than neighbors in the same age group, and they also reported having marginally more social contacts with friends.

Discussion

This article has provided a summary of the results of the Vitamin G research program, with an emphasis on the mechanisms behind the relationship between greenspace and health. The studies within the program provide converging evidence for a positive relationship between greenspace in the residential environment and health, using different self-reported and objective greenspace and health indicators. Stress reduction and social cohesion were found to be the most important mechanisms.

The relationships between greenspace and health are consistent with other epidemiological studies conducted in England (Mitchell and Popham 2007), Denmark (Nielsen and Hansen 2007), and Sweden (Björk et al. 2008). However, a recent study in New Zealand found no relationship between greenspace and mortality (Richardson et al. 2010). The authors explained this in terms of diminishing marginal returns at the large amount of greenspace, characteristic for New Zealand residential neighborhoods.

Greenspace was found to be especially important for mental health. In the national study that used physician-assessed morbidity, the strongest relationships were found for anxiety disorders and depression. This is in line with findings from previous studies that have also consistently reported stronger positive impacts of contact with greenspace on mental than on physical health (e.g., Sugiyama et al. 2008).

Also consistent with previous research (e.g., Mitchell and Popham 2008), the overall results suggest that people with lower SES, younger people, and the elderly benefit more from green areas in their residential environment. Although exposure to greenspace was not measured, these findings might be tentatively explained by the fact that these groups spend more time in the vicinity of their homes and thus are more exposed to characteristics of their residential environment. Moreover, those who spend less time in the residential environment might be additionally affected by a lack of greenspace as is typical for many work environments (Kwan 2009).

The mediation analyses in the neighborhood study suggest that stress reduction and social cohesion, in that order, are the most important mechanisms in explaining the relationship between greenspace and health. These findings are consistent with findings of previous research that have consistently shown strong links between greenspace and stress and social cohesion and only weak and mixed evidence for links between greenspace and physical activity. Few studies, however, have directly tested mediational models. As a notable exception, a survey in Australia showed that recreational walking could fully explain the link between greenness and physical health, whereas the relationship between greenness and mental health was partly accounted for by recreational walking and social cohesion (Sugiyama et al. 2008). These results are consistent with the finding in the Vitamin G program that social cohesion is an important mediator of greenspace–health relationships. The findings of the Vitamin G program with respect to the mediational role of physical activities (e.g., recreational walking) were more mixed and inconclusive, however.

Strengths and Limitations

Contrary to many other studies, the Vitamin G program encompassed different geographical scales and levels of urbanity, taking into account different subjective as well as objective health indicators and various indicators of greenspace. Most of the Vitamin G studies also used large data sets and state-of-the-art statistical (multilevel and mediational) analyses. Same-source bias was avoided by using objective indicators for greenspace along with observations. These aspects contribute to the validity and practical applicability of the results.

The most important limitations relate to exposure and causality, probably the two most difficult issues to address in studies of environment and health. The length and intensity of residents' exposure to green elements was not directly measured. With regard to causality, selection cannot be ruled out as an alternative explanation for our results but can only be made less plausible by using adequate control variables in the analyses.

Other limitations include the fact that we investigated only three mechanisms, although there are several other possible candidates (e.g., connectedness to nature, increased vitality; cf. Hartig et al. 2011). Furthermore, we have only looked at relatively nearby greenspace. For some purposes, people might seek out more distant areas. Moreover, our studies were done in a densely populated country with very little (wild) nature. As a consequence, the importance of nearby greenspace for health might have been overestimated as compared to other countries with more abundant nature (Richardson et al. 2010).

Future Research and Practical Implications

Future research should address these limitations. Better measurements of exposure are possible through Global Positioning System (GPS) measurements, as already used in transport geography. Using such measurements could provide detailed information on the time people spend in particular environments during daily activities (Kwan 2009). Direct selection effects can be ruled out with longitudinal studies, which include geocoded information.

With respect to implications for spatial planning and public health, it is important to study in more detail which aspects of greenspace are related to which mechanisms and which health outcomes. An important general guideline that can be derived from the Vitamin G program is that the quality of urban greenspace, besides quantity, has an additional beneficial effect on health. Proximity (comparing distances of 1 and 3 km) did not make much difference for self-rated health, but nearby green areas seem to be more important for physician-assessed morbidity, whereas less proximate areas seem to be more important for buffering stress. Moreover, we still do not know much about the relative contribution of greenspace compared to other environmental characteristics, both socioeconomic and physical.

More insight is also needed in possible international variation in the greenspace–health relationship. Research in northwestern Europe indicates that the relationship exists, irrespective of large differences in population density and types of greenspace, but as greenspace might affect different individuals in different ways, more attention to the interaction between individuals and their environment is required. By answering these questions, future research could provide a fuller understanding of the intricacies of the greenspace–health relationship and help to design effective and health-promoting greenspace policies.

References

Berto, R., S. Massaccesi, and M. Pasini. 2008. Do eye movements measured across high and low fascination photographs differ? Addressing Kaplan's fascination hypothesis. *Journal of Environmental Psychology* 28: 185–91.

Björk, J., M. Albin, P. Grahn, H. Jacobsson, J. Ardö, J. Wadbro, P.-O. Östergren, and E. Skärbäck. 2008. Recreational values of the natural environment in relation to neighbourhood satisfaction, physical activity, obesity and wellbeing. *Journal of Epidemiology and Community Health* 62 (e2): 1–7.

Coley, R. L., F. E. Kuo, and W. C. Sullivan. 1997. Where does community grow? The social context created by nature in urban public housing. *Environment and Behaviour* 29:468–94.

De Vries, S., T. Classen, S.-M. Eigenheer-Hug, K. Korpela, J. Maas, R. Mitchell, and P. Schantz. 2011. Contributions of natural environments to physical activity; Theory and evidence base. In *Forests, trees and human health*, ed. K. Nilsson, M. Sangster, C. Gallis, S. De Vries, K. Seeland, and J. Schipperijn, 205–44. New York: Springer.

De Vries, S., S. M. E. Van Dillen, P. P. Groenewegen, and P. Spreeuwenberg. Forthcoming. Streetscape greenery and human health: Stress, social cohesion and physical activity as possible mediators.

De Vries, S., R. A. Verheij, P. P. Groenewegen, and P. Spreeuwenberg. 2003. Natural environments—Healthy environments? An exploratory analysis of the relationship between greenspace and health. *Environment and Planning A* 35:1717–31.

Duncan, M., and K. Mummery. 2005. Psychosocial and environmental factors associated with physical activity among city dwellers in regional Queensland. *Preventive Medicine* 40:363–72.

Ellaway, A., S. Macintyre, and X. Bonnefoy. 2005. Graffiti, greenery, and obesity in adults: Secondary analysis of European cross sectional survey. *British Medical Journal* 31:611–12.

Grahn, P., and U. A. Stigsdotter. 2003. Landscape planning and stress. *Urban Forestry and Urban Greening* 2:1–18.

Groenewegen, P. P., A. E. Van den Berg, S. De Vries, and R. A. Verheij. 2006. Vitamin G: Effects of green space on health, well-being, and social safety. *BMC Public Health* 6:149.

Hartig, T., M. Mang, and G. W. Evans. 1991. Restorative effects of natural environment experiences. *Environment and Behavior* 23:3–27.

Hartig, T., A. E. Van den Berg, C. M. Hagerhall, M. Tomalak, N. Bauer, R. Hansmann, A. Ojala, et al. 2011. Health benefits of nature experiences: Psychological, social and cultural processes. In *Forests, trees and human health*, ed. K. Nilsson, M. Sangster, C. Gallis, S. De Vries, K. Seeland, and J. Schipperijn, 127–68. New York: Springer.

Hillsdon, M., J. Panter, C. Foster, and A. Jones. 2006. The relationship between access and quality of urban green space with population physical activity. *Journal of the Royal Institute of Public Health* 120:1127–32.

Hoehner, C. M., L. K. Brennan Ramirez, M. B. Elliott, S. L. Handy, and R. C. Brownson. 2005. Perceived and objective environmental measures and physical activity among urban adults. *American Journal of Preventive Medicine* 28:105–16.

Howard, E. [1902] 1946. *Garden cities of tomorrow*. London: Faber and Faber.

Kaplan, R. 2001. The nature of the view from home—Psychological benefits. *Environment and Behavior* 33:507–42.

Kaplan, S. 1995. The restorative benefits of nature: Toward an integrative framework. *Journal of Environmental Psychology* 15:169–82.

Kuo, F. E., W. C. Sullivan, and A. Wiley. 1998. Fertile ground for community: Inner-city neighborhood common spaces. *American Journal of Community Psychology* 26:823–51.

Kwan, M.-P. 2009. From place-based to people-based exposure measures. *Social Science and Medicine* 69:1311–13.

Kweon, B. S., W. C. Sullivan, and A. Wiley. 1998. Green common spaces and the social integration of inner-city older adults. *Environment and Behavior* 30: 823–58.

Maas, J., S. M. E. Van Dillen, R. A. Verheij, and P. P. Groenewegen. 2009. Social contacts as a possible mechanism behind the relation between green space and health: A multilevel analysis. *Health & Place* 15:586–92.

Maas, J., R. A. Verheij, S. De Vries, P. Spreeuwenberg, F. Schellevis, and P. P. Groenewegen. 2009. Morbidity is related to a green living environment. *Journal of Epidemiology and Community Health* 63:967–73.

Maas, J., R. A. Verheij, P. P. Groenewegen, S. De Vries, and P. Spreeuwenberg. 2006. Green space, urbanity and health: How strong is the relation? *Journal of Epidemiology and Community Health* 60:587–92.

Maas, J., R. A. Verheij, P. Spreeuwenberg, and P. P. Groenewegen. 2008. Physical activity as a possible mechanism behind the relationship between green space and health: A multilevel analysis. *BMC Public Health* 8:206.

McGinn, A. P., K. R. Evenson, A. H. Herring, and S. L. Huston. 2007. The relationship between leisure, walking and transportation activity with the natural environment. *Health & Place* 13:588–602.

Mitchell, R., and F. Popham. 2007. Greenspace, urbanity and health: Relationships in England. *Journal of Epidemiology and Community Health* 61:681–83.

———. 2008. Effect of exposure to natural environment on health inequalities: An observational population study. *The Lancet* 372 (9650): 1655–60.

Nielsen, T. S., and K. B. Hansen. 2007. Do green areas affect health? Results from a Danish survey on the use of green areas and health indicators. *Health and Place* 13:839–50.

Ottosson, J., and P. Grahn. 2005. A comparison of leisure time spent in a garden with leisure time spent indoors: On measures of restoration in residents in geriatric care. *Landscape Research* 30:23–55.

Owen, N., N. Humpel, E. Leslie, A. Bauman, and J. Sallis. 2004. Understanding environmental influences on walking: Review and research agenda. *American Journal of Preventive Medicine* 27:67–76.

Pikora, T., B. Giles-Corti, F. Bull, K. Jamrozik, and R. Donovan. 2003. Developing a framework for assessment of the environmental determinants of walking and cycling. *Social Science and Medicine* 56:1693–1703.

Pretty, J., J. Peacock, R. Hine, M. Sellens, N. South, and M. Griffin. 2007. Green exercise in the UK countryside: Effects on health and psychological well-being, and implications for policy and planning. *Journal of Environmental Planning and Management* 50:211–31.

Prezza, M., M. Amici, T. Roberti, and G. Tedeschi. 2001. Sense of community referred to the whole town: Its relations with neighboring, loneliness, life satisfaction, and area of residence. *Journal of Community Psychology* 29:29–52.

Richardson, E., J. Pearce, R. Mitchell, P. Day, and S. Kingham. 2010. The association between green space and cause-specific mortality in urban New Zealand: An ecological analysis of green space utility. *BMC Public Health* 10:240.

Sugiyama, T., E. Leslie, B. Giles-Corti, and N. Owen. 2008. Associations of neighbourhood greenness with physical and mental health: Do walking, social coherence and local social interaction explain the relationships? *Journal of Epidemiology and Community Health* 62:e9.

Takano, T., K. Nakamura, and M. Watanabe. 2002. Urban residential environments and senior citizens' longevity in megacity areas: The importance of walkable green spaces. *Journal of Epidemiology and Community Health* 56:913–18.

Ulrich, R. S. 1993. Biophilia, biophobia and natural landscapes. In *The biophilia hypothesis*, ed. S. R. Kellert and E. O. Wilson, 73–137. Washington, DC: Island.

Ulrich, R. S., R. F. Simons, B. D. Losito, E. Fiorito, M. A. Miles, and M. Zelson. 1991. Stress recovery during exposure to natural and urban environments. *Journal of Environmental Psychology* 11:201–03.

U.S. Department of Health and Human Services. 1996. *Physical activity and health: A report of the Surgeon General.* Atlanta, GA: U.S. Department of Health and Human Services, Centers for Disease Control and Prevention, National Center for Chronic Disease Prevention and Health Promotion.

Van den Berg, A. E., and M. H. G. Custers. 2011. Gardening promotes neuroendocrine and affective restoration from stress. *Journal of Health Psychology* 16:3–11.

Van den Berg, A. E., S. L. Koole, and N. Y. Van der Wulp. 2003. Environmental preference and restoration: How are they related? *Journal of Environmental Psychology* 23:2135–46.

Van den Berg, A. E., J. Maas, R. A. Verheij, and P. P. Groenewegen. 2010. Green space as a buffer between stressful life events and health. *Social Science and Medicine* 70:1203–10.

Van den Berg, A. E., M. Van Winsum-Westra, S. De Vries, and S. Van Dillen. 2010. Allotment gardening and health: A comparative survey among allotment gardeners and their neighbors without an allotment. *Environmental Health* 9:74.

Van Dillen, S. M. E., S. De Vries, P. P. Groenewegen, and P. Spreeuwenberg. 2011. Greenspace in urban neighbourhoods and residents' health: Adding quality to quantity. *Journal of Epidemiology and Community Health* (epub ahead of print).

Völker, B., H. D. Flap, and S. Lindenberg. 2007. When are neighbourhoods communities? Community in Dutch neighbourhoods. *European Sociological Review* 23:99–114.

Westert, G. P., F. G. Schellevis, D. H. De Bakker, P. P. Groenewegen, J. M. Bensing, and J. Van der Zee. 2005. Monitoring health inequalities through general practice: The second Dutch national survey of general practice. *European Journal of Public Health* 15: 59–65.

Integration of Spatial and Social Network Analysis in Disease Transmission Studies

Michael Emch,* Elisabeth D. Root,† Sophia Giebultowicz,* Mohammad Ali,‡
Carolina Perez-Heydrich,§ and Mohammad Yunus#

*Department of Geography and Carolina Population Center, University of North Carolina at Chapel Hill
†Department of Geography and Institute of Behavioral Sciences, University of Colorado at Boulder
‡International Vaccine Institute, Seoul, Korea
§Carolina Population Center and Department of Biostatistics, University of North Carolina at Chapel Hill
#International Centre for Diarrhoeal Disease Research, Bangladesh

This study presents a case study of how social network and spatial analytical methods can be used simultaneously for disease transmission modeling. The article first reviews strategies employed in previous studies and then offers the example of transmission of two bacterial diarrheal diseases in rural Bangladesh. The goal is to understand how diseases vary socially above and beyond the effects of the local neighborhood context. Patterns of cholera and shigellosis incidence are analyzed in space and within kinship-based social networks in Matlab, Bangladesh. Data include a spatially referenced longitudinal demographic database that consists of approximately 200,000 people and laboratory-confirmed cholera and shigellosis cases from 1983 to 2003. Matrices are created of kinship ties among households using a complete network design and distance matrices are also created to model spatial relationships. Moran's I statistics are calculated to measure clustering within both social and spatial matrices. Combined spatial effects and spatial disturbance models are built to simultaneously analyze spatial and social effects while controlling for local environmental context. Results indicate that cholera and shigellosis always cluster in space and only sometimes within social networks. This suggests that the local environment is most important for understanding transmission of both diseases, although kinship-based social networks also influence their transmission. Simultaneous spatial and social network analysis can help us better understand disease transmission and this study offers several strategies on how.

这项研究提出了一个有关社会网络和空间分析方法是怎样可同时用于疾病的传播模型的案例研究。文章首先回顾了应用于以往研究中的策略，然后提供了两种细菌性腹泻病在孟加拉国农村传播的例子。我们的目标是了解疾病如何随社会不同，并超出当地社区范围的影响。在空间和亲属关系为基础的社会网络内，用 Matlab 分析孟加拉国的霍乱和志贺氏菌的发病模式。数据包括引用空间纵向人口数据库，它由约 20 万人和从 1983 年至 2003 年实验室确诊的霍乱和志贺氏菌病例组成。使用一个完整的网络设计创建家庭之间的血缘关系矩阵，也创建距离矩阵来模拟空间关系。在社会和空间的矩阵内，计算 Moran's I 统计值。建立综合空间效果和空间干扰模型，同时控制当地环境背景。结果表明，霍乱和志贺氏菌总是在空间上集聚，但只是有时在社交网络集聚。这表明，当地的环境是了解这两种疾病传播的最重要因子，虽然以血缘关系为基础的社会网络也影响其传输。空间和社会网络分析一起，可以帮助我们更好地了解疾病传播。这项研究提供了几个相关战略。*关键词：孟加拉国，霍乱，邻里效应，志贺氏菌病，社交网络。*

Este artículo presenta un estudio de caso sobre la manera de utilizar simultáneamente los métodos de redes sociales y el análisis espacial para modelar la transmisión de enfermedades. En el artículo se revisan primero las estrategias utilizadas en estudios anteriores, para luego ofrecer un ejemplo de transmisión de dos enfermedades diarreicas bacteriales en las zonas rurales de Bangladesh. El objetivo es entender cómo varían socialmente las enfermedades más allá de los efectos del contexto del vecindario local. Los patrones de la incidencia del cólera y la shigellosis se analizaron espacialmente y dentro de las redes sociales basadas en parentesco en Matlab, Bangladesh. La información incluye una base de datos demográfica longitudinal referenciada espacialmente, consistente en cerca de 200.000 personas y casos de cólera y shigellosis confirmados por exámenes de laboratorio entre 1983 y 2003. Las matrices se crearon a partir de lazos de parentesco entre hogares que utilizan un diseño de red completo y también se hicieron matrices de distancia para modelar las relaciones espaciales. Se calcularon estadísticas I de Moran para medir agrupamientos dentro de ambas matrices, la social y la espacial. Los efectos espaciales y los modelos de alteración espacial combinados se construyeron para analizar simultáneamente los efectos espaciales y sociales mientras se ejercía control del contexto ambiental local. Los resultados indican que el cólera y la shigellosis siempre se agrupan en el espacio y solo a veces dentro de las redes sociales. Esto sugiere que el entorno local es lo más importante para comprender la transmisión de ambas enfermedades, aunque las redes sociales basadas en parentesco también influyen en su transmisión. El análisis espacial simultáneo con el de las redes sociales puede ayudarnos a entender mejor la transmisión de enfermedades y este estudio ofrece varias estrategias sobre cómo hacerlo.

This study demonstrates how combined social network and spatial analytical methods can be used simultaneously for disease transmission modeling. Separately, these two methods have been extensively used to incorporate physical and social context into health studies, but researchers are only just beginning to explore using them simultaneously. Infectious disease transmission can be influenced by neighborhood-level environmental circumstances, as well as personal interactions within social networks. Individuals with social ties are more likely to interact with one another and share common household assets, thereby increasing the likelihood of transmitting infectious diseases. This article first reviews strategies employed in previous studies to examine the effect of social ties and geographic proximity on disease transmission and then presents a case study of transmission of two bacterial diarrheal diseases in rural Bangladesh. This study investigates cholera and shigellosis transmission within social networks while controlling for spatial context in rural Bangladesh from 1983 to 2003. Cholera and shigellosis have been linked to the local environment and both have been shown to cluster in space (Miller, Feachem, and Drasar 1985; Emch and Ali 2001; Carrel et al. 2009; Ruiz-Moreno et al. 2010). They also can be spread through person-to-person contact, however, and transmission is thus also controlled by social processes (Collins 1998). The use of integrated social and spatial analysis provides evidence as to whether local environmental transmission is more important than person-to-person diarrheal disease transmission.

Background

Cholera and Shigellosis

Cholera is a bacterial disease, characterized by watery diarrhea and dehydration. It has been linked to the aquatic environment and is a significant cause of morbidity and mortality in much of Asia and Africa (Colwell et al. 1990; Islam, Drasar, and Sack 1993;

Colwell 1996). A study in rural Bangladesh reported a case-fatality rate (CFR) of 4 percent, but studies have reported wide ranges (Siddique et al. 1992). In 2009, 221,226 cases of cholera were reported to the World Health Organization (WHO), with a CFR of 2.24 percent (WHO 2009). In Somalia, where the public health system has virtually collapsed, a CFR of 13.6 percent was reported in a recent epidemic (Reuters 2007). The etiologic agent of cholera, *Vibrio cholerae*, can persist in brackish, coastal, and freshwater environments for significant periods of time. Transmission occurs via the fecal–oral route, through ingestion of contaminated surface water that is used for bathing, washing clothes and dishes, cooking, and sometimes drinking (Hoque et al. 1996). Person-to-person transmission has been reported during social events and through contamination of water and food in households (St. Louis et al. 1990; Swerdlow et al. 1992). The infective dose is very high at approximately 100,000 to 1 million bacteria.

Shigellosis is an infectious disease caused by various species of bacteria in the genus *Shigella*: *S. boydii*, *S. dysenteriae*, *S. flexneri*, and *S. sonnei*. The bacteria are transmitted from an infected person to another, usually through the fecal–oral route. The infective dose for shigellosis is very low, at only 10 to 200 organisms. Humans are the natural host and the bacteria are present in the diarrheal stools of infected persons while they are ill and for a week or two afterward. Most shigellosis infections are the result of the bacteria passing from stools or soiled fingers of one person to the mouth of another person. This happens when basic hygiene and hand-washing habits are inadequate. Breastfeeding status, age, and nutritional status are the dominant predictors of pediatric shigellosis. Family and environmental variables can also have an impact on the risk of shigellosis. Children are at lower risk for shigellosis if the sentinel patient is an adult, because sick adults might have less personal contact with young children (Ahmed et al. 1997). The CFR varies significantly but can be as high as 5 to 15 percent for some strains in areas where medical care is poor (Rahaman et al. 1975; Kotloff et al. 1999).

Social Network Analysis

Social network analysis is used to measure relationships between social entities (Wasserman and Faust 1994; Hanneman 2001) and is particularly useful for measuring social relationships that influence disease outcomes or health interventions (Morris 2004). In this study, we measure kinship relationships between households and how those relationships affect cholera and shigellosis disease incidence. Relationships with kin lay the foundation for most social interaction in certain societies and might illuminate the risk of transmission through situations such as visits and shared meals. In social network analysis, linkages between social actors are modeled using empirical data. In our study the actors are people with a kinship relationship that fosters interactions between physical residences. There are different types of networks and thus there are different analytical procedures based on how network data are collected. Complete (or full) networks require information about each actor's ties with all other actors (Hanneman 2001) instead of just a sample of ties as is done with local or partial network designs (Morris 2004). The field research area for this study offers a unique opportunity to build a complete social network because all kinship ties are meticulously recorded and internal migration tracked over time. We can therefore build ties between individuals who live in the same households as well as households with kinship ties to an individual who has migrated to another area. Well-known examples of complete network designs include the National Longitudinal Study of Adolescent Health (AddHealth; Bearman and Burns 1998; Bearman et al. 2004) and the Nang Rong, Thailand, study (Faust et al. 1999; Rindfuss et al. 2004; Entwisle et al. 2007).

Several studies have considered social networks and spatial context simultaneously (e.g., Faust et al. 1999), but few have investigated social networks and geographic context to examine health outcomes. Shared geographic space functions the same way as shared religious beliefs or ethnicity; actors in closer proximity use less effort to connect to one another than to those located further away and might influence and affect each other to a greater degree (McPherson, Smith-Lovin, and Cook 2001). Early research examining networks and space primarily focused on the influence of distance in relationship formation, using survey data on affiliations and measuring distance between associated units. Ties between students in common housing communities were shown to be related to the spatial proximity of dwellings (Caplow and Forman 1950; Festinger,

Schachter, and Back 1950), and individuals of different races who lived closer to one another were shown to be more likely to form friendships and thus decrease interracial hostility (Deutsch and Collins 1951). Studies on communication technology have found that despite technological advances that make communication easier, relationships are often formed and maintained with those in closer proximity (Wellman, Carrington, and Hall 1988; Wellman 1996; Mok, Wellman, and Basu 2004).

Christakis and Fowler (2007) incorporated social as well as spatial distance into their analysis of 5,124 individuals from the Framingham Heart Study. They studied social ties among family members, friends, and neighbors and the probability of obesity within those networks. Using a geographic information system (GIS), residential data, and geocoding techniques, spatial distance between the ego (individual of interest) and alter (the connected individual) was measured as a Euclidian distance. They found that although social proximity affected the likelihood of obesity in both parties, geographic distance appeared to have no effect. Radil, Flint, and Tita (2010) studied Los Angeles gang networks and the location of each gang's "turf," combining social and spatial methods to better understand violence and youth behavior. They analyzed two types of "embeddedness" in relation to geographic patterns of violence, namely, how gangs are socially positioned within the network and their relative geographic location. Liu, King, and Bearman (2010) examined autism in California, finding that children living in closer proximity to children with autism were more likely to be diagnosed with the disorder. The driving factor behind spatial clustering and autism prevalence was social influence and information sharing among families connected in space, therefore influencing the prevalence of diagnoses.

Several infectious disease studies have shown spatial clustering of individuals who are socially connected, often through a physical or sexual relationship. For example, sexual partnerships between individuals are shown to correspond to spatial proximity and common neighborhoods. Wylie, Cabral, and Jolly (2005) identified distinct geographic clusters of different chlamydia strains in Manitoba that were transmitted to sexual partners. Rothenberg et al. (2005) showed geographic and social clustering of HIV/AIDS in Colorado Springs intravenous drug networks. Other studies have used physical locations as nodes in a network or used measures of centrality and geography as variables related to an outcome. Klovdahl (2001) found social settings such as bars and clubs to be points of tuberculosis transmission in a network where actors consisted of both patients and places they frequented in the Houston metropolitan area. Bates et al. (2007) found that diarrheal disease risk in rural Ecuador was many times higher in more socially connected and densely populated communities.

This study is one of the first to simultaneously examine social and spatial clustering of disease. We first measure each form of clustering separately to understand what patterns exist before controlling for the other. Social clustering observed on its own can be influenced by the shared environment; we therefore also use a spatial effects–spatial disturbance model to estimate social effects while controlling for both known confounding variables and unknown underlying spatial effects. Specific study questions include the following:

1. Do shigellosis and cholera cluster in kinship networks?
2. Do shigellosis and cholera cluster geographically?
3. After controlling for known covariates and potential spatial autocorrelation, do shigellosis and cholera cluster within kinship networks?

Because the primary transmission pathway for cholera is environmental, whereas shigellosis is commonly transmitted through person-to-person contact, we hypothesize that cholera will exhibit stronger spatial clustering and shigellosis will cluster more within social networks.

Data

This study was conducted at the International Centre for Diarrhoeal Disease Research, Bangladesh (ICDDR, B). The research site for the ICDDR, B and for this project is called Matlab because the Centre's hospital is located in Matlab Town (Figure 1). Matlab is in south-central Bangladesh, approximately 50 km southeast of Dhaka, adjacent to where the Ganges River meets the Meghna River, forming the Lower Meghna River. A demographic surveillance system (DSS) has recorded all vital events of the study area population since 1966; the study area population is approximately 200,000. People are visited monthly by community health workers and if they have severe diarrhea they are treated at a hospital run by the Centre. The Matlab DSS is the most comprehensive longitudinal demographic database of a large population in the developing world. The people in the study area live in clusters of patrilineally related groups of households called baris that have been mapped and

included in a comprehensive GIS of the study area (Figure 1; Emch 1999; Ali et al. 2001). Other spatial features included in the GIS database include the roads, rivers, ponds, and the location and depths of tube wells used for drinking water. Individual-level study data include the bari of residence of all Matlab residents between 1 January 1983 and 31 December 2003, dates of in- and outmigration, and all laboratory-confirmed cholera and shigellosis cases at the ICDDR, B diarrhea hospital.

Methods

The methods employed for this study include (1) building a kinship-based complete network for all baris in Matlab, (2) creating spatial neighborhoods at multiple scales, (3) calculating Moran's I autocorrelation coefficients using the social network and spatial neighborhoods separately, and (4) integrating the social and spatial components by building a spatial effects–spatial disturbance model.

Network and Matrix Creation

The Matlab DSS was used to construct the kinship-based network. It records all kinship ties and contains the exact dates each person resided in different baris over time; therefore, a person can be traced from bari to bari over the course of his or her life in the Matlab study area. Migrations between baris are primarily kinship based; for example, due to marriage. The actors in the network are thus individuals with some kinship relationship that leads to relocation from one physical residence to another. The kinship-based network was created and analyzed under the assumption that when an individual moves, he or she maintains contact with the previous bari of residence at least for a short period of time. The mutual interaction between the old and new baris forms a nondirectional social connection. Kinship-based relations are appropriate in this study population, given that these types of networks are an integral component of social interaction in rural settings in the developing world (Hollinger and Haller 1990; Guest and Chamratrithirong 1992). Traditional customs such as shared meals and family visits encourage social and physical interaction between kin. In these settings, transmission of infectious disease is possible, depending on sanitary conditions and practices, the infective dose of the disease, and individual characteristics. Cholera and shigellosis are both able to spread through contaminated food and water, potentially raising the risk for populations living

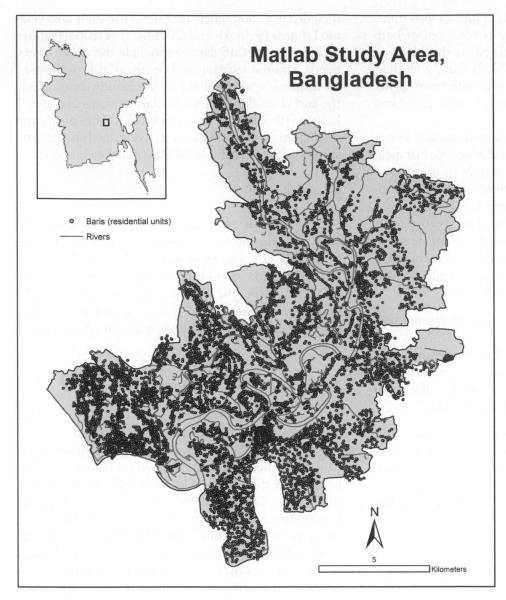

Figure 1. Matlab study area and GIS database. (Color figure available online.)

or interacting with infected individuals. The rationale for employing social network methods in this case is that these models will capture social and physical contact both within and beyond the household.

The kinship network is based on individual-level migrations linking *baris*, which are the "nodes," or units of analysis in the network. Each individual-level migration from *bari* x to *bari* y creates a social linkage between those two *baris*. These linkages are nondirectional, meaning that the connection is mutual, rather than one-directional. Each linkage of this type is called a *dyad*. A complete list of all dyads, or an edgelist, can be represented as an $n \times n$ matrix, where n equals the number of actors. Graphs are another form of visualizing networks. For example, the eight *baris* included

in Figure 2 show the family relationships between all *baris*. *Bari* 5 has no family ties to any of the other 7 *baris*, whereas *baris* 1, 2, and 3 are related, as are *baris* 6 and 7 and *baris* 4 and 8. We can represent relationships not only as graphs that can be visually inspected but also in the form of matrices, which allow mathematical and computational methods to be employed that summarize and find patterns (Wasserman and Faust 1994; Hanneman 2001). A network matrix is a rectangular arrangement of a set of elements represented as cells that are organized within rows and columns. Table 1 shows a hypothetical matrix of relationships between *baris* 1 through 8 and is the matrix representation of the graph shown in Figure 2. The lines between nodes in Figure 2 have arrows pointing toward each node because the

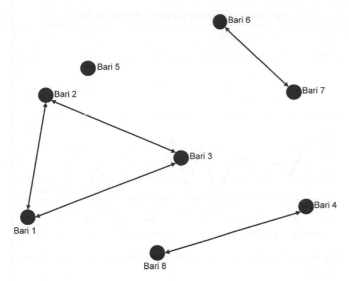

Figure 2. Network graph of eight extended household units.

social relationships between family members are nondirectional. In a social adjacency matrix, 1 represents the presence of a single, nondirectional social connection between two *baris* and 0 represents no social connection. In Table 1, a value of 1 is given if there is a family relationship, and a value of 0 denotes that there is no relationship.

For this study, multiple social networks were created and constructed for each year independent of others; that is, the linkages between *baris* for 1983 were not considered when constructing the social network for *baris* in 1984, and so forth. This was to account for uncertainty regarding how long an active social linkage based on a migration might last. Each year was then analyzed independently to investigate trends over time, using matrices created from the dyads of *bari*-to-*bari* connections.

It is important to note that we restrict our social network analyses to kinship networks due to logistical difficulties in obtaining more refined network data, such as

Table 1. Binary network matrix of eight extended household units

	Bari 1	Bari 2	Bari 3	Bari 4	Bari 5	Bari 6	Bari 7	Bari 8
Bari 1	0	1	1	0	0	0	0	0
Bari 2	1	0	1	0	0	0	0	0
Bari 3	1	1	0	0	0	0	0	0
Bari 4	0	0	0	0	0	0	0	1
Bari 5	0	0	0	0	0	0	0	0
Bari 6	0	0	0	0	0	0	1	0
Bari 7	0	0	0	0	0	1	0	0
Bari 8	0	0	0	1	0	0	0	0

relationships associated with school or work networks. Given the transmission dynamics of both cholera and shigellosis, namely, that both can be transmitted via fecal–oral routes, we would expect family ties to exposed or infected households to be associated with an increased risk of infection. Higher expected frequency of contacts among related households would reflect a greater propensity for individuals to be inadvertently exposed to respective pathogens through limited hygiene or food preparation practices of family members.

Moran's I

Spatial processes were expected to contribute significantly to cholera and shigellosis incidence because of the importance of environmental transmission to the disease dynamics. For instance, because water sources can be shared by neighboring *baris*, contamination of a water source will likely expose all *baris* within a neighborhood to cholera, shigellosis, or both. Thus, environmental transmission of these pathogens is likely to be a spatially controlled process. Clustering of cholera and shigellosis in space was examined using the Moran's I statistic at various neighborhood scales to compare the effects of the local environment (Moran 1950). To measure spatial clustering of cholera and shigellosis, for each *bari* all other *baris* located within a 500-, 1,000-, and 2,000-m buffer were identified. This was done to make three distance-band spatial matrices of all *baris*, where 1 represented a common neighborhood between two *baris* and 0 represented no common neighborhood. The three different buffers, or "neighborhoods," were used to compare spatial clustering at various scales. The matrix created from the social network was based on the existence of a single tie or more in a given year, with 1 representing a connection. The total number of *baris* evaluated in both the social and spatial analysis was 8,873. The dependent variable of interest was the rate of cholera or shigellosis in a *bari* during a specific year, aggregated from all individual recorded cases. For the entire twenty-one-year study period, there were 8,765 cases of cholera and 5,492 cases of shigellosis in Matlab. For each year, the four 8,873 × 8,873 matrices, one of social adjacency and three representing the different shared spatial neighborhoods, were row-standardized into weight matrices. This gave both social affiliates and spatial neighbors equal "weight" in terms of their influence on a particular *bari*. The matrices could then each be multiplied by the $n \times 1$ vector of disease rates per *bari*, generating a lag operator that represents the average rate of disease in neighboring *baris* or those

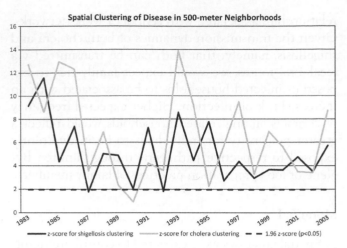

Figure 3. Moran's I z scores within 500 m for shigellosis and cholera.

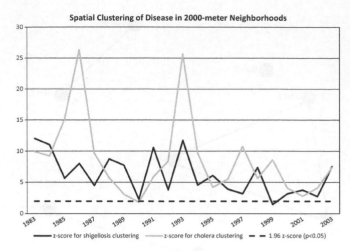

Figure 5. Moran's I z scores within 2,000 m for shigellosis and cholera.

either socially affiliated (social lag) or spatially connected (spatial lag).

The global Moran's I statistic was applied to cholera and shigellosis rates to identify both the social and spatial clustering. Typically used as a measure of spatial autocorrelation, Moran's I can also be applied to detect clustering of other elements, such as language or cultural variables, within other types of networks, such as friendships or kinship (Dow 2007). Z scores for hypothesis tests were derived using 10,000 runs of Monte Carlo simulations, under a null hypothesis of no network autocorrelation either in geographic or social space. The test was run for each of the twenty-one years using the social connectivity matrix as well as the three spatial distance matrices. Each separate analysis produced both the coefficient representing the extent of clustering and a z

score for each year of the data. The analyses were done in Stata 10 (StataCorp 2007) and MATLAB 7.7.0.

Spatial Effects–Spatial Disturbance Models

Social clustering might be due to spatial clustering; that is, individuals who are socially connected are more likely to live close to one another and thus are more likely to be affected by the same environmental risk factors. Not all of the critical spatial factors are always known, however. In such cases, it is beneficial to also acknowledge the error, or disturbances, that exist in a spatial context but are not captured by a model. We therefore utilized maximum likelihood estimation procedures and spatial interaction estimation methods outlined by Doreian (1980, 1981). A combined linear spatial effects–spatial disturbance model was built to estimate social effects while controlling for both known independent variables and unknown underlying spatial effects (Doreian 1980, 1981; Anselin 1988). In other words, a disturbance model can measure the social

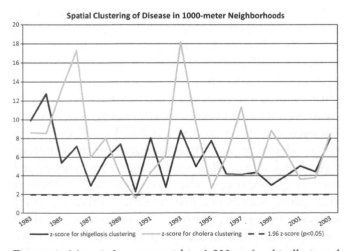

Figure 4. Moran's I z scores within 1,000 m for shigellosis and cholera.

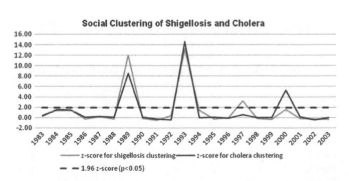

Figure 6. Moran's I z scores for social network matrix for shigellosis and cholera.

effect of each disease above and beyond the neighborhood effect, whether those variables are known or not.

One social network matrix (W_1) and another representing the shared spatial neighborhood at a 1,000-m scale (W_2) were row-standardized into weight matrices. This granted social affiliates and spatial neighbors equal levels of influence on a particular *bari*. The matrices were then each multiplied by the $n \times 1$ vector (Y) of either cholera or shigellosis rates per *bari* to generate a lag operator representing the average rate of each disease in *baris* either socially affiliated (social lag) or spatially connected (spatial lag). For each *bari*, control variables (defined by the design matrix X) were created based on proximity to environmental features (i.e., rivers, ponds, and tube wells) that have been found to be associated with cholera or shigellosis transmission in previous research (Emch 1999; Emch, Ali, and Yunus 2008). Distance to roads, which could facilitate social interaction, was also included as a covariate. We also included depth of the nearest tube well as a covariate, because tube well depth is expected to be inversely associated with disease risk (Escamilla et al. 2011). Using the preceding data, the spatial disturbance model was of the form $Y = \rho_1 W_1 Y + \rho_2 W_2 Y + X\beta$. The social effect ($\rho_1$) was assessed in terms of both existence and strength, whereas the spatial disturbance (ρ_2) was used to correct the bias potentially created by autocorrelation at the neighborhood level. Using the social and spatial weights matrices, the model was run for each year using MATLAB 7.7.0 and the LeSage Econometrics Toolbox (LeSage 1999; MATLAB 2010).

Results

Test statistics associated with the Moran's I analyses indicated that across most years cholera and shigellosis cases were significantly clustered in space (Figures 3–5); however, except for four time periods, these diseases failed to demonstrate significant clustering across social networks (Figure 6). In the 500-m neighborhood analysis there was significant spatial clustering of both shigellosis and cholera over the course of twenty-one years, with the exception of 1990 for cholera, when clustering was not significant (Figure 3). Similarly, at the 1,000-m scale, both diseases continued to cluster significantly in space during all years except 1990 (Figure 4). Figure 5 shows significant spatial clustering for all years for both diseases at the 2,000-m scale. The spatial clustering of both diseases was thus robust across

Table 2. Spatial error and significance, social effect and significance, and additional environmental variables that affect cholera rate

Year	Spatial error	Significance	Social effect	Signifiance	Other variables
1983	0.47	**	0.01	**	Pond,** tube well*
1984	0.48	**	0.04		Road*
1985	0.58	**	0.05	**	Road,* pond*
1986	0.61	**	−0.01		Road**
1987	0.54	**	0.00		
1988	0.53	**	0.00		
1989	0.46	**	0.28	**	
1990	0.34	**	0.00		
1991	0.42	**	−0.01		
1992	0.52	**	−0.01		Pond,* tube well*
1993	0.56	**	0.15	**	
1994	0.53	**	−0.01		
1995	0.34	**	−0.01		
1996	0.45	**	0.00		Tube well depth**
1997	0.54	**	0.03		
1998	0.42	**	−0.01		River,* tube well depth**
1999	0.57	**	−0.01		
2000	0.54	**	0.14	**	
2001	0.43	**	0.00		Road**
2002	0.45	**	−0.01		Tube well*
2003	0.51	**	−0.01		

*$p \leq 0.05$.
**$p \leq 0.01$.

Table 3. Spatial error and significance, social effect and significance, and additional environmental variables that affect shigellosis rates

Year	Spatial error	Significance	Social effect	Significance	Other variables
1983	0.52	**	−0.02		Road*
1984	0.60	**	0.03		Tube well*
1985	0.43	**	−0.01		Road*
1986	0.49	**	−0.01		
1987	0.42	**	0.02	**	
1988	0.44	**	0.02		
1989	0.55	**	0.02	**	Road*
1990	0.38	**	−0.01		
1991	0.49	**	−0.02		Road,* pond*
1992	0.40	**	0.00		
1993	0.46	**	0.21	**	
1994	0.42	**	−0.01		
1995	0.51	**	−0.01		
1996	0.46	**	0.00		Road*
1997	0.40	**	0.06	**	
1998	0.45	**	−0.01		
1999	0.41	**	0.00		
2000	0.44	**	0.02	**	
2001	0.39	**	0.39	**	
2002	0.51	**	0.00		
2003	0.59	**	0.00		

*$p \leq 0.05$.
**$p \leq 0.01$.

multiple spatial scales. Figure 6 shows that both shigellosis and cholera clustered significantly in the social network in 1989 and 1993, cholera clustered significantly in 1997, and shigellosis in 2000. This variation between the two diseases in 1997 and 2000 might indicate changes in epidemic trends and higher overall rates, which are then affected by network transmission.

The results of the combined social effects–spatial error model corroborated those found through the Moran's I cluster analyses. Namely, disease incidence could consistently be accounted for through spatial processes, and the impact of social ties was minimal across most years (Table 2). The spatial error (ρ_2) was significant for every year at an α level of 0.01 for both diseases. The parameter ρ_2 represents the extent to which the clustering of cholera and shigellosis rates not explained by measured independent variables or the social effect can be accounted for by the clustering of the error term. In other words, unaccounted-for variables related to similarity within the local environment were significant in all years. For cholera, when this underlying spatial error was controlled for, the social effects parameter (ρ_1) was significant at an α level of 0.01 for five out of twenty-one years, specifically for 1983, 1985, 1989, 1993, and 2000. The values represent the extent

to which cholera rates clustered in the network; the lower coefficients are a result of the small number of overall cholera cases given the population size. The environmental control variables showed varying levels of significance in different years (Table 2). For shigellosis, there were significant social effects in six years, one more than for cholera and, similar to cholera, the significant environmental control variables were different from year to year (Table 3).

Conclusions

This study shows that incidences of cholera and shigellosis are mostly associated with the local environment, but social networks also influence the transmission of both diseases in certain situations. The methods employed in this study compare the occurrence of social and spatial clustering separately and also examine the joint effects of spatial and social networks. Although there is a small but well-regarded body of prior literature examining the influence of social networks on diseases that have a behavioral component (e.g., sexually transmitted infections, obesity, and smoking), very few studies have examined the effect of social ties on infectious diseases such as cholera or shigellosis. In addition, only a very few studies have simultaneously examined the effect of both social connectedness and spatial proximity on disease transmission dynamics. This is one of the first to utilize a joint social–spatial model to examine disease incidence.

Calculating Moran's I spatial autocorrelation on both spatial and social matrices demonstrated that the local environment is consistently important in the transmission of shigellosis and cholera, and social networks are intermittently important. The social effects–spatial disturbance model accounts for the spatial autocorrelation of omitted predictor variables or the autocorrelation of the error term. The model resulted in five (cholera) and six (shigellosis) years during which there was a significant social effect above and beyond spatial effects, which were present in all years. During those years, processes related to kinship-based social networks influenced transmission and produced similar rates in connected extended households. In a recent paper investigating multiple transmission pathways for waterborne diseases, Tien and Earn (2010) noted that the relative contribution of a given transmission pathway can be outbreak dependent. In other words, under endemic conditions of cholera, shigellosis, or both, the primary route of transmission is environmental, through ingestion of contaminated water. Outbreaks of these diseases can be driven by other transmission pathways, however, such as through direct contact. Social processes, although relatively unimportant for explaining endemic disease processes, can be instrumental for driving episodic outbreaks. Thus, the observed peaks in the importance of social ties to cholera and shigellosis dynamics during four to five years could be attributed to outbreak, rather than endemic disease processes. Socially driven transmission of cholera, shigellosis, or both could be the result of actual physical transmission of the pathogen through direct person-to-person transmission or could also be due to similar behaviors across related *baris* (i.e., hygiene or water storage practices) that either increase or reduce collective risk.

Significant spatial error parameters estimated by the model for all years suggest the importance of unidentified spatial components in producing common disease rates among socially connected *baris*, but the spatial error parameter might "capture" those social interactions not included by the kinship network. Individuals will interact to some degree with their neighbors. As noted earlier, social networks were defined only through kinship linkages, which is a limitation of the study. Kinship relationships, however, are important social outlets in Matlab, and activities that can lead to cholera and shigellosis disease transmission, such as labor and meals, often take place with other family members (White 1997; Amin 1998). If cholera is spread via consumption of food or water contaminated by others, there is a significant chance that the transmission occurred between family members. In our analysis, the social effects term captures potential disease transmission attributable to familial interactions. Of course, individuals also interact with friends, many of whom live in close spatial proximity. The poor quality of roads and overall lack of transportation leads to limited mobility in Matlab, so an individual's activity space is rather limited. Because the primary source of exposure to cholera and shigellosis is through contaminated food and water, we would expect neighbors, especially those who share water resources, to be affected by the oral–fecal–oral transmission process. In our analysis, the spatial error term captures this aspect of disease transmission, the risk attributable to spatial proximity. Although it is a limitation that we cannot explicitly examine nonkinship social connections, we suggest that including the spatial error term captures many of these social ties because most social interactions in this study area take place within a small geographic area.

As discussed in the introduction, we expected to find stronger social clustering of shigellosis, due to the hypothesized predominantly person-to-person transmission dynamics of the disease. It is clear that our results do not support this hypothesis, which might also be related to the fact that we could not examine nonkinship social ties. Using an incomplete contact network could bias results because the level of exposure might be underestimated. We also expected that cholera would cluster more in space and our results supported this hypothesis. These results support the idea that cholera is an environmentally driven disease and that contact with contaminated water bodies is by far the most important mode of transmission.

This study explores ways to conduct integrated social network and spatial analysis. Both have been used to understand disease transmission separately, although spatial context is an important part of social interaction. This study provides a framework with which to address the relative contributions of social and spatial processes on infectious disease transmission dynamics, and can be widely applied to other disease systems for which social and spatial processes are likely to interact.

Acknowledgments

This study was conducted with the support of ICDDR, B donors that provide unrestricted support to the Centre for its operations and research. Current donors providing unrestricted support include the Australian International Development Agency, the Government of Bangladesh, the Canadian International Development Agency, the Embassy of the Kingdom of The Netherlands, the Swedish International Development Cooperative Agency, and the Department for International Development, UK. This study was also supported by the following grants: NIAID R03-AI076748 and NSF BCS-0924479. Funding for Sophia Giebultowicz was provided through the NSF IGERT program at the Carolina Population Center. Funding for Carolina Perez-Heydrich was provided by NIEHS T32ES007018.

References

Ahmed, F., J. D. Clemens, M. R. Rao, M. Ansaruzzaman, and E. Haque. 1997. Epidemiology of shigellosis among children exposed to cases of *Shigella* dysentery: A multivariate assessment. *American Journal of Tropical Medicine Hygiene* 56 (3): 258–64.

Ali, M., M. Emch, C. Ashley, and P. K. Streatfield. 2001. Implementation of a medical geographic information system: Concepts and uses. *Journal of Health, Population, and Nutrition* 19 (2): 100–10.

Amin, S. 1998. Family structure and change in rural Bangladesh: Implications for women's roles in the family. *Development and Change* 28 (2): 213–33.

Anselin, L. 1988. *Spatial econometrics: Methods and models.* Boston: Kluwer Academic.

Bates, S. J., J. Trostle, W. T. Cevallos, A. Hubbard, and J. N. S. Eisenberg. 2007. Relating diarrheal disease to social networks and the geographic configuration of communities in rural Ecuador. *American Journal of Epidemiology* 166 (9): 1088–95.

Bearman, P., K. Stovel, J. Moody, and L. Thalji. 2004. The structure of sexual networks and the National Longitudinal Study of Adolescent Health. In *Network epidemiology: A handbook for survey design and data collection,* ed. M. Morris, 201–19. Oxford, UK: Oxford University Press.

Bearman, P. S., and L. Burns. 1998. Adolescents, health and school: Early findings from the National Longitudinal Study of Adolescent Health. *NASSP Bulletin* 82: 601–23.

Caplow, T., and R. Forman. 1950. Neighborhood interaction in a homogeneous community. *American Sociological Review* 15:357–66.

Carrel, M., M. Emch, P. K. Streatfield, and M. Yunus. 2009. Spatio-temporal clustering of cholera: The impact of flood control in Matlab, Bangladesh, 1983–2003. *Health & Place* 15 (3): 771–82.

Christakis, N. A., and J. H. Fowler. 2007. The spread of obesity in a large social network over 32 years. *The New England Journal of Medicine* 357:370–79.

Collins, A. 1998. *Environment, health and population displacement: Development and change in Mozambique's diarrhoeal disease ecology.* Aldershot, UK: Ashgate.

Colwell, R. R. 1996. Global climate and infectious disease: The cholera paradigm. *Science* 274 (5295): 2025–31.

Colwell, R. R., M. L. Tamplin, P. R. Brayton, A. L. Gauzens, B. D. Tall, D. Herrington, M. M. Levine, S. Hall, A. Huq, and D. A. Sack. 1990. Environmental aspects of Vibrio cholerae in transmission of cholera. In *Advances in research on cholera and related diarrhoea,* 7th ed., S. Kuwahara and N. F. Pierce, 327–43. Tokyo: KTK Scientific.

Deutsch, M., and M. E. Collins. 1951. *Interracial housing: A psychological evaluation of a social experiment.* Minneapolis: University of Minnesota Press.

Doreian, P. 1980. Linear models with spatially distributed data: Spatial disturbances or spatial effects? *Sociological Methods & Research* 9:29–60.

———. 1981. Estimating linear models with spatially distributed data. *Sociological Methodology* 12:359–88.

Dow, M. M. 2007. Galton's problem as multiple network autocorrelation effects: Cultural trait transmission and ecological constraint. *Cross-Cultural Research* 41 (4): 336–63.

Emch, M. 1999. Diarrheal disease risk in Matlab, Bangladesh. *Social Science and Medicine* 49:519–30.

Emch, M., and M. Ali. 2001. Spatial and temporal patterns of diarrheal disease in Matlab, Bangladesh. *Environment and Planning A* 33:339–50.

Emch, M., M. Ali, and M. Yunus. 2008. Risk areas and neighborhood-level risk factors for *Shigella dysenteriae* 1 and *Shigella flexneri*: Implications for vaccine development. *Health & Place* 14 (1): 96–105.

Entwisle, B., K. Faust, R. Rindfuss, and T. Kaneda. 2007. Networks and contexts: Variation in the structure of social ties. *American Journal of Sociology* 112 (5): 1495–1533.

Escamilla, V., B. Wagner, M. Yunus, P. K. Streatfield, A. van Geen, and M. Emch. 2011. Impact of deep tubewells on childhood diarrhea in Bangladesh. *Bulletin of the World Health Organization* 89 (7): 521–27.

Faust, K., B. Entwisle, R. R. Rindfuss, S. J. Walsh, and Y. Sawangdee. 1999. Spatial arrangement of social and economic networks among villages in Nang Rong District, Thailand. *Social Networks* 21 (4): 311–37.

Festinger, L., S. Schachter, and K. Back. 1950. *Social processes in informal groups.* Stanford, CA: Stanford University Press.

Guest, P., and A. Chamratrithirong. 1992. The social context of fertility decline in Thailand. In *Fertility transitions, family structure, and population policy,* ed. C. Goldscheider, 67–99. Boulder, CO: Westview.

Hanneman, R. A. 2001. *Introduction to social network methods.* Riverside: Department of Sociology, University of California at Riverside.

Hollinger, F., and M. Haller. 1990. Kinship and social networks in modern societies: A cross-cultural comparison among seven nations. *European Sociological Review* 6:103–24.

Hoque, B. A., T. Juncker, R. B. Sack, M. Ali, and K. M. Aziz. 1996. Sustainability of a water, sanitation and hygiene education project in rural Bangladesh: A 5-year follow-up. *Bulletin of the World Health Organization* 74 (4): 431–37.

Islam, M. S., B. S. Drasar, and R. B. Sack. 1993. The aquatic environment as a reservoir of *Vibrio cholerae*: A review. *Journal of Diarrhoeal Disease Research* 11 (4): 197–206.

Klovdahl, A. S. 2001. Networks and tuberculosis: An undetected community outbreak involving public places. *Social Science & Medicine* 52 (5): 681–94.

Kotloff, K. L., J. P. Winickoff, B. Ivanoff, J. D. Clemens, D. L. Swerdlow, P. J. Sansonetti, G. K. Adak, and M. M. Levine. 1999. Global burden of *Shigella* infections: Implications for vaccine development and implementation of control strategies. *Bulletin of the World Health Organization* 77 (8): 651–66.

LeSage, J. P. 1999. Applied econometrics using MATLAB. http://www.spatial-econometrics.com (last accessed 10 November 2010).

Liu, K. Y., M. King, and P. Bearman. 2010. Social influence and the autism epidemic. *American Journal of Sociology* 5:1387–1434.

MATLAB. 2010. Version 7.7.0. The MathWorks, Inc., Natick, MA.

McPherson, M., L. Smith-Lovin, and J. M. Cook. 2001. Birds of a feather: Homophily in social networks. *Annual Review of Sociology* 27:415–44.

Miller, C. J., R. G. Feachem, and B. S. Drasar. 1985. Cholera epidemiology in developed and developing countries: New thoughts on transmission, seasonality, and control. *Lancet* 8423:261–63.

Mok, D., B. Wellman, and R. Basu. 2004. Does distance matter for relationships? Presentation at SUNBELT International Social Network Conference, Portoroz, Slovenia.

Moran, P. A. P. 1950. Notes on continuous stochastic phenomena, *Biometrika* 37:17–33.

Morris, M. 2004. *Network epidemiology: A handbook for survey design and data collection*. Oxford, UK: Oxford University Press.

Radil, S. M., C. Flint, and G. E. Tita. 2010. Spatializing social networks: Using social network analysis to investigate geographies of gang rivalry, territoriality, and violence in Los Angeles. *Annals of the Association of American Geographers* 100.307–26.

Rahaman, M. M., M. M. Khan, K. M. S. Aziz, M. S. Islam, and A. K. Kibriya. 1975. An outbreak of dysentery caused by *Shigella dysenteriae* type 1 on a Coral Island in the Bay of Bengal. *The Journal of Infectious Diseases* 132 (1): 15–19.

Reuters. 2007. Somalia: Cholera kills 82 in central region. *Alertnet* 5 February. http://www.alertnet.org/thenews/newsdesk/IRIN/289d0e116c1f0700790157eb152ae7ca.htm (last accessed 30 September 2009).

Rindfuss, R. R., A. Jampaklay, B. Entwistle, Y. Sawangdee, K. Faust, and P. Prasartkul. 2004. The collection and analysis of social network data in Nang Rong, Thailand. In *Network epidemiology: A handbook for survey design and data collection*, ed. M. Morris, 175–99. Oxford, UK: Oxford University Press.

Rothenberg, R., S. Q. Muth, S. Malone, J. J. Potterat, and D. E. Woodhouse. 2005. Social and geographic distance in HIV risk. *Sexually Transmitted Diseases* 32 (8): 506–12.

Ruiz-Moreno, D., M. Pascual, M. Emch, and M. Yunus. 2010. Spatial clustering in the spatio-temporal dynamics of endemic cholera. *BMC Infectious Diseases* 10:51.

Siddique, A. K., K. Zaman, A. H. Baqui, K. Akram, P. Mutsuddy, A. Eusof, K. Haider, S. Islam, and R. B. Sack. 1992. Cholera epidemics in Bangladesh: 1985–1991. *Journal of Diarrhoeal Disease Research* 10 (2): 79–86.

StataCorp. 2007. *Stata Statistical software: Release 10*. College Station, TX: StataCorp LP.

St. Louis, M., J. D. Porter, A. Helal, K. Drame, N. Hargrett-Bean, J. G. Wells, and R. V. Tauxe. 1990. Epidemic cholera in West Africa: The role of food-handling and high-risk foods. *American Journal of Epidemiology* 131 (4): 719–28.

Swerdlow, D. L, K. D. Greene, R. V. Tauxe, J. G. Wells, N. H. Bean, A. A. Ries, P. A. Blake, et al. 1992. Waterborne transmission of epidemic cholera in Trujillo, Peru: Lessons for a continent at risk. *The Lancet* 340: 28–32.

Tien, J. H., and D. J. D. Earn. 2010. Multiple transmission pathways and disease dynamics in a waterborne pathogen model. *Bulletin of Mathematical Biology* 72 (6): 1506–33.

Wasserman, S., and K. Faust. 1994. *Social network analysis: Methods and applications*. Cambridge, UK: Cambridge University Press.

Wellman, B. 1996. Are personal communities local? A Dumptarian reconsideration. *Social Networks* 18: 347–54.

Wellman, B., P. Carrington, and A. Hall. 1988. Networks as personal communities. In *Social structures: A network approach*, ed. B. Wellman and S. D. Berkowitz, 130–84. Cambridge, UK: Cambridge University Press.

White, S. C. 1997. Men, masculinities, and the politics of development. *Gender and Development* 5 (2): 14–22.

World Health Organization. 2009. World health statistics report 2009, part III: Selected infectious diseases. http://www.who.int/gho/publications/world_health_statistics/2009/en/index.html (last accessed 21 March 2012).

Wylie, J. L., T. Cabral, and A. M. Jolly. 2005. Identification of networks of sexually transmitted infection: A molecular, geographic, and social network analysis. *Journal of Infectious Diseases* 191 (6): 899–906.

Modeling Individual Vulnerability to Communicable Diseases: A Framework and Design

Ling Bian,* Yuxia Huang,† Liang Mao,‡ Eunjung Lim,§ Gyoungju Lee,# Yan Yang,* Murray Cohen,¶ and Deborah Wilson¥

*Department of Geography, University at Buffalo, State University of New York
†School of Engineering and Computing Sciences, Texas A&M University–Corpus Christi
‡Department of Geography, University of Florida
§Department of Geospatial Sciences, University of Maryland
#Department of Urban Engineering, Korea National University of Transportation, Korea
¶Consultants in Disease and Injury Control, Inc., Atlanta, Georgia
¥Office of Research Safety, National Institutes of Health

Reports on dangerous communicable diseases, such as severe acute respiratory syndrome (SARS) and H1N1 flu, have repeatedly stressed the importance of individuals in disease transmission. Still in its infancy, individual-based modeling faces many challenges. From the perspective of modeling approaches, this article explores (1) the framework of a three-population (daytime, nighttime, and pastime population) and two-scale (local and societal scale) social network; (2) a design that can represent heterogeneous, mobile, interacting individuals and their individualized vulnerability to infection; and (3) a simulation of individuals' vulnerability to influenza in an urban area in the Northeastern United States to illustrate the proposed framework and design. Simulation results correspond well to the reported epidemic information. The findings offer a valuable platform to devise much-needed spatially and temporally oriented control and intervention strategies for communicable diseases.

危险性传染病如严重急性呼吸道综合征（SARS）和 H1N1 禽流感的报告，一再强调个人在疾病传播中的重要性。基于个体的建模还处于起步阶段，并面临着许多挑战。这篇文章从建模方法的角度，探讨 (1) 三类人群（白天，夜间和休闲人群）的框架和两个规模（地方和社会的规模）的社会网络；(2) 能代表异类，移动，交互的个人和他们的个体易受感染性的设计；以及 (3) 模拟美国东北部市区个体的流感脆弱性，来说明拟议的框架和设计。仿真结果很符合所报道的疫情信息。这些发现提供了一个宝贵的平台，该平台为传染病制定所急需的面向时空的控制和干预策略。关键词：白天，夜间和休闲人群，基于个人的流行病学模型，流动性，社会网络。

Los informes relacionados con enfermedades contagiosas graves, tales como el síndrome respiratorio severo agudo (SARS) y la gripa N1N1, insisten sobre la importancia que corresponde a los individuos para la transmisión del mal. Todavía en su infancia, la modelización basada en el individuo enfrenta muchos retos. Desde la perspectiva de los enfoques modeladores, el presente artículo explora, (1) el marco de una red social integrada por una población vista desde tres dimensiones temporales (población diurna, nocturna y pasada) y a doble escala (escala local y social); (2) un diseño que puede representar individuos heterogéneos, móviles e interactivos y su vulnerabilidad a la infección individualizada, y (3) una simulación de la vulnerabilidad de los individuos a la influenza en un área urbana de los Estados Unidos del nordeste, para ilustrar el marco y diseño propuestos. Los resultados de la simulación se corresponden bien con la información epidemiológica disponible. Los descubrimientos hechos son una valiosa plataforma para desarrollar unos muy necesitados controles, orientados espacial y temporalmente, y las estrategias de intervención en el problema de las enfermedades contagiosas.

Communicable diseases are transmitted from individual to individual following a network of contacts in a population. Through this network, diseases spread across space and time. Reports on dangerous communicable diseases, such as severe acute respiratory syndrome (SARS), anthrax, bird flu, and, most recently, H1N1 flu, have repeatedly stressed the critical role of individuals in the spread of these diseases (McKenzie 2004). An understanding of who might be at risk, where and when the risk occurs, and with whom these at-risk individuals might be in contact are invaluable to the development of health policies to control and prevent the spread of these diseases.

Many types of simulation models have been devised to understand how communicable diseases propagate in a population (McKenzie 2004). Individual-based epidemiology models have generated great interest in recent years (Eubank et al. 2004; Ferguson et al. 2005; Longini et al. 2005; Cooley et al. 2008; Yang, Atkinson, and Ettema 2008). These models attempt to assess health risks at the individual level. Insights gained into how diseases are dispersed among individuals and how their vulnerability varies in space and time cannot be easily matched by other types of models (Riley 2007). In these models, a large number of individuals have been simulated. Not all of them, however, fully consider the uniqueness of individuals, especially their individualized mobility and interactions, although these are critical in assessing individual vulnerability. Further, many models focus on nighttime population at residential locations, although populations at workplaces and service places are equally vulnerable. Even fewer models have compared their results to reported epidemic information (Cooley et al. 2008).

Still in its infancy, individual-based modeling faces many challenges, including conceptual, design, and implementation issues. From the perspective of modeling approaches, this article addresses two aspects of individual-based epidemiology models: their framework and design. Specifically, this article explores (1) a framework of a three-population (daytime, nighttime, and pastime population) and two-scale (local and societal scale) social network; (2) a design that can represent heterogeneous, mobile, and interacting individuals and their heterogeneous vulnerability to infection; and (3) a simulation of individual vulnerability to influenza during an epidemic in an area in the Northeastern United States to illustrate the application of the model framework and design.

Framework

This research explores and extends a two-population and two-scale framework initially proposed by Bian (2004). The two populations are a nighttime population at home and a daytime population at workplaces. The two scales are a local scale and a societal scale. The daily movements of individuals between their homes and workplaces are represented as individual lifelines (Hägerstrand 1970; Miller 2005). These lifelines intersect at homes and workplaces, forming local-scaled networks. Travel between homes and workplaces by these individuals links the local networks into a society-scaled network. Through this two-population and two-scale network, diseases can spread by local infections and long-distance dispersions. Local infections occur through interactions within local networks (homes or workplaces), whereas long-distance dispersions occur when individuals travel between local networks (Bian 2004).

This framework explicitly accounts for heterogeneous and mobile individuals, their individualized interactions that vary by location and time, and the heterogeneity in an individual's exposure and infection. The separation between the nighttime and daytime populations fully accounts for two equally important sources of transmission in an urban environment. The two-scale social network allows for spatially and temporally dynamic interactions between individuals that result in the transmission of disease. Although lacking detail beyond a conceptual outline, this framework offers a guide for the design and implementation of individual-based models and is deemed the most appropriate for this discussion.

For this research, the original two-population and two-scale social network is extended to three populations while retaining the two scales. The third population is the pastime population at service places, which is an important daily activity of an urban population (Kwan 1999, 2002). The pastime population includes those who engage in activities outside home and workplaces, such as dining, shopping, and recreation. Note that the two scales refer to the structure of the social network. The actual modeling involves three scales: individuals, social groups (family, workplace, and service place), and the society-wide population.

Design

The design discussed here is intended to be generic for modeling individual vulnerability in the transmission of communicable diseases, whereas the basic design components and parameters can be modified to suit particular applications. The design specifies the representation of unique individuals, spatial–temporal scales of disease transmission, the structure of the social network, and individual vulnerability in the transmission of diseases, as described next.

Individuals

Individuals are represented and distinguished between one another by their unique attributes and behaviors. The design assigns four sets of attributes to each

Table 1. List of individual attributes and behaviors specified in the design

Attributes	Value/Unit
1. Demographic attributes	
Identity	1–245,842
Age	Child, youth, adult, senior
Occupation	Categories in NAICS
2. Spatial–temporal attributes	
Location	Home, workplace, service place
Time period	Daytime, nighttime, pastime
Date	Date
3. Affiliation attribute	
Identity of social group	Home ID, workplace ID, service place ID
Identity of individual in a social group	1–245,842
4. Infection attribute	
Infection probability	%
Infection stage	Latent period, incubation period, infectious period
Length of infection stages	Number of days
Infection status	Vaccinated, susceptible, latent, infectious, symptomatic, recovered

Behaviors (and associated attributes to implement behaviors)
1. Travel
    Travel
    Stay at the same location
    (Location, time period, date)
2. Interaction
    Interaction
    (Identity of social group, identity of individuals in the social group, infection status of these individuals, location, time period, date)
3. Infection
    Being infected
    (Infection probability, infection status)
4. Protection
    Vaccination
    (Infection status)
    Seeking medical treatment
    (Infection status)
    Staying home after developing symptoms
    (Identity of social group, identity of individuals in the social group)

Note: NAICS = North America Industry Classification System.

individual (Table 1). The first set is demographic characteristics (e.g., age, occupation) used to represent an individual's unique identity and offer a basis to prepare for the representation of his or her individual vulnerability to infection. The second set includes spatial–temporal attributes (e.g., location, time period in a day) that represent the spatial–temporal hetero-geneity in individuals. The third set considers the affiliations an individual has with the three types of social groups (i.e., family, workplace, and service place; e.g., the identity of a social group, the individuals in the group). Combined with the spatial–temporal attributes, this set helps to establish the individualized interactions that vary by location and time. These two sets are combined to prepare for the representation of disease transmission through these interactions. The fourth set of attributes is related to infection (e.g., infection probability, infection status) to estimate individual vulnerabilities during disease transmission.

Four sets of behaviors are also assigned to each individual (Table 1). These include mobility (e.g., traveling to different locations), interaction (e.g., interacting with other individuals at a given location and time period), disease transmission (e.g., being infected by others), and protective actions (e.g., vaccination, staying home once developing symptoms). Whereas an individual's attributes are assigned, their behaviors are represented by changes in those attributes that are relevant to a behavior. For example, travel can be represented by the change in the location attribute of an individual. Similarly, infection can be represented by changing the infection status from susceptible to infectious.

The four sets of attributes and four sets of behaviors are considered the model's primary parameters to distinguish them from a large number of secondary attributes and behaviors necessary for the modeling. These primary and secondary attributes and behaviors are deemed necessary and sufficient to support the intended modeling. Further, they can be implemented using data accessible to the public.

The design concept of unique individuals differs notably from classic and recently developed models, although the latter also considers individuals to various degrees. Classic population-based models divide a population into a number of population segments, each containing a large number of individuals who share a common health status, such as being susceptible to infections. The core of the classic models is the "homogeneous mixing" assumption that all individuals in a population segment are homogeneously mixed, mirrored by a uniform parameter value for all individuals in that segment. This assumption presents a design concept that is fundamentally different from the heterogeneous individual design used in this research and is most effective for modeling pandemics over a large area.

Population-based models are not appropriate for assessing individual vulnerability, but two other types of models do focus on unique individuals. One is the individual-based cellular automata model (Holmes 1997). Its design consists of unique individuals located in regular cells, one in each cell. Diseases spread through both local interactions between neighboring cells and by leapfrogging to distant locations to establish new foci of infection. The design concept of immobile individuals inherent in this type of model is not comparable to the individualized mobility design employed in this research and is perhaps intended for immobile individuals (e.g., plants).

Another type is the individual-based aspatial model (Keeling 1999; Koopman and Lynch 1999), in which unique individuals and their individualized interactions in a social network are explicitly represented but their spatial representation is not always explicitly considered. The design of these models is similar to the proposed design described in this research. Developed primarily for sexually transmitted diseases, these models emphasize individual social behaviors, not spatial ones. Individual-based models that incorporate individualized mobility are few in number. Their current status was discussed earlier and is not further elaborated here.

Spatial–Temporal Scale

The behaviors discussed earlier occur at three types of point locations—homes, workplaces, and service places—and follow transportation routes between them. A vector data model is effective in supporting the spatial scale at the point locations and the links between them.

The temporal scale in this design is defined as three mutually exclusive time periods of a day: daytime, pastime, and nighttime. The division of a day into these three periods is based on three factors—location, time, and activity—as each time period is bound to a location and a type of a social group. This scale is perhaps more meaningful than the arbitrarily defined time steps used in many models. Further, an infection process consists of a number of stages, such as the latent period and the infectious period. Each could last more than a day (Heymann 2004). The three-period division offers a sufficiently fine temporal scale for representing the length of the different stages of an infection process.

Social Network

Based on the representation of heterogeneous, mobile, and interacting individuals, the following design creates the three-population and two-scale social network. Several individuals join a social group by sharing the same location, time, and affiliation attributes of the social group. These groups represent social networks at the local scale. Their representation is supported by the multiple sets of attributes and behaviors listed in Table 1 (i.e., spatial–temporal attributes, affiliation attributes, and interaction behaviors). The three possible affiliations of all individuals (some might have less than three) collectively represent the three populations at the societal scale, along with the ties between them. This social network design accommodates the proposed framework (Section 2) and for this reason differs from those found in the current literature.

Generic network research evaluates the structure of a network and properties of its nodes and links to assess its performance in transmitting various tangible or intangible properties. Social network research also deals with these issues but in the context of social relationships and the variation in social influences through different relationships (Borgatti et al. 2009). Most social network research focuses on a finite number of individuals who share common social attributes. In addition, most social networks are static, meaning that the network structure and properties of its nodes and links do not change. Recent literature has explicitly studied dynamics in social networks (Carley 2003). Individual nodes can be added to or eliminated from the network without disturbing the majority of the established links. Further, most social networks are aspatial without considering the location and mobility of individuals.

In comparison, the proposed network is first a "census" social network that encompasses all individuals in a population, instead of the finite number typically used in social network research. Second, as opposed to a static network, the proposed network is dynamic in that an individual node alternates its link between other nodes on a daily basis (e.g., to coworkers during the day and to family members at night). Third, in contrast to the stable-link and dynamic-node design, the proposed network presents dynamic links while maintaining stable nodes. Finally, this social network is explicitly registered to spatial locations, and mobility is part of the construct of a network. This design is deemed effective in representing the daily routine contacts in a population and in assessing individual vulnerability, which is spatially and temporally heterogeneous.

Disease Transmission

Transmission of communicable diseases depends on three considerations: the contact of susceptible individuals with an infectious individual, the infection probability of these susceptible individuals, and the types of protective actions they take once they develop symptoms. These considerations can be implemented by the previously discussed interaction, infection, and protection behaviors, respectively. Those individuals who take preinfection protective actions (e.g., vaccinations) are excluded from acquiring an infection. Other individuals who are in direct contact with an infectious individual in the three social groups are candidates for acquiring infection. These candidates are identified by their affiliation attributes shared with the infectious individual. The probability of an infection depends on an individual's demographic attributes and associated infection attributes (Table 1). The resultant newly infectious individuals are sources of the next generation of infection in two possible ways. The first is where the infectious individuals continue to interact with the three possible social groups and infect others. The second is where the infectious individuals take protective actions (e.g., staying home), thus reducing the number of social groups with whom they interact and infect.

This design of disease transmission is built on the representation of unique individuals and their individualized interactions at different locations and time periods in a census social network. These design concepts allow the model to assess the identity of individuals exposed to and affected by an infection, the location and time of their infection, and its consequence (either recovery or death). The following simulation illustrates the application of the design in a working model.

Simulation

Individual vulnerability to seasonal influenza is simulated for an urban area in the Northeastern United States. Although the simulation encompasses a specific urban area, the discussion attempts to address common issues encountered in such an implementation, such as the types of data required and their sources, parameters and their values, and the types of process functions required to implement the model. Due to space concerns, only the key steps are reported here.

Each simulated individual is assigned to a household, a workplace, and multiple service places to establish the three-population and two-scale social network. These assignments are based on information drawn from multiple data sets. The transmission of influenza is subsequently simulated to assess the vulnerability of the simulated individuals.

Simulation of Individuals at Home

The nighttime population at home is represented by a residential area within an approximately 2-mile radius of a center point where the first infectious case is introduced. According to 2000 census data, this residential area has cohesive social–economic characteristics in terms of a high population density, mixed racial characteristics, and moderate incomes. Within this area are a total of 227 block groups containing 104,000 households and 245,842 individuals.

Three data sets are used to assign the individuals to a household: the 2000 Census Summary File 3 (SF3) data at the block group level (U.S. Census Bureau), a household data set obtained from ReferenceUSA, Inc. (ReferenceUSA Inc.), and Topologically Integrated Geographic Encoding and Referencing (TIGER) census block group boundary data (U.S. Census Bureau). Five sets of statistics in the SF3 file provide key information. Three sets are individual statistics (age, gender, and the relationship to the householder), and two are household statistics (size and type of household; see Appendix). Individuals are assigned to households according to the statistical distribution of these variables using a Monte Carlo method. This step populates the 104,000 households with 245,842 individuals. Those variables that are used to assign individuals to households now become the first set of home-related attributes of the simulated individuals (see Appendix). Most of these attributes are secondary information necessary for intermediate modeling processes or to derive the primary attributes listed in Table 1.

The resultant household and individual data created using the SF3 data are then merged with the ReferenceUSA household data set to obtain additional attributes for the simulated individuals. Most critical of these are the location and purchasing behavior of households, information that is necessary for the subsequent simulation. The merger is based on the attributes common to both data sets, including household size, characteristics of the householder, and the presence of children. The preceding processes build the nighttime population.

Simulation of Individuals at Workplaces

Now that individuals have home-related attributes, four sets of information are used to assign individuals to their respective workplaces. These are a workplace data set obtained from Environment Systems Research Institute Inc. (ESRI), additional SF3 data, the previously derived individual and household attributes, and the TIGER road network data (Appendix).

The ESRI workplace data set consists of 310,400 workplaces distributed over 5,434 block groups within a one-hour driving distance from the previously identified residential area. This one-hour driving distance is used because it encompasses approximately 99 percent of the travel-to-work distances of all workers in the residential area, according to the SF3 data. Each workplace contains a set of attributes. The most important are location and type of workplace. The workplace type is coded using the North America Industry Classification System (NAICS; U.S. Census Bureau).

Five sets of statistics in the additional SF3 data contribute to assigning individuals to a workplace: the number of workers in a household, the transportation mode and the travel time used for workers to travel to work, their occupation as coded by NAICS, and the number of vehicles in a household. Using the Monte Carlo method, each worker is assigned to a workplace based on the statistical distribution of these variables and two additional constraints. First, the assigned workplace is located within a zone between the minimum and maximum travel distances for all workers in a block group. The travel distance is estimated using the transportation mode and travel time in the SF3 data with the speed for each mode along the actual transportation routes being provided by the TIGER data. The second constraint is the match between a worker's occupation and the workplace type, both coded by NAICS. Consid-

ering that certain service-oriented workplaces are open during both daytime and pastime hours, two shifts of workers are assigned to these workplaces. School-aged individuals are assigned to the school nearest to their household identified in the ESRI workplace data.

The variables used to assign individuals to workplaces are in turn used as the second set, workplace-related attributes for individuals (see Appendix). This step builds the daytime population and links it to the previously created nighttime population.

Simulation of Individuals at Service Places

The 245,842 individuals who are affiliated with a home, most of whom are also affiliated with a workplace, are assigned to service places based on three sets of information. The first is a subset of workplaces identified as service places. The second is a travel diary obtained from the regional transportation authority agency for the area. The third is the home- and workplace-related attributes created previously.

According to NAICS, the 310,400 workplaces in the simulated area include 89,159 service places. The travel diary data is a survey of trips taken by individuals in selected households living in the area during the 1990s. This is the most recent survey of this kind for the study area and represents data for 39,934 trips taken by 9,256 individuals in 3,906 households. These data provide three sets of relevant information: the purpose of a trip (e.g., recreation), the type of origin and destination

Table 2. Key parameters, values, and sources used to implement the simulation

Parameter	Value	Source
Latent period	2 days	Heymann (2004)
Incubation period	3 days	Heymann (2004)
Infectious period	Children ($\leq$ 5 years): 7 days Youth and adults (6–64 years): 4 days Senios (> 64 years): 4 days	Heymann (2004)
Infection rates	Children: 0.05 Youth and adults: 0.03 Seniors: 0.05	Calibrated based on R_0 between 0.9 and 1.3 (Chowell, Miller, and Viboud 2007)
Probability of developing symptoms once infected	0.5	Ferguson et al. (2006)
Background immunity from vaccination	Children: 17.9% Adults: 15.6% Seniors: 62.7%	Euler et al. (2005) Molinari et al. (2007)
Withdrawal to home after developing symptoms	Children: 100% Adults: 33% Seniors: 100%	Metzger et al. (2004) Stoller, Forster, and Portugal (1993)
Probability of seeking Medical treatment	0.38	Metzger et al. (2004)

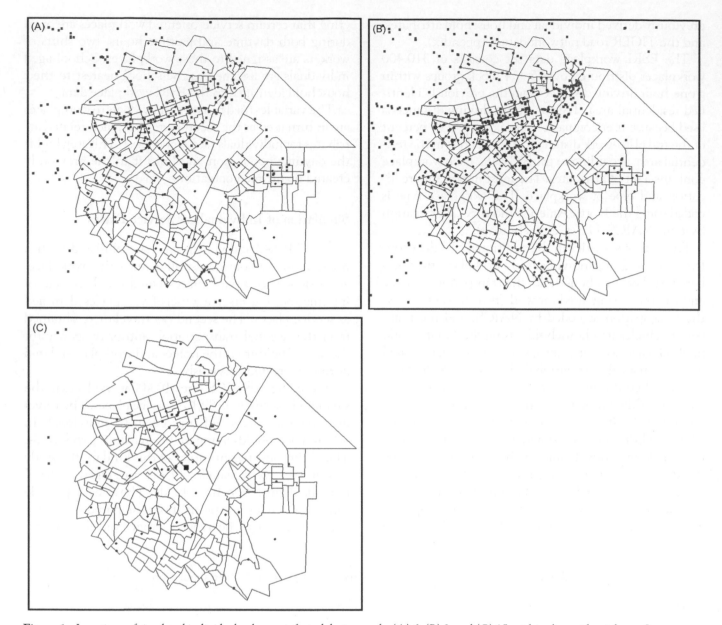

Figure 1. Locations of simulated individuals who are infected during weeks (A) 6, (B) 9, and (C) 15, within the residential area. Large square = the center point of the residential area; small square = home locations; diamond = workplace locations; red = a death case. (Color figure available online.)

(e.g., home–workplace), and the statistical distribution of these trips for both households and individuals (see Appendix).

According to the statistical distribution of these variables, the Monte Carlo method is used to assign individuals to a service place. For different dates during the epidemic, one of the twenty nearest service places in an NAICS category is randomly selected as the destined service place. Weekday and weekend service activities are differentiated as well. Weekends see an increased number of service trips and no workplace trips, except for those individuals who are assigned to work at service places.

The three populations and links among them are now established. Each individual now possesses three of the four sets of attributes (demographic, spatial–temporal, and affiliation attributes) and two sets of behaviors (travel and interaction). These prepare for the subsequent simulation of disease transmission and individual vulnerability.

Simulation of Individual Vulnerability

The infection attributes (Tables 1 and 2) and infection and protection behaviors are assigned to individuals. The simulation of disease transmission follows

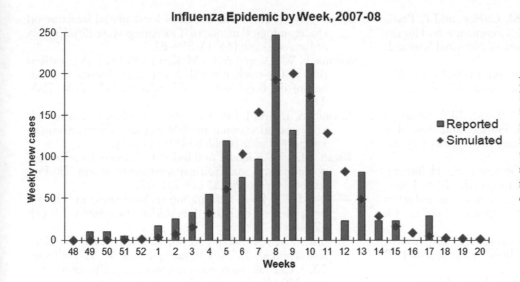

Figure 2. A comparison of the simulated weekly number of newly infected individuals as an average of the 100 realizations to the actual number of weekly influenza-like illness cases reported to the Centers for Disease Contol for the simulated area during the 2007–2008 influenza season. (Color figure available online.)

the design specified earlier using the relevant primary attributes (demographic and infection attributes) and behaviors (interaction, infection, and protection). The first case of infection is introduced into the center of the residential area. The infection propagates through the population following the three-population and two-scale social network discussed earlier. The Monte Carlo method is used in all steps of the transmission simulation that involve, for example, selecting individuals to perform the infection or protection behavior according to their corresponding probabilities (Table 2).

The output of the simulation estimates individual vulnerabilities, including the identity of the infected individual and the location, time, and consequence of the infection. Also estimated is the collective health outcome of the population, expressed as the daily number of newly infected individuals over a 200-day period. The simulation is run for 100 realizations. In each realization, those individuals who receive a vaccination, those who are in direct contact with the first infectious individual (and their infection probability), and those who take protective actions are randomized.

Figure 1 displays the simulated spatial distribution of individuals who are infected during weeks 6, 9, and 15 to represent the rise, peak, and decline of the epidemic, respectively. The pattern is typical of the 100 realizations. Figure 2 displays the simulated weekly number of newly infected individuals compared to the actual number of weekly influenza-like illness cases reported to the Centers for Disease Control and Prevention (CDC) during the 2007–2008 influenza season when the simulated area was affected by a record influenza epidemic (CDC 2008). The estimated numbers are averages of 100 realizations. The shape, timing, and magnitude of

the peak of the predicted curve correspond well with those of the reported epidemic. To our knowledge, this level of correspondence has rarely been achieved by other individual-based models. The three-population and two-scale model provides a valuable foundation to devise much-needed spatially and temporally oriented control and intervention strategies for communicable diseases.

It is indeed difficult to resist citing the popular statement by George Box that all models are wrong but some are useful (Box and Draper 1987). Simulation models provide a useful platform to systematically evaluate gaps in our knowledge and support effective health policy decisions. The framework, design, and implementation issues discussed in this article offer such a platform.

Acknowledgments

This research was funded in part by the Department of Health and Human Services under Contract Award number HHSN263999900824B. The valuable comments and suggestions provided by the anonymous reviewers are gratefully acknowledged.

References

Bian, L. 2004. A conceptual framework for an individual-based spatially explicit epidemiological model. *Environment and Planning B* 31 (3): 381–95.

Borgatti, S. P., A. Mehra, D. J. Brass, and G. Labianca. 2009. Network analysis in the social sciences. *Science* 323 (5916): 892–95.

Box, G. E. P., and N. R. Draper. 1987. *Empirical model-building and response surfaces.* Hoboken, NJ: Wiley.

Carley, K. M. 2003. Dynamic network analysis. In *Dynamic social network modeling and analysis: Workshop summary and papers*, ed. R. Breiger, K. M. Carley, and P. Pattison, 133–45. Washington, DC: Committee on Human Factors, National Research Council, National Research Council.

Centers for Disease Control and Prevention (CDC). 2008. http://wonder.cdc.gov/ (last accessed 30 November 2011).

Chowell, G., M. A. Miller, and C. Viboud. 2007. Seasonal influenza in the United States, France, and Australia: Transmission and prospects for control. *Epidemiology and Infection* 136 (6): 852–64.

Cooley, P., L. Ganapathi, G. Ghneim, S. Holmberg, W. Wheaton, and C. R. Hollingsworth. 2008. Using influenza-like illness data to reconstruct an influenza outbreak. *Mathematical and Computer Modeling* 48: 929–39.

Environment Systems Research Institute, Inc. http://www.esri.com/data/esri_data/business.html/ (last accessed 30 November 2011).

Eubank, S., H. Guclu, V. S. A. Kumar, M. V. Marathe, A. Srinivasan, Z. Toroczkai, and N. Wang. 2004. Modelling disease outbreaks in realistic urban social networks. *Nature* 429 (6988): 180–84.

Euler, G. L., C. B. Bridges, C. J. Brown, P. J. Lu, J. Singleton, S. Stokley, S. Y. Chu, M. McCauley, M. W. Link, and A. H. Mokdad. 2005. Estimated influenza vaccination coverage among adults and children—United States, September 1, 2004–January 31, 2005. *Morbidity and Mortality Weekly Report* 54 (12): 304–07.

Federal Highway Administration. 2000. *Manual on uniform traffic control devices, millenium edition*. Washington, DC: Federal Highway Administration.

Ferguson, N. M., D. A. T. Cummings, S. Cauchemez, C. Fraser, S. Riley, A. Meeyai, S. Iamsirithaworn, and D. S. Burke. 2005. Strategies for containing emerging influenza pandemic in Southeast Asia. *Nature* 437 (7056): 209–14.

Ferguson, N. M., D. A. Cummings, C. Fraser, J. C. Cajka, P. C. Cooley, and D. S. Burke. 2006. Strategies for mitigating an influenza pandemic. *Nature* 442 (7101): 448–52.

Hägerstrand, T. 1970. What about people in regional science? *Papers of the Regional Science Association* 24 (1): 7–21.

Heymann, D. L. 2004. *Control of communicable diseases manual*. 18th ed. Washington, DC: American Public Health Association.

Holmes, E. E. 1997. Basic epidemiological concepts in a spatial context. In *Spatial ecology*, ed. D. Tilman and P. Kareiva, 111–36. Princeton, NJ: Princeton University Press.

Keeling, M. J. 1999. The effects of local spatial structure on epidemiological invasions. *Proceedings of the Royal Society of London B* 266 (1421): 859–67.

Kermack, W. O., and A. G. McKendrick. 1927. A contribution to the mathematical theory of epidemics. *Proceedings of the Royal Society of London A* 115 (772): 700–21.

Koopman, J., and J. Lynch. 1999. Individual causal models and population system models in epidemiology. *American Journal of Public Health* 89 (8): 1170–74.

Kwan, M. 1999. Gender and individual access to urban opportunities: A study using space-time measures. *The Professional Geographer* 51 (2): 211–27.

———. 2002. Time, information technologies, and the geographies of everyday life. *Urban Geography* 23 (5): 471–82.

Longini, I. M., A. Nizam, S. Xu, K. Ungchusak, W. Hanshaoworakul, D. A. Cummings, and M. E. Halloran. 2005. Containing pandemic influenza at the source. *Science* 309 (5737): 1083–87.

McKenzie, F. E. 2004. Smallpox models as policy tools. *Emerging Infectious Diseases* 10 (11): 2044–47.

Metzger, K. B., A. Hajat, M. Crawford, and F. Mostashari. 2004. How many illnesses does one emergency department visit represent? Using a population-based telephone survey to estimate the syndromic multiplier. *Morbidity and Mortality Weekly Report* 53:106–11.

Miller, H. J. 2005. A measurement theory for time geography. *Geographical Analysis* 37 (1): 17–45.

Molinari, N. A. M., I. R. Ortega-Sanchez, M. L. Messonnier, W. W. Thompson, P. M. Wortley, E. Weintraub, and C. B. Bridges. 2007. The annual impact of seasonal influenza in the US: measuring disease burden and costs. *Vaccine* 25 (27): 5086–96.

ReferenceUSA, Inc. http://www.referenceusa.com/ (last accessed 30 November 2011).

Riley, S. 2007. Large-scale spatial transmission models of infectious disease. *Science* 316 (5829): 1298–1301.

Stoller, E. P., L. E. Forster, and S. Portugal. 1993. Self-care responses to symptoms by older people: A health diary study of illness behavior. *Medical Care* 31 (1): 24–42.

Thompson, D. C., V. Rebolledo, R. S. Thompson, A. Kaufman, and F. P. Rivara. 1997. Bike speed measurements in a recreational population: Validity of self reported speed. *Injury Prevention* 3 (1): 43–45.

U.S. Census Bureau. http://www.census.gov (last accessed 30 November 2011).

Yang, Y., P. Atkinson, and D. Ettema. 2008. Individual space–time activity-based modelling of infectious disease transmission within a city. *Journal of the Royal Society Interface* 5 (24): 759–72.

Appendix Individual attributes, their values, and source data used to implement the simulation

Home-related attributes	Source data
Individual attributes	
1. Identity	Simulated
2. Gender (male, female)	SF3
3. Age (≤ 5, 6–15, 16–17, 18–64, ≥ 65)	SF3
4. Relationship to the householder (householder, spouse, own children, other children, other adults in the household)	SF3
5. Occupation (27 categories coded in NAICS)	SF3
Household attributes	
1. Home ID	Simulated
2. Household size (number of individuals)	SF3
3. Household type (family households: married couple, single-father and single-mother families; nonfamily households: living alone, not living alone)	SF3
4. Number of children	SF3
5. Gender composition (number of individuals for each gender)	SF3
6. Income level (18 categories defined by RefUSA household data)	RefUSA
7. Number of vehicles	SF3
8. Number of workers	SF3
9. Home address	RefUSA
10. Block group ID	SF3
11. Census tract ID	SF3

Workplace-related attributes	Source data
SF3, Home-related attributes	
1. Workers (identity, gender, age, occupation)	SF3, Home-related attributes
2. Travel mode (walk, bike, bus, car, and subway)	SF3, Home-related attributes
3. Travel time	SF3
4. Travel speed	Thompson et al. (1997); Federal Highway Administration (2000); U.S. Census Bureau
5. Travel distance	SF3, TIGER
6. Travel route	TIGER
Workplace attributes	
1. Workplace ID	Simulated
2. Workplace address	ESRI
3. Workplace type (27 categories coded in NAICS)	ESRI
4. Number of employees	ESRI

Service place-related attributes	Source data
Individual attributes	
1. Identity (age, gender, occupation, home ID, workplace ID)	Home-related and workplace-related attributes
2. Travel characteristics (mode, time, speed, distance, route)	Workplace-related attributes
3. Number of daily trips	Travel diary
4. Sequence of daily trips	Home-related and workplace-related attributes
5. Trip purpose (pick up or drop off, shopping, eating out, recreation, personal, and social)	Travel diary
6. Origin–destination (home-to-service, work-to-service, and service-to-service)	Travel diary
7. Day of week (weekday, weekend)	Simulated
8. Time period of trip (daytime, pastime)	Simulated
Service place attributes	
1. Service place ID	Simulated
2. Service place address	ESRI
3. Service place type (10 categories coded in NAICS)	ESRI
4. Number of employees	ESRI
5. Shift of employees (daytime, pastime)	Simulated

Note: SF3 = 2000 U.S. Census Summary File 3; NAICS = North America Industry Classification System; RefUSA = ReferenceUSA household data set; TIGER = Topologically Integrated Geographic Encoding and Referencing U.S. Census block group data; ESRI = Environmental Systems Research Institute workplace data set.

Population Movement and Vector-Borne Disease Transmission: Differentiating Spatial–Temporal Diffusion Patterns of Commuting and Noncommuting Dengue Cases

Tzai-Hung Wen,* Min-Hau Lin,† and Chi-Tai Fang‡

*Department of Geography, College of Science, National Taiwan University
†Graduate Institute of Health Policy and Management, College of Public Health, National Taiwan University
‡Graduate Institute of Epidemiology and Preventive Medicine, College of Public Health, National Taiwan University

Commuters who acquire dengue infections could be an important route for the transmission of the virus from their homes to workplaces. Understanding the effects of routine human movement on dengue transmission can be helpful in identifying high-risk areas for effective intervention. This study investigated the effects of local environmental and demographic characteristics to clarify the role of the daily commute in dengue transmission. We analyzed the clustering patterns of space–time distances between commuting and noncommuting dengue cases from June 2007 to January 2008 in Tainan City, Taiwan. We also analyzed the network topology of space–time distances to identify possible key individuals and conducted time-to-event analysis for geographic diffusion through commuting versus noncommuting dengue cases. Our significant findings indicate that most of the space–time distances of noncommuting cases clustered within 100 m and one week, whereas commuting cases clustered within 2 to 4 km and one to five weeks. Analysis of the temporality of the geographical diffusion by villages showed that commuting cases diffuse more rapidly across villages than noncommuting cases in the late epidemic period. The role of commuting was identified as a significant risk factor contributing to epidemic diffusion (hazard ratio: 3.08, p value < 0.05). Local neighborhood characteristics (number of vacant grounds and empty houses) are independent facilitating factors for diffusion through both noncommuting cases and commuting cases (hazard ratio: 1.035 and 1.022, respectively, both $p < 0.05$). Higher population density is a significant risk factor only for diffusion through commuters (hazard ratio: 1.174). In summary, noncommuters, mostly elderly adults and housewives, might initiate local outbreaks, whereas commuters carrying the virus to geographically distant areas cause large-scale epidemics.

携带登革热感染病的通勤者可能是该病毒从他们的家到工作场所的一个重要传播途径。了解人们的日常活动对登革热传播的影响，可能有助于确定可有效干预的高风险地区。本研究调查当地的环境和人口特征的影响，以澄清每天上下班通勤对登革热传播的作用。我们分析台湾的台南市在 2007 年 6 月至 2008 年 1 月间，通勤与非通勤之间的时空距离的聚类模式。通过通勤与非通勤人群中登革热病例，我们还分析了时空距离的网络拓扑结构，以识别可能的关键个人，和进行地理扩散的时间与事件分析。最重要的研究结果表明，大多数非通勤的情况下，病例聚集在 100 米和 1 个星期的时空距离，而对通勤的情况，病例聚集于 2 至 4 公里和 1 到 5 个星期内。由乡村地理扩散的时间性分析表明，通勤比不通勤的病例，在后期流行期间更迅速地弥漫到整个村庄。通勤的作用被确定为一个造成疫情扩散（危险比：3.08，p 值<0.05）的重要风险因素。当地社区的特点（空置地块和空房子的数量）是通勤和不通勤病例通用的扩散的独立促进因素（危险比分别为 1.035 和 1.022，均 p <0.05）。更高的人口密度，只是通勤情况中的一个重要的扩散危险因素（危险比：1.174）。总之，非通勤者，大多是老年人和家庭主妇，可能引发局部暴发，而乘客携带病毒到相距遥远的地区则造成大规模的流行。*关键词：通勤，登革热，疾病传播的风险，空间扩散，台湾。*

Los viajeros pendulares, o *conmutantes*, que se infecten de dengue podrían ser una vía importante para la transmisión del virus desde sus hogares hasta los lugares de trabajo. La comprensión de los efectos que tienen los movimientos humanos rutinarios sobre la transmisión del dengue puede ayudar mucho en la identificación de las áreas de alto riesgo para que allí pueda ejercerse una intervención efectiva. Este estudio investigó los efectos que tienen las características ambientales y demográficas locales para aclarar el papel que cumple el desplazamiento diario en la transmisión del dengue. Analizamos los patrones de agrupamiento registrados en contexto de distancias espacio–tiempo entre los casos de dengue cuyos orígenes se deben a viajeros pendulares y no conmutantes, entre junio de 2007 y enero de 2008 en la Ciudad de Tainan, Taiwán. También analizamos la topología de red de las distancias espacio–tiempo para identificar posibles individuos claves y realizamos análisis del tiempo transcurrido hasta el evento para la difusión geográfica en casos de dengue originados por viajes pendulares contra los transmitidos de otras maneras. Nuestros hallazgos significativos indican que la mayoría de las distancias espacio–tiempo de los casos no asociados con viajes pendulares se agruparon dentro de 100 m y una semana, en tanto que los casos difundidos mediante ese tipo de desplazamiento se agruparon dentro de 2 a 4 km y una a cinco semanas. El análisis de la temporalidad de la difusión geográfica por aldeas mostró que el contagio originado en viajes pendulares se difunde más rápidamente en las aldeas que en el caso de los no conmutantes en el período epidémico tardío. El papel de los viajes pendulares se identificó como un factor de riesgo significativo que contribuye a la difusión epidémica (razón de amenaza: 3.08, valor $p < 0.05$). Las características locales del vecindario (número de terrenos vacíos y casas desocupadas) son factores facilitadores independientes para la difusión tanto en los casos de no conmutantes como los de conmutantes (razón de amenaza: 1.035 y 1.022, respectivamente, ambos $p < 0.05$). Una mayor densidad de población es un factor de riesgo significativo solo para la difusión a través de conmutantes (razón de amenaza: 1.174). En resumen, los no conmutantes, principalmente adultos viejos y amas de casa, podrían iniciar brotes locales, mientras los viajeros pendulares que transportan el virus a áreas geográficas distantes causan epidemias a gran escala.

Dengue fever (DF) is one of the world's most widely spread mosquito-borne diseases and threatens millions of people in the two thirds of the world where the disease is commonly found. The clinical manifestations of dengue include DF, dengue hemorrhagic fever (DHF), and the most severe and potentially fatal dengue shock syndrome (DSS; Guzman et al. 2010). Infected cases are mainly distributed in tropical and subtropical areas in accordance with vector habitats for *Aedes aegypti* and *Aedes albopictus* (Halstead 2008). As currently there are no effective vaccines or antiviral drugs available for prevention of dengue infection, mosquito control and environmental management are regarded as the main intervention strategies for dengue control. Mosquito control relies on pesticide application to reduce adult and larval populations; environmental management, including campaigns and public education, focuses on larval habitat elimination (Guzman et al. 2010).

Identifying characteristics of local environments with larval habitats can be beneficial for targeting dengue control in these areas. Thammapalo et al. (2008) found that DF and DHF incidence in Thailand was strongly associated with environmental risk factors, including the percentages of shop-houses, brick-made houses, and houses with poor garbage disposal. Inadequate and badly covered storage of in-house water containers could also cause the proliferation of *Aedes aegypti* (de Mattos Almeida et al. 2007). Vacant spaces and empty houses are other possible mosquito breeding sites (Tan 1997; Ang and Singh 2001).

Due to the limited flight range (around 250 m) of mosquitoes in search of a blood meal (Gu et al. 2006), the causes of large-scale dengue diffusion might include other factors in addition to local environmental factors. Hales et al. (1999) found temporal correlations between monthly reports of DF cases on different islands of the South Pacific, with the propagation of infection spreading from larger islands to their smaller neighbors. Therefore, population density and human travel might play important roles in modulating or speeding up dengue transmission (Kan et al. 2008; Adams and Kapan 2009; Wen et al. 2010). Stoddard et al. (2009) also found that common exposure sites (e.g., markets, schools, or restaurants) have higher dengue transmission risks than local neighborhoods. Therefore, understanding the effects of human movement on dengue transmission can be helpful in identifying important demographic characteristics and high-risk areas in the transmission of pathogens and in effective dengue control (Vazquez-Prokopec et al. 2009). The objective of this study is to compare the epidemiology, diffusion patterns, and possible determinants between commuting and noncommuting cases to clarify the role of the commute in dengue transmission.

Data and Methods

Study Areas and Study Populations

Tainan City, one of the most densely populated cities in southern Taiwan, covers approximately 175.6 square

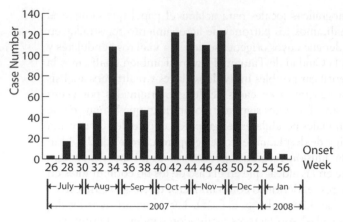

Figure 1. Epidemic curve of the dengue epidemic in Tainan City.

kilometers and has 771,900 inhabitants. Its population density is ranked fifth highest in Taiwan (Ministry of the Interior 2007). The city suffered a severe epidemic of dengue from June 2007 to January 2008 (Figure 1). We analyzed the 1,403 laboratory-confirmed dengue cases in this large epidemic. All of the dengue cases were reported by Taiwan Centers for Disease Control and confirmed in the laboratory by molecular identification, serological diagnosis, or virus isolation. In

our data set there are 1,017 confirmed dengue cases, with the number of illness onset by week, complete residential addresses, and working places (72.5 percent of total cases). We aggregated these cases in villages by their residences and mapped the cumulative incidence (case count divided by population) of each village in Figure 2. There are 548 cases whose residences are identical to their working places, categorized as non-commuting cases. The remainder of the dengue cases (469 cases) can be categorized as commuters. There are only twenty-nine cases (6.1 percent of commuting cases) whose working places are in their residential neighborhoods. Most commuters routinely travel across districts. A geographic information system (GIS), Arc-GIS 9.3, was used in this study for mapping and spatial analysis.

Exploratory Space–Time Analysis of the Spreading Dynamics

Because the week number of illness onset and residential location of each confirmed case are recorded in the database, we can calculate case-to-case distances in space and time. Based on the concept of the Knox

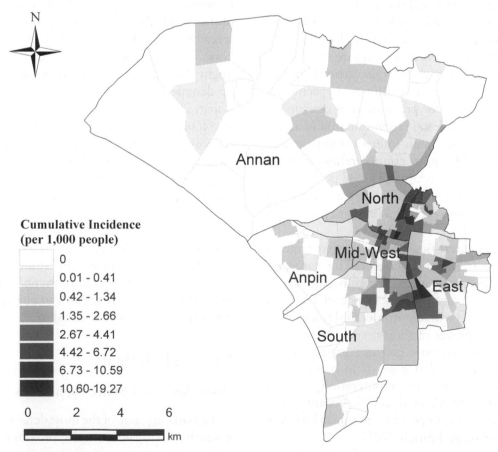

Figure 2. Spatial distribution of dengue incidence in Tainan City.

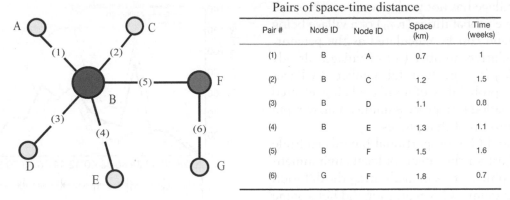

Pairs of space-time distance

Pair #	Node ID	Node ID	Space (km)	Time (weeks)
(1)	B	A	0.7	1
(2)	B	C	1.2	1.5
(3)	B	D	1.1	0.8
(4)	B	E	1.3	1.1
(5)	B	F	1.5	1.6
(6)	G	F	1.8	0.7

Figure 3. A hypothesized example of network topology of pairs of space–time distances. The circle size is proportional to the degree centrality of each node.

statistic (Knox 1964; Kulldorff and Hjalmars 1999), cumulative pairs of points calculated at a given space-distance (in 100 m) and time-distance (in weeks) are counted. We then compared the clustering patterns of space–time distances between commuting and non-commuting cases at these distances within the same intervals. MATLAB R2008 was used for analyzing cumulative pairs of space–time distances, and SigmaPlot 10.0 was used for mapping the diagrams of space–time distances.

Network Analysis of Pairs of Space–Time Distances

The larger numbers of cumulative pairs of points calculated at a given space-distance and time-distance can be regarded as space–time clustering of cases (Kulldorff and Hjalmars 1999). The outlier detection technique is used to determine the significantly large numbers of cumulative pairs of points. The upper fence as the threshold in this study is defined as the third quartile plus three times the interquartile range (Triola 2009). These pairs of space–time distances were selected and their spatial distribution mapped when the number of these cumulative pairs of points was larger than the upper fence. The link-node network structure composed of these pairs was further analyzed. As shown in Figure 3, one node that exists in many pairs represents a person who has many connections with others in time and space, and he or she could be regarded as a possible key individual. The concept is identical to degree-based centrality (the number of links that a node has), which is a commonly used measure in social network analysis (Freeman 1978). Therefore, based on the network topology of space–time distances, we determined the degree-based centrality of each dengue case to identify possible key individuals. Flowpy was used to establish geo-referenced

link-node networks in the GIS environment and MAT-LAB R2008 was used to analyze the network structure.

Time-to-Event Analysis of Geographical Diffusions

Time-to-event analysis, also called survival analysis in biostatistical and clinical medicine literature, examines the determinants of temporality in processes by analyzing time-to-event data. We applied time-to-event analysis to study the diffusion processes of dengue epidemic across geographical areas. The smallest local administrative division, designated as the village, was used as the unit for analysis. There are a total of 266 villages in Tainan City. Two events were modeled: (1) the occurrence of the first commuting case in a village and (2) the occurrence of the first noncommuting cases in a village. The week of the first dengue case in Tainan City (twenty-sixth week of 2007) was used as the zero time point. The cumulative probabilities for a village being affected by dengue via commuting cases versus that via noncommuting cases over time were computed and plotted using the Kaplan–Meier method and compared by log-rank test (Kleinbaum and Klein 2005). In brief, the cumulative probabilities ($R_{t(j)}$) of a village being affected by dengue at a given time point ($t_{(j)}$) were computed by the following formula:

$$R(t_{(j)}) = 1 - \hat{S}(t_{(j)})$$

$$= 1 - \prod_{i=1}^{j} \Pr(T > t_{(i)} | T \geq t_{(i)}) = \frac{n_{t(j)}}{N}, \quad (1)$$

where $\Pr(T > t_{(i)} | T \geq t_{(i)})$ is the conditional probability of a village being not affected during week i ($t_{(i)}$), given that the village has not yet been affected by week *i*. The cumulative probabilities ($R_{t(j)}$) of a village being affected by dengue can be calculated as the percentage of the cumulative number (n_j) of villages already affected in week j ($t_{(j)}$) in the total number of villages (*N*). The higher probability of a village being affected in a given time period by a process implies a more rapid geographical diffusion of that process.

We further used Cox proportional hazards multiple regression to adjust for the effects of local environmental and demographic factors in facilitating the diffusion of the dengue epidemic. Cox regression did not assume any specific parametric distribution on the temporality of the two modeled events. The only assumption is that the ratios of instantaneous incidence rate (hazard) of a village being affected by dengue between villages with or without a characteristic did not change over time (Kleinbaum and Klein 2005). The regression model included one dummy variable (1 = geographical diffusion via commuting cases; 0 = geographical diffusion via noncommuting cases) and five numeric variables: the numbers of schools, markets, parks, vacant grounds, and empty houses in the village, as well as the population density.

Results

Descriptive Analysis

Demographics. We divided the dengue cases into two groups—the commuting group and the noncommuting group—for comparisons of demographics and spatial-temporal patterns. As shown in Table 1, most patients in the noncommuting group are middle-aged or elderly adults. In the noncommuting group, 52 percent of males and 38.1 percent of females are older than sixty years old and 31.9 percent of males and 49.2 percent of females are forty to sixty years old. On the other hand,

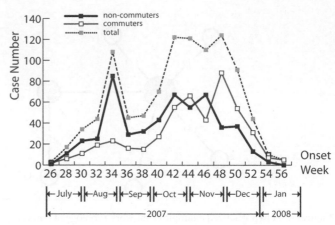

Figure 4. Epidemic curves of commuting and noncommuting dengue cases.

most patients in the commuting group are young adults or teenagers. In the commuting group, 62.8 percent of males and 63.3 percent of females are younger than forty years old. In general, the average ages of men and women in the noncommuting group are 59.5 and 54.1, respectively, whereas in the commuting group these averages are 32.1 and 33.2, respectively. The average age of a noncommuting case is statistically and significantly older than the average age of a commuting case (*p* value < 0.001).

Temporal and Spatial Distribution. Figure 4 shows the temporal progression of these two groups. The first epidemic wave (in August and early September) was induced by noncommuters, and most cases were noncommuters during this period. The patients in the second (major) wave (from late September to December) were mixed between commuters and noncommuters. The number of noncommuters in the beginning of the second wave (September) was still slightly larger than the number of commuters, but commuting cases

Table 1. Demographics of commuting and noncommuting dengue cases

Age	Noncommuting dengue cases			Commuting dengue cases		
	Males	Females	Total	Males	Females	Total
<20	9 (3.3%)	5 (2.0%)	14	**87(32.3%)**	57 (28.6%)	144
20–39	35 (12.8%)	27 (10.6%)	62	82 (30.5%)	**69 (34.7%)**	151
40–60	87 (31.9%)	**125(49.2%)**	212	86 (32.0%)	62 (31.2%)	148
>60	**142(52.0%)**	97 (38.1%)	239	14 (5.2%)	11 (5.5%)	25
Mean age[a]	59.5	54.5	57.1	32.1	33.2	32.6

Note: The largest number of each column is shown in bold.

[a]The mean age of noncommuting dengue cases was significantly older than commuting dengue cases by *t* test (*t* statistic = 22.34, *p* < 0.0001).

144

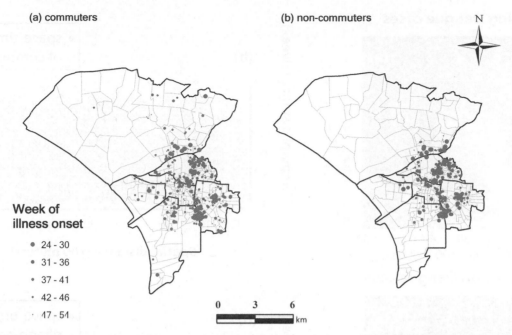

Figure 5. Spatial distribution of commuting and noncommuting dengue cases with the week numbers of illness onset in Tainan City.

dominated during the last epidemic period (from late November to December).

Spatial distributions of noncommuting and commuting cases with the week number of illness onset are shown in Figure 5. The spatial extent of commuting cases (Figure 5B) was more scattered than that of non-commuting cases (Figure 5A). The larger circle sizes of the map express the earlier stages of the epidemic. The central and southern areas of the city had a mix of the commuting and noncommuting cases in the beginning of the epidemic.

Exploratory Space–Time Analysis

Figure 6 shows the cumulative pairs of space–time distances for noncommuting and commuting cases, respectively. Deeper red coloring indicates the presence of more cumulative pairs in specific space–time distance intervals, which also indicates spatial–temporal clustering or interaction. The space–time clusters are further identified as red dots by the outlier detection technique in Figure 6B and 6D. Different space–time clustering patterns for noncommuting and commuting cases were identified. Most space–time pairs of non-commuting cases clustered within 100 m and one week, whereas commuting cases clustered within 2 to 4 km and one to five weeks. This means that on average the interval of illness onset between two noncommuting cases was within one week, and the residential distance

between the two was within 100 m. On the other hand, the interval of illness onset between two commuting cases was between one and five weeks, and the residential distance between the two was around 2 to 4 km.

Network Analysis for Identifying Possible Key Individuals

We then mapped space–time pairs within the distance of 100 m and one week illness onset interval for noncommuting cases and pairs with 2- to 4-km distances and one- to five-week illness onset intervals for commuting cases, as shown in Figure 7. The links between commuting cases covered most areas of the city, whereas noncommuting cases clustered in specific areas. The network density of these pairs showed that one intense spot of noncommuting cases occurred in the southeast of the city, with others scattered across the city, whereas commuting cases occurred in the center of the city (Figure 8).

Degree centrality in the network analysis was then calculated to identify the most influential and potentially key persons spreading the virus. The circles in Figure 9 indicate the residential locations of dengue cases, with larger circles representing higher degrees of centrality. Figure 9A shows higher degrees of centrality for noncommuting cases focused in the southeast part of the city, at the outbreak location of the first wave, whereas commuting cases are focused in the center of the city, at the epidemic foci of the second wave (Figure 9B).

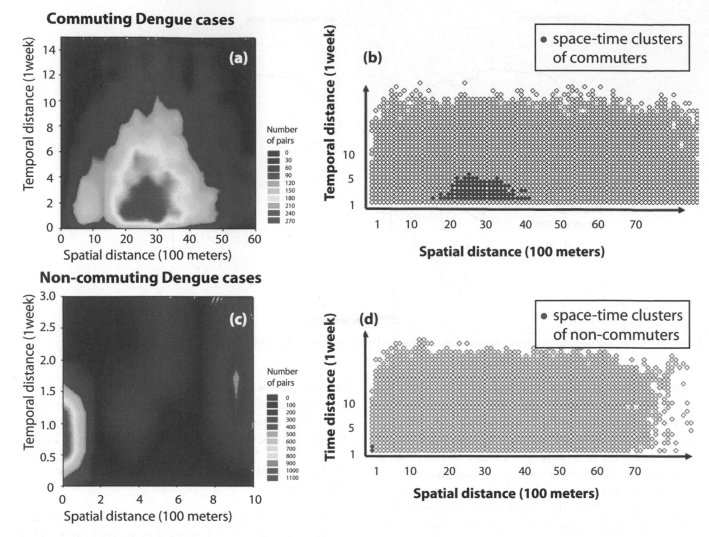

Figure 6. Patterns of space–time distance diagrams and space–time clusters of noncommuting dengue cases and commuting dengue cases. (Color figure available online.)

Kaplan–Meier Analysis for Geographic Diffusions

Figure 10 shows significantly higher cumulative probabilities for a village being affected by dengue via commuting cases in comparison with that via noncommuting cases during the epidemic period (log rank test, p value < 0.001). The difference becomes apparent after week forty, which indicated that the geographical diffusion via commuting cases is more rapid than the geographical diffusion via noncommuting cases.

Identifying Determinants for Geographic Diffusions

Cox proportional hazards regression showed that, after adjusting for the effects of other variables, geographical diffusion via commuting cases is more rapid than the

geographical diffusion via noncommuting cases (hazard ratio 3.080, p < 0.05; Table 2). We then separately analyzed factors in facilitating diffusion of dengue via noncommuting cases versus that via commuting cases. In univariate analysis, local neighborhood characteristics, including the number of vacant grounds, empty houses, and the local population density, are significant facilitating factors for the diffusion through both commuting cases and noncommuting cases (Tables 3 and 4). In multivariate analysis, local environmental risk factors (numbers of vacant grounds and empty houses) are independent facilitating factors for diffusion through both noncommuting and commuting cases (hazard ratio: 1.035 and 1.022, respectively, both p < 0.05). Meanwhile, higher population density is a significant risk factor only for the diffusion through commuting cases (hazard ratio: 1.174, p value < 0.05) but not for noncommuting cases.

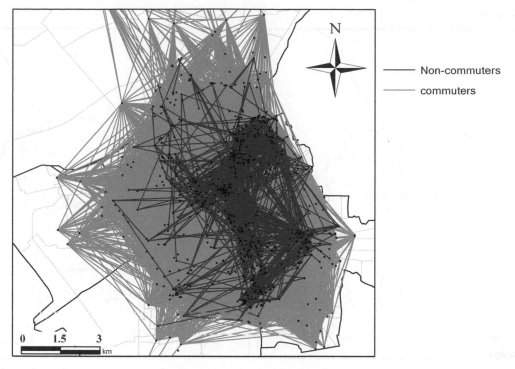

Figure 7. Network topology of noncommuting and commuting pairs of space–time distances.

Discussion

This study compared spatial–temporal and demographic patterns of commuting and noncommuting dengue patients and the determinants for disease diffusion to identify the effects of commuting on disease transmission. The analytical results suggest that commuting and noncommuting patients have different

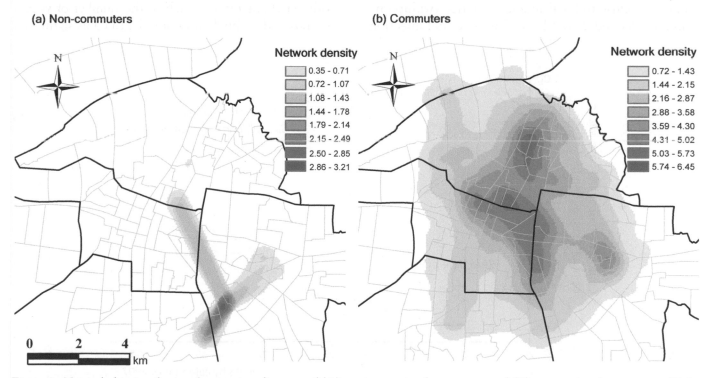

Figure 8. Network density of pairs of space–time distances of (A) noncommuting dengue cases and (B) commuting dengue cases. (Color figure available online.)

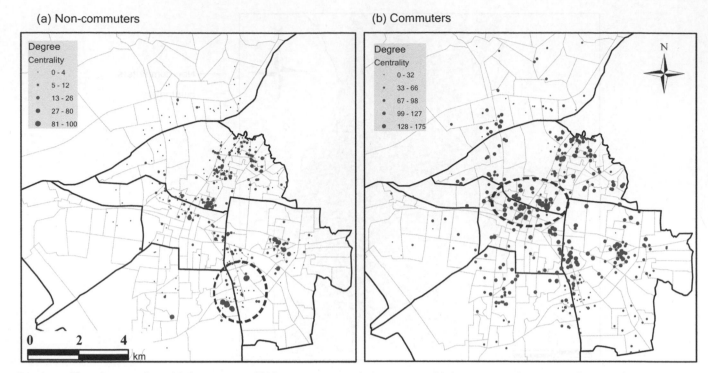

Figure 9. Identification of possible key persons of (A) noncommuting dengue cases and (B) commuting dengue cases by using degree centrality in network analysis. The circle size is proportional to the degree centrality of each node.

space–time clustering patterns (Figure 6). Most non-commuters clustered within a small range in time and space (within one week and around 100 m), whereas commuters clustered within a larger range (within one to five weeks and 2 to 4 km). The potential infection

sources of the first and second epidemic wave were also identified, at different locations, by using degree centrality in network analysis (Figure 9). Local neighborhood characteristics such as the number of vacant grounds and empty houses are identified as significant

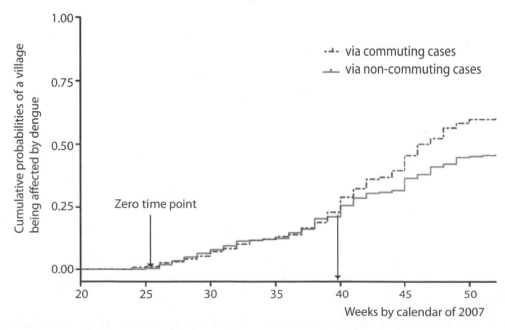

Figure 10. Kaplan–Meier curves for geographical diffusion via (A) noncommuting dengue cases and (B) commuting dengue cases.

Table 2. The results of Cox's regression analysis

Variables	Hazard ratios[a]
Commute	**3.080 (2.46, 3.85)**
No. of markets	1.038 (0.78, 1.38)
No. of vacant grounds and empty houses	**1.026 (1.01, 1.04)**
No. of schools	0.917 (0.81, 1.04)
No. of parks	1.008 (0.89, 1.14)
Population density	**1.098 (1.04, 1.15)**

Note: The numbers within the parentheses are the lower and upper limits of the 95 percent confidence interval. Statistically significant values are shown in bold (*p* value < 0.05).
[a]For each unit of increase in the variable.

risk factors for dengue diffusion and have larger effects on dengue patients who are noncommuters than those who are commuters (Table 2).

The methodological significance of this study is to propose the analytical procedures to quantify the geographical diffusion process by using the space–time distance diagrams and identify possible key persons of diffusion by determining the space–time clusters and the centrality of an individual. The residence and living environment of possible key persons could be the source of infection. Global space–time statistical methods, such as the Knox index and the Mantel index, identify the overall spatial–temporal clustering patterns (Kulldorff and Hjalmars 1999). Beyond the global patterns, this study further analyzed the significant space–time pairs from the space–time distance diagrams and identified the localized potential sources of diffusion. Epidemiologists often used the techniques of contact tracing to investigate the source of infection (Mitruka et al. 2011), but contact tracing usually relies

Table 3. Determinants of diffusion through dengue-infected noncommuters

Variables	Hazard ratios[a]	
	Univariate analysis	Multivariate analysis
No. of markets	1.121 (0.71, 1.77)	0.949 (0.59, 1.53)
No. of vacant grounds	**1.028 (1.00, 1.05)**	**1.035 (1.01, 1.06)[b]**
No. of empty houses	**1.104 (1.07, 1.14)**	
No. of schools	0.931 (0.75, 1.15)	0.826 (0.66, 1.04)
No. of parks	0.961 (0.76, 1.22)	0.993 (0.79, 1.25)
Population density	**1.124 (1.05, 1.21)**	1.077 (0.99, 1.17)

Note: The numbers within the parentheses are the lower and upper limits of the 95 percent confidence interval. Statistically significant values are shown in bold (*p* value < 0.05). Dependent variable is time to the first non-comuting dengue case.
[a]For each unit of increase in the variable.
[b]For a total number of vacant grounds and empty houses.

Table 4. Determinants of diffusion through dengue-infected commuters

Variables	Hazard ratios[a]	
	Univariate analysis	Multivariate analysis
No. of markets	1.209 (0.81, 1.81)	0.940 (0.60, 1.47)
No. of vacant grounds	**1.036 (1.02, 1.06)**	**1.022 (1.00, 1.04)[b]**
No. of empty houses	**1.075 (1.04, 1.11)**	
No. of schools	1.045 (0.88, 1.25)	0.913 (0.75, 1.11)
No. of parks	1.093 (0.93, 1.28)	1.127 (0.97, 1.32)
Population density	**1.206 (1.13, 1.28)**	**1.174 (1.09, 1.27)**

Note: The numbers within the parentheses are the lower and upper limits of the 95 percent confidence interval. Statistically significant values are shown in bold (*p* < 0.05). Dependent variable is time to the first commuting dengue case.
[a]For each unit of increase in the variable.
[b]For a total number of vacant grounds and empty houses.

on comprehensive contact records and epidemiological investigation. It is not feasible when response time and intervention resources are limited. Space and time of a patient's illness onset are important clues to infection source identification. These space–time relationships could provide insights of possible contacts and disease transmission. Instead of comprehensive contact investigation, our study demonstrates the feasibility of identifying possible infection sources by analyzing space–time relationships of patients' illness onset and differentiating these patterns between dengue-infected commuters and noncommuters.

This study suggested that commuting patients have larger ranges of space–time distances of diffusion than noncommuters (Figures 7 and 8) and clarified the role of commuting as a significant risk factor contributing to epidemic diffusion (Table 2), implying that routine human movement could be an important behavioral factor that facilitates the large-scale diffusion of dengue virus. Dengue-infected commuters during the period of viremia (defined as four to five days after the date of illness onset but potentially extending to twelve days; Halstead 2008) could be spreading the virus around their residences and workplaces through mosquito bites. Dengue-infected commuters could have a greater capability to spread the virus across geographical areas than dengue-infected noncommuters (Stoddard et al. 2009). Therefore, a surveillance system should have different levels of alarm that reflect the potential diffusion of someone who is reported as a dengue-infected case and is a commuter or noncommuter.

The possibly key noncommuting cases identified in this study were those of the first epidemic wave (in August and September, weeks thirty to thirty-six), which was induced by a cluster of infected elders at the elder center (Figure 9A). The difference in cumulative incidence between noncommuting and commuting cases was not significant during this period (Figure 10), implying that the virus circulated around both commuters and noncommuters. The second epidemic wave indicated that commuters have a significantly higher cumulative incidence than noncommuters after week thirty-nine (Figure 10). Most of the possibly key commuting cases are located at the center of the city (Figure 9B), implying that localized areas could be a source of diffusion and the commute could have induced the epidemic to become more widespread, transforming transmission from a local or small scale to a larger scale (Figures 8 and 10).

Local neighborhood characteristics such as the number of vacant grounds and empty houses in the neighborhood are identified as significant facilitating factors in this study (Tables 2 and 3). The reason for this result could be that the lifestyle of most of those infected with dengue primarily involves regular activities around their residences. Vacant grounds and empty houses are regarded as potential mosquito breeding areas (Tan 1997; Ang and Singh 2001). Therefore, noncommuters could have increased probabilities of human–mosquito contact around their residences. On the other hand, population density on an epidemic triggered by commuting cases is also a statistically significant factor. This could explain why, with an increase in the proportion of infected commuters, the second epidemic wave shifted to the center of the city. In addition, an epidemic triggered by commuters also has higher hazard ratios for common exposure sites, including markets, schools, and parks in a village. These parameters are not statistically significant, however. Our results suggest the spatial targeting of interventions should not only focus on the residences of infected commuters but also on common exposure sites and densely populated areas.

This study had notable limitations. First, we did not incorporate local entomological data (e.g., the mosquito density or Breteau index) into our study. Due to dengue virus transmission by mosquito bites, routine entomological surveillance data could assist in confirming that the possible sources of infection identified by this study would be the potential risk areas with higher mosquito density. Entomological surveillance was not established to systematically collect mosquito data with a standardized procedure in time and space, however. Therefore, we did not use entomological data because they lacked consistency and were unable to adequately represent the locations covered in our study.

Second, the study assumed that people were infected either at their workplaces or at their residences. Although epidemiological investigation recorded other places the patients visited before their illness onset dates to trace the source of infection, most of these recorded places were coded by place names rather than addresses or geographic coordinates, and therefore these places could not be identified and geocoded for further spatial analysis. The integration of epidemiological investigation with location-based technology will make it possible to construct a more comprehensive picture of disease spread (Vazquez-Prokopec et al. 2009). More detailed space–time distance diagrams could be constructed if information about locations each patient visited before and after illness onset could be geocoded. Multiple purposes of people movement such as commuting, leisure (shopping), or recreation, could be analyzed further and more possible sources of diffusion could also be identified.

Finally, mosquito control and intervention strategies could also affect the diffusion patterns of an epidemic. This study focused on the initiation of the epidemic in each village and assumed that there is no geographic heterogeneity of mosquito control measures during the epidemic. The spatial and temporal variations of the epidemic were assumed to be caused by local environmental and demographic characteristics.

Conclusion

An increase in the number of international travelers and local commuters has caused a surge of global and regional epidemics in the past several decades (Warren, Bell, and Budd 2010). In this study, the effects of routine commutes on disease diffusion are visualized and measured quantitatively. Commuting and noncommuting patients have different diffusion patterns and determinants in a dengue epidemic. The noncommuters, mostly elderly adults and housewives, might initiate a local epidemic, whereas commuters carrying the virus to geographically distant areas cause a large-scale epidemic. The rapid notification and diagnosis of noncommuting cases is important for providing earlier warning signals of emerging local dengue epidemics. Along with comprehensive tracking of commuting cases, these early alerts allow for appropriate interventions and faster response times, preventing subsequent large-scale epidemics.

Acknowledgments

This research was supported by grants from the National Science Council (NSC 98-2410-H-002-168-

MY2) and financial support provided by the Infectious Diseases Research and Education Center, Department of Health, and National Taiwan University. We would like to thank the health bureaus of the Tainan City government and the Centers for Disease Control in Taiwan for contributing dengue epidemiological data. We also thank Professor Chwan-Chuen King for her constructive suggestions.

References

Adams, B., and D. D. Kapan. 2009. Man bites mosquito: Understanding the contribution of human movement to vector-borne disease dynamics. *PLoS One* 4:e6763.

Ang, K. T., and S. Singh. 2001. Epidemiology and new initiatives in the prevention and control of dengue in Malaysia. *Dengue Bulletin* 25:7–14.

ArcGIS, Version 9.3. Redlands, CA: ESRI, Inc.

de Mattos Almeida, M. C., W. T. Caiaffa, R. M. Assuncao, and F. A. Proietti. 2007. Spatial vulnerability to dengue in a Brazilian urban area during a 7-year surveillance. *Journal of Urban Health* 84:334–45.

Flowpy. Santa Barbara, CA: ENJ.

Freeman, L. C. 1978. Centrality in social networks conceptual clarification. *Social Networks* 1 (3): 215–39.

Gu, W. D., J. L. Regens, J. C. Beier, and R. J. Novak. 2006. Source reduction of mosquito larval habitats has unexpected consequences on malaria transmission. *Proceedings of the National Academy of Sciences of the United States of America* 103:17560–63.

Guzman, M. G., S. B. Halstead, H. Artsob, P. Buchy, J. Farrar, D. J. Gubler, E. Hunsperger, et al. 2010. Dengue: A continuing global threat. *Nature Reviews Microbiology* 8:S7–S16.

Hales, S., P. Weinstein, Y. Souares, and A. Woodward. 1999. El Nino and the dynamics of vectorborne disease transmission. *Environmental Health Perspectives* 107: 99–102.

Halstead, S. B. 2008. *Dengue*. London: Imperial College Press.

Kan, C. C., P. F. Lee, T. H. Wen, D. Y. Chao, M. H. Wu, N. H. Lin, S. Y. Huang, et al. 2008. Two clustering diffusion patterns identified from the 2001–2003 dengue epidemic, Kaohsiung, Taiwan. *American Journal of Tropical Medicine and Hygiene* 79:344–52.

Kleinbaum, D. G., and M. Klein. 2005. *Survival analysis: A self-learning text*. New York: Springer-Verlag.

Knox, E. G. 1964. The detection of space–time interactions. *The Royal Statistical Society Series C: Applied Statistics* 13:25–29.

Kulldorff, M., and U. Hjalmars. 1999. The Knox method and other tests for space–time interaction. *Biometrics* 55:544–52.

MATLAB. 2008. Mathworks, Inc., Natick, MA.

Ministry of the Interior. 2007. *Taiwan area population statistics*. Taipei, Taiwan: Ministry of the Interior.

Mitruka, K., J. E. Oeltmann, K. Ijaz, and M. B. Haddad. 2011. Tuberculosis outbreak investigations in the United States, 2002–2008. *Emerging Infectious Diseases*. 17 (3): 425–31.

SigmaPlot, 10.0. San Jose, CA: Systat Software Inc.

Stoddard, S. T., A. C. Morrison, G. M. Vazquez-Prokopec, V. Paz Soldan, T. J. Kochel, U. Kitron, J. P. Elder, and T. W. Scott. 2009. The role of human movement in the transmission of vector-borne pathogens. *PLoS Neglected Tropical Diseases* 3:e481.

Tan, B. T. 1997. Control of dengue fever/dengue hemorrhagic fever in Singapore. *Dengue Bulletin* 21: 30–34.

Thammapalo, S., V. Chongsuvivatwong, A. Geater, and M. Dueravee. 2008. Environmental factors and incidence of dengue fever and dengue haemorrhagic fever in an urban area, Southern Thailand. *Epidemiology and Infection* 136:135–43.

Triola, M. F. 2009. *Elementary statistics*. Boston: Addison Wesley.

Vazquez-Prokopec, G. M., S. T. Stoddard, V. Paz-Soldan, A. C. Morrison, J. P. Elder, T. J. Kochel, T. W. Scott, and U. Kitron. 2009. Usefulness of commercially available GPS data-loggers for tracking human movement and exposure to dengue virus. *International Journal of Health Geographics* 8:68.

Warren, A., M. Bell, and L. Budd. 2010. Airports, localities and disease: Representations of global travel during the H1N1 pandemic. *Health & Place* 16: 727–35.

Wen, T. H., N. H. Lin, D. Y. Chao, K. P. Hwang, C. C. Kan, K. C. Lin, J. T. Wu, S. Y. Huang, I. C. Fan, and C. C. King. 2010. Spatial-temporal patterns of dengue in areas at risk of dengue hemorrhagic fever in Kaohsiung, Taiwan, 2002. *International Journal of Infectious Diseases* 14:e334–43.

Climate Change and Risk Projection: Dynamic Spatial Models of Tsetse and African Trypanosomiasis in Kenya

Joseph P. Messina,* Nathan J. Moore,[†] Mark H. DeVisser,* Paul F. McCord,* and Edward D. Walker[‡]

*Department of Geography, Center for Global Change and Earth Observations, and AgBioResearch, Michigan State University
[†]Department of Geography and Center for Global Change and Earth Observations, Michigan State University, and Department of Environmental and Resource Sciences, Zhejiang University, Hangzhou, China
[‡]Department of Entomology and Department of Microbiology and Molecular Genetics, Michigan State University

African trypanosomiasis, otherwise known as *sleeping sickness* in humans and *nagana* in animals, is a parasitic protist passed cyclically by the tsetse fly. Despite more than a century of control and eradication efforts, the fly remains widely distributed across Africa and coextensive with other prevalent diseases. Control and planning are hampered by spatially and temporally variant vector distributions, ecologically irrelevant boundaries, and neglect. Tsetse are particularly well suited to move into previously disease-free areas under climate change scenarios, placing unprepared populations at risk. Here we present the modeling framework ATcast, which combines a dynamically downscaled regional climate model with a temporally and spatially dynamic species distribution model to predict tsetse populations over space and time. These modeled results are integrated with Kenyan population data to predict, for the period 2050 to 2059, exposure potential to tsetse and, by association, sleeping sickness and nagana across Kenya.

非洲锥虫病，否则称为人体昏睡病或动物那加那病，是一种通过采采蝇传播循环的寄生性原生动物。尽管作了超过一个世纪的控制和根除的努力，苍蝇仍然广泛的在非洲分布并和其他流行疾病共存。控制和规划被它们具时空变异性的向量分布，生态无关的界限，和忽视等因素所阻碍。在气候变化情景下，采采蝇特别适合移动到以前无病的地区，给毫无准备的人群带来风险。在这里，我们提出 ATcast 的的建模框架，它结合了动态的降尺度的区域气候模型与时空动态物种分布模型，在时空上预测采采蝇种群。这些模拟结果与肯尼亚的人口数据相结合，预测了肯尼亚在 2050 年到 2059 期间暴露于采采蝇和相关的昏睡病和那加那病的可能性。关键词：气候变化，肯尼亚，风险预测，空间模型，采采蝇。

La tripanosomiasis africana, también conocida como *enfermedad del sueño* en los humanos y *nagana* en los animales, es una protista parásita transmitida cíclicamente por la mosca tsé-tsé. A pesar de más de un siglo de control y esfuerzos por erradicarla, la mosca permanece ampliamente distribuida a través de África y concurre con otras enfermedades prevalentes. Los procesos de control y planificación son obstaculizados por las variables distribuciones espaciales y temporales del vector, límites ecológicamente irrelevantes y por el descuido. Las moscas tsé-tsé están particularmente bien equipadas para desplazarse hacia áreas que antes estaban libres de la enfermedad gracias a nuevos escenarios derivados del cambio climático, colocando en riesgo a poblaciones que no estaban preparadas para enfrentarlo. Aquí presentamos la modelización de un escenario AT, que combina un modelo de clima regional dinámicamente reducido en escala con un modelo de distribución de especies, temporal y espacialmente dinámicas, para predecir las poblaciones de tsé-tsé en el espacio y el tiempo. Estos resultados modelados se integran con los datos de la población keniana para predecir, para el período 2050 a 2059, el potencial de exposición a la tsé-tsé y por asociación el potencial de enfermedad del sueño y nagana a través de Kenia.

African trypanosomiasis (AT), a neglected tropical disease, is a zoonotic, parasitic infection of wildlife, domesticated animals, and humans. Its causative agents (parasites of the *Trypanosoma brucei* species complex) are transmitted by the bite of the tsetse fly (genus *Glossina*). Approximately 8.5 million km² in thirty-seven sub-Saharan Africa countries are infested with tsetse (Allsopp 2001), resulting in approximately 70 million people with exposure risk (World Health Organization [WHO] 2010). Two major epidemics occurred in the first half of the twentieth century, one between 1896 and 1906 and the other in 1920 (WHO 2010). By the mid-1960s, human African trypanosomiasis (HAT) appeared to be under control. By the mid-1970s, however, HAT reemerged due to a breakdown in surveillance and control programs compounded by drug resistance, genetic changes in the parasite, civil conflict, and anthropogenic (land use and cover changes) and natural (climate) environmental change. In the mid-1990s it was estimated that at least 300,000 cases were underreported due to lack of surveillance capabilities, diagnostic expertise, and health care access (WHO 2010). In response to these limitations, the WHO, with public and private partnerships, initiated a new surveillance and elimination program, under which approximately 25,000 new cases were reported with an annual estimated rate of 50,000 to 70,000 cases (Weekly Epidemiological Record 2006). In 2010, the number of reported human cases of the disease dropped to 7,131, leading some to hope for eventual complete control of the human disease complex (WHO 2011). The disease is also considered one of the most important economically debilitating diseases in sub-Saharan Africa, with animal African trypanosomiasis (AAT) reducing livestock productivity by 20 percent to 40 percent in tsetse areas (Hursey 2001). In Kenya, where agriculture accounts for roughly a quarter of gross domestic product, the economic burden of African trypanosomiasis is acutely felt at both local and national scales (Bourn et al. 2001).

We hypothesize that climate change and anthropogenic activity combine to modify the environment to enhance or degrade habitat suitability for tsetse. We know that tsetse occupy environmental niches in Kenya that, based on existing biophysical data, they should not, and we know that tsetse are missing from areas in Kenya where they should exist in large and stable populations. These findings make cost-effective surveillance, control, and intervention efforts extremely difficult and traditional epidemiological prediction almost impossible. Many studies of tsetse exist (cf. Welburn, Maudlin, and Simarro 2009), but there are no disease vector studies that integrate fundamental niche models, species movement models, and climate change data. Some studies for other vectors have shown promise (i.e., Peterson [2009] for malaria; González et al. [2010] for leishmaniasis), but validation is largely limited to existing data sets (Kulkarni, Desrochers, and Kerr 2010) and often only linked to temporally static, biophysical variables. Although consistently reported to be important, few studies have empirically linked human activities, ecological stressors, vector responses, and disease emergence interacting at multiple spatiotemporal scales. In response, we quantify systematically the space–time distribution of tsetse across Kenya and deterministically predict the changing tsetse distributions expected to emerge with a changing climate. Future tsetse distributions are then placed and discussed within the social framework of Kenya.

Tsetse Control

Today, the international public health strategy for many vector-borne diseases has changed from eradication to ecological perspective vector control (Torr, Hargrove, and Vale 2005; Ferguson et al. 2010). A wide variety of techniques exist to control tsetse populations, including insecticide spraying, wild host culling, and land cover modification. The combination of ineffective application, emergent resistance, and environmental concerns, however, motivated the search for alternative strategies (Grant 2001). The sterile insect technique, one such alternative that has received widespread publicity for controlling tsetse in Zanzibar, has been ineffective in other locations. Anthropogenic landscape modification (autonomous control) involves the removal of tsetse habitat through the natural expansion of human settlement and cropland. This indirect method of control is the most effective and least expensive of the land clearing methods; however, it is difficult to quantify the extent to which tsetse habitat is reduced and it is not a practical policy solution (Bourn et al. 2001). The most frequently used ecologically friendly techniques are point-source control methods, which attract the flies with a combination of visual and odor cues to traps or targets (Leak, Ejigu, and Vreysen 2008). Although cost effective, widespread implementation of point-source control methods has been difficult and never effective over the long term. In East Africa, governments generally lack the infrastructure to manage and sustain the traps over large, diffuse affected areas and often situate traps or targets with only anecdotal evidence of tsetse presence (Hide 1999).

Tsetse and Trypanosomiasis in Kenya

Eight species of tsetse are found in Kenya, covering roughly 25 percent of Kenya's land area, including 60 percent of productive rangeland. Tsetse occupy diverse habitats in distinct "fly belts": the North and South belts near Mt. Kenya, the South Rift, Lake Victoria basin, Central Kenya, Trans Mara-Narok, and the Coastal belts. These belts are infested with one or more tsetse species with distributional limits that are set by intersections of physical, biological, and anthropogenic boundaries (Ford 1971). The most common habitats

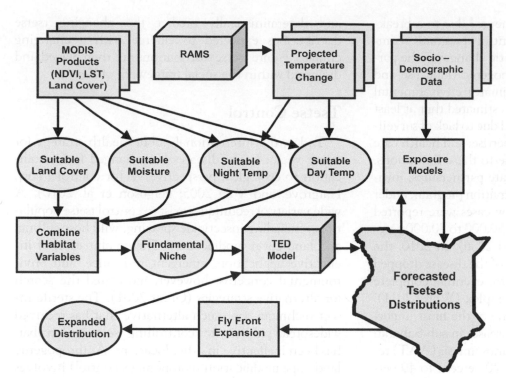

Figure 1. The African Trypanosomiasis Forecasting System (ATcast) modeling framework. MODIS = Moderate Resolution Imaging Spectroradiometer; NDVI = Normalized Difference Vegetation Index; LST = Land Surface Temperature; RAMS = Resional Atmospheric Modeling System; TED = Tsetse Ecological Distribution.

are riparian vegetation and woody savannah. Typical environmental limits for tsetse are day temperatures below 17°C and above 36°C, annual rainfall less than 300 mm, and lack of suitable resting sites (Jordan 1986; Terblanche et al. 2008). Different tsetse fly species can coexist in the same areas, making it difficult to assess quickly the causative agent in human or animal epidemics (Hide 1999). Tsetse flies are one of the few insect K-strategists, with long life expectancy (average of ninety days per female), high survival rates (>90 percent daily survivorship in adults), and low reproduction rates of one live pupa deposited in a suitable soil every six to nine days. The tsetse fly vector carries the parasites to different animal hosts, allowing cyclical transmission, but the primary animal reservoirs are wild ungulates and domestic cattle. Humans might also contribute to the reservoir pool (WHO 2010), and both animals and humans contribute to *Trypanosoma* genetic exchange (Hide 1999). Taxonomically, tsetse exist as three distinct clades. We focus on the *morsitans* or savannah group.

Methods

Our modeling environment, ATcast (for African Trypanosomiasis Forecasting System), is an integrated space–time projection ecological model (Figure 1). The Tsetse Ecological Distribution (TED) model (DeVisser

et al. 2010), a spatially explicit dynamic subcomponent of ATcast, predicts tsetse distributions at 250-m spatial and sixteen-day temporal resolution and can be described in two parts: (1) a spatially explicit fundamental niche model that identifies suitable tsetse habitat and (2) a fly movement model that integrates tsetse distributions and fly movement rates. The fundamental niche model uses four Moderate Resolution Imaging Spectroradiometer (MODIS) data sets: (1) the MODIS Terra Normalized Difference Vegetation Index (NDVI) Vegetation Indices 250 m V005 (MOD13Q1) product as a surrogate for available moisture (Williams et al. 1992), (2) the MODIS Terra Day Land Surface Temperature (LST) 1 km V005 product (MOD11A2), (3) the MODIS Terra Night LST 1 km V005 product (MOD11A2), and (4) the 1 km MODIS type 1 Global Land Cover product. Each of the four data sets is classified to a suitable versus unsuitable habitat classification scheme and combined to create a tsetse fundamental niche map every sixteen days. Temperature, moisture, and land cover parameterizations vary by species, but here are set for G. *morsitans*, commonly known as the woody savannah species and the most spatially extensive tsetse in Kenya (Bourn et al. 2001).

Habitat suitability alone is a poor predictor of tsetse presence, so a fly movement model was developed to identify potential tsetse distributions within the fundamental niche. The fly movement model does not predict

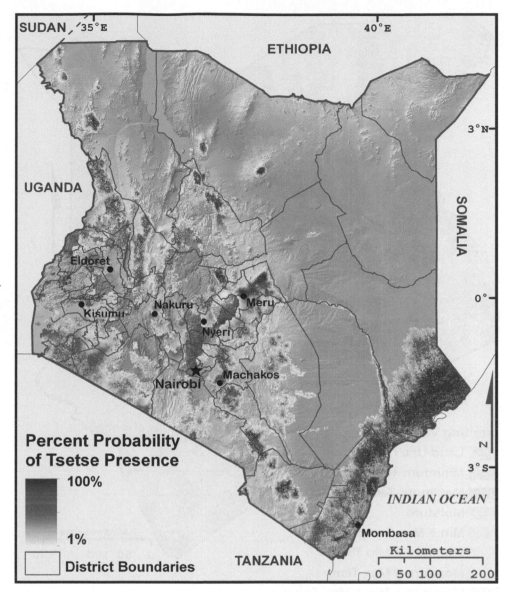

Figure 2. The Tsetse Ecological Distribution (TED) model percentage probability map overlaid on a physiographic map of Kenya. The location of several of the largest cities and district boundaries are included as indicators of higher human population densities. (Color figure available online.)

the movement rates of individuals but rather models tsetse distributions as a dynamic population-scale "fly front" (Hargrove 2000). If ecologically suitable tsetse habitat predicted by the fundamental niche model is encountered by the expanded tsetse distributions, then tsetse are allowed to persist in that location. This results in individual binary presence–absence maps at sixteen-day intervals, from 1 January 2001 to the acquisition date of the most recently available MODIS data products used in the model (here 15 October 2010). The probability of tsetse presence is the product of the sum of the binary distribution maps divided by the total number of scenes. The maximum extent of tsetse distributions is also produced and identifies any cells in which tsetse were predicted present during the time period analyzed (Figure 2). The formal validation of

the TED model is presented in DeVisser et al. (2010). Figure 3 is the limiting variable map constructed by using one landscape variable (i.e., day LST, night LST, NDVI, or Land Cover) at a time to model sensitivity of the tsetse fundamental niche. The maximum extents of tsetse distributions were then calculated for each of the four maps and compared to the normal TED model maximum extent map, identifying locations where one or more landscape variables are limiting tsetse distributions. The limiting variable map helps identify those areas immediately susceptible to change (Figure 3). For example, if under a future climate scenario moisture were to increase, these areas might become suitable for tsetse. Alternatively, areas where land cover is the limiting variable would likely become suitable if the land was abandoned. The minimum and maximum

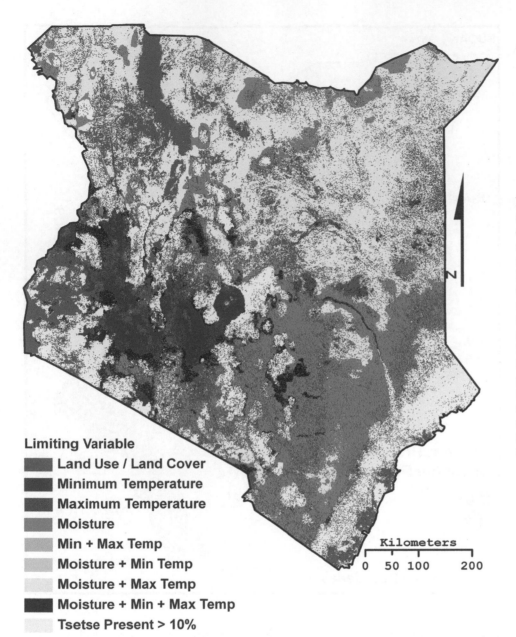

Limiting Variable

- Land Use / Land Cover
- Minimum Temperature
- Maximum Temperature
- Moisture
- Min + Max Temp
- Moisture + Min Temp
- Moisture + Max Temp
- Moisture + Min + Max Temp
- Tsetse Present > 10%

Figure 3. The spatial location where an environmental variable (i.e., land cover, minimum temperature, maximum temperature, and moisture) limits tsetse distributions within the Tsetse Ecological Distribution (TED) model. Combination classes indicate locations where more than one variable limits tsetse distributions (e.g., moisture + maximum temperature predicted to be both too hot and too dry at some point in the year for tsetse distributions to persist). The tsetse present >10% class indicates the locations where the TED model percentage probability map predicts a greater than 10 percent probability of tsetse always being present. (Color figure available online.)

temperature limiting variables are explicitly modeled in the following climate projection section.

To model climate change, we loosely coupled the Regional Atmospheric Modeling System (RAMS) version 4.4 (Cotton et al. 2003) with TED. The RAMS model is a state-of-the-art atmospheric model that numerically solves the fully compressible nonhydrostatic equations described by Tripoli and Cotton (1982) and captures exchanges of heat, momentum, and radiation between the surface and atmosphere. The modeled spatial extent spanned Kenya, Tanzania, Uganda, Rwanda, and Burundi and the vertical domain had thirty-three levels stretching to 32,581 m high. Surface and vegeta-

tion dynamics were governed by the LEAF-2 submodel (Walko et al. 2000). Land cover was taken from Torbick et al. (2006) with land cover parameters including albedo and fractional cover linked to appropriate global land cover classes. The RAMS parameterization was similar to that described in Moore et al. (2010), which documents extensive validation against observation and explains how MODIS vegetation time-series spline functions replaced more generic latitude–longitude functions to represent leaf area index and fractional cover for East Africa. For validation, six-hourly boundary conditions were obtained from the National Centers for Environmental Prediction (Kalnay et al. 1996).

The RAMS-derived climate simulations spanned two decades: 2000–2009 and 2050–2059, driven with boundary conditions supplied from the Community Climate System Model 3.0 SRES Scenario A1B (cf. Gent 2006); CO_2 levels with RAMS were updated in concordance with the boundary conditions. To explore possible changes in tsetse distributions given potential climate change, projected change in mean monthly minimum and maximum temperature between 2001 and 2009 and 2051 and 2059 from RAMS were added to the mean MODIS LST data and used as inputs in the TED model. The change in mean monthly minimum temperature was added to the mean nighttime LST data under the assumption that minimum temperatures most often occur at night. The change in mean monthly maximum temperature was subsequently added to the mean daytime LST data. Given the use of mean LST data, mean NDVIs in conjunction with the mode land use–land cover (LULC) class of each cell were used in lieu of projected NDVI and LULC data. The tsetse percentage probability map and maximum extent map from the TED model using the RAMS-projected changes in mean monthly temperature were then compared to the maps generated using the mean day and night LST, mean NDVI, and mode LULC data from 2001 to 2009, and potential changes in tsetse distributions were then identified.

At mesoscales, model assessment and validation were divided into design evaluation, sensitivity, and application error and uncertainty (cf. Santner, Williams, and Notz 2003). The projection components were particularly challenging to evaluate. There are very few studies that address uncertainty attendant to regional climate model choice due to the complexities of multiple comparable simulations (model intercomparison projects excepted). The few that have indicate that the choice of regional climate model has a large impact on measures of uncertainty, particularly in areas where parameterizations are poorly characterized. Because climate uncertainty propagates into forecast models, this additional constraint allowed for defining the limits of the ATcast projections (cf. Moore and Messina 2010). To reduce model-driven uncertainty in the results, the climate projections focused on the use of temperature data alone. Moisture, population, and land cover projections could all be used, but the cone of uncertainty would surely exceed the parameter space of even this deterministic model implementation.

Results and Discussion

The risk of acquiring AT is largely dependent on the intensity and duration of tsetse exposure and the susceptibility of the host population. Figure 4 presents those areas with emerging tsetse populations and those areas that should experience a decline in tsetse, with the important Highlands region identified. Slightly less than 20 percent of Kenya's land is considered high- or medium-potential agricultural land (Alila and Atieno 2006). Because most of Kenya's croplands are concentrated in the higher and historically more reliable rainfall zones of the Highlands, Lake Victoria basin, and a narrow strip along the coast, cropping and mixed farming have a distinct spatial distribution across Kenya. In fact, more than 90 percent of Kenya's croplands are found in these areas, and specifically for all areas identified as impacted by changing tsetse distributions in Figure 4, agriculture is the predominant occupation (World Resources Institute et al. 2007). Figure 5 is a bivariate plot map comparing current and future tsetse distributions, focusing on the Highlands region, with human population quantiles derived from the 2008 gridded global population data (Landscan 2008).

Following van de Steeg et al.'s (2010) Highland districts definition, Table 1 identifies districts of greatest tsetse expansion and, of these, all but Koibatek and Elgeyo are found in the Highlands. Further, of the districts where expansion is taking place, only Narok and Koibatek have a population density lower than Kenya's population average. Of the districts where the greatest amount of contraction is taking place, only Machakos is found in the Highlands and only Makueni and Machakos have population densities greater than Kenya's average population density.

Of the districts with the greatest net change (Table 2), only Kitui and Kajiado are experiencing a net decrease in tsetse distributions, both rangeland districts with relatively low human population densities. Of the districts with the greatest net change, all districts experiencing a net increase in tsetse distributions are found in the Highlands (notice that the two non-Highland districts where expansion was taking place—Koibatek and Elgeyo—are no longer listed). Additionally, apart from Narok, all districts with a net increase have a population density greater than Kenya's average population density. Figure 6 presents an alternative view of the distribution of expansion and contraction zones. Although the Highland region is not

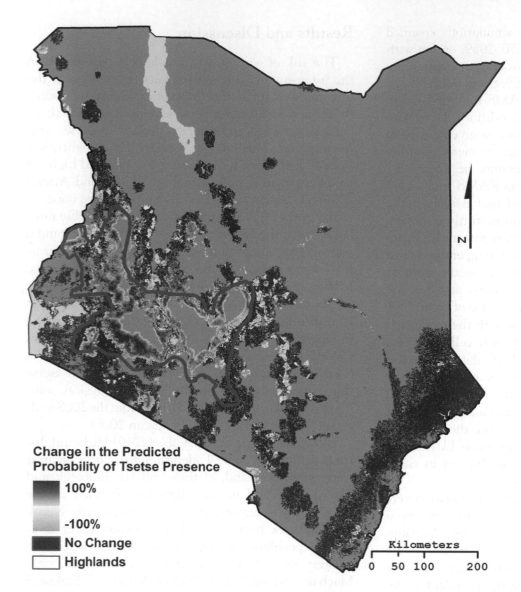

Change in the Predicted Probability of Tsetse Presence

- 100%
- -100%
- No Change
- Highlands

Kilometers

0   50 100   200

Figure 4. The Tsetse Ecological Distribution (TED) model–projected change in the percentage probability of tsetse occurrence based on projected change in temperature from Regional Atmospheric Modeling System. An expansion of tsetse indicates a location where tsetse were not predicted in 2001–2009 but are projected by 2051–2059. A contraction in tsetse indicates a location where tsetse were predicted in 2001–2009 but not projected as present by 2051–2059. A positive value indicates a location where the probability of tsetse occurrence is projected to increase by the displayed percentage. Conversely, a negative value indicates a location where the probability of tsetse occurrence is projected to decrease by the displayed percentage. The region referred to as the Kenyan Highlands is outlined in blue. (Color figure available online.)

all at similarly high elevations, it represents a general physiographic description of the region. That the link between elevation change and expansion or contraction is strong is not surprising given that the dynamic variable was temperature. It does provide additional support for the concern that tsetse will expand into the economically critical higher elevation areas of Kenya.

Susceptible human populations are those who live in proximity or travel into ecosystems suitable for tsetse habitation in which there are inadequate control or population preventative measures. It is likely that males and females within different age groups have different occupation and mobility patterns, resulting in similar levels of risk within ecosystems but differential levels of risk across ecosystems. Further, infected populations with lower incomes are at higher risk of debilitating sequelae because of limited access to preventative and curative resources. Finally, varying levels of population density are associated with varying levels of human risk depending on the threshold level of the ecosystem and tsetse prevention and control efforts. Specifically, emerging epidemic or endemic ecosystems require different thresholds of tsetse vectors and infective reservoirs to support the transmission cycle. For example, in densely populated Highland areas trypanosomiasis transmission might have high potential in emerging ecosystems (i.e., new tsetse niches) if there is a growing human-infection reservoir base in addition to animal reservoirs and prevention and control measures have not been initiated.

At the contextual level, political commitment, the transportation network, and the socioeconomic infrastructure in which populations reside and travel can directly or indirectly impact the transmission cycle

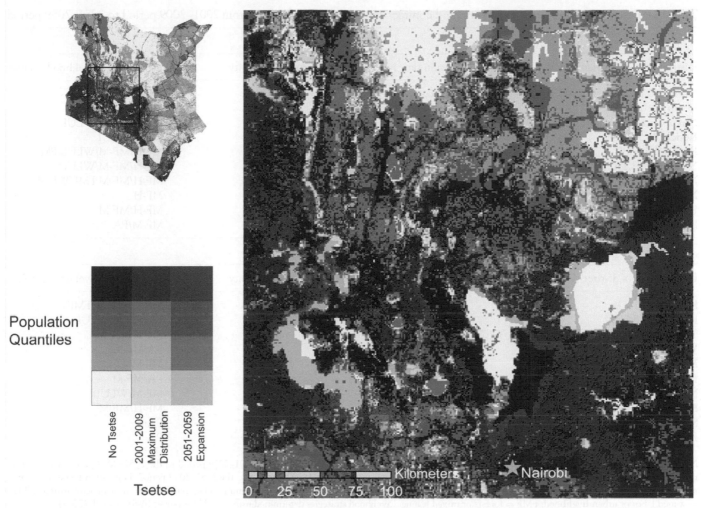

Figure 5. Present tsetse distributions (2001–2009) and future tsetse expansion areas (2051–2059) with corresponding distributions of human population. The future expansion of tsetse distributions will take place primarily in the Highlands, the most populated areas of Kenya. Human population in areas of no tsetse is represented with colors ranging from light gray to black. Human population in areas of present tsetse distributions is represented with colors ranging from light purple to dark purple. Human population in areas of future tsetse expansion is represented with colors ranging from light red to dark red. Ancillary data sources: Population (Landscan 2008), lakes (World Resources Institute et al. 2007). (Color figure available online.)

Figure 6. The frequency of cells with projected expanded/contracted maximum extent tsetse distributions from the Tsetse Ecological Distribution (TED) model using projected change in temperature from Regional Atmospheric Modeling System, plotted against elevation in 1-m increments.

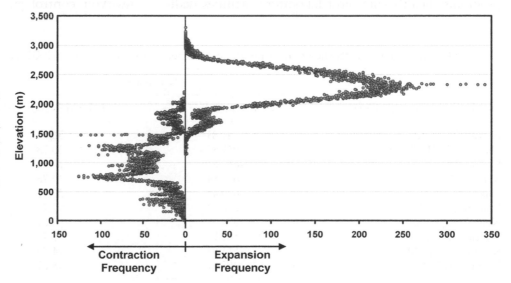

Table 1. Districts with greatest expansion and contraction in tsetse distributions from 2001–2009 period to 2051–2059 period

	Expansion		
District/area of district (km^2)	Expansion (km^2)	Population density	Predominant livelihood strategy
Narok/17,731.80	1,866.41	33.39	PA/WLUL/MF-H
Uasin Gishu/3,373.89	962.35	232.35[a]	MF-H/WLUL
Nakuru/7,605.95	708.68	180.00[a]	MF-H/MF-M/WLUL
Kericho/2,581.04	638.84	285.50[a]	MF-H/WLUL
Bomet/2,369.87	600.74	247.72[a]	MF-H/MF-M/WLUL/FMF
Nyandarua/3,270.53	574.98	170.68[a]	MF-H/MF-M/WLUL
Nyeri/3,370.60	567.72	219.09[a]	MF-H/MF-M/FMF/WLUL
Nandi/2,873.14	485.81	218.81[a]	MF-H
Koibatek/2,996.59	394.29	53.88	MF-H/MF-M
Elgeyo/1,450.21	278.84	113.31[a]	MF-M/PA
	Contraction		
District/area of district (km^2)	Contraction (km^2)	Population density	Predominant livelihood strategy
Kitui/30,391.70	662.81	30.88	MF-M/WLUL/FMF
Kajiado/21,847.20	511.92	18.63	PA/MF-M/WLUL
Turkana/61,037.30	397.87	8.3	PA/WLUL
West Pokot/9,284.93	303.62	36.86	PA/MF-M
Makueni/8,281.20	206.32	105.92[a]	MF-M
Baringo/7,943.17	187.35	34.72	PA/MF-M
Machakos/6,021.20	167.22	176.78[a]	MF-M/WLUL
Koibatek/2,996.59	144.88	53.88	MF-H/MF-M
Samburu/21,189.30	131.52	7.96	PA
Isiolo/25,114.10	115.09	4.65	PA

Note: Expansion represents the potential increase in tsetse distributions within a district from the 2001–2009 period to the 2051–2059 period. Contraction represents the potential decrease in tsetse distributions within a district from the 2001–2009 period to the 2051–2059 period. Population figures calculated using Landscan (2008) product for Kenya. MF-H = mixed farming—high potential; MF-M = mixed farming—marginal; PA = pastoral or agropastoral; WLUL = waged labor or urban livelihood; FMF = forests or mixed fishing. Livelihood strategies determined using World Resources Institute et al. (2007).
[a]Indicates that district population is greater than the 2008 average population density of 68 people/km^2 (Landscan 2008).

at different scales. Structural factors at both national and local scales include policies related to resource allocation for trypanosomiasis prevention and control programs, health care, and laboratory facilities dedi- cated to the diagnosis and treatment of trypanosomiasis infection. Further, the proximity of major and minor towns (i.e., distance and time) to tsetse or animal reservoir control programs influences the likelihood

Table 2. Districts with the greatest net change in tsetse distributions from 2001–2009 period to 2051–2059 period

	Net change		
District/area of district (km^2)	Net change (km^2)	Infested % 2001–2009	Infested % 2051–2059
Narok/17,731.80	1,840.83	56.64	67.02
Uasin Gishu/3,373.89	957.35	20.78	49.15
Kitui/30,391.70	−662.81	19.39	17.21
Kericho/2,581.04	638.84	49.78	74.53
Nakuru/7,605.95	629.44	27.92	36.20
Bomet/2,369.87	600.74	57.72	83.07
Nyandarua/3,270.53	574.98	5.07	22.65
Nyeri/3,370.60	567.72	37.77	54.62
Kajiado/21,847.20	−511.92	14.06	11.72
Nandi/2,873.14	485.81	38.07	54.98

Note: Net change represents the districts with the greatest total change in potential tsetse distributions from the 2001–2009 period to the 2051–2059 period.

of correct diagnosis. Area poverty might increase the risk of tsetse exposure and human trypanosomiasis via the poor quality of the local environment (e.g., presence of wild animal reservoirs and overgrowth of brush). Finally, it is likely that future contextual factors (i.e., structural factors and area poverty) will indirectly impact tsetse exposure and trypanosomiasis infection by exacerbating population-level risk factors in certain ecosystems. These demographic and social variables and their direct and indirect interactive effects on the risk of human trypanosomiasis infection should drive specific risk-reduction control strategies.

Tsetse flies are a particularly attractive vector insect for disease ecology space–time modeling. First, tsetse fly populations of the various known species tend to have low vagility (i.e., low dispersal rate) and tend to cluster in favorable habitats. Second, tsetse population density varies substantially seasonally, and there is only modest evidence for density-dependent control of population density, suggesting that density-independent factors operate primarily in regulating population size (Rogers and Randolph 1985). Thus, both locally and regionally, populations can be expected to vary spatially in density as a function of the availability and dynamics of adequate habitat for adults and for their live-born progeny. This largely explains the mystery surrounding emergent and disappearing populations, but scalar climate impacts alter the ability of tsetse to move into historically occupied spaces and also open historically inhospitable spaces. This is not endogenously predictable and is completely missed by the traditionally relied upon presence-only sampling.

Climate change will alter many infectious disease systems and impact heretofore unsuspecting and vulnerable populations (cf. Sutherst 2004). We certainly agree with Rogers and Randolph's (2002) assertion that proposals for the eradication of tsetse typically ignore historical, political, and ecological precedents. Tsetse flies exist in a complex space–time dynamic directly driven by ecological and anthropomorphic conditions. Reflecting this, ATcast produces spatiotemporal map products identifying the realized niche for tsetse and populations at risk given the complex multiscale interactions of climate, people, and the environment.

Tsetse and AT are very likely to (re)emerge under climate change scenarios as significant disease challenges. Planning, mapping, and monitoring efforts benefit from collaborations among interested organizations, although much remains to be done (Simarro et al. 2010). The probable expansion of tsetse into the Kenyan Highlands directly threatens the core of the agricultural dairy industry and places what is currently a large population of people at new exposure risk. By effectively preparing the health care system, veterinary services, and control measures, these potentially serious impacts might be mitigated.

Acknowledgments

This research was supported by the National Institutes of Health, Office of the Director, Roadmap Initiative, and NIGMS Award No. RGM084704A.

References

Alila, P. O., and R. Atieno. 2006. Agricultural policy in Kenya: Issues and processes. Paper presented at the Future Agricultures Consortium Workshop, Institute of Development Studies, University of Sussex, UK.

Allsopp, R. 2001. Options for vector control against trypanosomiasis in Africa. *Trends in Parasitology* 17 (1): 15–19.

Bourn, D., R. Reid, D. Rogers, B. Snow, and W. Wint. 2001. *Environmental change and the autonomous control of tsetse and trypanosomiasis in sub-Saharan Africa.* Oxford, UK: Information Press.

Cotton, W. R., R. A. Pielke, Sr., R. L. Walko, G. E. Liston, C. Tremback, H. Jiang, R. L. McAnelly, J. Y. Harrington, and M. E. Nicholls. 2003. RAMS 2001: Current status and future directions. *Meteorology Atmospheric Physics* 82 (1–4): 5–29.

DeVisser, M. H., J. P. Messina, N. J. Moore, D. P. Lusch, and J. Maitima. 2010. A dynamic species distribution model of Glossina subgenus Morsitans: The identification of tsetse reservoirs and refugia. *Ecosphere* 1 (1): Art. 6.

Ferguson, H. M., A. Dornhaus, A. Beeche, C. Borgemeister, M. Gottlieb, M. S. Mulla, J. E. Gimnig, D. Fish, and G. F. Killeen. 2010. Ecology: A prerequisite for malaria elimination and eradication. *PLoS Medicine* 7 (8): e1000303.

Ford, J. 1971. *The role of the trypanosomiases in African ecology: A study of the tsetse fly problem.* Oxford, UK: Oxford University Press.

Gent, P. R., ed. 2006. Special issue on Community Climate System Model (CCSM). *Journal of Climate* 19 (11).

González, C., O. Wang, S. E. Strutz, C. González-Salazar, V. Sánchez-Cordero, and S. Sarkar. 2010. Climate change and risk of leishmaniasis in North America: Predictions from ecological niche models of vector and reservoir species. *PLoS Neglected Tropical Diseases* 4 (1): e585.

Grant, I. F. 2001. Insecticides for tsetse and trypanosomiasis control: Is the environmental risk acceptable? *Trends in Parasitology* 17 (1): 10–14.

Hargrove, J. W. 2000. A theoretical study of the invasion of cleared areas by tsetse flies (Diptera: Glossinidae). *Bulletin of Entomological Research* 90:201–09.

Hide, G. 1999. History of sleeping sickness in East Africa. *Clinical Microbiology Reviews* 12 (1): 112–25.

Hursey, B. S. 2001. The programme against African trypanosomiasis: Aims, objectives and achievements. *Trends in Parasitology* 17 (1): 2–3.

Jordan, A. M. 1986. *Trypanosomiasis control and African rural development*. New York: Longman.

Kalnay, E., M. Kanamitsu, R. Kistler, W. Collins, D. Deavan, M. Iredell, S. Saha, et al. 1996. The NCEP/NCAR 40-year reanalysis project. *Bulletin of the American Meteorology Society* 77 (3): 437–71.

Kulkarni, M. A., R. E. Desrochers, and J. T. Kerr. 2010. High resolution niche models of malaria vectors in northern Tanzania: A new capacity to predict malaria risk? *PLoS One* 5 (2): e9396.

Landscan. 2008. *2008 global population database*. Oak Ridge, TN: UT-Battelle, operated by Oak Ridge National Laboratory.

Leak, S. G. A., D. Ejigu, and M. J. Vreysen. 2008. *Collection of baseline entomological baseline data for tsetse area-wide integrated pest management programmes*. Rome, Italy: Food and Agriculture Organization of the United Nations.

Moore, N., and J. Messina. 2010. A landscape and climate data logistic model of tsetse distribution in Kenya. *PLoS One* 5 (7): e11809.

Moore, N., N. Torbick, B. Pijanowski, B. Lofgren, J. Wang, D.-Y. Kim, J. Andresen, and J. Olson. 2010. Adapting MODIS-derived LAI and fractional cover into the RAMS model in East Africa. *International Journal of Climatology* 30 (13): 1954–69.

Peterson, A. T. 2009. Shifting suitability for malaria vectors across Africa with warming climates. *BMC Infectious Diseases* 9:59.

Rogers, D. J., and S. E. Randolph. 1985. Population ecology of tsetse. *Annual Review of Entomology* 30:197–216.

———. 2002. A response to the aim of eradicating tsetse from Africa. *Trends in Parasitology* 18 (12): 534–36.

Santner, T. J., B. J. Williams, and W. I. Notz. 2003. *The design and analysis of computer experiments*. New York: Springer-Verlag.

Simarro, P. P., G. Cecchi, M. Paone, J. R. Franco, A. Diarra, J. A. Ruiz, E. M. Fèvre, F. Courtin, R. C. Mattioli, and J. G. Jannin. 2010. The atlas of human African trypanosomiasis: A contribution to global mapping of neglected tropical diseases. *International Journal of Health Geographics* 9:57.

Sutherst, R. W. 2004. Global change and human vulnerability to vector-borne diseases. *Clinical Microbiology Reviews* 17 (1): 136–73.

Terblanche, J. S., S. Clusella-Trullas, J. A. Deere, and S. L. Chown. 2008. Thermal tolerance in a south-east African population of the tsetse fly *Glossina pallidipes* (Diptera, Glossinidae): Implications for forecasting climate change impacts. *Journal of Insect Physiology* 54 (1): 114–27.

Torbick, N., D. Lusch, J. Qi, N. Moore, J. Olson, and J. Ge. 2006. Developing land use/land cover parameterization for climate and land modeling in East Africa. *International Journal of Remote Sensing* 27 (19): 4227–44.

Torr, S. J., J. W. Hargrove, and G. A. Vale. 2005. Towards a rational policy for dealing with tsetse. *Trends in Parasitology* 21 (11): 537–41.

Tripoli, G., and W. Cotton. 1982. The Colorado State University three-dimensional cloud/mesoscale model: Part I. General theoretical framework and sensitivity experiments. *Journal de Recherches Atmospheriques* 16:185–219.

Van de Steeg, J. A., P. H. Verburg, I. Baltenweck, and S. J. Staal. 2010. Characterization of the spatial distribution of farming systems in the Kenyan Highlands. *Applied Geography* 30 (2): 239–53.

Walko, R. L., L. E. Band, J. Baron, T. G. F. Kittel, R. Lammers, T. J. Lee, D. Ojima, et al. 2000. Coupled atmosphere-biophysics-hydrology models for environmental modeling. *Journal of Applied Meteorology* 39 (6): 931–44.

Weekly Epidemiological Record. 2006. Human African trypanosomiasis (sleeping sickness): Epidemiological update. *Weekly Epidemiological Record* 8 (81): 69–80.

Welburn, S. C., I. Maudlin, and P. P. Simarro. 2009. Controlling sleeping sickness—A review. *Parasitology* 136 (14): 1943–49.

Williams, B., D. Rogers, G. Staton, B. Ripley, and T. Booth. 1992. Statistical modeling of georeferenced data: Mapping tsetse distributions in Zimbabwe using climate and vegetation data. In *Modelling vector-borne and other parasitic diseases*, ed. B. D. Perry and J. W. Hansen, 267–80. Nairobi, Kenya: The International Laboratory for Research on Animal Diseases

World Health Organization (WHO). 2010. African trypanosomiasis (sleeping sickness). http://www.who.int/mediacentre/factsheets/fs259/en/ (last accessed 30 November 2010).

———. 2011. New cases of human African trypanosomiasis continue to drop: Decline strengthens prospects for elimination. http://www.who.int/neglected_diseases/disease_management/HAT_cases_drop/en/index.html (last accessed 7 July 2011).

World Resources Institute; Department of Resource Surveys and Remote Sensing, Ministry of Environment and Natural Resources, Kenya; Central Bureau of Statistics, Ministry of Planning and National Development, Kenya; and International Livestock Research Institute. 2007. *Nature's benefits in Kenya, an atlas of ecosystems and human well-being*. Washington, DC, and Nairobi, Kenya: World Resources Institute.

Spatial-Temporal Analysis of Cancer Risk in Epidemiologic Studies with Residential Histories

David C. Wheeler,* Mary H. Ward,[†] and Lance A. Waller[‡]

*Department of Biostatistics, School of Medicine, Virginia Commonwealth University
[†]Occupational and Environmental Epidemiology Branch, Division of Cancer Epidemiology and Genetics, National Cancer Institute, National Institutes of Health, U.S. Department of Health and Human Services
[‡]Department of Biostatistics and Bioinformatics, Rollins School of Public Health, Emory University

Exploring spatial-temporal patterns of disease incidence identifies areas of significantly elevated risk and can lead to discoveries of disease risk factors. One popular way to investigate patterns in risk over space and time is spatial-temporal cluster detection analysis. The identification of significant clusters could lead to etiological hypotheses to explain the pattern of elevated risk and to additional epidemiologic studies to explore these hypotheses. Several methodological issues and data challenges that arise in space–time cluster analysis of chronic diseases, such as cancer, include poor spatial precision of residence locations, long disease latencies, and adjustment for known risk factors. This article reviews the key challenges faced when performing cluster analyses of chronic diseases and presents a spatial-temporal analysis of non-Hodgkin lymphoma (NHL) risk addressing these challenges. Residential histories, collected as part of a population-based case-control study of NHL (the National Cancer Institute [NCI] Surveillance, Epidemiology, and End Results [SEER] NHL study) in four SEER centers (Detroit metropolitan area, Los Angeles, Seattle metropolitan area, and Iowa) were geocoded. In this analysis, we explored previously detected spatial-temporal clusters and adjusted for exposure to polychlorinated biphenyls (PCBs) and genetic polymorphisms in four genes, previously found to be associated with NHL, using a generalized additive model framework. We found that the genetic factors and PCB exposure did not fully explain previously detected areas of elevated risk.

探索疾病发病率的时空模式，识别显著的高风险区域，可能发现导致疾病的危险因素。调查时空风险模式的一种流行方式是时空群集检测分析。本文确定重大的集群可能会产生病因的假说来解释高风险的模式，和产生探讨这些假说的额外的流行病学研究。在对慢性疾病，如癌症的时空聚类分析的过程中，出现了几个方法问题和数据挑战，包括居住地点较差的空间精度，长的疾病潜伏期，和对已知危险因素的调整。本文综述在进行慢性病的聚类分析时所面临的关键挑战，并提出了 个有关非霍奇金淋巴瘤 (NHL) 风险的时空分析来应对这些挑战。对四个 SEER 中心（底特律大都市区，洛杉矶，西雅图大都市区，和爱荷华）的住宅历史（国家癌症研究所[NCI]的监测，流行病学和 NHL 研究的最终结果[SEER]）进行了地理编码，把它们作为以人口为基础的 NHL 病例对照研究的一部分进行收集。在这个分析中，我们使用广义相加模型框架，探讨了先前侦测到的时空集群和调整暴露在多氯联苯 (PCB) 和四个基因中的遗传多态性，它们以前被发现与 NHL 相关。我们发现，遗传因素和 PCB 暴露并不能完全解释先前侦测到的高风险地区。关键词：癌症，病例对照研究，聚类分析，流行病学，广义相加模型，非霍奇金淋巴瘤。

La exploración de los patrones espacio-temporales de la incidencia de una enfermedad sirve para identificar las áreas de riesgo significativamente alto, y puede llevar al descubrimiento de factores de riesgo de la enfermedad. Una manera popular para investigar los patrones de riesgo a través del espacio y el tiempo es el análisis de detección de agrupamiento temporal en el espacio. La identificación de agrupamientos significativos podría conducir a hipótesis etiológicas que expliquen el patrón de riesgo elevado y a estudios epidemiológicos adicionales que exploren estas hipótesis. Varios aspectos metodológicos y retos sobre datos que surgen en el análisis del agrupamiento espacio-temporal de enfermedades crónicas, cáncer por ejemplo, incluyen una precisión espacial muy pobre sobre la localización de las residencias, largas latencias de la enfermedad y ajuste de factores de riesgo conocidos. Este artículo revisa los retos claves que se enfrentan cuando se efectúan análisis de agrupamiento para enfermedades crónicas, y presenta un análisis espacio-temporal de riesgo del linfoma no Hodgkin (NHL), que aboca estos retos. Se geo-referenciaron las historias residenciales, recogidas como parte de un estudio de caso para control basado en población de NHL (el estudio de Vigilancia, Epidemiología y Resultados Finales [SEER] del Instituto

Nacional de Cáncer [NCI]), realizado en cuatro centros SEER (área metropolitana de Detroit, Los Ángeles, área metropolitana de Seattle, e Iowa). En este análisis exploramos agrupamientos espacio-temporales previamente detectados y ajustados por exposición a bifenilos policlorados (PCB) y polimorfismos genéticos en cuatro genes, que previamente habían sido hallados asociados con NHL, utilizando un marco de modelo aditivo generalizado. Descubrimos que los factores genéticos y la exposición a los PCB no explicaban completamente las áreas de alto riego previamente detectadas.

E xploring spatial-temporal patterns of disease incidence has proven to be beneficial for identifying areas of significantly elevated risk and discovering significant factors associated with risk. Particularly for cancer, there is a long history of research analyzing geographic patterns in disease incidence and mortality with the objective of discovering environmental determinants of disease (Fraumeni and Blot 1977). Examples of risk factors revealed by analytic epidemiologic studies that followed on observations of geographic patterns of cancer include exposure to asbestos from shipyards as a risk factor for lung cancer among men along the southeastern United States seaboard (Blot et al. 1979) and chronic use of snuff as a risk factor for oral and pharyngeal cancer among women in the southern United States (Winn et al. 1981).

Although there have been success stories in pursuing leads from analyzing geographic patterns of disease, most early studies of spatial patterns of cancer were ecological studies, using data on disease and the population at risk aggregated to areal units, such as counties. Ecological studies have a number of inherent analytic challenges (Beale et al. 2010) that limit their role in etiologic research. These challenges include spatial inaccuracy of data, exposure misclassification, and ecological bias (Wakefield and Elliott 1999; Elliott and Savitz 2008). In addition, analyses in ecological studies are usually based on administrative geographic boundaries that are not inherently meaningful for studying disease. These studies lack information on residential history and risk factors for individuals. Furthermore, environmental exposure data of interest typically will have been collected on different spatial scales.

For establishing causal factors in chronic diseases, studies should collect individual-level data (Elliott and Savitz 2008). In public health research, individual-level data are the foundation of case-control and cohort studies. Often, these epidemiologic studies contain spatial information at the individual level through residential addresses, which might include the address at time of diagnosis for a case or time of study enrollment for controls or cohort subjects. Increasingly, residential histories over long periods of a participant's lifetime are available (collected directly from participants); hence,

it is possible to consider residential mobility and disease latency when analyzing disease patterns. In addition, with the increasing accessibility of geographic information systems and geocoding technology, it is possible to analyze epidemiologic data at a finer spatial scale than in the past.

One approach to analyzing geographic patterns in disease that makes use of individual-level data is the detection of spatial clusters or areas of significantly elevated risk. The identification of clusters in space and time can lead to the development of hypotheses to explain the pattern of elevated risk and reveal important clues about disease etiology. We note the distinction in goals between detecting an individual cluster (or clusters) and approaches to describe general clustering of disease, the general tendency for cases to occur nearer other cases than one might expect under equal risk (Besag and Newell 1991; Waller and Gotway 2004). The incorporation of temporal data further refines the analysis by linking cases that are coincident in both time and space.

Our discussion focuses on cluster detection for individual-level epidemiologic studies with residential histories. In the remainder of this article, we discuss the challenges that epidemiologic studies with residential histories present for existing approaches in cluster detection and then present a spatial-temporal analysis of non-Hodgkin lymphoma (NHL) risk, addressing these challenges through a statistical analysis approach that evaluates residential histories and adjusts for known risk factors. Our motivating interest centers on evaluation of whether or not previously detected areas of significantly elevated NHL risk in a case-control study could be explained by adjusting for additional environmental and genetic risk factors.

Cluster Analysis Approaches for Epidemiologic Studies with Residential Histories

There are several existing methods for spatial cluster detection within individual-level data. Among the most commonly used methods are local scan statistics

(Kulldorff 1997, 2006), kernel density ratio estimation (Bithell 1990; Kelsall and Diggle 1995), Q-statistics (Jacquez et al. 2005), and generalized additive models (Kelsall and Diggle 1998; Vieira et al. 2005; Webster et al. 2006), and we limit our discussion to these methods. Few cluster detection methods are designed to fully evaluate the multidimensional, spatial, and temporal data that are increasingly available within epidemiologic studies. To improve the power to detect unexplained clusters when exploring spatio-temporal patterns of disease in individual-level data, analysis approaches should explicitly consider residential patterns that change over time due to migration and adjust for known risk factors.

In many cluster studies, the residential locations of study subjects at time of diagnosis are typically the only address information available and are assumed to be a reasonable surrogate for unmeasured environmental exposures, defined broadly to include lifestyle factors as well as pollutants. Due to residential mobility, the residence at time of diagnosis of disease might not accurately reflect the most relevant environmental exposures for diseases with long latencies, such as cancer. We define latency as the number of years between exposure to a relevant risk factor and the diagnosis of disease. For diseases with long latencies, migration must be considered. Researchers in public health and geography (Jacquez et al. 2005; Vieira et al. 2005; Sabel et al. 2009) have recognized the importance of migration when studying patterns and etiology of disease. Ignoring migration when studying health outcomes with long latencies can lead to exposure misclassification, diminished study power, and biased risk estimates (Tong 2000). Migration bias can occur when there is differential migration related to a factor of interest among study population groups (Tong 2000). The factor of interest is typically space in a cluster detection study.

Among existing methods, only Q-statistics were designed to adjust for migration. Q-statistics can consider the entire residential history of study subjects but require an a priori knowledge of the number of nearest neighbor subjects to use in defining the relevant cluster size when searching over space for clusters. Unfortunately, one often does not know which number of nearest neighbors is relevant for a particular study. The local scan statistics and the kernel density ratios were designed to model only one relevant location for each study subject. Analyses with these methods typically use the residential address at the time of diagnosis, which makes the implicit assumption that individuals do not migrate (at least between the time of the relevant ex-

posure and the diagnosis of disease) or that the latency between causal exposures and diagnosis of disease is negligible (Jacquez 2004). Generalized additive models (GAMs) have the potential to model several residential locations in a residential history, but studies to date using GAMs have either assumed one latency period with little empirical justification or have included all historical residential locations for each subject in one statistical model, violating the model assumption of independent observations with potentially biased model parameters (Vieira et al. 2005; Webster et al. 2006). Vieira et al. (2008) explored latency while estimating disease risk spatially using GAMs in overlapping time periods but included multiple addresses per subject in each time period. An adjustment is needed to include several records for a subject in a GAM, or the data must be structured in a way to include only one record per subject to have an unbiased model.

As the goal of cluster detection is hypothesis generation through identification of geographic areas of unusually high risk, any known risk factor that could explain a detected cluster should be adjusted for in the analysis. Any cluster observed after adjustment for known risk factors could be potentially explained by a yet unknown spatially or temporally patterned risk factor. Local scan statistics and kernel density methods do not allow for simple adjustment for risk factors with individual-level data. With Q-statistics, adjustment for risk factors is done separately from cluster detection. In contrast, GAMs can adjust for risk factors and test for clusters within a unified statistical framework. We next present an investigation of areas of significantly elevated NHL risk illustrating a GAM-based approach that simultaneously adjusts for risk factors, considers latency periods, and tests for significant clusters in one unified statistical framework.

Spatial-Temporal Analysis of Non-Hodgkin Lymphoma Risk

Study Population

Since 1975 in the United States, the annual age-adjusted incidence rate of NHL has increased more than 75 percent from 11.1 to 19.8 per 100,000 person years (Ries et al. 2003). The cause for this increase is largely undetermined and little is known about the etiology of NHL, except for established risk factors that include certain viral infections, immune suppression, and a family history of hematolymphoproliferative cancers

(Chatterjee et al. 2004). Incidence of NHL increases with age, is higher in men, and is 40 to 70 percent higher in whites compared to blacks (Jemal et al. 2004). NHL incidence has also been associated with specific genetic polymorphisms (Morton et al. 2008) and environmental risk factors, including pesticides (Zahm et al. 1990), insecticides such as chlordane (Colt et al. 2006), and polychlorinated biphenyls (PCBs; Colt et al. 2005). Taken together, the established risk factors account for only a small proportion of the total annual NHL cases.

A previous analysis of NHL risk in the National Cancer Institute (NCI) Surveillance, Epidemiology, and End Results (SEER) NHL study, a population-based case-control study of NHL at four SEER centers (Detroit, Los Angeles, Seattle, and Iowa), revealed unexplained areas of significant risk in Detroit, Los Angeles, and Iowa after adjusting for the risk factors age, race, gender, education, and home treatment for termites before 1988, a surrogate for exposure to chlordane (Wheeler et al. 2011). The analysis also explored the latency period that might be relevant for environmental exposures for NHL and found that a lag time of twenty years before diagnosis was most associated with risk of NHL. Here, we perform an analysis of NHL risk in the NCI-SEER study to adjust for additional risk factors of exposure to PCBs and presence of specific genetic polymorphisms available for a subset of subjects to ascertain whether or not they explain the previously detected areas of elevated risk.

Details of the NCI-SEER study have been reported previously (Chatterjee et al. 2004; Morton et al. 2008; Wheeler et al. 2011). NHL cases aged twenty to seventy-four years were identified between 1 July 1998 and 30 June 2000. Participants provided lifetime residential histories that were then matched to geographic address databases. In addition to demographic data and select risk factors available for all subjects, carpet dust samples from used vacuum cleaner bags were collected in homes to measure residential exposure to PCBs for 58 percent of cases and 56 percent of controls, and genotyping was conducted from DNA samples for 89 percent of cases and 93 percent of controls. Details on the collection and analysis of the dust samples are available in Colt et al. (2005). Based on the previous findings of a significant association between NHL incidence and residential levels of PCB congener 180 (Colt et al. 2005; Morton et al. 2008), we used concentrations of this PCB in our analysis. We used a binary measure of PCB 180 exposure, defined as 1 if PCB 180 was $\geq$ 44.4 ng/g in dust, where this level was the lower bound for the

highest category of exposure in Morton et al. (2008). Based on findings of increased risk of NHL with genetic polymorphisms for the genes FCGR2A, RAG1, TNF, and XRCC1 (Morton et al. 2008), we included these in our analysis.

To be consistent with the analysis of Wheeler et al. (2011) while exploring the elevated areas of NHL risk, this analysis included only study participants with a complete twenty-year residential history within one of the three study centers that contained areas of significantly elevated risk. A total of 671 cases (67 percent) and 516 controls (68 percent) met this criterion. Among these subjects, there were 521 cases and 404 controls with complete genetic data and 305 cases and 212 controls with complete genetic and PCB data.

GAM Analysis with Lag Times

We used GAMs (Hastie and Tibshirani 1990) to model spatially the probability that an individual was diagnosed with NHL. The methods are further detailed in Wheeler et al. (2011). Given the coordinates (s_1, s_2) for residential locations (s) at a particular time t, the odds of being a case are modeled as

$$\text{logit}[p(s_1, s_2)] = \alpha + \beta'x + Z_t(s_1, s_2), \qquad (1)$$

where the left side of the equation is the log of the disease odds at location s, α is an intercept, β is a vector of regression coefficients, x is a vector of covariates observed at location s, and $Z_t(s_1, s_2)$ is a function of the residential locations at a particular time point. This function provides spatial smoothing of the locations and models spatial variation not explained by the covariates. The spatially smoothed term can be considered a surrogate for unmeasured environmental factors at a specified time. The technique of smoothing over residential locations is used to measure the density of cases relative to controls over space. This approach models cases and controls as a marked heterogeneous Poisson point process with intensity $\lambda(s) = \lambda_1(s) + \lambda_0(s)$, where $\lambda_1(s)$ denotes the intensity of cases and $\lambda_0(s)$ the intensity of controls.

Within this framework, we can evaluate several lag times in years before diagnosis—for example, twenty years and ten years—through Z_t and select the one that best explains the risk of disease. Analysis of deviance (ANODEV) can be used to evaluate the significance of the lag times by testing differences in deviances between nested models, with and without Z_t. The difference in deviances for two nested models approximately follows a chi-square distribution with an associated p value. A significantly lower deviance from a model with a lag time of k years indicates that using the smoothed pattern of residential locations from k years before diagnosis significantly explained overall disease risk. This model specification does not consider the duration spent at each residence but rather the pattern of residences at any time t.

For the form of the spatial smoothing function, we used loess, or locally weighted scatterplot smoothing (Cleveland 1979), and smoothed over both spatial dimensions. The smoothing function has a span parameter that controls the amount of smoothing. The span parameter must be estimated, and we selected the span that minimized the Akaike information criterion (Akaike 1973) over a large range of span values. We estimated the GAM model parameters in the statistical analysis software R (R Development Core Team 2010) using the GAM package, version 1.03.

To assess the variation in risk of disease over space, we plotted the local odds ratios (ORs) using the model specified in Equation 1. To produce a map of local ORs, we first estimated all parameters for the model expressed in Equation 1 using the study data. We then predicted the log odds over a rectangular grid placed over the study area using the estimated model parameters. To provide an interpretable OR map, we used the entire study population as the reference and divided the odds from the spatial model at each grid point by the odds from the null model.

For inference on clusters, we identified areas of significantly elevated risk using Monte Carlo randomization. This procedure compares the observed local ORs to distributions of local ORs under the null hypothesis that case status does not depend on location (Waller and Gotway 2004). We used 999 Monte Carlo samples to build the permutation distribution of ORs at each grid location, using the optimal span from the observed data for the permutations. We identified

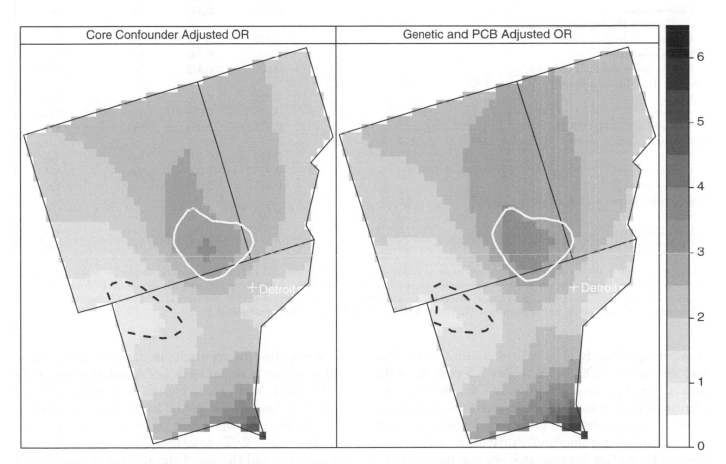

Figure 1. Local odds ratios (ORs, scale at right) for non-Hodgkin lymphoma adjusted for the core covariates (age, gender, race, education, home termite treatment) and additionally for four genetic polymorphisms and PCB 180 exposure at a residential lag time of twenty years in the Detroit study area using the model with missing variable coding. Areas of statistically significant elevated ORs are identified with a solid white line and statistically significant lowered ORs are identified with a dashed black line. PCB = polychlorinated biphenyl.

Table 1. Sample size, estimated span parameter, approximate p value for the spatial term, and presence of significantly elevated or lowered risk areas of non-Hodgkin lymphoma for crude and adjusted models for several sets of data for three centers

Model	Cases	Controls	Total	Span	p value	High-risk cluster	Low-risk cluster
Detroit: Missing indicator	214	144	358				
Crude				0.600	0.071	Yes	Yes
Core covariates				0.600	0.072	Yes	Yes
Core + genes and PCB 180				0.600	0.093	Yes	Yes
Detroit: Genes subset	128	91	219				
Crude				0.625	0.051	Yes	Yes
Core covariates				0.625	0.093	Yes	Yes
Core + genes				0.625	0.086	Yes	Yes
Detroit: Genes + PCB subset	65	41	106				
Crude				0.700	0.022	Yes	Yes
Core covariates				0.700	0.021	Yes	Yes
Core + genes and PCB 180				0.700	0.019	Yes	No
Iowa: Missing indicator	267	211	478				
Crude				0.625	0.211	Yes	No
Core covariates				0.625	0.144	Yes	No
Core + genes and PCB 180				0.625	0.204	Yes	No
Iowa: Genes subset	233	186	419				
Crude				0.625	0.337	Yes	No
Core covariates				0.625	0.225	Yes	No
Core + genes				0.625	0.318	Yes	No
Iowa: Genes + PCB subset	133	109	242				
Crude				0.600	0.378	No	No
Core covariates				0.600	0.346	No	No
Core + genes and PCB 180				0.600	0.422	No	No
Los Angeles: Missing indicator	190	161	351				
Crude				0.275	0.003	Yes	Yes
Adjusted				0.275	0.029	Yes	Yes
Adjusted + Genes, PCB 180				0.275	0.024	Yes	Yes
Los Angeles: Genes subset	160	127	287				
Crude				0.375	0.001	Yes	Yes
Core covariates				0.375	0.009	Yes	Yes
Core + genes				0.375	0.009	Yes	Yes
Los Angeles: Genes + PCB subset	107	62	169				
Crude				0.500	0.132	No	Yes
Core covariates				0.500	0.057	Yes	Yes
Core + genes and PCB 180				0.425	0.038	Yes	Yes

Note: For each center, the first set is all subjects with complete twenty-year residential histories and uses missing indicator coding for the genetic factors and PCB exposure, the second set is only subjects with complete genetic data, and the third set is only subjects with complete genetic and PCB data. The core covariate adjusted model includes age, gender, race, education, and home treatment for termites before 1988. The other adjusted models also include genetic polymorphisms in four genes, as well as exposure to PCB 180 in some models. PCB = polychlorinated biphenyl.

areas of significantly elevated risk as those areas that had an observed OR in the upper 2.5 percent of the ranked permutation distribution of ORs. Similarly, we identified areas of significantly lowered risk of disease as those having an observed OR in the lower 2.5 percent of the ranked permutation distribution. Clusters of either elevated or lowered risk are significant at the 0.05 level (assuming a two-tailed distribution). We mapped the local ORs and highlighted the significant areas of risk for disease simultaneously.

We applied the approach just described in the three study centers (Detroit, Iowa, Los Angeles) in the NCI-SEER NHL study where clusters were previously detected (Wheeler et al. 2011). In the previous adjusted models, the most significant lag time was twenty years in Detroit ($p = 0.07$), Iowa ($p = 0.14$), and Los Angeles ($p = 0.03$), and clusters of elevated risk were detected at a time lag of twenty years in all three study areas; therefore, we focused on this time lag in our analysis. We fitted separate models for each center. We fitted

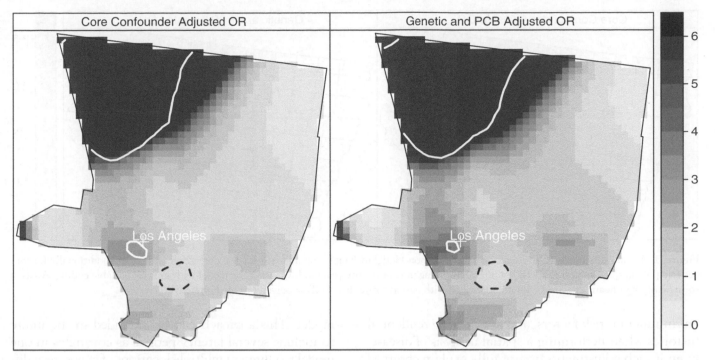

Figure 2. Local odds ratios (ORs, scale at right) for non-Hodgkin lymphoma adjusted for the core confounders and additionally for four genetic polymorphisms and PCB 180 at a residential lag time of twenty years in the Los Angeles study area using the model with missing variable coding. Areas of statistically significant elevated ORs are identified with a solid white line and statistically significant lowered ORs are identified with a dashed black line. PCB = polychlorinated biphenyl.

crude models; models adjusted for the core covariates age at enrollment, gender, race, education, and home treatment for termites before 1988; and models additionally adjusted for PCB 180 and genetic polymorphisms for the four previously mentioned genes. The covariates did not vary over time in the models. We fitted the models for three sets of data for each center. We used a set of all subjects with complete residential histories and for whom missing values for PCB 180 and genetic risk factors were coded with a missing indicator, a set of subjects with complete genetic data, and a set with complete genetic and PCB data.

Results

The analyses from all three subsets of study subjects showed that adjusting for PCB 180 and the four genetic factors made little difference in Detroit, Iowa, and Los Angeles at a lag time of twenty years before diagnosis in terms of the significance of the spatial term in the models and the presence of significantly elevated or lowered areas of NHL risk (Table 1). The only change in significant clusters due to adjusting for the genetic factors and PCB 180 occurred in Detroit with the model that included those with complete genetic and PCB data, where an area of significantly lowered risk was explained

by these factors. Adjusting for the core covariates was adequate to detect an area of significantly elevated risk that was not found by the crude model in Los Angeles in the genetic and PCB subset of data.

The locations of the areas of significantly elevated and lowered NHL risk remained the same across adjusted models, although the shape of the detected areas changed slightly for some models. In Detroit, the delineation of the area of significant elevated risk (in southeast Oakland County) was consistent when adjusting for the core covariates and additionally for the genetic factors and PCB exposure (Figure 1). In Los Angeles, an area of elevated risk (West Hollywood) decreased slightly in size after adjusting for PCB exposure and the genetic factors (Figure 2). The other region of elevated risk contains sparsely populated areas. The area of significantly elevated risk in Iowa, including parts of Wayne County and Appanoose County, also decreased in size after adjusting for PCB exposure and the genetic polymorphisms (Figure 3). The risk overall decreased in Iowa after adjusting for the additional genetic and PCB risk factors, and the approximate p values for the spatial term increased with the adjustment in each of the three sets of data.

Our study demonstrates the importance of adjusting for suspected risk factors, including genetic and

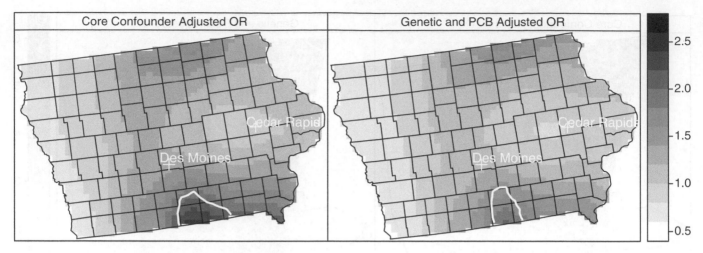

Figure 3. Local odds ratios (ORs, scale at right) for non-Hodgkin lymphoma adjusted for the core confounders and additionally for four genetic polymorphisms and PCB 180 at a residential lag time of twenty years in Iowa using the model with missing variable coding. Areas of statistically significant elevated ORs are identified with a solid white line. PCB = polychlorinated biphenyl.

environmental risk factors, and considering residential histories when performing a spatial analysis of disease. Even if such adjustments do not fully explain observed patterns, they refine the hypotheses generated by the analyses. Our study also highlights the potential heterogeneity in risk factors for NHL incidence across different geographic areas. Additional efforts are required to identify risk factors that might explain areas of significant risk in Detroit, Iowa, and Los Angeles.

Conclusions

In this article, we first reviewed the challenges encountered when performing spatial cluster analysis of chronic diseases and then presented an analysis that addressed these challenges. We investigated previously detected areas of significantly elevated risk of NHL in a case–control study to see if PCB exposure and genetic risk factors could explain the clusters. We found that adjusting for genetic factors and PCB 180 levels in homes did not explain significantly elevated risk areas in Detroit, Iowa, and Los Angeles but did explain an area of significantly lowered risk in Detroit in one model. Our analysis demonstrates an approach to spatial-temporal analysis of disease for epidemiologic studies with individual-level suspected risk factors and residential histories. Our approach is based on the well-established GAM, which provides a unified statistical framework for adjusting for risk factors, estimating disease risk spatially, and assessing significance of elevated risk for individual-level epidemiologic studies. A strength of our approach is that it is straightforward to evaluate several latency periods within an unbiased

model. This approach can be extended in the future to include several latency periods as covariates in one model to represent multiple locations of exposure at different time points. It should also be possible to include duration at each residence in a model. A limitation of this approach is that a latency period not considered in the model might be the most relevant for a particular disease. The selection of latency period candidates to evaluate in an exploratory analysis might be somewhat arbitrary. In addition, not all available residential locations would typically be evaluated in this approach. Extending existing approaches to include all address information is an area for methodological development in the future. Another limitation, common to all cluster detection approaches, is that detection of a cluster only identifies a location of potential exposure but does not identify the nature of any exposure. In summary, our application of this approach serves as an illustrative example for those interested in performing space–time cluster analysis of chronic diseases with suspected latencies and limited risk factor information to assist in the generation of new hypotheses about potential risk factors. The approach is especially applicable to other chronic diseases with suspected environmental causes.

References

Akaike, H. 1973. Information theory and an extension of the maximum likelihood principle. In *Second international symposium on information theory*, ed. B. Petran and F. Csaaki, 267–81. Budapest, Hungary: Akademi Kiado.

Beale, L., S. Hodgson, J. Abellan, S. LeFevre, and L. Jarup. 2010. Evaluation of spatial relationships between health and the environment: The rapid inquiry facility. *Environmental Health Perspectives* 118:1306–12.

Besag, J., and J. Newell. 1991. The detection of clusters in rare diseases. *Journal of the Royal Statistical Society, Series A* 154:143–55.

Bithell, J. 1990. An application of density estimation to geographical epidemiology. *Statistics in Medicine* 9:691–701.

Blot, W., J. Fraumeni, Jr., T. Mason, and R. Hoover. 1979. Developing clues to environmental cancer: A stepwise approach with the use of cancer mortality data. *Environmental Health Perspectives* 32:53–58.

Chatterjee, N., P. Hartge, J. Cerhan, W. Cozen, S. Davis, N. Ishibe, J. Colt, L. Goldin, and R. Severson. 2004. Risk of non-Hodgkin's lymphoma and family history of lymphatic, hematologic, and other cancers. *Cancer Epidemiology Biomarkers and Prevention* 13:1415–21.

Cleveland, W. 1979. Robust locally weighted regression and smoothing scatterplots. *Journal of the American Statistical Association* 74:829–36.

Colt, J., S. Davis, R. Severson, C. Lynch, W. Cozen, D. Camann, E. Engels, A. Blair, and P. Hartge. 2006. Residential insecticide use and risk of non-Hodgkin's lymphoma. *Cancer Epidemiology Biomarkers and Prevention* 15(2):251–57.

Colt, J., R. Severson, J. Lubin., N. Rothman, D. Camann, S. Davis, J. Cerhan, W. Cozen, and P. Hartge. 2005. Organochlorines in carpet dust and non-Hodgkin lymphoma. *Epidemiology* 16:516–25.

Elliott, P., and D. Savitz. 2008. Design issues in small-area studies of environment and health. *Environmental Health Perspectives* 116:1098–1104.

Fraumeni, J., Jr., and W. Blot. 1977. Geographic variation in esophageal cancer mortality in the United States. *Journal of Chronic Diseases* 30:759–67.

Hastie, T., and R. Tibshirani. 1990. *Generalized additive models*. London: Chapman & Hall.

Jacquez, G. 2004. Current practices in the spatial analysis of cancer: Flies in the ointment. *International Journal of Health Geographics* 3:22.

Jacquez, G., A. Kaufmann, J. Meliker, P. Goovaerts, G. AvRuskin, and J. Nriagu. 2005. Global, local and focused geographic clustering for case-control data with residential histories. *Environmental Health* 4 (4): 1–19.

Jemal, A., R. Tiwari, T. Murray, A. Ghafoor, A. Samuels, E. Ward, E. Feuer, and M. Thun. 2004. Cancer statistics. *CA: A Cancer Journal for Clinicians* 54:8–29.

Kelsall, J., and P. Diggle. 1995. Non-parametric estimation of spatial variation in relative risk. *Statistics in Medicine* 14:2335–42.

———. 1998. Spatial variation in risk of disease: A non-parametric binary regression approach. *Applied Statistics* 47:559–73.

Kulldorff, M. 1997. A spatial scan statistic. *Communications in Statistics: Theory and Methods* 26:1487–96.

———. 2006. *SaTScan: Software for the spatial and space–time scan statistics*. Silver Spring, MD: Information Management Services.

Morton, L., S. Wang, W. Cozen, M. Linet, N. Chatterjee, S. Davis, R. Severson, et al. 2008. Etiologic heterogeneity among non-Hodgkin lymphoma subtypes. *Blood* 112:5150–60.

R Development Core Team. 2010. *R: A language and environment for statistical computing*. Vienna, Austria: R Foundation for Statistical Computing.

Ries, L., M. Eisner, C. Kosary, B. Hankey, B. Miller, L. Clegg, A. Mariotto, M. Fay, E. Feuer, and B. Edwards. 2003. *SEER cancer statistics review, 1975–2000*. Bethesda, MD: National Cancer Institute.

Sabel, C., P. Boyle, G. Raab, M. Loytonen, and P. Maasilta. 2009. Modelling individual space–time exposure opportunities: A novel approach to unravelling the genetic or environmental disease causation debate. *Spatial and Spatio-Temporal Epidemiology* 1:85–94.

Tong, S. 2000. Migration bias in ecologic studies. *European Journal of Epidemiology* 16:365–69.

Vieira, V., T. Webster, J. Weinberg, and A. Aschengrau. 2008. Spatial-temporal analysis of breast cancer in upper Cape Cod, Massachusetts. *International Journal of Health Geographics* 7:46.

Vieira, V., T. Webster, J. Weinberg, A. Aschengrau, and D. Ozonoff. 2005. Spatial analysis of lung, colorectal, and breast cancer on Cape Cod: An application of generalized additive models to case-control data. *Environmental Health* 4 (11): 1–18.

Wakefield, J., and P. Elliott. 1999. Issues in the statistical analysis of small area health data. *Statistics in Medicine* 18:2377–99.

Waller, L., and C. Gotway. 2004. *Applied spatial statistics for public health data*. New York: Wiley.

Webster, T., V. Vieira, J. Weinberg, and A. Aschengrau. 2006. Method for mapping population-based case-controls studies: An application using generalized additive models. *International Journal of Health Geographics* 5 (26): 1–10.

Wheeler, D., A. De Roos, J. Cerhan, L. Morton, R. Severson, W. Cozen, and M. Ward. 2011. Spatial-temporal cluster analysis of non-Hodgkin lymphoma in the NCI-SEER NHL Study. *Environmental Health* 10:63.

Winn, D., W. Blot, C. Shy, L. Pickle, A. Toledo, and J. Fraumeni, Jr. 1981. Snuff dipping and oral cancer among women in the southern United States. *New England Journal of Medicine* 304:745–49.

Zahm, S., D. Weisenburger, P. Babbitt, R. Saal, J. Vaught, K. Cantor, and A. Blair. 1990. A case-control study of non-Hodgkin's lymphoma and the herbicide 2,4-dicholorophenoxyacetic acids (2,4-D) in eastern Nebraska. *Epidemiology* 1:349–56.

An Examination of Spatial Concentrations of Sex Exchange and Sex Exchange Norms Among Drug Users in Baltimore, Maryland

Karin Elizabeth Tobin,* Laura Hester,† Melissa Ann Davey-Rothwell,* and Carl Asher Latkin*

*Department of Health, Behavior and Society, Johns Hopkins Bloomberg School of Public Health
†Department of Epidemiology, Johns Hopkins Bloomberg School of Public Health

Baltimore, Maryland, consistently ranks highest nationally in rates of sexually transmitted diseases and sexually transmitted infections (STDs/STIs) and human immunodeficiency virus (HIV) infection. Prior studies have identified geographic areas where STI and HIV infection in the city is most prevalent. It is well established that sex exchange behavior is associated with HIV and STIs, yet it is not well understood how sex exchangers are spatially distributed within the high-risk areas. We sought to examine the spatial distribution of individuals who report sex exchange compared to those who do not exchange. Additionally, we examined the spatial context of perceived norms about sex exchange. Data for the study came from a baseline sample of predominately injection drug users ($n = 842$). Of these, 21 percent reported sex exchange in the prior ninety days. All valid baseline residential addresses of recruited participants living within Baltimore City boundaries were geocoded. The multidistance spatial cluster analysis (Ripley's K-function) was used to separately calculate the K-functions for the addresses of recruited participants reporting sex exchange versus non-sex exchange. Evidence of spatial clustering of sex exchangers was observed and norms aligned with these clusters. Of particular interest was the high density of sex exchangers in one specific housing complex of East Baltimore, which happens to be the oldest in Baltimore. These findings can inform targeted efforts for screening and testing for HIV and STIs and placement of both individual and structural-level interventions that focus on increasing access to risk reduction materials and changing norms about risk behaviors.

马里兰州巴尔的摩的性传播疾病和性传播感染 (STDs/STIs) 以及人类免疫缺陷病毒 (HIV) 感染率始终在全国排名最高。此前的研究已经确定 STI 和 HIV 感染病例在该城市的哪些地理区域最盛行。虽然性交换行为与 HIV 和 STIs 相关已被普遍公认，但它不能很好地解释性交换者在高风险地区的空间分布。我们通过对比那些所报告的有性交换行为的人群和那些没有性交换的人群，力求来审查他们的空间分布。此外，我们审查性交换者的自觉规范的空间环境。我们的研究数据来自一个以注射吸毒者 ($n = 842$) 为主的基准样品。其中，21% 的人报告了在过去 90 天有性交换的行为。我们对居住在巴尔的摩市边界内的，所招募的参与者的有效基准住址进行了地理编码。使用多距离空间聚类分析（里普利的 K-函数），分别计算所记录的有性交换和无性交换招募参与者的地址的 K 函数值。观察性交换者空间聚类的证据和与这些聚类集群所对应的规范。特别感兴趣的是，东巴尔的摩的一个存在最高密度性交换行为的特定建筑群，恰好是巴尔的摩最古老的建筑群。这些结果可启发我们有针对性地筛查和检查 HIV 和 STIs，以及帮助我们制定个人和结构水平的干预措施，这些干预措施着重于增强对降低风险的物质的获取和改变有关危险行为的规范。*关键词：规范，里普利的K-功能，性交换，空间聚类。*

Consistentemente, la ciudad de Baltimore, Maryland, se ubica en el grupo con tasas nacionales más elevadas de enfermedades transmitidas sexualmente y de infecciones de transmisión sexual (ETS/ITS), y de infección por el virus de inmunodeficiencia humana (VIH). Estudios anteriores han identificado áreas geográficas de la ciudad en donde las ITS y la infección con VIH tienen mayor prevalencia. Está bien establecido que la conducta del intercambio sexual se asocia con VIH e ITS, pero no se entiende bien de qué manera están distribuidos espacialmente dentro de las áreas de alto riesgo quienes se involucran en este intercambio. Nosotros buscamos examinar la distribución espacial de individuos que informan de intercambio sexual en comparación con aquellos que no lo practican. Examinamos también el contexto espacial de cómo se perciben las normas sobre intercambio sexual. Los datos para este estudio provienen de una muestra de base constituida predominantemente por usuarios de drogas inyectables ($n = 842$). De estos, el 21 por ciento reportaron intercambio sexual en los noventa días anteriores. Se geo-referenciaron todas las direcciones residenciales de los participantes reclutados que viven dentro de los límites de la ciudad de Baltimore. El análisis de aglomeración espacial de distancia múltiple (la función K de Ripley) se utilizó para calcular separadamente las funciones K de las direcciones de los participante reclutados que reportaron intercambio sexual versus los que no lo reportan. Se observó evidencia de aglomeración espacial de quienes tuvieron intercambio sexual y las normas estuvieron en línea con estos aglomerados. De particular interés fue el registro de alta densidad de intercambios sexuales en un complejo residencial específico de Baltimore Oriental, que resultó ser el más viejo de esa ciudad. Estos hallazgos puede informar esfuerzos específicamente orientados a chequeos y pruebas de laboratorio para VIH e ITS y asignación de intervenciones, tanto individuales como de nivel estructural, enfocadas a aumentar el acceso a materiales para reducción de riesgo y cambio de normas sobre conductas riesgosas.

I t is well established that exchanging sex for money or drugs is associated with increased risk of infection with human immunodeficiency virus (HIV) and sexually transmitted infections (STIs; Astemborski et al. 1994; Rietmeijer et al. 1998; Doherty et al. 2000). Recent attention to sex risk environments (e.g., bathhouses, sex resorts, commercial sex areas) has gained attention in public health research (Rhodes 2002; Wylie, Shah, and Jolly 2007). These environments facilitate social interactions between individuals and have been shown to be associated with infectious disease transmission (Wohl et al. 2010), which underscores the value and importance of including geographic dimensions in the study and analysis of HIV risk behavior.

Baltimore consistently ranks among the most burdened cities in rates of HIV and STIs in the United States (Centers for Disease Control and Prevention 2009, 2010; Center for Sexually Transmitted Infection Prevention, DHMH, Baltimore City Health Department, and Maryland Office of Planning 2011). Prior geographic studies conducted in Baltimore have identified core areas of gonorrhea transmission (Zenilman et al. 1999; Jennings et al. 2005), chlamydia (Hardick et al. 2003), and HIV infection (Towe et al. 2010). These studies have demonstrated that STIs are not evenly distributed throughout the city but are found in areas that correspond with poor social cohesion (Ellen et al. 2004), poverty, crimes, and drug use. Less is known about the spatial distribution of individuals who exchange sex relative to this area characterized by high sex risk. Identifying locations where sex exchangers cluster can refine prevention and treatment efforts.

One mechanism through which neighborhood or residential location has been hypothesized to influence health is through facilitating social interactions and formation and perpetuation of social norms. It has been argued that culture and social norms are rooted in place (Mills et al. 2001; Gesler and Kearns 2002). Descriptive norms are perceptions of an individual about their peers engaging in a specific behavior. Injunctive norms are perceptions about other approval of a behavior (Cialdini, Reno, and Kallgren 1990). Studies with a wide range of populations and ethnic groups have reported positive associations between both types of norms and HIV risk behaviors such as condom use (Albarracin et al. 2001; Hart, Peterson, and Community Intervention Trial for Youth Study Team 2004), use of shooting galleries (Tobin, Davey-Rothwell, and Latkin 2010), sharing injection equipment (Andia et al. 2008), and sex exchange (Davey-Rothwell and Latkin 2008). Little attention has been paid to the spatial context of social norms and their association with behavior. Understanding the geographies of norms could inform placement of interventions designed to change norms.

The aims of this study were, first, to examine the spatial distribution of individuals who report sex exchange compared to those who do not exchange; second, to examine the spatial distribution of descriptive and injunctive norms on sex exchange; and third, to examine associations between spatial variables and sex exchange. Attention to spatial concentrations of poor health behaviors is important because it can inform resource allocation, public health screening, and interventions.

Methods

Study Population and Recruitment

Data for this study came from the baseline of the STEP into Action study. Recruitment and assessment methods have been described in detail elsewhere (Tobin et al. 2010). In brief, recruited study participants, herein referred to as index participants, were individuals aged eighteen years and older who self-reported injection drug use in the prior six months or were drug users or sex partners that the index group recruited into the study. Index participants were recruited by trained field staff, through participant word of mouth, and with advertisements. All eligible participants provided written informed consent and completed an assessment on sex risk behavior using audio computer-assisted self-interview software. Trained research assistants then administered a survey to collect sociodemographics, drug use history, perceived norms, and residential address data. All participants were remunerated for the study visit. The Johns Hopkins Bloomberg School of Public Health Institutional Review Board approved this study.

Measures

Sex exchange was operationalized as having sex in exchange for drugs, money, food, or shelter in the prior ninety days. To assess descriptive norms, participants were asked "How many of your friends have sex or turn tricks in exchange for money or drugs?" A dichotomous variable was created (0 = none, 1 = a few to all). Injunctive norms were determined using the question "How many of your friends would disapprove if you had sex for money or drugs?" (0 = none and 1 = a few to all). Census data came from 2000 U.S. Census File 1 and File 3 (U.S. Census Bureau 2010). Variables included were median household income per census block group, percentage of the census block group population that is minority race (black, Asian, Hawaiian/Pacific Islander, Native American, other), percentage of households (both vacant and occupied) that are available for rent, and percentage in the labor force.

Violent crime data came from the 2005 Baltimore City Police Department Report, which was classified using the U.S. Department of Justice Bureau of Justice Statistics and was cleaned and geocoded by the Johns Hopkins University Eisenhower Library.

Spatial Methods

All valid, baseline residential addresses of STEP participants living within Baltimore city boundaries were geocoded using Environmental Systems Research Institute's (ESRI) Streetmap U.S.A. extension (ESRI 2009). The geocoding match was 99 percent with an average score of 97. Using the residential addresses of STEP participants, the kernel density of sex exchange behavior, descriptive norms, and injunctive norms was calculated and displayed using ArcMap 9.3 (Silverman, 1998; ESRI 2009). The spatial locations of high-density regions of sex exchange and injunctive and descriptive norms were identified and compared to choropleth maps of the selected demographic, socioeconomic, and structural census variables.

The Ripley's K-function was calculated to evaluate clustering of baseline residential addresses for participants reporting either sex exchange or no sex exchange in comparison with a randomly distributed population. We used the weighted K-function ($K(h)$) proposed by Ripley (1976):

$$K(h) = \hat{\lambda}^{-1} \sum_{i=1}^{N} \sum_{i=1}^{N} w_{ij}\delta(d(i.j) < h)$$

where N is the number of residences for participants in the group of interest, i is the sex exchanger's or non-sex exchanger's residential location, λ is the number of residences in Baltimore City divided by the area of Baltimore City, w_{ij} is the conditional probability that a sex exchanger or non-sex exchanger lives within the Euclidean distance between i and another location j, and h is the distance between the randomly selected residence and the boundary.

The K-function defined the probability of observing a participant reporting certain norms or sex exchange behaviors at a location. To visually evaluate clustering for each group, the K-function was transformed into an $\hat{L}$ function to allow comparison of an estimated K-function to a straight line. The $\hat{L}$ function value was plotted over a range of distance for each group. Upper and lower 99 percent confidence bands were calculated for each $\hat{L}$ plot using 999 Monte Carlo simulations in the Baltimore City land boundaries. At distance ranges where the plot exceeded the confidence bands, residential clustering was significantly greater than what was expected if the residence had been randomly spatially distributed (Waller and Gotway 2004).

To evaluate whether sex exchangers cluster more than non-sex exchangers, the ratio of the log intensity function was calculated using a Bernoulli model, where the distribution of residences of participants reporting sex exchange were compared to the distribution of residences of participants reporting no sex exchange. This gave the spatial variation in risk of sex exchange. We also calculated the spatial variation in risk of no sex exchange, with the distribution of residences with sex exchangers acting as the control. All spatial statistics calculations were made using the *splancs* package in the R open-source software environment (R Development Core Team 2008; Rowlingson et al. 2010). A surface plot of the spatial variation in risk was created to visually identify high-risk areas using CrimeStat III (Levine 2010). Bivariate associations between sex exchange and census variables were examined using *t* tests.

Table 1. Sample characteristics of 751 geocoded study participants

Variable	No sex exchange[a] n (%)	Sex exchange[b] n (%)	p value
Mean age (SD)	43.9 (7.53)	40.2 (7.70)	< 0.001
Gender			
Male	375 (66%)	47 (31%)	
Female	194 (34%)	104 (69%)	< 0.001
Sexual identity			
Gay	14 (2%)	6 (4%)	
Straight	542 (95%)	116 (77%)	
Bisexual	13 (2%)	29 (19%)	< 0.001
Unemployed past 6 months			
No	58 (10%)	8 (5%)	
Yes	511 (90%)	143 (95%)	0.08
Homeless past 6 months			
No	408 (72%)	75 (50%)	
Yes	161 (28%)	76 (50%)	< 0.001
Incarcerated past 6 months			
No	409 (72%)	106 (70%)	
Yes	160 (28%)	45 (30%)	0.69
Injected drugs past 6 months			
No	102 (18%)	10 (7%)	
Yes	467 (82%)	141 (93%)	< 0.001
Smoked crack past 6 months			
No	196 (34%)	32 (21%)	
Yes	373 (66%)	119 (79%)	< 0.01
How many of your friends have sex for money or drugs			
None	197 (35%)	10 (7%)	
Some or all	372 (65%)	141 (93%)	< 0.001
How many of your friends would disapprove if you were to have sex for money or drugs			
None	134 (24%)	58 (38%)	
Some or all	435 (76%)	93 (62%)	< 0.001

[a]n = 569.
[b]n = 151.

Results

The final sample included 720 participants who reported a valid address during their baseline survey. Of these, 151 (20 percent) reported sex exchange. Table 1 presents comparisons of baseline characteristics and norms comparing sex exchangers to non-sex exchangers. A greater proportion of sex exchangers were younger, were female, were bisexual, were homeless, had smoked crack, and had injected drugs in the past six months. A greater proportion of sex exchangers perceived that a few to all of their friends exchange sex ($p < 0.0001$) compared to nonexchangers. A smaller proportion of sex exchangers perceived that a few to all of their friends would disapprove of sex exchange (injunctive norms, $p < 0.0001$). No census variables were statistically associated with sex exchange behavior in bivariate analysis (Table 2).

Table 2. Bivariate associations of census variables and sex exchange

Variable	No sex exchange[a]	Sex exchange[b]	p value
Median household income	22,324	21,396	0.34
Percentage minority	0.85 (0.23)	0.87 (0.19)	0.42
Percentage rental	0.55 (0.24)	0.58 (0.23)	0.19
Percentage in labor force	0.50 (0.11)	0.50 (0.11)	0.97
Violent crimes (log transform)	3.94 (0.73)	3.91 (0.78)	0.67

[a]n = 569.
[b]n = 151.

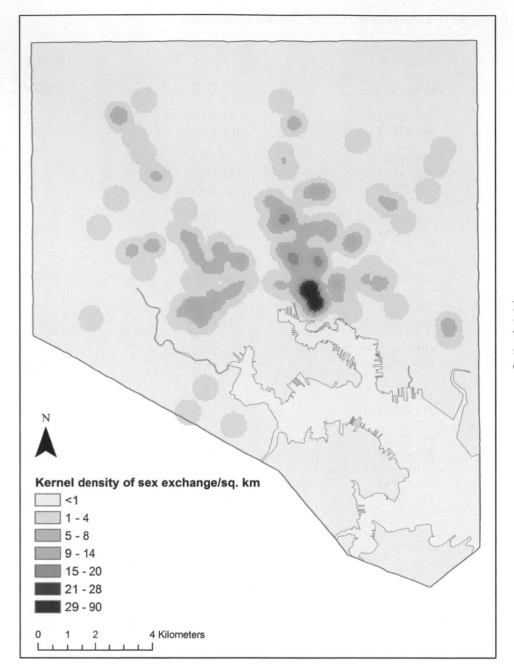

Figure 1. Spatial distribution of participants who report sex exchange, The STEP Into Action study, Baltimore, Maryland. (Color figure available online.)

Figure 1 displays the kernel density of participants who reported sex exchange. Those reporting sex exchange behaviors were most concentrated in an area that corresponds to a subsidized housing development in East Baltimore. The two other high-density areas included another public housing development in West Baltimore and a residential section in East Baltimore. Figure 2 displays the kernel density of non-sex exchange participants. Residences of non-sex exchangers were more spatially distributed across Baltimore and were not as densely located in subsidized housing areas. The K-function suggests that the observed sex ex-change and non-sex exchange behaviors both cluster significantly more than expected. When the ratio of spatial intensity is used to compare the spatial variation in risk of sex exchange versus non-sex exchange, however, participants reporting sex exchange behaviors cluster significantly more.

The regions with a high kernel density of sex ex-change spatially aligned with regions that had a high kernel density of participants reporting that a few to all friends engage in sex exchange (descriptive norms) and no friends would disapprove of sex exchange (injunctive norms; maps not shown). Further, this density

Figure 2. Spatial distribution of participants who did not report sex exchange, The STEP Into Action study, Baltimore, Maryland. (Color figure available online.)

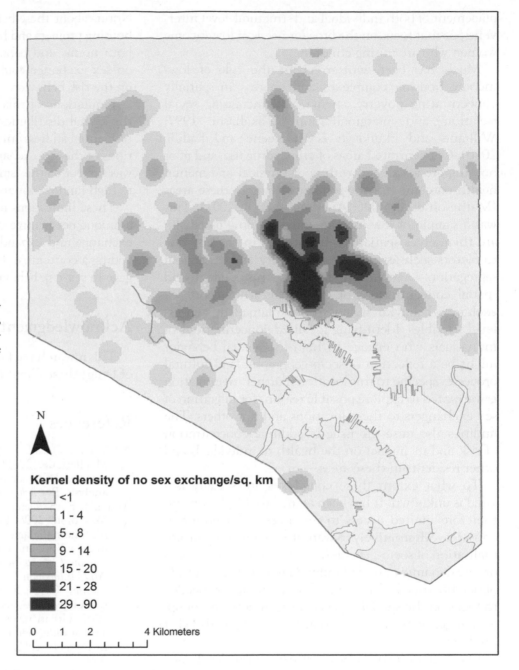

N

Kernel density of no sex exchange/sq. km

- <1
- 1 - 4
- 5 - 8
- 9 - 14
- 15 - 20
- 21 - 28
- 29 - 90

0   1   2   4 Kilometers

was highly concentrated in the housing development in East Baltimore.

Discussion

This study found evidence of greater spatial clustering of individuals who report sex exchange behaviors compared to individuals who did not report sex exchange after adjusting for the distribution of the sample. Specifically, clustering was observed in two low-income housing complexes and one low-income residential neighborhood. Of particular interest was the high density of sex exchangers in one specific housing complex in East Baltimore, which is the oldest in Baltimore. According to U.S. Census data, the demographic profile of this housing complex is over one-third-single female households with children under the age of eighteen, compared to another housing complex in East Baltimore that did not have a clustering of sex exchangers, where the proportion was 19 percent (U.S. Census Bureau 2000). These findings can inform targeted efforts for screening and testing for HIV and STIs and placement of both individual and structural-level interventions that focus on the broader needs of low-income women who are raising children.

Much has been written about the role of low-income housing complexes and ghettos in spatially concentrating poverty, stigma, and increasing social isolation and marginalization (Takahashi 1997; Williams and Ekundayo 2001; Keene and Padilla 2010). Concentrated areas of crime, drug use, and poor housing are associated with poor physical and mental health outcomes for the residents living in these areas (Matheson et al. 2006; Aneshensel et al. 2007). This was a sample of low-income, minority drug users who are themselves spatially concentrated in the city due to factors such as housing availability and residential segregation. Accounting for this, this study found spatial concentrations of individuals who report sex exchange behavior that is not explained by census-level variables. Identifying localized concentrations of individuals who engage in high-risk sexual behavior might be a reflection of socio-geographic and cultural processes specific to this location. Further investigation is warranted to explore possible reasons for migration of sex exchangers to these locations and not others. The findings also raise the issue of additive concentration of risk and its impact on the health of individuals and other residents in these areas.

To what extent these complexes are socially isolated is unknown. It has been argued that isolation and therefore limited access to resources is important to consider. Alternatively, Kwan et al.'s (2008) conceptualization of socio-geographic context includes a focus on the dynamic nature of individual movement outside of neighborhood of residence. Future studies are needed to focus on the spatial aspects of social networks of sex exchangers within and external to the neighborhood of residence.

Our lack of associations between census-level variables and sex exchange might reflect an issue of scale. That is, census-level or ZIP code–based variables did not adequately or precisely explain socio–spatial processes that occur in places such as housing complexes. Employing ethnographic or other field-based methods would enable a more in-depth exploration of the lived experiences of individuals who live in these places and the pertinent geographic factors that are in play regarding sex exchange.

As expected, we found that norms were aligned with reports of behavior. Norms are a social process that has been found to predict behavior and might be one pathway through which place impacts health behavior.

Norms about these behaviors might be more public in housing projects and hence have a greater influence on both norms and behaviors. Concentrations of norms on sex exchange could challenge efforts to intervene on the risk behavior.

Limitations of this analysis should be noted. First, the spatial distribution was based on the self-reported residential address only and we lacked data on the locations of the sex exchange behaviors. Also, the analysis was subject to the "small numbers problem" that leads to high random variability (McLafferty 2008).

These limitations notwithstanding, this study offers a unique perspective on the geographic context of sex exchange and expands behavioral health research by adding a contextual dimension to individuals who engage in sex-risk behavior.

Acknowledgment

This research was funded through National Institutes of Drug Abuse Grant R01 DA016555.

References

Albarracin, D., B. T. Johnson, M. Fishbein, and P. A. Muellerleile. 2001. Theories of reasoned action and planned behavior as models of condom use: A meta-analysis. *Psychological Bulletin* 127 (1): 142–61.

Andia, J. F., S. Deren, R. R. Robles, S. Y. Kang, and H. M. Colon. 2008. Peer norms and sharing of injection paraphernalia among Puerto Rican injection drug users in New York and Puerto Rico. *AIDS Education and Prevention: Official Publication of the International Society for AIDS Education* 20 (3): 249–57.

Aneshensel, C. S., R. G. Wight, D. Miller-Martinez, A. L. Botticello, A. S. Karlamangla, and T. E. Seeman. 2007. Urban neighborhoods and depressive symptoms among older adults. *The Journals of Gerontology, Series B: Psychological Sciences and Social Sciences* 62 (1): S52–S59.

Astemborski, J., D. Vlahov, D. Warren, L. Solomon, and K. E. Nelson. 1994. The trading of sex for drugs or money and HIV seropositivity among female intravenous drug users. *American Journal of Public Health* 84 (3): 382–87.

Center for Sexually Transmitted Infection Prevention, DHMH, Baltimore City Health Department and Maryland Office of Planning. 2011. 2010 gonorrhea distribution by jurisdictions and age groups. http://ideha.dhmh.maryland.gov/OIDPCS/CSTIP/CSTIPDocuments/2010_GC_Table_AgeGroup_by_Jurisdiction (last accessed 10 April 2012).

Centers for Disease Control and Prevention (CDC). 2009. *HIV/AIDS surveillance report, 2007.* Atlanta, GA: U.S. Department of Health and Human Services, CDC.

———. 2010. *Sexually transmitted disease surveillance 2009.* Atlanta, GA: U.S. Department of Health and Human Services.

Cialdini, R. B., R. R. Reno, and C. A. Kallgren. 1990. A focus theory of normative conduct: Recycling the concept of norms to reduce littering in public places. *Journal of Personality and Social Psychology* 58 (6): 1015–26.

Davey-Rothwell, M. A., and C. A. Latkin. 2008. An examination of perceived norms and exchanging sex for money or drugs among women injectors in Baltimore, MD, USA. *International Journal of STD & AIDS* 19 (1): 47–50.

Doherty, M. C., R. S. Garfein, E. Monterroso, D. Brown, and D. Vlahov. 2000. Correlates of HIV infection among young adult short-term injection drug users. *AIDS (London, England)* 14 (6): 717–26.

Ellen, J. M., J. M. Jennings, T. Meyers, S. E. Chung, and R. Taylor. 2004. Perceived social cohesion and prevalence of sexually transmitted diseases. *Sexually Transmitted Diseases* 31 (2): 117–22.

Environmental Systems Research Institute (ESRI). 2009. *ESRI and maps ArcGIS 9.3 media kit.* Redlands, CA: ESRI.

Gesler, W. M., and R. A. Kearns. 2002. Culture, place, and health. In *Critical geographies*, ed. T. Skelton and G. Valentine, 11–35. London and New York: Routledge.

Hardick, J., Y. H. Hsieh, S. Tulloch, J. Kus, J. Tawes, and C. A. Gaydos. 2003. Surveillance of chlamydia trachomatis and neisseria gonorrhoeae infections in women in detention in Baltimore, Maryland. *Sexually Transmitted Diseases* 30 (1): 64–70.

Hart, T., J. L. Peterson, and Community Intervention Trial for Youth Study Team. 2004. Predictors of risky sexual behavior among young African American men who have sex with men. *American Journal of Public Health* 94 (7): 1122–24.

Jennings, J. M., F. C. Curriero, D. Celentano, and J. M. Ellen. 2005. Geographic identification of high gonorrhea transmission areas in Baltimore, Maryland. *American Journal of Epidemiology* 161 (1): 73–80.

Keene, D. E., and M. B. Padilla. 2010. Race, class and the stigma of place: Moving to "opportunity" in eastern Iowa. *Health & Place* 16 (6): 1216–23.

Kwan, M., R. D. Peterson, C. R. Browning, L. A. Burrington, C. A. Calder, and L. J. Krivo. 2008. Reconceptualizing sociogeographic context for the study of drug use, abuse, and addiction. In *Geography and drug addiction*, ed. Y. F. Thomas, D. Richardson, and I. Cheung, 437–46. New York: Springer.

Levine, N. 2010. *CrimeStat: A spatial statistics program for the analysis of crime incident locations.* Version 3.3. Houston, TX, and Washington, DC: Ned Levine & Associates, and the National Institute of Justice.

Matheson, F. I., R. Moineddin, J. R. Dunn, M. I. Creatore, P. Gozdyra, and R. H. Glazier. 2006. Urban neighborhoods, chronic stress, gender and depression. *Social Science & Medicine* 63 (10): 2604–16.

McLafferty, S. 2008. Placing substance abuse: Geographical perspectives on substance use and addiction. In *Geography and drug addiction*, ed. Y. F. Thomas, D. Richardson, and I. Cheung, 1–16. New York: Springer.

Mills, T. C., R. Stall, L. Pollack, J. P. Paul, D. Binson, J. Canchola, and J. A. Catania. 2001. Health-related characteristics of men who have sex with men: A comparison of those living in "gay ghettos" with those living elsewhere. *American Journal of Public Health* 91 (6): 980–83.

R Development Core Team. 2008. *R: A language and environment for statistical computing.* Vienna, Austria: R Foundation for Statistical Computing.

Rhodes, T. 2002. The "risk environment": A framework for understanding and reducing drug-related harm. *International Journal of Drug Policy* 13 (2): 85–94.

Rietmeijer, C. A., R. J. Wolitski, M. Fishbein, N. H. Corby, and D. L. Cohn. 1998. Sex hustling, injection drug use, and non-gay identification by men who have sex with men: Associations with high-risk sexual behaviors and condom use. *Sexually Transmitted Diseases* 25 (7): 353–60.

Ripley, B. D. 1976. The second order analysis of stationary point patterns. *Journal of Applied Probability* 13:225–66.

Rowlingson, B., P. Diggle, R. Bivand, G. Petris, and S. Eglen. 2010. *Splancs: Spatial and space-time point pattern analysis.* Vol. R package version 2.

Silverman, B. W. 1998. *Density estimation for statistics and data analysis.* London: Chapman & Hall/CRC.

Takahashi, L. M. 1997. The socio-spatial stigmatization of homelessness and HIV/AIDS: Toward an explanation of the NIMBY syndrome. *Social Science & Medicine* 45 (6): 903–14.

Tobin, K. E., M. Davey-Rothwell, and C. A. Latkin. 2010. Social-level correlates of shooting gallery attendance: A focus on networks and norms. *AIDS and Behavior* 14 (5): 1142–48.

Tobin, K. E., S. J. Kuramoto, M. A. Davey-Rothwell, and C. A. Latkin. 2010. The STEP into action study: A peer-based, personal risk network-focused HIV prevention intervention with injection drug users in Baltimore, Maryland. *Addiction* 106 (2): 366–75.

Towe, V. L., F. Sifakis, R. M. Gindi, S. G. Sherman, C. Flynn, H. Hauck, and D. D. Celentano. 2010. Prevalence of HIV infection and sexual risk behaviors among individuals having heterosexual sex in low income neighborhoods in Baltimore, MD: The BESURE study. *Journal of Acquired Immune Deficiency Syndromes* 53 (4): 522–28.

U.S. Census Bureau. 2000. *Profile of general demographic characteristics: 2000.* Baltimore, MD: Baltimore City Department of Planning.

———. 2010. U.S. census summary file 1 (SF 1) and summary file 3 (SF3) 100-percent data. Online database. http://www.census.gov/prod/cen2000/doc/sf1.pdf (last accessed 10 October 2010).

Waller, L. A., and C. A. Gotway. 2004. *Applied spatial statistics for public health.* Hoboken, NJ: Wiley Interscience.

Williams, P. B., and O. Ekundayo. 2001. Study of distribution and factors affecting syphilis epidemic among inner-city minorities of Baltimore. *Public Health* 115 (6): 387–93.

Wohl, D. A., M. R. Khan, C. Tisdale, K. Norcott, J. Duncan, A. M. Kaplan, and S. S. Weir. 2010. Locating the places people meet new sexual partners in a southern US city to inform HIV/STI prevention and testing efforts. *AIDS and Behavior* 15 (2): 283–91.

Wylie, J. L., L. Shah, and A. Jolly. 2007. Incorporating geographic settings into a social network analysis of injection drug use and bloodborne pathogen prevalence. *Health & Place* 13 (3): 617–28.

Zenilman, J. M., N. Ellish, A. Fresia, and G. Glass. 1999. The geography of sexual partnerships in Baltimore: Ap-plications of core theory dynamics using a geographic information system. *Sexually Transmitted Diseases* 26 (2): 75–81.

Five Essential Properties of Disease Maps

Kirsten M. M. Beyer,* Chetan Tiwari,† and Gerard Rushton‡

*Institute for Health and Society, Medical College of Wisconsin
†Department of Geography, University of North Texas
‡Department of Geography, The University of Iowa

We argue that as the disease map user group grows, disease maps must prioritize several essential properties that support public health uses of disease maps. We identify and describe five important properties of disease maps that will produce maps appropriate for public health purposes: (1) Control the population basis of spatial support for estimating rates, (2) display rates continuously through space, (3) provide maximum geographic detail across the map, (4) consider directly and indirectly age–sex-adjusted rates, and (5) visualize rates within a relevant place context. We present an approach to realize these properties and illustrate it with small-area data from a population-based cancer registry. Users whose interests are in selecting areas for interventions to improve the health of local populations will find maps with these five properties useful. We discuss benefits and limitations of our approach, as well as future logical extensions of this work.

我们认为随着使用疾病地图用户群的增长，这些地图必须优先考虑支持公众健康的地图所使用的几个基本属性。我们识别和描述五种疾病地图的重要属性，它们将以公共健康为目的产生适当的地图：（1）控制空间化的人口基础以估计比率，（2）连续显示穿越时空的比率，（3）在跨越整幅地图的范围提供最大的地理细节，（4）直接和间接地考虑年龄，性别调整后的比率，（5）在一个相关的背景下，使比率可视化。我们提出了实现这些特性的一种方法，并以来自人口为基础的癌症登记资料的小面积数据来说明该方法。那些对选择干预地区并对改善当地居民的健康感兴趣的用户，会发现具有这五个属性的地图很有用。我们讨论了该方法的优点和局限性，以及今后对这项工作的逻辑拓展。关键词：癌症，疾病制图，空间分析。

Sostenemos que por cuanto el grupo de usuarios del mapa crece, los mapas sobre enfermedades deben priorizar varias propiedades esenciales que favorecen los usos de los mapas para fines de salud pública. Identificamos y describimos cinco propiedades importantes de los mapas de enfermedades que pueden hacerlos apropiados para los propósitos de la salud pública: (1) Controlan la base poblacional de apoyo espacial para calcular las tasas, (2) despliegan las tasas de manera continua a través del espacio, (3) proveen el máximo de detalle geográfico a través del mapa, (4) consideran directa e indirectamente las tasas ajustadas edad–sexo, y (5) visualizan las tasas dentro de un contexto relevante de lugar. Presentamos un enfoque para comprender estas propiedades e ilustrarlas con datos sobre un área pequeña de un registro de cáncer basado en población. Los usuarios cuyos intereses están en la selección de áreas para hacer intervenciones que mejoren la salud de las poblaciones locales encontrarán útiles los mapas que exhiban estas cinco propiedades. Discutimos los beneficios y limitaciones de nuestro enfoque, lo mismo que las lógicas extensiones futuras de este trabajo.

The last decade has seen great growth in the production of disease and mortality maps throughout the world. Although this growth has been accompanied by research on the fundamentals of map design and production, some essential properties of these maps have received relatively little attention, particularly from geographers.

Map properties that have received attention include selection of color schemes (Brewer et al. 1997; Brewer and Pickle 2002), best units for display (Boscoe and Pickle 2003), spurious variability of disease rates (Morris and Munasinghe 1993), brushing techniques (MacEachren and Kraak 1997), micromaps (Carr, Wallin, and Carr 2000), and continuous representation (Tyczynski et al. 2006). Additional attention focuses on modeling the spatial pattern of rates by adjusting for patterns of spatial autocorrelation (Goovaerts 2006) and detection and visualization of disease clusters (Kulldorff and Nagarwalla 1995; Openshaw, Charlton, and Craft 1988; Chen et al. 2008).

Considerable effort has continued the choropleth disease mapping approach, which uses administratively defined boundaries for rate calculation and representation. The problems with this approach are well known, including (1) the *small numbers problem*, whereby rates based on small populations are more variable—and less reliable—than rates based on large populations; (2) the *modifiable areal unit problem*, whereby the choice of bounded or areal unit can change the observed spatial pattern; (3) the *limited variability problem*, where mapping a rate by a predetermined unit prevents the exploration of the true spatial variation within it; and (4) the *visual impact of large areas problem*, where areal units defined by large areas of geographic space carry more visual weight than smaller areas, despite the fact that these larger areas are often characterized by low population density and thus are more likely to have unstable rates. These problems have been described many times (Openshaw 1984; Gelman and Price 1999; Lawson 2001; Waller and Gotway 2004), and a significant amount of work attempts to adjust rates shown on choropleth maps, including headbanging (Mungiole, Pickle, and Simonson 1999), Bayesian adjustments (Clayton and Kaldor 1987; Bernardinelli and Montomoli 1992), and smoothing (Kafadar 1996; Shi et al. 2007).

Other approaches seek to address these issues by avoiding them at the outset. These methods prioritize the definition of both the geographic space and population on which a rate will be based, delineating this space before the rate is calculated to minimize the variance in the map. Techniques of density estimation and spatial filtering describe this approach. These techniques conceptualize disease rates as varying continuously over geographic space and aggregate data in such a way that a stable disease rate is calculated by overlapping windows over the map and is presented as a continuous surface (Brunsdon 1995; Rushton and Lolonis 1996; Talbot et al. 2000; Tiwari and Rushton 2005; Beyer and Rushton 2009; Shi 2009, 2010; Davies and Hazelton 2010).

Others have aggregated areal units based on contiguity or population characteristics before calculating rates, seeking to maintain the use of areal units for presentation of rates but also attempting to delineate bounded areas in such a way as to address the small numbers problem (Morris and Munasinghe 1993; Mu and Wang 2008). In addition, developments in geostatistics have been brought to bear on the small numbers problem by searching for spatial correlations in disease patterns and using that knowledge to model spatial patterns. Kriging methods as applied to disease mapping have summarized the nature of spatial autocorrelation in disease rates with a semivariogram and then used that semivariogram to model disease rates for locations where no information has been observed (Kyriakidis 2004; Goovaerts 2006). Some have combined mapping approaches, including using kriging to analyze regional estimates of disease

risk already smoothed using empirical Bayes techniques (Berke 2004).

As the practice of mapping disease has developed, the disease map user group has grown beyond health geographers, spatial epidemiologists, and spatial statisticians to include disease prevention and control organizations, advocacy organizations, and the public. If disease maps are to be used by these groups to result in improvements in population health, it is important that disease maps prioritize several essential properties that support public health uses of disease maps, including surveillance, resource allocation, and communication.

In this article, we identify and describe five important properties of disease maps that will produce maps appropriate for public health purposes. We present an approach to realize these five properties and illustrate it with small-area data from a population-based cancer registry. Users whose interests are in selecting areas for interventions to improve the health of local populations will find maps with these five properties useful.

Our Approach

We developed a methodology for creating disease maps and implemented it in a Web-based disease mapping system we call WebDMAP (http://www.webdmap.com). This system implements the method known as *adaptive spatial filtering*, which uses an adaptive bandwidth filter of uniform shape that increases in size inversely with population density for disease rate calculation (Brunsdon 1995; Rushton and Lolonis 1996; Tiwari and Rushton 2005; Beyer and Rushton 2009; Shi 2009, 2010; Davies and Hazelton 2010; Talbot et al. 2000).

In adaptive spatial filtering, a grid is laid over the study area, and for each grid point a rate is calculated by using a circular filter that expands to obtain data from multiple geocoded locations until it obtains enough observations to calculate a stable rate. The user defines a threshold value to guide the filter bandwidth size, such as an expected number of disease events or a population size that will result in a stable rate calculation. The area in between the grid points is interpolated to create a continuous surface representation of a disease burden. The geographic scale of the measured rates will vary across the map as the density of the population at risk varies.

We utilized this system to map burdens of cancer in Iowa from 2000 to 2005; examples from this work are illustrated in the next section, and a large selection of maps can be found on the Internet (http://www.uiowa.edu/iowacancermaps/). The adaptive spatially filtered maps are based on disease and population data aggregated from the 949 Iowa five-digit ZIP codes; data were geocoded to a single point within each ZIP code—the centroid of the largest incorporated area (e.g., city, town, or village) polygon in each ZIP code. The maps use ZIP code data from the Iowa Cancer Registry for 2000 to 2005 linked to U.S. population data for Iowa ZIP Code Tabulation Areas (ZCTAs) from the 2000 Census. Rates were adjusted to take into account the age and sex of populations in each ZIP code. As cancer data are for ZIP codes for the years 2000 to 2005, and population data are for ZCTAs for the year 2000, our maps are subject to issues of spatial and temporal misalignment between numerator and denominator.

Five Essential Properties of Disease Maps

Property 1: Control the Population Basis of Spatial Support for Estimates of Rates

Many of the problems of conventional disease maps stem from failing to control the spatial basis of support for rate calculation, defined as the population residing within the geographic area used as the analysis unit for the calculation of a disease rate (Gotway and Young 2002; Waller and Gotway 2004). Although administratively defined spatial supports—such as census geographies—might increase the ease with which a rate can be mapped, they might not represent appropriate supports for the phenomenon of interest and could result in rates that are unreliable for interpretation. We consider the careful control of spatial support to be the first essential property of a disease map.

In Figure 1, we illustrate the difference between the use of an administratively defined spatial support and a support controlled to be appropriate to the phenomenon of interest. Figure 1 is intended to present the "best map" that can be made using each of three different mapping approaches and should be interpreted as a comparison of mapping approaches, not a comparison of spatial patterns. There are a number of key distinctions among the three maps, including data from different geographic units (county, ZIP code) and data for different age groups, depending on the age–sex adjustment strategy used. These differences are necessary and specific to each approach and illustrate the idea that the best map that can be made with each approach might in fact be a different map of slightly different

information, because what can be achieved with each method is different.

As shown in Figure 1, if we map colorectal cancer incidence for the state of Iowa using adaptive spatial filtering and ZIP code-level data and compare the result to a map that uses county boundaries for rate calculation and representation, we observe a very different map. In Figures 1 and 2, we use quantile breaks with fifteen classes to show a gradual transition and three main categories; rates that are higher than expected are shown in shades of red, whereas rates lower than expected are shown in shades of blue, with darker shades indicating the most extreme rates. Areas experiencing rates close to statewide rates are not colored.

In Figure 1, we see that whereas the spatial basis of support for counties is based on predetermined administrative areas, the spatial basis of support for the adaptive spatially filtered maps is based on the spatial pattern of the underlying population distribution. Although the size and shape of county boundaries in Iowa are relatively similar, the populations represented by these boundaries vary widely, resulting in varying levels of statistical confidence across the map.

It is important that a discussion of spatial support not be limited to statistical questions. There also exists the challenge of taking into account the nature of territorial organization, which brings with it a landscape of established social, economic, and other characteristics that might contribute to the spatial distribution of disease.

Property 2: Display Rates Continuously through Space

Very closely related to the control of spatial support is the importance of presenting a disease rate visually so that it best estimates what a true disease risk pattern might look like on the ground. Despite the fact that bounded units are often used for disease rate representation, there is little reason to suspect that the actual spatial pattern of risk follows such boundaries. A more likely assumption is that risk surfaces are more continuous in nature, varying along with biological, social, and environmental characteristics of populations. Maps that utilize bounded units for representation prevent the observation of any variation of risk that might take place within boundaries.

In general, the spatial point pattern of disease events over a region can be regarded as the outcome of a spatial stochastic process in which the disease cases occur continuously across geographic space and independently of

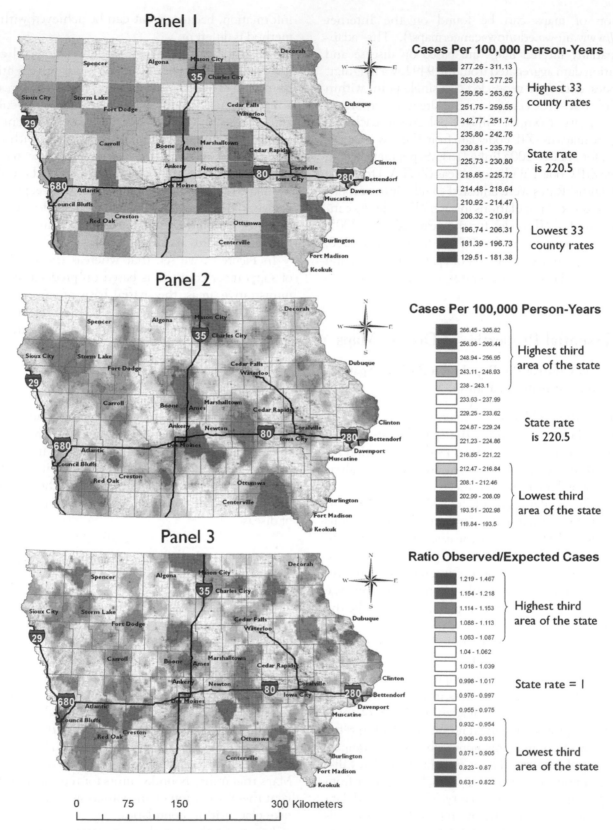

Figure 1. Three maps of colorectal cancer (CRC) incidence in Iowa, 2000–2005. Panel 1: CRC incidence rates by county, using direct age–sex adjustment. Adjustments were made based on male and female populations in the following three age groups: 50–69, 70–79, and 80 and older. Panel 2: CRC incidence mapped using adaptive spatial filtering and direct age–sex adjustment. Adjustments were made based on male and female populations in the following three age groups: 50–69, 70–79, and 80 and older. Panel 3: CRC incidence mapped using adaptive spatial filtering and indirect age–sex adjustment. This map was created using five-year age groups and was not restricted to ages 50 and older. (Color figure available online.)

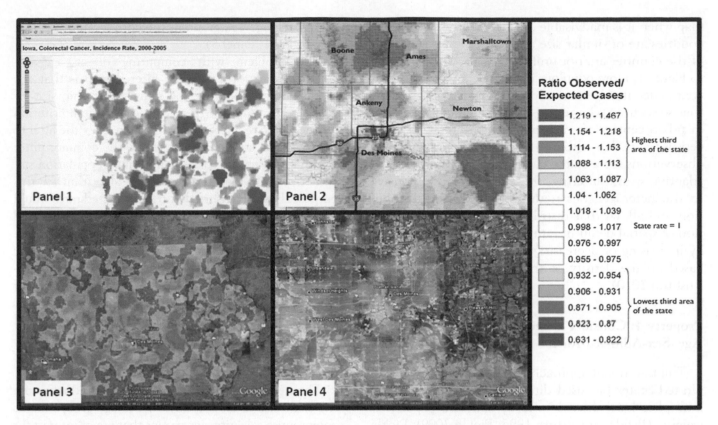

Figure 2. Visualization and place context. Panel 1: Raster layers created in WebDMAP are overlaid on OpenStreetMap data using the OpenLayers Web mapping interface. Panel 2: For presentation as a printed map, a number of place context features can be added to the disease layer, including county boundaries, place names, major highways, and a base map showing the locations of settlements, water bodies, and other topographical features of interest. Panel 3: Raster layers created in a geographic information system software package can be exported for visualization in other applications, such as the Google Earth mapping service. Panel 4: In the Google Earth mapping service, a user can easily zoom in to a particular locality of interest to explore the burden of disease at a fine geographic scale. (Color figure available online.)

one another. This pattern is often described using an intensity function. Kernel density estimation methods, including the adaptive spatial filter method, have been used previously to represent this intensity function as a continuous surface (Diggle, Besag, and Gleaves 1976).

Of course, not all disease maps utilizing a continuous surface presentation are created equally. Some maps that present disease rates as varying continuously are created through the "smoothing" (e.g., the adjustment of a disease rate based on other information, such as an empirical Bayes prior or rates from contiguous counties) of precalculated rates for large areas (Kafadar 1999; Tyczynski et al. 2006) as opposed to the iterative aggregation of numerator and denominator information before calculating and directly reporting a disease rate (Rushton and Lolonis 1996; Talbot et al. 2000).

Again, considering Figure 1, we see that instead of a spatial pattern that stops abruptly at the boundaries of an administratively defined areal unit (the county), the spatially filtered maps vary continuously over geo-graphic space, eliminating representation problems associated with maps that employ bounded units.

It is important to note that features of some landscapes could temper the benefit of a continuous representation. Significant geographical features (e.g., mountains, racial segregation) might disrupt geographical space in such a way that a different approach might be necessary. A distinction should be made between artificial breaks in space, imposed by predetermined geographical boundaries, and more natural breaks that could actually be relevant in interrupting the spatial distribution of the disease rate.

Property 3: Provide Maximum Geographic Detail across the Map

A central criticism that can be made of disease maps that use administratively defined boundaries is that they allow those boundaries to remove spatial detail from the map, when it is available, and force spatial detail into a map when it is inadvisable due to small numbers. Iowa counties are of similar size and shape, but populations of the counties are not uniform. Again with reference to Figure 1, more populous counties can provide intracounty detail. Alternatively, choosing a spatial support that seeks to maximize the geographic detail available in urban centers often means using a unit (ZIP code, census tract) that is then too small to contain enough observations for a stable calculation in rural areas. In adaptive spatial filtering, the mapmaker is responsive to the heterogeneous nature of population distribution and allows that distribution to guide the definition of spatial detail in rate mapping. This is achieved by increasing or decreasing the number of filters used based on the density of the population (Tiwari and Rushton 2010).

Property 4: Consider Directly and Indirectly Age–Sex-Adjusted Rates

The traditional approach to disease mapping in the United States has used direct age adjustment, which multiplies local disease rates by standard population weights (Pickle and White 1995; Pickle 2009). Pickle (2009) cited twelve U.S. atlases of mortality published since 1975; all used the direct standardization method. Many international atlases, however, have used the indirect method. Walter and Birnie (1991) found that 59 percent of the forty-six international health atlases they examined used indirect adjustment. The direct method shows the measure of disease that would prevail in a local area if it had the same age–sex composition as the standard population. This is useful when the purpose of the map is to compare rates from place to place because it removes effects due to differences in the age–sex composition of local areas. When the purpose of the map is to compare the impact of different disease rates on local populations, however, indirect adjustment should be used. This method applies standard rates to local population distributions. In addition to measuring the burden of disease rates on local populations, indirectly age-adjusted rates have the advantage of rate stability. They also maximize geographic detail because rates can be more reliably computed for small-area populations.

We use adaptive spatial filtering to map both indirectly and directly age–sex-adjusted rates. To map indirectly age–sex-adjusted rates, our adaptive filters obtain observed and expected numbers of cancer cases for each ZIP code, where the expected number of cases was calculated by multiplying the statewide cancer incidence rate for each age–sex group by the number of individu-

als in each age–sex group in the corresponding ZCTA (see Figure 1, Panel 3).

The problem with computing direct age–sex-adjusted rates for a spatially filtered map is that the spatial filters should be the same geographic size for each age stratum, yet to meet the minimum criterion for the adaptive filter it is necessary to apply the filter to each age–sex stratum to determine the common filter size that will ensure that the minimum population size is met. We developed a three-step algorithm for this purpose. First, to ensure that the computed rates would be based on the same-size spatial filters, we established combinations of age groups that for the state as a whole had approximately equal numbers of colorectal cancer cases. Second, for each grid point we computed the filter size needed to estimate a stable local rate for each of these age groups. We then used the largest of these filter sizes to compute the rates at the grid point and weighted each rate by its associated population weight in the standard population. The sum of the weighted rates is the directly age-adjusted rate (see Figure 1, Panel 2).

The need to ensure that directly adjusted maps use the same-size filters for each of the age groups might compromise selecting age groups that are of interest for biological or medical reasons. This compromise is necessary for valid implementation of the direct-adjustment method to a common geography on different age groups. The direct approach risks the calculation of unstable local rates, as it requires the computation of age–sex-specific rates for small areas. It follows, also, that direct age-adjusted maps will routinely have less detail in their geography and in their age structure.

Both adjustment strategies are valid; the purpose of the map should determine which of the two should be selected. If the purpose is to compare disease rates across space, direct age adjustment should be used, whereas if the purpose is to compare the needs of regions for resources to address health disparities, indirect age adjustment should be used.

Property 5: Visualize Rates within a Relevant Place Context to Enhance Interpretation

It is increasingly recognized that disease maps cannot be produced in a vacuum, hidden from public audiences, but should be available to public health practitioners and the public so that they can be used in improving population health (Bell et al. 2006; Beyer and Rushton 2009). Along with this imperative comes the necessity to increase the readability and effectiveness of communication of a disease map, including its ability to "place" its assertion of relative disease risks (MacEachren, Brewer, and Pickle 1998; Boscoe et al. 2003).

Figure 2 illustrates some of the approaches that could be taken to enhance visualization. Panel 2 shows a map designed for print. The background used for these maps is the 1:100,000 scale topographic map from the U.S. Geological Survey, which shows information such as the locations of cities and towns and physical landmarks such as rivers and lakes. The major highways, county boundaries, and names of some cities are also included to assist in interpretation. Panels 1, 3, and 4 show visualization options in the widely available Google Earth mapping service and OpenLayers application.

Conclusions

Disease maps are increasingly providing the context within which policy questions are raised and resolved (Wennberg and Gittelsohn 1973; Dartmouth Medical School 2008; Epstein 2010). Patterns of disease have become the focus of many who are asking for actions to reduce health inequalities. The prevailing approach to disease mapping, however, begins with a table of data with defined areas as rows and disease attributes of those areas as columns. The effective application of the principles outlined in this article will move disease mapping away from this assumption and will be important to developing scientific knowledge that can inform disease prevention and control efforts.

Future work should explore the implications of disease maps in terms of measures of statistical confidence across the map and ways in which these measures could be represented (Krieger et al. 2004). In addition, there is no literature discussing the special issues that arise in making directly adjusted maps with spatially adaptive filters; building on the approach used in this article, more formal comparisons of direct and indirect adjustment are needed.

Our approach is characterized by a number of limitations. Although disease risk varies with population and environmental characteristics across geographic space, methods used for producing spatially continuous representations of that risk do not themselves consider such place-based differences. Disease rates in rural areas with small numbers of people might be based on information from relatively distant places to achieve rate stability; further, the information might come from a restricted number of directions if a rural area is located in a corner of the study area for which data are available. Although the problem of combining data from distant locations, and therefore diluting the characterization of risk, can be partially addressed by imposing a distance-based restriction limiting the maximum size of the spatial filter, there is a need for further improvements.

In practice, this limitation can be considered in terms of two questions for future research: (1) Can filters be constrained to address the potential problem of including information from distant locations or different (e.g., urban vs. rural) settings? (2) Does this limitation constitute an inability to examine the role of local environmental processes on risk? Future work should explore the possibility of placing constraints on filters to restrict information they include and to identify whether these restrictions change or improve estimation. Regardless of which strategy is adopted to constrain the spatial filter definition, it is critical that the mapmaker is given information to analyze and visualize the spatial basis of support that was used to compute every rate that is represented on the map.

In conclusion, it is important first to consider the purpose of the desired map before determining the most appropriate mapping approach. Disease maps might seek to describe the spatial distribution of disease, identify unknown risk factors, verify hypotheses (e.g., sources of exposure, social determinants), or communicate risk. Each purpose might call for a different map and a different approach, including consideration of geographic scale, cartographic representation, and methodology.

We recognize the diversity of disease mapping approaches and appreciate the advantages and disadvantages of different approaches. We hope that this article can initiate a discussion of the future of disease mapping to maximize the utility of disease maps for both an increased understanding of the causes of geographic variations in health and the benefit of disease prevention and control.

Acknowledgments

The authors wish to thank colleagues Marc Armstrong, David Bennett, Qiang Cai, Martha Carvour, David Haynes, Soumya Mazumdar, and Alberto Segre (The University of Iowa); Charles Lynch and Michelle West (State Health Registry of Iowa); David Stinchcomb (National Cancer Institute); James Cucinelli and Dan Kavan (Information Management Services, Inc.); Sam LeFevre and Benjamin Goodrich (Utah Department of Health); and Kevin Henry (New Jersey

Department of Health and Senior Services) for their collaboration and suggestions. We would also like to thank three anonymous reviewers. The work described here was funded in part by the National Cancer Institute, SEER Rapid Response Surveillance Study N01-PC-31543.

References

Bell, B., R. Hoskins, L. Pickle, and D. Wartenberg. 2006. Current practices in spatial analysis of cancer data: Mapping health statistics to inform policymakers and the public. *International Journal of Health Geographics* 5:49.

Berke, O. 2004. Exploratory disease mapping: Kriging the spatial risk function from regional count data. *International Journal of Health Geographics* 3:18.

Bernardinelli, L., and C. Montomoli. 1992. Empirical Bayes versus fully Bayesian analysis of geographical variation in disease risk. *Statistics in Medicine* 11:983–1007.

Beyer, K., and G. Rushton. 2009. Mapping cancer for community engagement. *Preventing Chronic Disease* 6:A03.

Boscoe, F. P., C. McLaughlin, M. J. Schymura, and C. L. Kielb. 2003. Visualization of the spatial scan statistic using nested circles. *Health Place* 9:273–77.

Boscoe, F., and L. Pickle. 2003. Choosing geographic units for choropleth rate maps, with an emphasis on public health applications. *Cartography and Geographic Information Science* 30:237–49.

Brewer, C., A. MacEachren, L. Pickle, and D. Herrmann. 1997. Mapping mortality: Evaluating color schemes for choropleth maps. *Annals of the Association of American Geographers* 87:411–38.

Brewer, C., and L. Pickle. 2002. Evaluation of methods for classifying epidemiological data on choropleth maps in series. *Annals of the Association of American Geographers* 92:662–81.

Brunsdon, C. 1995. Estimating probability surfaces for geographical point data: An adaptive kernel algorithm. *Computers & Geosciences* 21:877–94.

Carr, D., J. Wallin, and D. Carr. 2000. Two new templates for epidemiology applications: Linked micromap plots and conditioned choropleth maps. *Statistics in Medicine* 19:2521–38.

Chen, J., R. E. Roth, A. T. Naito, E. J. Lengerich, and A. M. MacEachren. 2008. Geovisual analytics to enhance spatial scan statistic interpretation: An analysis of U.S. cervical cancer mortality. *International Journal of Health Geographics* 7:57.

Clayton, D., and J. Kaldor. 1987. Empirical Bayes estimates of age-standardized relative risks for use in disease mapping. *Biometrics* 43:671–81.

Dartmouth Medical School. 2008. *Tracking the care of patients with severe chronic illness: The Dartmouth atlas of health care 2008.* Dartmouth, NH: The Dartmouth Institute for Health Policy and Clinical Practice.

Davies, T. M., and M. L. Hazelton. 2010. Adaptive kernel estimation of spatial relative risk. *Statistics in Medicine* 29:2423–37.

Diggle, P. J., J. Besag, and J. T. Gleaves. 1976. Statistical analysis of spatial point patterns by means of distance methods. *Biometrics* 32:659–67.

Epstein, A. M. 2010. Geographic variation in Medicare spending. *New England Journal of Medicine* 363:85–86.

Gelman, A., and P. N. Price. 1999. All maps of parameter estimates are misleading. *Statistics in Medicine* 18: 3221–34.

Goovaerts, P. 2006. Geostatistical analysis of disease data: Accounting for spatial support and population density in the isopleth mapping of cancer mortality risk using area-to-point Poisson kriging. *International Journal of Health Geographics* 5:52.

Gotway, C. A., and L. J. Young. 2002. Combining incompatible spatial data. *Journal of the American Statistical Association* 97:632–48.

Kafadar, K. 1996. Smoothing geographical data, particularly rates of disease. *Statistics in Medicine* 15:2539–60.

———. 1999. Simultaneous smoothing and adjusting mortality rates in U.S. counties: Melanoma in white females and white males. *Statistics in Medicine* 18:3167–88.

Krieger, N., P. Waterman, J. Chen, D. Rehkopf, and S. Subramanian. 2004. *Geocoding and monitoring U.S. socioeconomic inequalities in health: An introduction to using area-based socioeconomic measures—The Public Health Disparities Geocoding Project monograph.* Cambridge, MA: Harvard School of Public Health. http://www.hsph.harvard.edu/thegeocodingproject/ (last accessed 15 December 2010).

Kulldorff, M., and N. Nagarwalla. 1995. Spatial disease clusters: Detection and inference. *Statistics in Medicine* 14:799–810.

Kyriakidis, P. 2004. A geostatistical framework for area-to-point spatial interpolation. *Geographical Analysis* 36:259–90.

Lawson, A. 2001. *Statistical methods in spatial epidemiology.* New York: Wiley.

MacEachren, A., C. Brewer, and L. Pickle. 1998. Visualizing georeferenced data: Representing reliability of health statistics. *Environment and Planning A* 30:1547–62.

MacEachren, A., and M. Kraak. 1997. Exploratory cartographic visualization: Advancing the agenda. *Computers & Geosciences* 23:335–43.

Morris, R., and R. Munasinghe. 1993. Aggregation of existing geographic regions to diminish spurious variability of disease rates. *Statistics in Medicine* 12:1915–29.

Mu, L., and F. Wang. 2008. A scale-space clustering method: Mitigating the effect of scale in the analysis of zone-based data. *Annals of the Association of American Geographers* 98:85–101.

Mungiole, M., L. W. Pickle, and K. H. Simonson. 1999. Application of a weighted head-banging algorithm to mortality data maps. *Statistics in Medicine* 18:3201–3209.

Openshaw, S. 1984. The modifiable areal unit problem. *Concepts and Techniques in Modern Geography* 38:41.

Openshaw, S., M. Charlton, and A. Craft. 1988. Searching for leukaemia clusters using a geographical analysis machine. *Papers of the Regional Science Association* 64:95–106.

Pickle, L. 2009. A history and critique of U.S. mortality atlases. *Spatial and Spatio-temporal Epidemiology* 1:3–17.

Pickle, L. W., and A. A. White. 1995. Effects of the choice of age-adjustment method on maps of death rates. *Statistics in Medicine* 14:615–27.

Rushton, G., and P. Lolonis. 1996. Exploratory spatial analysis of birth defect rates in an urban population. *Statistics in Medicine* 15:717–26.

Shi, X. 2009. A geocomputational process for characterizing the spatial pattern of lung cancer incidence in New Hampshire. *Annals of the Association of American Geographers* 99:521–33.

———. 2010. Selection of bandwidth type and adjustment side in kernel density estimation over inhomogeneous backgrounds. *International Journal of Geographical Information Science* 24:643–60.

Shi, X., E. Duell, E. Demidenko, T. Onega, B. Wilson, and D. Hoftiezer. 2007. A polygon-based locally-weighted-average method for smoothing disease rates of small units. *Epidemiology* 18:523.

Talbot, T. O., M. Kulldorff, S. P. Forand, and V. B. Haley. 2000. Evaluation of spatial filters to create smoothed maps of health data. *Statistics in Medicine* 19:2399–2408.

Tiwari, C., and G. Rushton. 2005. Using spatially adaptive filters to map late stage colorectal cancer incidence in Iowa. In *Developments in spatial data handling*, ed. P. Fisher, 665–76. Berlin: Springer.

———. 2010. A spatial analysis system for integrating data, methods and models on environmental risks and health outcomes. *Transactions in GIS* 14:177–95.

Tyczynski, J. E., K. Pasanen, H. J. Berkel, and E. Pukkala. 2006. *Atlas of cancer incidence and mortality in Ohio*. Dayton, OH: Cancer Prevention Institute.

Waller, L., and C. Gotway. 2004. *Applied spatial statistics for public health data*. New York: Wiley-Interscience.

Walter, S., and S. Birnie. 1991. Mapping mortality and morbidity patterns: An international comparison. *International Journal of Epidemiology* 20:678.

Wennberg, J., and A. Gittelsohn. 1973. Small area variations in health care delivery. *Science* 182:1102–08.

Migrant Workers in Home Care: Routes, Responsibilities, and Respect

Kim England* and Isabel Dyck†

*Department of Geography, University of Washington
†School of Geography, Queen Mary, University of London

We consider the increasingly common provision of home-based health care by migrant care workers. In particular, we explore the racial division of paid reproductive care and ideas about embodied work to show that although (im)migrants tend to fall to the bottom of the hierarchy of care work, the reasons are multifaceted and complex. We draw on interview data from a larger study of long-term home care in Ontario to explore the lived experience of care work by migrant workers, emphasizing their social agency. We organize our discussion around the themes of routes, responsibilities, and respect and emphasize the embodied and power-inflected care work relation. Through these themes we explore the different routes the migrants took into care work—how they found their jobs and what role those jobs play in their lives. Then we address the responsibilities of different home care jobs and the relational dynamic of how job responsibilities are actually practiced. Finally, the theme of respect examines how the workers try to treat their clients with dignity but sometimes the work relation is marked by racism and friction over what counts as "good" care. We show that care work is constructed and experienced through a complex interweaving of embodiment, labor market inequalities, and the province's regulatory mechanisms of care provision.

我们认为，由移民保健工作者提供的，以家庭为基础的医疗保健，已越来越普遍。特别是，我们通过探索带薪生育保健的种族分化和其所体现的工作思路来表明，虽然移民倾向于从事底层护理工作，其原因是多方面的和复杂的。我们借鉴位于安大略省所进行的一个长期家庭护理的大型研究采访数据，探讨移民工人的护理工作的生活经验，强调他们的社会机构。我们围绕路线，责任和尊重的主题，并强调所体现的和受权力所驱的护理工作关系。通过这些主题，我们探讨移民所采取的不同的加入护理工作行业的方法，即他们怎样找到的工作，和这些工作在他们的生活中发挥了什么作用。然后，我们探讨不同家庭护理工作的责任，以及这些岗位职责的关系动态是如何实行的。最后针对尊重的主题，探讨虽然这些工人如何尝试有尊严地对待他们的客户，但是有时工作关系被打上种族主义的标志，并在什么是"好"的护理标准上存在分歧。我们的研究表明，护理工作是通过错综复杂的原因交织，劳动力市场的不平等，和该省护理服务的监管机制来构建和进行的。*关键词：护理，体现，移民，种族化，工作。*

En este estudio consideramos la provisión cada vez más común de servicios de salud a domicilio por trabajadores migrantes de salud. En particular, exploramos la división por raza de atención reproductiva paga e ideas acerca del trabajo personificado, para mostrar que aunque los inmigrantes quedan ubicados en el fondo de la jerarquía del trabajo de atención, las razones son multifacéticas y complejas. Nos basamos en entrevistas realizadas en un estudio mayor de atención de largo plazo en hogares de Ontario, para explorar la experiencia vivida en el trabajo de atención por trabajadores migrantes, haciendo énfasis en su agencia social. Nuestra discusión la organizamos alrededor de los temas de rutas, responsabilidades y respeto, y enfatizamos la relación del trabajo de atención personificada y ajena a intereses de poder. A través de estos temas exploramos las diferentes rutas que siguieron los migrantes para llegar a este tipo de trabajo—cómo consiguieron este empleo y qué papel juega en sus vidas este trabajo. Luego abocamos las responsabilidades de los diferentes empleos de atención domiciliaria y la dinámica relacional sobre cómo se practican realmente las responsabilidades del trabajo. Por último, el tema del respeto examina la manera como los trabajadores intentan tratar a sus clientes con dignidad, aunque algunas veces la relación de trabajo esté marcada por racismo y fricción sobre qué es lo que cuenta como "buena" atención. Mostramos que el trabajo de atención se construye y experimenta a través de un complejo entretejido de personificación, inequidades del mercado laboral y los mecanismos reguladores de la provincia sobre provisión de cuidados de la salud.

The ongoing restructuring of health care systems greatly impacts the working conditions and experience of health care workers. In Canada a growing proportion of those health care workers are foreign-born, as is the case in the United Kingdom and the United States. Much attention centers on the transnational migration of doctors and nurses or on migrants employed in the low-paid, nonprofessionalized health care occupations in institutionalized workplaces like hospitals and nursing homes. There is relatively little work, however, on the growing proportion of migrants working in community-based home care. In this article we focus on this particular and increasingly common group of health care workers. We bring together threads from existing literature on theorizing social reproduction, gendered and racialized inequalities in paid employment, and the marginalization of care in capitalist economies. Sifting through that scholarship in light of our empirical findings, we identified three overarching themes: routes (immigration process, job search, and credentials), responsibilities (tasks, rules, and family), and respect (being valued, discrimination, and work dynamics). We focus on the lived experience of home care workers and address these themes in a fashion that emphasizes their social agency. Too often (im)migrant care workers are positioned as passive participants, trapped into accepting lower wages and undesirable jobs. We advocate a more complex interpretation whereby numerous processes and practices come together to produce an evolving labor market that seems to slot women into particular jobs but where women also have initiative and make "rational" choices.

We use Zimmerman, Litt, and Bose's (2006, 3–4) definition of care work as "the multifaceted labor that produces the daily living conditions that make basic human health and well-being possible." This particular definition appeals precisely because it is multifaceted. It includes the most common understanding of care work—the care of children, the elderly, and people with illnesses and disabilities. These equate to the relational, responsive, face-to-face aspects of care work noted in the care ethics literature that emphasizes the emotional and power-inflected dynamic of interdependent care relations (Kittay 1999; Tronto 2005; Bondi 2008). Zimmerman, Litt, and Bose's definition, however, also includes the more instrumental, less relational activities of housekeeping and domestic tasks such as cleaning floors, grocery shopping, and cooking.

We draw on the literature about the social organization of jobs in care work. Glenn (1992) was among the first to identify a "racial division of paid reproductive labor"—a hierarchy where white women tend to hold face-to-face supervisory and professional positions (e.g., nurses and social workers), and women of color do the "heavy, dirty, 'back-room' chores of food preparation and cooking, cleaning, changing bed pans and the like" (20). More recently, Glenn's racial division of paid reproductive labor has been taken global by scholars placing it in the context of the transnational migration of care workers. For example, Parreñas (2001) and Hondagneu-Sotelo (2007) expanded Glenn's framework to argue that low-wage migrant women of color are found at the bottom of the racialized hierarchy of paid reproductive labor and the increase in migrant domestic workers in the global north equates to an international transfer of caregiving from the global south. In geography, too, this transnational migration of domestic workers is an important research topic. Much of that work has looked at employer employee relations in the live-in caregiver situation, where the empirical focus is on paid care in a single home (e.g., Pratt 2004; Cox 2006). We know less about the experience of care workers hired by agencies working shifts in several different homes (but see Meintel, Fortin, and Cognet 2006). These workers receive instructions from their agencies and then go into people's homes, where the actual care provided is also the result of negotiations with the client and depends on the materiality of each home.

The racialized migrant division of paid reproductive labor underscores the physicality of the actual bodies doing care work. In the literature on embodiment, the concept of "body work" is used to describe the close, intimate, often messy work carried out on other people's bodies (Twigg 2000; McDowell 2009; England and Dyck 2011). Curiously, the corporeal aspect of paid work is often overlooked, even though work is embodied and the corporeality of difference is played out in the divisions of labor. Some scholars address the embodied practices of low-wage workers in other labor market sectors, suggesting that work practices literally embody global capital labor processes: cleaning hotel rooms, cooking and serving restaurant meals, and cleaning the offices of global elites (e.g., Aguiar and Herod 2006). In the case of home care, there is a dual embodiment because not only is the worker's body the direct instrument of care, but also his or her labor focuses on other people's bodily functions, such as toileting and catheter management. Moreover, the particular worksites of labor processes also play a constitutive part in the embodiment of political economy. In our study these multiple embodiments of care work occur in the putatively private homes of care recipients, and as our focus is on international migrants, some falling to the bottom of the racialized division of paid reproductive labor, global processes are also brought into the homes of care recipients.

The Study and Data

The context of our research is the neoliberalized restructuring of the Canadian public health care system. Health care provision is a provincial responsibility and in Ontario neoliberalized health care means shorter hospital stays and patients moving from hospitals into community care sooner than in the past. It is also cheaper for older people to be attended to via publicly funded home care rather than in hospitals or nursing homes. Thus, a large proportion of home care services are provided to people who have chronic illnesses, physical disabilities, or age-related frailties. Ontario now provides home care through a process of managed competition. The province is divided into regional Community Care Access Centers (CCACs), which assess individuals' care needs and determine their level of service, and nonprofit and for-profit agencies compete for contracts to deliver services within a particular CCAC. The managed competition process introduced stricter eligibility requirements and reductions in the number and length of home visits by home care workers (see England et al. 2007).

Our analysis is based on an ethnographic, multidisciplinary study of home care in Ontario. Data were collected from seventeen cases recruited from across the province via our partnering CCACs. Each case included interviews with the care recipient, the paid care workers, and (if relevant) the family caregiver. All care recipients were receiving services from agencies that won contracts from their CCAC, and the paid care workers were employed by those agencies.[1] In this article we restrict our analysis to the eight cases in which one of the paid care workers is an immigrant. All were women who migrated to Canada as adults. Three were born in Jamaica, and one each came from Chile, Germany, Eritrea, Kenya, and Morocco.[2] Ivy (Jamaican) had lived in Canada the longest (twenty-six years) and Irma (Kenyan) the least (six years). We do not intend this to be a representative sample; rather, they allow us to trace general processes and make theoretical points about migrant home care workers in relation to the broad categories of routes, responsibilities, and respect. In some instances the migrants' narratives echo those of Canadian-born workers, such as seeking dignity through their work. In other instances, the im-

migration process makes a significant difference, such as influencing why they went into and stay in home care.

Routes

The theme of routes came up in various ways in the care workers' narratives. Obviously, as immigrants, all of the women had taken international routes to wind up in home care jobs; some arrived in Canada as sponsored family members, whereas others came as refugees. Some came directly to Canada from their county of birth, and others came via other countries. Jocie, a registered nurse (RN), elected to come to Canada from Morocco via the United Kingdom; Rahma (attendant) fled Eritrea and lived in Kenya for a decade before she, her husband, and their younger children moved to Canada. Jocie's route most closely fits within the frame of the transnational migration of skilled workers (Kofman and Raghuram 2006). She went to England to train as a nurse. As a French speaker, she then immigrated first to Québec and later to Ontario.

There were other routes taken into care work. Most of the personal support workers (PSWs) and attendants in the larger study (but not the RNs) found their way into home care work through personal contacts. This echoes Hanson and Pratt's (1995) finding that women in female-dominated work tend to find jobs through other women in their social networks, rather than through formal channels. For example, Brenda (attendant, Jamaica) got her first job twelve years earlier when "a friend told me about [the agency]; I called and started working two hours a day. At first I didn't think it would last [laughs]." Of course, this is one way that occupational segregation is reproduced, because information comes from "strong ties" (i.e., friends, family, and neighbors), as in Brenda's case. As a job search technique, however, informal methods provide an insider's view about what to expect on the job and information about the norms around tasks, pay, vacation time, and so on. From the viewpoint of the home care agencies, word-of-mouth recruiting is a cheap and quick way to fill PSW and attendant positions and might mean less turnover than among those hired through formal means. Perhaps, then, it is not surprising that Brenda's employer interviewed her within twenty-four hours (others talked of similarly fast-moving hiring processes).

The educational achievements of the migrant women are highly varied and cannot be conveniently mapped onto the figure of the "poorly educated immigrant woman" channeled into paid care jobs as a last resort. Valentina (attendant, Chile) held a BA from Chile; Brenda (attendant, Jamaica) had some postsecondary education; and Rahma, an Eritrean refugee, had less than a high school education (most of the Canadian-born home care workers had some postsecondary education; two had BAs). Valentina exemplifies the evident frustrations of immigrants who find that they are unable to translate their professional skills into the Canadian labor market. "Well, I am immigrant," she said. "In Canada I never worked [as a librarian] because in my country the degree is not a master's, it's a BA. Here they're looking for a master's degree for a librarian."

Certainly, a growing literature points to the negative labor market and poverty implications for recent immigrants, like Valentina, whose credentials and work experience from their countries of birth are not recognized (Preston, Lo, and Wang 2003; Creese, Dyck, and McLaren 2008). Irma had similar difficulties to Valentina but experienced a different outcome. When she first arrived in Canada, Irma had already trained as a nurse in Kenya and was keen to find a nursing job, but she had to get licensed before she could be an RN. In the meantime, Irma remarked, "All I could do was work in home care, and I worked too in a nursing home as an aide. I could not do any more until I qualified as an RN." When the required documentation of her credentials was finally retrieved from Kenya there were still issues with her qualifying in Ontario. In the end, the wife of one of her care recipients took up the cause and, Irma said, the wife "went to the College and complained. I was afraid; I did not want to make trouble. But after she spoke with them I was allowed to write." Finally, five years after her arrival, Irma was able to do training "to make sure I know the procedures here, to learn the proper terminologies." She found the training easy, "similar to back home in Kenya," and had recently passed the exam. Yet, she said:

> Apart from the high technology here, I don't feel it is very different. You still must know how to look after the patient's comfort. The physical care is still the same. It doesn't change if you are in a different country. It doesn't mean that the patient has different symptoms. It's only the high technology that is new.

When asked how they got attracted into home care, the immigrant women offered different explanations than the Canadian-born women. Home care work was more readily available to them than other jobs because agencies were less concerned about their lack of work experience in Canada, gaps in their paid-work history, and, for the non-English speakers, their limited language skills. Home care work could usually be fitted around their family responsibilities; for example, Jocie (RN, Morocco) worked part time as her children were young. Others talked of going into care work to bolster their family's standard of living or in response to their husband's unemployment or job insecurity. Rahma (attendant, Eritrea) spoke often of her family's financial struggles and repeatedly said she was happy to be working, even though she knew she was not well paid and sometimes had to work six days a week: "I'm helping my husband and helping my family because things are expensive for us." Like many of the women, Rahma was also sending remittances to support her extended family. Thus for many, despite its shortcomings, care work is desirable work; it offers some job security and provides workers' families with a critical means of economic survival. Even poorly paid care work is understood by many of those in it as offering important opportunities for women from a range of backgrounds to contribute to their family's income (Giles and Preston 1996).

Responsibilities

In Ontario, 30 percent of the 2006 employed labor force was made up of immigrants, whereas immigrants make up 28 percent of RNs, 22 percent of registered practical nurses (RPNs), and 38 percent of PSWs and attendants. Although only subsets of these are employed in the home care sector, the census statistics give a flavor of the migrant division of health care work. Across all seventeen cases in our home care study, twenty-nine paid care workers were interviewed, of whom twenty-one were Canadian-born. Of those, nine were RNs and eight were PSWs or attendants (and two others were RPNs and one was a physiotherapist). On the other hand, six of the eight migrant workers were PSWs or attendants, and the two others were RNs, Jocie (Morocco) and Leah (Jamaica). This admittedly small group of migrant care workers does reflect the racialized divisions of paid reproductive labor described in the literature (see Meintel, Fortin, and Cognet [2006] for an example from another Canadian province). On closer analysis, however, we find more complexity. For example, Jocie is at the higher end of the care hierarchy. She is a "high-tech" nurse, with advanced training in intravenous technology management. These highly sought-after skills make her, as she puts it, "part of a specialized team" responsible for numerous clients across an extensive territory.

Increasingly, nurses (in all work settings) are "high tech, low touch," dealing with assessments and the more technical aspects of care. Reflecting Twigg's (2000) observations about hierarchies in nursing, the work of RNs is marked by distance from the direct, intimate care of bodies. The more labor-intensive, high-touch body work is usually the responsibility of PSWs or attendants. Their responsibilities include the personal care of their clients' bodies, often carried out in the most intimate spaces of home (bedrooms and bathrooms). Their jobs transgress the bodily and domestic boundaries of normative social interactions. The care workers' discussion of intimate body work (bathing, toileting, catheters) points up crucial elements involved in the actual practice of home care work, often demonstrating the skill involved in this devalued, supposedly unskilled work. Brenda (attendant, Jamaica) describes how she bathes one care recipient:

> So we talk and talk until we reach upstairs and she's in bathtub. You have to talk with them because they don't really want to have a bath. And by the time the conversation is over she's getting lathered with soap, and says, "Oh you tricked me." All the time she doesn't realize what's happening yet. She's a darling; I like that one, even though sometimes she don't want to have the bath.

Specific job responsibilities are laid out by provincial health care policy guidelines that then inform the formation of home care relations and how care is actually put into practice in the clients' homes (England and Dyck 2011). Care agency constraints on time allocated to tasks and the legal demarcation of job category boundaries further impact home care work practices. For instance, Irma (Kenya) is employed part time as an attendant for a home care agency, but she also holds a second, more lucrative job working night shifts as an RN in a hospital. Irma's experience demonstrates how provincial policy directives and agency regulations reinforce occupational divisions of care work. When asked if there were rules about what tasks she was allowed to perform, she responded:

> I am working here as a homemaker, but if there is a dressing [needed] I am not allowed to do that. They have to get a nurse. Even though I am [qualified] to do it, it has to be done by a nurse. [But] I do that in the hospital, because there I work as a registered nurse.

Irma's embodied labor involves her crossing the divisions of paid care work on a daily basis and shows how specific care sites (home or hospital) co-constitute the corporeal practices of care work.

The relational dynamic of care work involves a careful weaving together of negotiating boundaries, embodiment, and the care worker's own sense of appropriate care (what Datta et al. [2010] described as a "migrant ethic of care" wherein their life experience and cultural values shape their perceptions of what constitutes good care). For example, care recipients seek to maintain their modesty and dignity and at the same time the care workers negotiate "the rules" of their employer and shape their own work practices to help their clients. We did find that the Canadian-born care workers are more likely to bend the rules further or push back against the agencies in ways that risked their jobs. Generally the immigrants were more circumspect. For example, Rahma (attendant, Eritrea) said, "I tell her now it's more than two hours but I'll finish [the laundry], and tomorrow when I come, I cut the time from her." Rahma was the most concerned about calling the office for permission to do certain things. For her, keeping her job was critical for her family's finances, so she was not prepared to jeopardize that. She also pointed out, however, that she used the rules as a strategy for maintaining clear boundaries with clients: "I say 'I don't have permission, I can't do this,'" when asked to do tasks beyond her job description or to stay beyond the allocated time.

Respect

Respect is central to an individual's feeling of self-worth and self-esteem and is a fundamental part of the dynamics of paid work relations. Obviously the amount of respect experienced by care workers (and the care recipient) is highly variable and nuanced, dependent on the specificity of the relationship, along with the comfort level about negotiating material care practices in a particular situation. As Bondi (2008) showed, the care work relationship is inflected by both power and affect or emotion on both sides of the relationship. For example, Irma (attendant, Kenya) explained that William is ashamed of being naked before strangers, yet his daily home care is about bathing and dressing. His modesty, she explains, is "his proud character" and his embarrassment about his loss of bodily control: "That is why he hates a different provider coming to him every day. That is why it must be the same person." And that "same person" is Irma. Although she is now an RN, she continues as William's attendant on a part-time basis.

Irma's particular relationship can be seen as an example of the "international transfer of caregiving" that Parreñas (2001) described, although in this case it is transferred to an elderly, incontinent man with limited motor skills rather than a child. Irma describes her relationship with William and his wife as being like family. She does not use this label unreflexively, as she said that being part of a family means you observe behavior over a long period of time and come to understand people very intimately: "It is not all sweetness, though you care about them very much." Respect, individuality, being mindful of their client's privacy (within their home, but also at the scale of the body), and treating them with dignity were common threads in the care workers' interviews.

Another aspect of respect is about language skills contributing to difficult working conditions. We found this thread in the interviews less often than we expected. Language skills were not raised in any substantive way by Alexa (attendant, Germany) or Jocie (RN, French-speaking Moroccan), nor by their care recipients. It came up in passing in Rahma's case (attendant, Eritrea), who speaks broken English and is working to develop her language skills. The issue of accented English came up in the case of the native English-speaking women, however. In their study of African immigrants in Vancouver, Creese and Kambere (2003) found that despite coming from English-speaking former British colonies where the education system is also British based, their "accent" and race mark them as immigrants. Ivy, originally from Jamaica, migrated to Canada twenty-six years earlier. Ivy's client Bernice remarked that "I can't understand her half the time—what she's talking about? She's been here for twenty-five years . . . you know how thick her accent is." Bernice linked this directly to what she described as Ivy's laziness and said "I can't take any more of (Ivy's) stupidity." Bernice also devalues Ivy's care work, for example, complaining that her cleaning "is slipshod. . . . Like today I asked her to clean the bathroom; she stands in the hall and shoots Lysol through the door and that's the bathroom clean." Bernice's remarks were the clearest example from our data of the intermingling of anti-immigrant and racist discourses: "Common-sense discourses [that] construct people of color as immigrants and immigrants as people of color" (Creese and Kambere 2003, 566). Bernice pulls on a racist script to describe her changing care provision situation, when it is likely that Ivy's allocated time for Bernice has been cut because of home care restructuring. This also echoes experiences of racism and discrimination reported by home care workers in Québec (Meintel, Fortin, and Cognet 2006), and migrant nurses in the United States, suggesting that "foreign" care workers are scapegoated when the effects of economic and labor force restructuring reverberate on a global scale (Ball 2004).

Changes in publicly funded home care also help explain the reactions of another care recipient. Hannah said that until Valentina (attendant, Chile):

> Every single caregiver [the agency] sent was from the Caribbean. They were sending a different person every single day, a different one. It was SO hard for me. The majority of them were lazy. They came in with these huge bags, God knows what they took.

In addition to accusations of "laziness," Hannah raises another theme that emerged in several cases (including those with Canadian-born workers): thinly veiled accusations of theft. Again troubling in what this reveals about racism and lack of respect for care workers, these unpleasant remarks could also be understood as the vulnerability arising from the lack of continuity of care and the difficulty of "opening your home to strangers" expressed by several care recipients.

Hannah was increasingly fearful that Valentina would resign from her job as an attendant. In her interview, Valentina seemed to be struggling with her own identity and self-respect: "I'm a librarian, not a housekeeper," she said at one point in her interview. In addition, after several months of helping Hannah bathe, cleaning her home, and preparing meals (Valentina complained of peeling lots of potatoes and carrots), she believed that her own health was beginning to suffer (wrist pain). Hannah said Valentina had discussed this with her but then remarked that Valentina was "too cheap to buy some carpal tunnel splints." The relational production of respect (or perhaps demise of it!) in this particular situation flags a second set of vulnerabilities. Narratives of vulnerable bodies abound but usually only in reference to care recipients. In our more relational understanding of care work, there is a dual vulnerability of both the paid care workers' and the care recipients' bodies—a trusted caregiver starts suffering from work-related health problems (e.g., carpal tunnel syndrome) that impacts her ability to work with a client, and this might mean that the "vulnerable" client faces problems with continuity of care, which in turn impacts the client's health. In a context where home care reform means cost cutting and rationing care, recipients feel vulnerable to having their service cut. Care workers, on the other hand, are concerned about their job security and new work practices that leave them questioning how (or if) they can continue providing quality care that allows them to maintain their own health, fits within their ethic of care, and still adheres to rules about what is allowed and paid for by the system.

Conclusion

Although these data are from a Canadian study, with particular health care policies and agency directives affecting the routes taken by migrant workers, their job responsibilities, and their sense of respect, our article reflects and informs other scholarship on migrant workers in multicultural settings in the global north. The globalization of care work links together the transnational flows of women and a range of care jobs that are frequently poorly paid and undervalued. There is also recognition of the sometimes demanding and demeaning conditions under which paid care workers perform their work. Certainly our study suggests that there is a racialized migrant division of paid reproductive labor in Ontario, but we have deliberately highlighted the social agency and "rational" choices of migrant care workers in our discussion of the routes, responsibilities, and respect associated with home care jobs. Although the association of paid care work as "women's work" endures, we show that it is a viable, even attractive, option for women with limited avenues for economic survival.

We show that the social organization of labor serving capital interests has effects at the scale of the body, including the bodies of those providing care. That the care workers in our study are immigrants further complicates the work relation and actual practices of care work. Moreover, our embodied approach draws on racialized divisions of labor to underscore the significance of an intersectional approach that embeds care work within power hierarchies of gender, race, and class. As racialized workers, the women in this study experience their access to care work and the day-to-day experiences with care recipients in a context inflected by their global migration, intertwined with labor market inequalities along with provincial policy directives and agency rules. Thus, they might have to endure racism in particular care work situations, and many apply a "migrant ethic of care" to their work rooted in their own life experiences and the skills they bring to the workplace. In these instances, the global enters the intimate spaces of Canadian homes via migrant care workers, which can transform the very meanings and practices of care work. Most obviously the racialized occupational hierarchy within the care workforce is evident in the overrepresentation of migrant women in the more "dirty work" of care work, compared with their Canadian-born counterparts. Despite this occupational clustering, we argue that even for those at more marginalized intersections within multiple hierarchies of power there is still social agency, resistance, and choice.

Notes

1. The participants' names are pseudonyms. The qualitative research was funded by the Social Sciences and Humanities Research Council of Canada. The research team was led by principal investigator Patricia McKeever and included co-investigators J. Angus, M. Chipman, A. Dolan, I. Dyck, J. Eakin, K. England, D. Gastaldo, and B. Poland and research coordinator K. Osterlund. We thank the clients, their families, and the care workers for their time, energy, and enthusiasm in participating in this research.
2. The Canadian-born nurses who volunteered their ethnic background were of northern European (primarily British) origin or in one case, Ukrainian descent.

References

Aguiar, L. M., and A. Herod, eds. 2006. *The dirty work of neoliberalism: Cleaners in the global economy.* Oxford, UK: Blackwell.

Ball, R. E. 2004. Divergent development, racialised rights: Globalised labour markets and the trade of nurses—The case of the Philippines. *Women's Studies International Forum* 27 (2): 119–33.

Bondi, L. 2008. On the relational dynamics of caring: A psychotherapeutic approach to emotional and power dimensions of women's care work. *Gender, Place and Culture* 15 (3): 249–65.

Cox, R. 2006. *The servant problem: Paid domestic work in a global economy.* London: I. B. Tauris.

Creese, G., I. Dyck, and A. T. McLaren. 2008. The "flexible" immigrant? Human capital discourse, the family and labour market strategies. *Journal of International Migration and Integration* 9 (3): 269–88.

Creese, G., and E. N. Kambere. 2003. What colour is your English? *Canadian Review of Sociology and Anthropology* 40 (5): 565–73.

Datta, K., C. McIlwaine, Y. Evans, J. Herbert, J. May, and J. Wills. 2010. A migrant ethic of care? Negotiating care and caring among migrant workers in London's low pay economy. *Feminist Review* 94:93–116.

England, K., and I. Dyck. 2011. Managing the body work of home care. *Sociology of Health and Illness* 33 (2): 206–19.

England, K., J. Eakin, D. Gastaldo, and P. McKeever. 2007. Neoliberalizing home care: Managed competition and restructuring home care in Ontario. In *Neoliberalization: Networks, states, peoples,* ed. K. England and K. Ward, 169–94. Oxford, UK: Blackwell.

Giles, W., and V. Preston. 1996. The domestication of women's work: A comparison of Chinese and Portuguese immigrant workers. *Studies in Political Economy* 51:147–81.

Glenn, E. N. 1992. From servitude to service work: Historical continuities in the racial division of paid reproductive labor. *Signs* 18 (1): 1–43.

Hanson, S., and G. Pratt. 1995. *Gender, work, and space.* London and New York: Routledge.

Hondagneu-Sotelo, P. 2007. *Domestica: Immigrant workers cleaning and caring in the shadows of affluence.* Berkeley and Los Angeles: University of California Press.

Kittay, E. F. 1999. *Love's labor: Essays on women, equality and dependency*. London and New York: Routledge.

Kofman, E., and Raghuram, P. 2006. Women and global labour migrations: Incorporating skilled workers. *Antipode* 38 (2): 282–303.

McDowell, L. 2009. *Working bodies: Interactive service employment and workplace identities*. Oxford, UK: Wiley-Blackwell.

Meintel, D., S. Fortin, and M. Cognet. 2006. On the road and on their own: Autonomy and giving in home health care in Québec. *Gender, Place and Culture* 13 (5): 563–80.

Parreñas, R. S. 2001. *Servants of globalization: Women, migration and domestic work*. Stanford, CA: Stanford University Press.

Pratt, G. 2004. *Working feminism*. Philadelphia: Temple University Press.

Preston, V., L. Lo, and S. Wang. 2003. Immigrants' economic status in Toronto: Stories of triumph and disappointment. In *The world in a city*, ed. P. Anisef and M. Lanphier, 192–262. Toronto: University of Toronto Press.

Tronto, J. 2005. Care as the work of citizens. In *Women and citizenship*, ed. M. Friedmann, 130–47. Oxford, UK: Oxford University Press.

Twigg, J. 2000. *Bathing, the body and community care*. London and New York: Routledge.

Zimmerman, M. K., J. S. Litt, and C. E. Bose. 2006. *Global dimensions of gender and carework*. Stanford, CA: Stanford University Press.

Urban Politics and Mental Health: An Agenda for Health Geographic Research

Joseph Pierce,* Deborah G. Martin,† Alexander W. Scherr,‡ and Amelia Greiner§

*Geography Department, The Florida State University
†Graduate School of Geography, Clark University
‡School of Law, University of Georgia
§The Bloustein School of Planning and Public Policy, Rutgers University

Siting of mental health service facilities has often been subject to public opposition and political struggles. These processes have produced a landscape of mental health provision that is powerfully uneven and concentrated in economically and socially depressed areas. We argue that understanding this landscape requires an examination of the political processes that shape such siting decisions. Although health geographers (most importantly Dear and Wolch) have periodically engaged with politics, the important role of informal development politics in producing landscapes of health remains insufficiently examined. We introduce the case of residential social service facility ("group home") siting in central Massachusetts to explore the political dynamics of the production of health. Siting of group homes in Massachusetts is governed by a legal framework that provides social service agencies with legal protection and autonomy from local governments as they make siting choices. This exemption from local zoning ordinances often shifts local politics from formal to informal channels, leading to the application of many forms of soft influence over siting decisions. A comprehensive geographic analysis of mental health should include the social and political processes of siting.

精神健康服务设施的选址往往受到公众的反对并涉及政治斗争。这些过程已经造成了精神健康供应集中于经济和社会状况不景气地区的极不平衡的景观。我们认为，理解这一景观需要审查造成这样的选址决定的政治进程。虽然健康地理学者（最重要的是 Dear 和 Wolch）定期从事政治活动，非正式政治发展在产生健康景观中的重要作用仍然审查得不够。我们通过介绍座落在马萨诸塞州中部的住宅社会服务设施（"团体家屋"）情况，探索健康生产的政治动态。这些设施在马萨诸塞州的选址由一个法律框架所管制，它在作选址决定时给社会服务机构提供受法律保护的脱离地方政府的自主权。这种从当地分区条例的豁免，经常把地方政治从正式的转移到非正式的渠道，导致了多种形式的软性影响在选址决策中的应用。一个综合性的精神健康的地理分析应包括社会和政治进程的选址。关键词：团体家屋，健康政治，地方发展，心理健康，选址。

A menudo la ubicación de instalaciones para servicios de salud mental ha generado oposición pública y enfrentamientos políticos. Estos procesos han dado lugar a un paisaje de atención a la salud mental poderosamente desequilibrado y que se concentra en áreas económica y socialmente deprimidas. Nuestro argumento es que para comprender este paisaje se requiere un examen de los procesos políticos que dan lugar a tales decisiones locacionales. Aunque los geógrafos especialistas en temas de la salud (singularmente Dear y Wolch) periódicamente se involucran en política, el importante papel que juegan las políticas de desarrollo informal en la producción de paisajes de la salud sigue estando insuficientemente estudiado. Traemos a cuento el caso de las instalaciones del servicio social residencial ("hogar grupal") ubicadas en la parte central de Massachusetts para explorar la dinámica política de la producción de salud. La ubicación de los hogares grupales en Massachusetts está controlada por un marco legal que provee a las agencias de servicios sociales con protección legal y autonomía frente a los gobiernos locales en lo que concierne a seleccionar sitios de ubicación. Esta excepción al ordenamiento de la zonificación local a menudo desvía la política local de los canales formales a los informales, lo cual conduce a la aplicación de muchas formas de influencia suave en las decisiones de selección de sitios. Un análisis geográfico minucioso de los problemas de salud mental debe incluir los procesos sociales y políticos de la localización.

Health geographers over the past twenty years have contributed to an increasingly sophisticated understanding of landscapes of human health. Attention to health service provision in this literature has often been focused on the impact landscapes of provision could have on health outcomes; that is, on the causal relationship between particular distributions of sites and health outcomes in distributions of populations (Kearns and Moon 2002; Curtis 2004; Dummer 2008). Health geographers' increasing interest in and engagement with these complex landscapes of health suggest that there is a need for further investigation of the social and political processes that produce them. The work of Jennifer Wolch, Michael Dear, and others on the siting of group homes—sometimes called residential social service facilities, where a group of people live together in a setting supported by social service professionals (Dear and Taylor 1982; Dear and Wolch 1987; Wolch 1990; Dear 1992)—forms the bedrock of analyses of health siting in the geographic literature. Their work in the 1980s and early 1990s demonstrated that siting is often extensively contested through social and political processes. They argue that these conflicts exert a strong influence on resultant geographies of mental well-being. Although these findings have been widely acknowledged, the bulk of scholarly attention to their work comes from urban and political geographers who are not specifically focused on health or health outcomes but rather, the implications of a particular set of relations between the state and the nonprofit sector for a broader urban politics (M. P. Brown 1997; Lake and Newman 2002; Trudeau 2008).

We suggest that the multifaceted politics of health facility siting should be an explicit component of health geography analyses, particularly as a means for understanding the interface between geographies of health and urban development. We develop this argument through an examination of the case of residential social service facility ("group home") siting in central Massachusetts. Through interviews with community and social service agency (SSA) stakeholders, as well as analysis of legal documents and media accounts, we see evidence that, although formal structures to "depoliticize" siting exist (e.g., legal frameworks in Massachusetts), informal debates about the appropriateness of particular group home sites pervade settings such as public hearings and ad hoc citizen committees (Martin and Pierce forthcoming). These contestations provide an opportunity to examine the perceptions and dynamics that contribute to the production of particular health geographies in urban settings.

Because it invokes social norms and political–economic processes influencing land use, the siting of group homes offers a striking empirical case through which to work toward integrating geographic literatures of health, including recent explorations of landscapes

of well-being, with urban and political geographic insights about urban political economy. Our examination of the production of the group home landscape in central Massachusetts highlights a complex intersection of SSA siting practices and considerations of client well-being, local development politics and land use concerns, and formal state legal mechanisms that seek to provide SSA discretion for group home siting. In what follows, we explore key scholarship on health facility siting within public health, health geography, and urban political geography. We then draw on a case study of group home siting in central Massachusetts, before arguing for a research agenda that draws on health and urban political geographic research perspectives.

Literature Review: Siting Mental Health Provision

In the public and mental health literatures on group homes and supportive housing siting, much discussion has focused on successful institutional siting, where success is the ability to permit, build, and open a facility in the face of community opposition (Spreat and Conroy 2001; Scalzo 2008). Siting-oriented literature in public and mental health often addresses the varied responses that different group home populations receive. One study found the greatest opposition to substance users, followed by adults with severe mental illness (Galster et al. 2002); another study of adult attitudes toward varying facilities found that group homes for the mentally retarded were less controversial than other populations, such as people with mental disabilities, depression, or AIDS (Takahashi and Gaber 1998). Within this literature, community opposition to group homes is often taken as a given (Davidson 1981), and there is limited direct engagement with systematic or political critiques of siting practices.

When scholars within this community explore desirable characteristics for sites, their results tend to positively reflect on the choices that SSA administrators are already making (Zippay 1997; Zippay and Thompson 2007; Wong and Stanhope 2009), such as locating in (often low-income) areas with access to daily services and transportation. There is sometimes an acknowledgment of a lack of certainty of what makes for a good or bad neighborhood environment, however (Newman 2001; Wong and Stanhope 2009). Health geographers have extensively documented the uneven spatial distribution of health outcomes (Curtis and Jones 1998;

Eisenhauer 2001; Cummins 2007), pointing to the distribution of illness such as asthma (Loh and Sugerman-Brozan 2002) or AIDS (Arnold et al. 2009), as well as the correlation between such health outcomes and a variety of sociodemographic factors both within nations (Dummer 2008) and globally. In a number of these studies, the concentration of "a poverty of health" has been shown to correlate with low socioeconomic status (Smith and Easterlow 2005).

The past fifteen years have also yielded a growing discussion of more social and "place-based" analyses of landscapes of health and the local contexts of health services (Curtis and Jones 1998; Cutchin 1999; T. Brown and Duncan 2002; Milligan and Conradson 2006). In particular, scholars have explored the contexts and landscapes that might be conducive to increased well-being (Kearns and Gesler 1998; Curtis et al. 2007; Fleuret and Atkinson 2007). As in the public and mental health literature, these accounts emphasize therapeutic environments in which care ought to be provided (Gastaldo, Andrews, and Khanlou 2004; Gesler 2005). There is scholarship, drawing connections to urban planning and political geography, that suggests that planners should approach public facility siting by seeking to promote "service hubs" or "decentralized concentration" (DeVerteuil 2000; see also Nelson and Wolch 1985). Where studies have attempted to make this linkage, however, they have often been grounded in other subdisciplines, such as environmental justice (Maantay 2002; Sister, Wolch, and Wilson 2010) or a broader urban political economic analysis.[1]

Some health geographers have addressed the management and regulation of health services (Milligan and Conradson 2006; DeVerteuil and Wilton 2009). The specific issue of the process of siting health provision facilities, however, has received little systematic attention aside from the cluster of work involving Dear (Dear and Taylor 1982; Joseph and Hall 1985; Dear and Wolch 1987; Dear 1992) and Wolch (1990). Specifically, Dear, Wolch, and others (Dear and Taylor 1982; Dear and Wolch 1987; Wolch 1990; Dear 1992) addressed the role of the state in the provision of social services and the production of public health, including siting processes. Wolch's primary concern in the late 1980s was the devolution of the state from official involvement in health and welfare provision. She argued that, by shifting public responsibility for social welfare to nonprofit, nongovernmental SSAs, the state both abdicated its responsibilities for its citizens and removed from public view and debate formal oversight of services, creating a "shadow state" of nonprofit groups with services clustered in dense, low-income neighborhoods.

Although some recent work (as in Milligan and Conradson 2006) explored the implications of the hybridized state–voluntary sector for health geographers, by and large, Dear and Wolch's work has been engaged by scholars working within urban development or urban political frameworks. For example, Lake and Newman (2002) noted the absence of citizen involvement in the politics of non-profit-led urban redevelopment. Many urban scholars (M. P. Brown 1997; Mitchell 2001; Lake and Newman 2002; Hankins and Martin 2006) cited Wolch's (1990) "shadow state," highlighting a broader move toward neoliberalism and privatization of public (governmental) functions into market logics (Brenner and Theodore 2002; Harvey 2006; Leitner, Peck, and Sheppard 2006). Reincorporating such attention to "politics" in research on health can help better answer questions about the locus of decision making that produces health landscapes and outcomes such as definitions of well-being and health in urban social life.

By *politics* we mean processes of negotiation and contestation over how space is governed and behavior regulated: politics as negotiation over the ordering of society. For political scientists, this ordering occurs in the "institutional sphere" of government and its legitimating documents (Kernell and Jacobson 2006). Yet for many scholars, politics occur well beyond the institutional framework of states and structures of decision making. Nancy (2002, 22) articulated politics as "the site of detotalization," in which the incommensurability, or irreducible difference, of competing ideals and goals are made concrete, negotiated, and arbitrated. Laclau (1990, 35) similarly argued that politics expose "the moment of antagonism where the undecidable nature of the alternatives and their resolution through power relations becomes fully visible." Politics can include associations and negotiations in daily life (Amin 2006). Staeheli (2008), for example, stressed a distinction between "public" politics that are fully within and legitimated by governmental realms and those that are private but nonetheless represent broad social claims and antagonisms. For these scholars, politics permeate both formal and informal spheres of human action.

Although a systematic review of the theoretical literature on politics is beyond this article's scope, we highlight efforts to distinguish between formal politics of the state and those that proceed outside of it. For health geographers, these distinctions are important because they signal the boundaries of arenas of decision making regarding issues such as notions of well-being, suitability of programs for particular health outcomes, siting of services, and definitions of services themselves (e.g., over what constitutes mental vs. physical health, or whether obesity is fundamentally a medical or transportation problem). Understanding the locus of decision making, the mechanisms for participation, and the institutions and actors engaged in negotiation all offer a means for better conceptualizing and analyzing the production of geographies of health. Unpacking these political dimensions—who, what, and where—in the case of siting of group homes in Massachusetts helps to expose the production of particular health landscapes more generally and their ongoing reproduction in urban arenas.

The Case of Group Home Siting in Central Massachusetts

In exploring how local development politics shapes health landscapes, we consider a case drawn from research on group home siting in urban central Massachusetts. This research is part of a project investigating urban (re)development politics and legal frameworks regulating land use. We conducted interviews with twenty-eight stakeholders to explore the process of group home siting. Stakeholders included SSA administrators and employees, lawyers, local community advocates and opponents, and politicians. We also performed a media analysis of forty articles from local newspapers from January 2005 through December 2006, a period covering some of the most active and publicly visible contestations over group homes in central Massachusetts. Additionally, we conducted exploratory analysis of a spatial data set of the locations of group homes acquired from the City of Worcester, enhancing this data set with on-the-ground neighborhood observations of each identified group home site in the city.

Group homes provide critical service delivery to vulnerable populations (Wolch 1990). Although facilities provide varying services, we focus in particular on those offering short- to medium-term transitional care in the domains of mental health, addiction, and homelessness. Many communities initially resist the siting of group homes in their midst, using formal tools such as community planning boards and zoning laws as well as informal political, social, and economic relationships to displace group homes into other areas with less social and economic capital (Dear 1992). This opposition employs a politics; the particular form of this politics, and perhaps the degree of opposition itself, is structured by the local governing framework for siting itself.

In Massachusetts, however, many of the formal levers that regulate group home siting in other jurisdictions (e.g., zoning) have been explicitly demobilized under a legal framework that is commonly referred to as the Dover Amendment or simply Dover. Part of the state's zoning-enabling legislation, Dover exempts religious and educational institutions from local land use laws, excepting "reasonable regulations" applied equally to all area buildings (e.g., regulation of parking, bulk, setback, etc.; Massachusetts General Law chapter 40A § 3 2003). Massachusetts courts have interpreted the term *educational* broadly, including residential facilities like group homes that incorporate life skills training in their service programs. As a result, SSAs in Massachusetts have a great deal of power under the law to site according to their preferences. Although Dover's applicability to each site requires certification by a municipal building inspector, the process is administrative and requires no oversight or input by either community organizations or elected government bodies.

In the last decade, some new group home sites in central Massachusetts were hotly contested as inappropriate to their surroundings. In our interviews and in media coverage, Dover was repeatedly mentioned (by both opponents and proponents) as an essential tool allowing SSAs to site despite opposition. One social service administrator told us, "[The] Dover amendment is like a great shield; it was our biggest tool in our arsenal in this [siting] discussion . . . it sort of shut everyone up." Another administrator highlighted the importance of Dover generally, saying, "Dover exists because [previously] there was a lot of discrimination." A resident opposed to a particular site in Worcester also saw Dover as a tool but one that has been abused: "Dover is a pretty powerful law. . . . Dover was an excellent law that protected . . . needy people, and prevented discrimination, but unfortunately . . . it created these large social service organizations that used Dover . . . to avoid any opposition to anything."

Rather than simply tilting the political field of siting contestation toward SSAs, however, the "big stick" of Dover instead moves, or displaces, political contestation over siting out of formal spheres of institutional politics where it would be centered on municipal debates about land use zoning and into more informal realms, where consultation with city officials or elites might still be important but occurs informally. In the preceding quotes, Dover is alternately seen as the only means for successful siting or an inappropriate source of power; in either case, the law is powerful for siting and supersedes local laws. Yet the supposed supremacy of Dover does not eliminate the need to negotiate the relationships within the neighborhood, the city, or the state that sustain SSA's ability to provide services. A media account of siting conflicts points to the importance of these informal political relations: "[SSA] officials say they've learned their lesson. . . . [An official says that] 'We've learned how important it is to reach out to the vast majority of residents who have legitimate questions we need to respond to'" (Schaffer 2006). Although getting facts to resident neighbors might be useful to SSAs, the context of Dover clearly establishes that SSAs have no legal obligation to do so; their mandate to inform is instead social, moral, and, thus, informally political.

In addition, one of the social service administrators quoted earlier commented on relations with city and state officials: "I've seen other agencies do things a certain way and kind of disregard the political class or whatever . . . they're entirely from their rights, but . . . we have a good relationship with the political class [of the] city; we depended on them for some of our funding." Indeed, these funding decisions are sites of political negotiation in both formal and informal spheres; in one instance, a state senator (temporarily) removed state funding from one social service organization to protest what the senator called a lack of communication about the siting of a group home by the SSA involved (Hammel 2006). Less punitively, the city of Worcester convened a task force on group home location that asked (in the name of the city but without force of law) that SSAs systematically notify the city and nearby residents before opening such facilities (see Martin and Pierce forthcoming). This effort by municipal officials to reinsert the local state in a process that Dover has formally depoliticized reveals the ongoing highly political yet informal negotiations over social service provision in Massachusetts. Despite a formal legislative and political consensus regarding the autonomy of SSA decision making, then, we see informal political dynamics such as interpersonal relationships between city and agency officials critically driving the discussion over siting of group homes.

Our exploratory spatial analysis of the Worcester database of group home properties provides some empirical perspective on these competing—or awkwardly coexisting—accounts of siting. Our results suggest a Worcester group home landscape that is strongly clustered, with clusters that correlate strongly with poverty in the city. Worcester's postindustrial decline, like that of many U.S. cities, has disproportionately affected low-income neighborhoods (Wyly and Hammel 2004). The area, known as Main South, is defined by its relatively low average income, concomitant high rates of poverty, and high (for Worcester) populations of black, Hispanic, and foreign-born immigrant residents (U.S. Bureau of the Census 2000; see Figure 1). This area of the city overlaps substantially with areas zoned for industrial and manufacturing uses, many of which contain factories and warehouses that stand vacant but continue to harbor environmental toxins (Krueger 2007).

Group homes in Worcester are heavily concentrated in Main South (see Figure 2). They are highly clustered ($p < 0.005$), and the high-density cluster is centered at the north end of Main South amid Worcester's highest concentrations of poverty. Although Main South is also the most densely populated area of Worcester, the clustered pattern is still observable even when normalizing for population density. To confirm the implications of this pattern, we conducted a visual survey of all of the identified residential social service facilities in Worcester. Group homes are overwhelmingly located in contexts that suffer the social patterns from which facility residents are seeking to extricate themselves (e.g., poverty, drug addiction, and drug use).

Although many positive outcomes can emerge from services provided in low-income areas, as one provider indicated in an interview, it is harder to convince service participants that they do not need to do drugs when drug culture is right outside their door. One social service administrator was quite explicit about desirable locations: "It's good if there is public transportation ready at hand and we actually try to stay away from things like bars and liquor stores that are not so appealing or are good for our folks." Another administrator commented that in addiction treatment, "[Y]ou can't have [clients] in the environment where everybody is dealing and doing [drugs] . . . it's just too tempting, it's too overwhelming." The data in Worcester suggest, however, that group homes—with a range of services, not simply addiction treatment—are in fact in the same areas where drug dealing is widely known to be problematic. It is prima facie startling that, given the legal resources provided to SSAs in their siting processes, they have chosen to site facilities in contexts that they might not believe maximize the effectiveness of their services.

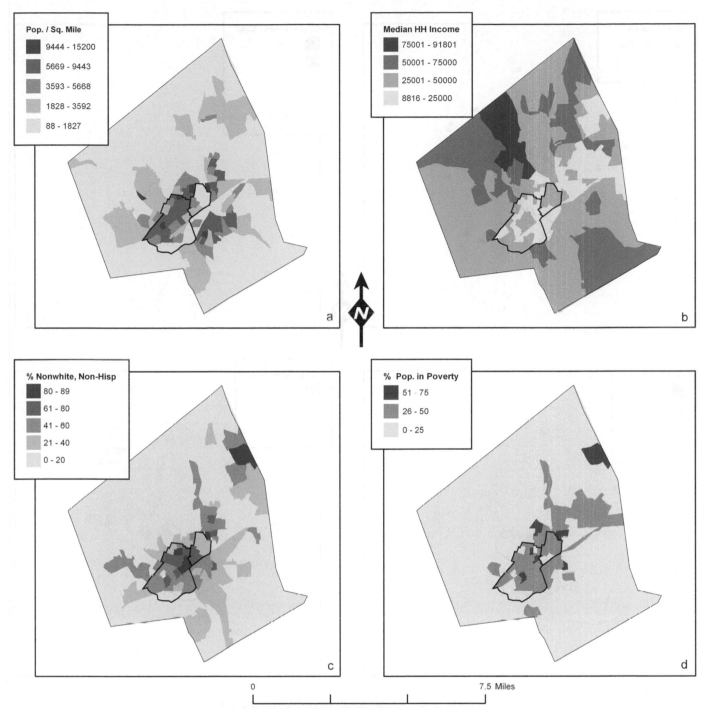

Figure 1. Contextualizing Worcester. (A) Population density. (B) Median household income. (C) Percentage of population non-white and non-Hispanic. (D) Percentage of population below the poverty line. Outlined neighborhoods are Main South (lower left) and central business district. *Sources:* U.S. Bureau of the Census (2000), City of Worcester Technical Services (2005, 2008).

Conclusion: The Political Production of Group Home Landscapes

The Massachusetts case illustrates that despite a powerful legal mechanism that allows SSAs wide nominal discretion to site without appeal to a political process, political forces outside the formal arena systematically shape the siting through the application of informal power and the (usually) implicit threat of governmental funding disruption to their organizational parents. The landscape of group homes that results from these complex processes of siting

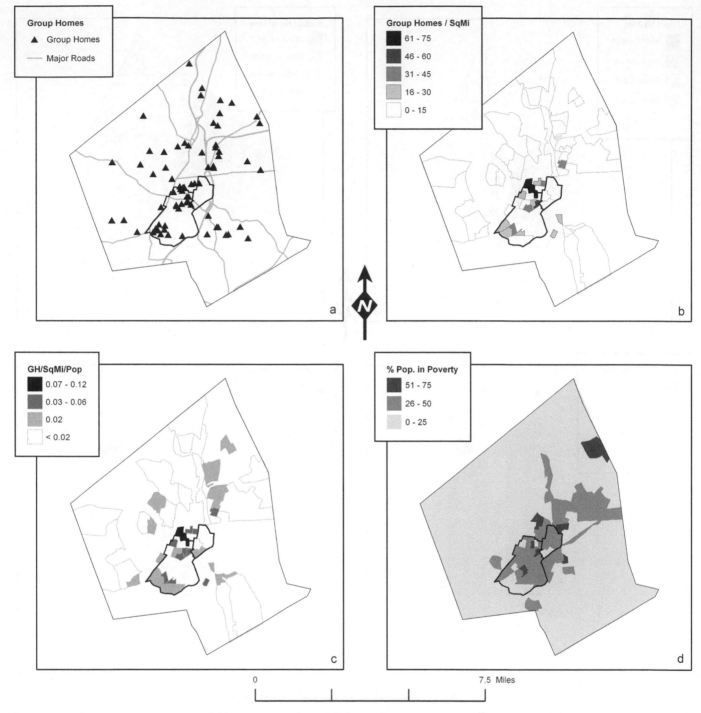

Figure 2. Group homes in Worcester. (A) Group home sites. (B) Group homes per square mile. (C) Group home density normalized by population. (D) Percentage population in poverty (repeated). Outlined neighborhoods are Main South (lower left) and central business distrcit. *Sources:* U.S. Bureau of the Census (2000), City of Worcester Technical Services (2005, 2008).

is highly uneven and concentrated in low-income neighborhoods with residents who have relatively less "soft" power to affect these siting choices.

Our analysis of interview data indicates that the produced landscape of group homes in central Massachusetts is not governed primarily by what makes a

site efficacious—although that is certainly a discourse that is used to shape siting outcomes—but also critically by what makes a site *politically possible* for SSA administrators. This finding reframes the analytical conversation about group home siting away from the intrinsic properties of a particular parcel of land as viewed by a particular organization's administrators and toward the social and political processes that assign acceptability or desirability to particular uses of parcels or neighborhoods. Put another way, focusing on siting as a political process is a way of insisting on the relevance of the informal relationships that shape local development in producing a particular mental health provision landscape.

The decades-long process of social service devolution articulated by Dear and Wolch in the 1980s and early 1990s has over the intervening decades altered the political context in which decision making about the production of health landscapes is made, moving the locus of responsibility for care from the state to distributed private institutions (Milligan and Conradson 2006). By more explicitly engaging with the politics of siting processes, health geographers can simultaneously (1) better understand the important causal relationships through which development politics shape mental health landscapes and (2) identify and focus on the informal political negotiations that critically shape siting choices in a contemporary neoliberal political context.

We see potential for a consolidation of efforts at the vanguard of geographic thinking about health landscapes in terms of various socially and politically constructed spaces (Milligan and Conradson 2006; DeVerteuil and Wilton 2009) with work in urban and political geography that continues to develop the implications of Wolch's (1990) understanding of the shadow state with contemporary explorations of how local development politics are prosecuted (Leitner, Sheppard, and Sziarto 2008; Staeheli 2008). Explicitly engaging with the politics of mental health siting provokes difficult questions. Which health priorities are foregrounded in political processes? How is what makes a "good site" negotiated? Offering analytical tools that link health landscapes with everyday processes of political decision making can help make the existing insights of health geographers more accessible to policymakers, directing research toward how health interrelates with other dimensions of urban development.

Note

1. Environmental justice analyses link health concerns to political struggles by examining the distribution of environmental inequity; conceptualizations of health are thus driven by this economic and social justice lens. Although a justice perspective parallels our concerns, length limitations lead us to focus on reintegrating contributions from political and urban–political scholarship more directly following Dear and Wolch.

References

Amin, A. 2006. The good city. *Urban Studies* 43 (5): 1009–23.

Arnold, M., L. Hsu, S. Pipkin, W. McFarland, and G. W. Rutherford. 2009. Race, place and AIDS: The role of socioeconomic context on racial disparities in treatment and survival in San Francisco. *Social Science & Medicine* 69 (1): 121–28.

Brenner, N., and N. Theodore. 2002. Cities and the geographies of "actually existing neoliberalism." *Antipode* 34 (3): 349–79.

Brown, M. P. 1997. *RePlacing citizenship: AIDS activism and radical democracy.* New York: Guilford.

Brown, T., and C. Duncan. 2002. Placing geographies of public health. *Area* 33 (4): 361–69.

Cummins, S. 2007. Commentary: Investigating neighbourhood effects on health—Avoiding the "local trap." *International Journal of Epidemiology* 36 (2): 355–57.

Curtis, S. 2004. *Health and inequality: Geographical perspectives.* London: Sage.

Curtis, S., W. Gesler, K. Fabian, S. Francis, and S. Priebe. 2007. Therapeutic landscapes in hospital design: A qualitative assessment by staff and service users of the design of a new mental health inpatient unit. *Environment and Planning C: Government and Policy* 25:591–610.

Curtis, S., and I. R. Jones. 1998. Is there a place for geography in the analysis of health inequality? *Sociology of Health & Illness* 20 (5): 645–72.

Cutchin, M. P. 1999. Qualitative explorations in health geography: Using pragmatism and related concepts as guides. *The Professional Geographer* 51 (2): 265–74.

Davidson, J. L. 1981. Location of community based treatment centers. *Social Service Review* 55 (2): 221–41.

Dear, M. 1992. Understanding and overcoming the NIMBY syndrome. *Journal of the American Planning Association* 58 (3): 288–300.

Dear, M., and S. M. Taylor. 1982. *Not on our street: Community attitudes toward the mentally ill.* London: Pion.

Dear, M., and J. Wolch. 1987. *Landscapes of despair: From deinstitutionalization to homelessness.* Princeton, NJ: Princeton University Press.

DeVerteuil, G. 2000. Reconsidering the legacy of urban public facility location theory in human geography. *Progress in Human Geography* 24 (1): 47–69.

DeVerteuil, G., and R. Wilton. 2009. Spaces of abeyance, care and survival: The addiction treatment system as a site of "regulatory richness." *Political Geography* 28:463–72.

Dummer, T. J. B. 2008. Health geography: Supporting public health policy and planning. *Canadian Medical Association Journal* 178 (9): 1177–80.

Eisenhauer, E. 2001. In poor health: Supermarket redlining and urban nutrition. *GeoJournal* 53:125–33.

Fleuret, S., and S. Atkinson. 2007. Wellbeing, health and geography: A critical review and research agenda. *New Zealand Geographer* 63:106–18.

Galster, G., K. Petit, A. Santiago, and P. Tatian. 2002. The impact of supportive housing on neighborhood crime rates. *Journal of Urban Affairs* 24 (3): 289–315.

Gastaldo, D., G. Andrews, and N. Khanlou. 2004. Therapeutic landscapes of the mind: Theorizing some intersections between health geography, health promotion and immigration studies. *Critical Public Health* 14 (2): 157–76.

Gesler, W. 2005. Therapeutic landscapes: An evolving theme. *Health & Place* 10 (2): 117–28.

Hammel, L. 2006. SMOC, O'Brien mum after talks. *Worcester Telegram and Gazette* 31 January. http://www.telegram.com (last accessed 18 December 2007).

Hankins, K. B., and D. G. Martin. 2006. Charter schools and urban regimes in neoliberal context: Making workers and new spaces in metropolitan Atlanta. *International Journal of Urban and Regional Research* 30 (3): 528–47.

Harvey, D. 2006. Neo-liberalism as creative destruction. *Geografiska Annaler B* 88 (2): 145–58.

Joseph, A., and G. Hall. 1985. The locational concentration of group homes in Toronto. *The Professional Geographer* 37 (2): 143–54.

Kearns, R., and G. Moon. 2002. From medical to health geography: Novelty, place and theory after a decade of change. *Progress in Human Geography* 26 (5): 605–25.

Kearns, R. A., and W. M. Gesler. 1998. Conclusion. In *Putting health into place: Landscape, identity, and well-being*, ed. R. A. Kearns and W. M. Gesler, 289–96. Syracuse, NY: Syracuse University Press.

Kernell, S., and G. C. Jacobson, eds. 2006. *The logic of American politics*. 3rd ed. Washington, DC: CQ Press.

Krueger, R. 2007. Making "smart" use of a sewer in Worcester, Massachusetts: A cautionary note on smart growth as an economic development policy. *Local Environment* 12 (2): 93–110.

Laclau, E. 1990. *New reflections on the revolution of our time*. London: Verso.

Lake, R. W., and K. Newman. 2002. Differential citizenship in the shadow state. *GeoJournal* 58:109–20.

Leitner, H., J. Peck, and E. S. Sheppard. 2006. *Contesting neoliberalism: Urban frontiers*. New York: Guilford.

Leitner, H., E. Sheppard, and K. Sziarto. 2008. The spatialities of contentious politics. *Transactions of the Institute of British Geographers* 33:157–72.

Loh, P., and J. Sugerman-Brozan. 2002. Environmental justice organizing for environmental health: Case study on asthma and diesel exhaust in Roxbury, Massachusetts. *Annals of the American Academy of Political and Social Science* 584 (1): 110–24.

Maantay, J. 2002. Zoning, law, health, and environmental justice: What's the connection? *The Journal of Law, Medicine, and Ethics* 30:572–93.

Martin, D. M., and J. Pierce. Forthcoming. Reconceptualizing resistance: Residuals of the state and democratic radical pluralism. *Antipode*.

Massachusetts General Laws (M.G.L.) c. 40A, § 3, 2003.

Milligan, C., and D. Conradson, eds. 2006. *Landscapes of voluntarism: New spaces of health, welfare and governance*. Bristol, UK: The Policy Press.

Mitchell, K. 2001. Transnationalism, neo-liberalism, and the rise of the shadow state. *Economy and Society* 30 (2): 165–89.

Nancy, J. 2002. Is everything political? (A brief remark). *The New Centennial Review* 2 (3): 15–22.

Nelson, C., and J. Wolch. 1985. Intrametropolitan planning for community-based residential care: A goals programming approach. *Socio-Economic Planning Sciences* 19:205–12.

Newman, S. J. 2001. Housing attributes and serious mental illness: Implications for research and practice. *Psychiatric Services* 52:1309–17.

Scalzo, V. 2008. *Suitable locations for a supportive housing facility*. Wilkes-Barre, PA: Luzerne-Wyoming Counties Mental Health/Mental Retardation Program.

Schaffer, N. 2006. How the NIMBYs win. *Worcester Magazine* 2 February 2006. http://worcestermag.com (last accessed 14 July 2008).

Sister, C., J. Wolch, and J. Wilson. 2010. Got green? Addressing environmental justice in park provision. *GeoJournal* 75 (3): 229–48.

Smith, S. J., and D. Easterlow. 2005. The strange geography of health inequalities. *Transactions of the Institute of British Geographers* 30:173–90.

Spreat, S., and J. W. Conroy. 2001. Community placement for persons with significant cognitive challenges: An outcome analysis. *The Journal of the Association for Persons with Severe Handicaps* 26 (2): 106–13.

Staeheli, L. 2008. Political geography: Difference, recognition, and the contested terrains of political claims-making. *Progress in Human Geography* 32 (4): 561–70.

Takahashi, L. M., and S. L. Gaber. 1998. Controversial facility siting in the urban environment: Resident and planner perception in the United States. *Environment and Behavior* 30 (2): 184–215.

Trudeau, D. 2008. Towards a relational view of the shadow state. *Political Geography* 27 (6): 669–90.

U.S. Bureau of the Census. 2000. *Census of population and housing: 2000 Census* [Computer file]. Geolytics version. CensusCD 2000. East Brunswick, NJ: Geolytics, Inc.

Wolch, J. 1990. *The shadow state: Government and voluntary sector in transition*. New York: The Foundation Center.

Wong, Y.-L. I., and V. Stanhope. 2009. Conceptualizing community: A comparison of neighborhood characteristics of supportive housing for persons with psychiatric and developmental disabilities. *Social Science & Medicine* 68 (8): 1376–87.

Wyly, E., and D. Hammel. 2004. Gentrification, segregation, and discrimination in the American urban system. *Environment and Planning A* 36:1215–41.

Zippay, A. 1997. Trends in siting strategies. *Community Mental Health Journal* 33 (4): 301–10.

Zippay, A., and A. Thompson. 2007. Psychiatric housing: Locational patterns and choices. *American Journal of Orthopsychiatry* 77 (3): 392–401.

Geographic Barriers to Community-Based Psychiatric Treatment for Drug-Dependent Patients

Jeremy Mennis,* Gerald J. Stahler,* and David A. Baron†

*Department of Geography and Urban Studies, Temple University
†Department of Psychiatry and Behavioral Sciences, University of Southern California

The World Health Organization has urged governments worldwide to implement evidence-based treatment services for drug addiction and mental health disorders, but the role of geographic characteristics in influencing treatment continuity for this population has been largely understudied. Here, we employ logistic regression ($N = 294$) to investigate how accessibility and neighborhood socioeconomic context influenced treatment continuity for a sample of 294 drug-dependent patients who received acute inpatient psychiatric treatment at a large, inner-city hospital in Philadelphia, Pennsylvania, and who were then referred to outpatient care. Results indicate that longer travel time to treatment, a high crime rate in the patient's home neighborhood, and traveling from a relatively lower to a higher crime neighborhood for treatment suppress treatment continuity. These contextual influences are moderated by ethnicity, where whites are influenced more strongly by travel time to treatment. This likely reflects the locations of treatment programs relative to patterns of residential segregation. African Americans both reside and attend treatment within the very highest crime areas, and this appears to have a particularly negative impact on treatment continuity for African Americans. This research highlights the need for more careful consideration of geographic issues in psychiatric treatment planning.

虽然世界卫生组织敦促全球范围内的各国政府为吸毒成瘾和心理健康失调患者实施以证据为基础的治疗服务，但是地理特色对影响这个群体的连续性治疗的作用，在很大程度上还没有被充分研究。在这里，我们采用逻辑回归分析 (N = 294)，来探讨交通方便性和邻里的社会经济环境如何影响治疗的连续性，我们使用了一个共 294 位对药物依赖患者的样本，这些患者在宾夕法尼亚州费城的市内医院先接受了急性精神科住院治疗，然后被转到了门诊。结果表明，治疗所需的较长的旅行时间，病人住处邻里的高犯罪率，从一个低犯罪率街区去一个犯罪率相对高的街区进行治疗等因素抑制治疗的连续性。这些情况的影响依种族而变，其中白人受治疗路上所费时间的影响更强。这可能反映了治疗机构相对居住隔离模式的位置。非裔美国人不但在犯罪率很高的地方居住而且也在那里接受治疗，这似乎对非裔美国人治疗的连续性有特别负面的影响。这项研究突出了在精神病治疗规划中需要慎重考虑更多的地理问题。关键词：交通方便性，成瘾，药物治疗，地点和健康，物质的使用。

La Organización Mundial de la Salud ha urgido a los gobiernos de todo el mundo a implementar servicios de tratamiento basados en evidencia por desórdenes de drogadicción y salud mental; sin embargo, el papel que tienen las características geográficas para influir en la continuidad del tratamiento de esta población ha sido, en general, muy poco estudiado. En nuestro caso utilizamos la regresión logística ($N = 294$) para investigar cómo influye en la continuidad del tratamiento la accesibilidad y el contexto socioeconómico del vecindario, para una muestra de 294 individuos adictos a la droga, quienes recibieron intenso tratamiento psiquiátrico como pacientes internados en un hospital mayor para áreas deprimidas, en Filadelfia, Pensilvania, para luego ser asignados para tratamiento como pacientes ambulatorios. Los resultados indican que un tiempo más largo de viaje para el tratamiento, una tasa alta de criminalidad en el vecindario de residencia del paciente, y el viajar para tratamiento de un vecindario relativamente más bajo en crimen a uno de mayor criminalidad, son condiciones que reprimen la continuidad del tratamiento. Estas influencias contextuales son moderadas por etnicidad, donde los blancos son influidos más intensamente por el tiempo de viaje al tratamiento. Esto probablemente refleje las localizaciones de programas de tratamiento relativas a los patrones de segregación residencial. Los afroamericanos residen y atienden al tratamiento dentro de las áreas con mayor criminalidad, lo cual parece tener un impacto particularmente negativo sobre la continuidad del tratamiento para este grupo étnico. Esta investigación destaca la necesidad de una más cuidadosa consideración de los aspectos geográficos en la planificación del tratamiento psiquiátrico.

According to the World Health Organization (WHO), among the greatest causes of disability throughout the world are addiction and mental health disorders, which have been estimated to account for over one third of all healthy life years lost through noncommunicable diseases (Wang et al. 2007). Mental and substance use disorders represent a greater burden than either cancer or cardiovascular disease. Estimates made by the United Nations Office of Drugs and Crime (UNODC) suggest that more than 205 million people in the world use illicit drugs and that depending on the country, 35 percent to 85 percent of serious cases do not receive treatment (WHO 2004). Because this is a major public health problem that affects both industrialized and developing countries, the UNODC and WHO have urged governments worldwide to implement evidence-based treatment services (UNODC–WHO 2008).

Given the chronic nature of drug dependence, considerable research has demonstrated the importance of continuing community-based care and aftercare following primary interventions (McLellan et al. 2000). Treatment accessibility mediates treatment availability, however, whether in terms of affordability, limits on eligibility, cultural relevance, waiting time, hours of operation, or transportation access (Winstanley et al. 2008; Xu et al. 2008). Several studies suggest that geographic factors also influence patient engagement in continuing care and treatment completion, as well as rehospitalization (Jacobson, Robinson, and Buthenthal 2007; Stahler et al. 2009), although little research has focused on the effect of these community-level variables on substance abuse treatment outcomes and engagement in treatment.

The purpose of this study is to investigate geographic barriers to continuity of care for dually diagnosed patients, those with cooccurring mental health and substance use disorders, discharged from acute inpatient psychiatric care. Compared to those with a substance use disorder only, these patients have particularly complex treatment needs that complicate their compliance with treatment, and their rates of relapse and treatment discontinuation tend to be particularly high (Bradizza, Stasiewicz, and Paas 2006).

In geographic barriers, we include both travel accessibility to treatment as well as neighborhood socioeconomic characteristics that we believe might inhibit treatment adherence. We propose that treatment continuity is enhanced by ease of access to treatment. Further, we contend that psychological and environmental stressors that tend to increase the likelihood of treatment discontinuity might be greater in areas that have greater crime, poverty, segregation, or residential mobility. To this end, we analyze a sample of 294 dually diagnosed patients discharged to community-based treatment in Philadelphia, Pennsylvania. We model the influence of geographic characteristics on treatment continuity, while controlling for individual demographic and diagnostic characteristics.

Place Effects on Mental Health and Drug Dependency

Place can be considered as encompassing spatially distributed characteristics of risk and protection that support either healthy or unhealthy behaviors and outcomes. For example, socioeconomically disadvantaged neighborhoods are posited to have relatively poor access to resources that promote health, such as access to health care, nutritious food, or recreational opportunities (Macintyre 2007). Such deprivation amplification is understood as a contextual mechanism that can exaggerate the already negative health effects of poverty at the individual or household level. Because of urban residential segregation, and attendant socioeconomic disadvantage for many minority populations in the United States, the lack of accessibility to health-promoting resources in impoverished areas is associated with racial and ethnic inequity in health outcomes.

Cummins et al. (2007) argued that place-based health research should focus on the dynamic interactions that occur between individuals and the social and physical aspects of places. Certain neighborhood socioeconomic characteristics, such as concentrated disadvantage and residential mobility, act to weaken social capital within neighborhoods, consequently affecting health behaviors (Carpiano 2006). Such a view extends the work of Wilson (1987) and others who argued that neighborhood-level structural forces of concentrated disadvantage and racial and economic segregation weaken social cohesion in the form of civic institutions and other social bonds, thus reinforcing a cycle of poverty (Rankin and Quane 2002; Sampson 2003). In the context of health, health behaviors are seen as embedded within a set of social norms that are held within the community, such that unhealthy behaviors are, in part, a product of peer influence that encourage those unhealthy behaviors, a lack of social support or resources for healthy behaviors, and a lack of social control at the community level more generally.

Recently, researchers have sought to extend these theories of place and health to substance use and mental health disorders by focusing on properties of the socioeconomic, social, and built environments (Stahler et al. 2007; McLafferty 2008; Thomas, Richardson, and Cheung 2008; Stahler et al. 2009). It is well known that family environment, psychological stress, and peer behavior influence mental health problems and substance use disorders (Stockdale et al. 2007; Mason et al. 2010). The presence of poverty, chronic unemployment, and violent crime at both the household and neighborhood levels are certainly potential causes of stress. These can negatively influence mental health and substance use through weakened social control over deviant behavior, larger concentrations of alcohol sales and advertising, and increased illegal drug markets that provide greater access to illicit drugs (Galea, Rudenstine, and Vlahov 2005).

Consequently, neighborhood characteristics have been linked to a variety of mental health and substance use outcomes (Latkin and Curry 2003; Leventhal and Brooks-Gunn 2003; Curtis et al. 2006; Mair, Diez Roux, and Galea 2008; Mason et al. 2009; Mennis and Mason 2011). Treatment adherence to community-based treatment, as well as rehospitalization in a psychiatric unit, has also been shown to be related to the presence of alcohol sales and illicit drugs (Stahler et al. 2007; Stahler et al. 2009). Other researchers have found moderating effects, where demographic characteristics moderate the influence of neighborhood and social contextual influences on substance use (Mennis and Mason 2012). Alternatively, neighborhood characteristics can act to moderate the effects of individual- and family-level influences on health behaviors, where, for instance, living in a disadvantaged neighborhood might mute the positive influence of the family environment (Snedker and Herting 2008).

In this study we theorize that commonly held aspects of neighborhood socioeconomic character, such as indicators of poverty, unemployment, segregation, and residential mobility, influence treatment continuity. We consider these socioeconomic neighborhood characteristics as markers of neighborhood social disorganization, where weak social capital, peer influence toward deviant behavior (i.e., criminal acts or substance use), and segregation from mainstream cultural norms, along with limited physical (i.e., travel) accessibility to treatment, might act to mitigate treatment adherence for dually diagnosed patients participating in community-based treatment.

Methods

Sample Selection

Data describing a sample of 294 patients diagnosed by board-certified psychiatrists with cooccurring mental health and substance use disorders were collected from hospital records for this study. These patients were all treated in an acute inpatient psychiatric unit in a large inner-city hospital in Philadelphia, Pennsylvania, between 30 September and 31 December 2003. The length of admission ranged between three and thirteen days, with an average stay of 11.4 days. Psychiatrists conducting comprehensive intake interviews determined diagnosis using *Diagnostic and Statistical Manual of Mental Disorders* (4th ed., text revision [DSM–IV–TR]; American Psychiatric Association 2000) criteria. All patients were seen daily by an attending psychiatrist, received daily individual and group psychotherapy, and participated in twelve-step groups. Patients were referred for ongoing outpatient mental health and drug treatment as clinically indicated on discharge from the hospital. All patients were given a referral for outpatient treatment to one of fifty-two community-based mental health programs located throughout Philadelphia.

Criteria for inclusion in the study included a diagnosis of cooccurring disorders (at least one mental disorder and a substance use disorder) as well as a positive urine drug screen for prototypical illicit drugs at admission. In addition, the patient must have been referred to an outpatient treatment program. Table 1 shows the demographic and clinical characteristics of the sample. All procedures for gathering information from medical records during chart review were done in compliance with Health Insurance Portability and Accountability Act guidelines, after the study was given exempt status by the first author's university's institutional review board.

Variables Used in the Analysis

The primary outcome variable is treatment continuity, as measured by whether the patient attended his or her first outpatient appointment within thirty days of discharge. A description of each of the explanatory variables used in this analysis is presented in Table 2. Individual variables were selected based on prior evidence, their theoretical relevance to treatment continuity, and their availability in medical charts. These included

Table 1. Demographic and clinical characteristics of the sample

	n	%
Race		
African American	183	62
Hispanic	65	22
White	43	15
Other	3	1
Gender		
Male	152	52
Female	142	48
Chief complaint (can be more than one)		
Suicidal ideation	121	41
Suicide attempt	37	13
Auditory hallucinations	91	31
Aggressive behavior	64	22
Depression	105	36
Bizarre behavior	49	17
Paranoia	49	17
Positive drug screen (can be more than one)		
Cocaine	177	60
Cannabis	106	36
Opioids	35	12
Benzodiazepines	47	16
Phencyclidine	35	12
Barbiturates	13	4
Amphetamines	6	2
Axis I diagnosis at discharge		
Depression	102	35
Schizophrenia	50	17
Bipolar disorder	46	16
Psychotic disorder	40	14
Substance-induced mood disorder	20	7
Other	36	11
Institutional residence at discharge		
Yes	61	21
No	233	79
Attended outpatient treatment appointment within 30 days of hospital discharge		
Yes	92	31
No	202	69

Note: N = 294.

race, gender, age, chief complaint, drugs of abuse, and psychiatric diagnosis. Concerning geographic variables, we considered several variables for measuring accessibility to treatment, including the Euclidean (as-the-crow-flies) distance between the patient's discharge address and treatment program, the distance along the road network, the time it would take to drive along the road network, and the time it would take to ride public transportation (trains and the subway, but excluding buses). All four of these accessibility measures are significantly correlated with one another and with the outcome variable (where, as expected, lower distance and time traveled are associated with treatment continuity). Of the four accessibility variables, however, the driving time variable had the strongest correlation with the outcome. We therefore use the driving time variable as the measure of accessibility in this analysis. Minutes of driving time was divided by ten prior to entry into the logistic regression to aid in the interpretation of the odds ratio.

Data on the neighborhood character of patient's discharge addresses were derived from 2000 U.S. Census block group socioeconomic data as well as 2000–2002 arrest data acquired from the Philadelphia Police Department. Sixteen variables (Table 3) were carefully chosen to reflect neighborhood characteristics of race, racial entropy (Apparicio, Petkevitch, and Charron 2008; a measure of ethnic diversity), poverty, crime, and residential mobility. These block-group-level data were entered into a factor analysis to identify the primary mechanisms of variation relating to neighborhood character. Five rotated (varimax rotation) factors were generated, accounting for 74.5 percent of the total variation in the data set (Mennis and Mason 2012). Table 4 lists the factor loadings for each variable. Based on these factor loadings, each factor was interpreted as a dimension of neighborhood character (Table 4). The first factor captures concentrated disadvantage, as it loads highly on educational attainment, employment, and public assistance income in the expected direction. The second factor we regard as a degree of ethnic diversity, as it loads highly on characteristics such as percentage foreign born, linguistic isolation, and racial entropy. The third factor appears to reflect the presence of Hispanic population, who are highly concentrated in certain neighborhoods in Philadelphia. The fourth factor clearly reflects the presence of crime (notably as a dimension distinct from concentrated disadvantage), and the fifth factor reflects the degree of residential mobility, as indicated by its loadings on the percentage of housing units occupied by renters and percentage of people who have lived in the same house for five years. Each factor was used as an explanatory variable in the analysis.

We also considered that a socioeconomic difference between the patient's discharge address and the treatment location might suppress treatment continuity. Theoretically, people tend to feel most comfortable in places to which they are accustomed, and abrupt changes in socioeconomic environment might be undesirable, discouraging patients from

Table 2. Variable names and definitions

Variable name	Definition
Demographic	
Age	Continuous variable indicating age of patient in years
Male	Dichotomous variable indicating if patient is male
African American	Dichotomous variable indicating if patient self-identifies as African American
Hispanic	Dichotomous variable indicating if patient self-identifies as Hispanic
Chief complaint	
Bizarre	Dichotomous variable indicating if chief complaint is bizarre behavior
Depression	Dichotomous variable indicating if chief complaint is depression
Drug screen	
Opiates	Dichotomous variable indicating urinary drug screen indicates opiates
Cocaine	Dichotomous variable indicating urinary drug screen indicates cocaine
Discharge place	
Institutional	Dichotomous variable indicating patient is discharged to a shelter, halfway house, or other institutional setting
Accessibility	
Driving time	Continuous variable indicating driving time (in minutes) from patient's discharge location to treatment location
Socioeconomic	
Disadvantage	Continuous variable indicating the level of concentrated disadvantage at patient's discharge location
Diversity	Continuous variable indicating the level of ethnic diversity at patient's discharge location
Hispanic	Continuous variable indicating the level of Hispanic character at patient's discharge location
Crime	Continuous variable indicating the level of crime at patient's discharge location
Mobility	Continuous variable indicating the level of residential mobility at patient's discharge location
Socioeconomic gradient	
Disadvantage gradient	Continuous variable indicating change in disadvantage from patient's discharge location to treatment location (discharge location disadvantage minus program location disadvantage)
Diversity gradient	Continuous variable indicating change in diversity from patient's discharge location to treatment location (discharge location diversity minus program location diversity)
Hispanic gradient	Continuous variable indicating change in Hispanic from patient's discharge location to treatment location (discharge location Hispanic minus program location Hispanic)
Crime gradient	Continuous variable indicating change in crime from patient's discharge location to treatment location (discharge location crime minus program location crime)
Mobility gradient	Continuous variable indicating change in mobility from patient's discharge location to treatment location (discharge location mobility minus program location mobility)

Table 3. Variables used in the block group factor analysis

Variable	Definition
% Hispanic	Percentage of the total population that self-identifies as Hispanic
% White	Percentage of the total population that self-identifies as white and non-Hispanic
% African American	Percentage of the total population that self-identifies as African American
% Other race	Percentage of the total population that does not self-identify as white, African American, or Hispanic
% Renter	Percentage of the total households that are renter occupied
% Same house	Percentage of the population over five years old who live in the same house they occupied five years ago
% Linguistic isolation	Percentage of the households with no one over five years old who speaks English well
% Foreign	Percentage of the total population who were born outside the United States
% High school	Percentage of the total population over age twenty-five with a high school diploma or equivalent
% Employed	Percentage of the total civilian population over age sixteen who are employed
% Public assistance	Percentage of the total population receiving public assistance income
% Female headed	Percentage of households female headed with children under sixteen
% Vacancy	Percentage of housing units vacant
Violent crime rate	Number of violent crimes per capita, 2000–2002
Property crime rate	Number of property crimes per capita, 2000–2002
Racial entropy	Index of racial diversity (see text for explanation)

Note: $N = 1,766$.
Source: Mennis and Mason (2012).

Table 4. Variable loadings on factors from factor analysis of block group data

Variable	Factor 1	Factor 2	Factor 3	Factor 4	Factor 5
% White	−.851	−.074	.345	−.018	.044
% African American	.675	−.169	−.663	−.021	−.089
% High school	−.693	−.025	−.385	−.035	.169
% Employed	−.617	.081	−.124	−.116	−.144
% Public assistance	.789	−.054	.287	−.008	.026
% Female headed	.746	−.046	.076	−.090	−.019
% Vacancy	.668	−.138	−.061	.234	.102
% Other race	−.076	.868	−.043	.205	.141
Racial entropy	−.077	.556	.405	−.019	.386
% Linguistic isolation	.153	.627	.558	−.026	.020
% Foreign	−.177	.873	.093	−.005	.089
% Hispanic	.337	.126	.817	−.012	.041
Violent crime rate	.201	.035	.009	.947	.027
Property crime rate	−.040	.117	−.015	.960	.033
% Renter	.174	.113	−.059	.062	.839
% Same house	.097	−.163	−.116	−.005	−.868
% of Variation	25%	15%	12%	11%	12%
Interpretation	Concentrated disadvantage	Ethnic diversity	Hispanic	Crime	Residential mobility

Note: For each variable, the highest loading factor is highlighted. $N = 1,766$.
Source: Mennis and Mason (2012).

attending treatment at program locations in neighborhoods very different from their own. To capture this effect, we created five new socioeconomic gradient variables that capture the change in neighborhood socioeconomic character from the patient's discharge address to his or her treatment program location. For each of the five socioeconomic factors, the factor value of the block group that contains the patient's treatment program is subtracted from the factor value of the block group that contains the patient's discharge address. For example, a positive crime gradient value indicates that the patient is traveling from a relatively high- to a relatively low-crime neighborhood for treatment, whereas a negative value indicates the patient is traveling to a neighborhood with a higher crime character than the one to which he or she was discharged.

Analytic Strategy

Logistic regression was employed to test whether there is a significant relationship of each independent variable with treatment continuity. The analytical procedure was carried out in five stages, where first the characteristics of the individual patient were entered into the regression equation, then the institutional setting of the discharge address was entered, then the accessibility measure was entered, followed by the five

neighborhood socioeconomic characteristics, and then the socioeconomic gradient variables. The area under the receiver operating curve diagnostic was used to assess the model fit. Interaction terms were then entered separately into the logistic regression equation to investigate whether the influence of accessibility, neighborhood socioeconomic, and socioeconomic gradient variables that were found to be significant predictors of treatment continuity differed among ethnic groups, where each interaction term is the product of the hypothesized focal and moderating variables. Although we are aware that other studies have employed multilevel modeling to investigate neighborhood effects on individual health behaviors (e.g., Subramanian 2010), such an approach is inappropriate here, as over half of the patients have discharge addresses in a census block group that contains only one patient discharge address, and 83 percent are within a block group that contains two or less discharge address locations.

Results

Results of the logistic regression of treatment continuity are presented in Table 5. Model 1 includes explanatory variables describing the demographic character, cocaine and opiate use (derived from the urinary drug screen data), and presence of bizarre behavior and depression (derived from the chief complaint data).

Table 5. Logistic regression of treatment continuity (whether a patient attended treatment within one month of hospital discharge)

Independent variable	Model 1	Model 2	Model 3	Model 4	Model 5
Demographic					
Age	1.02 (1.42)	1.02 (1.47)	1.01 (0.82)	1.01 (0.91)	1.01 (0.41)
Male	0.92 (0.09)	0.88 (0.21)	0.89 (0.18)	0.88 (0.22)	0.84 (0.39)
African American	0.93 (0.04)	0.92 (0.04)	0.90 (0.07)	0.98 (0.00)	1.40 (0.44)
Hispanic	1.08 (0.03)	1.09 (0.04)	0.99 (0.00)	0.90 (0.05)	0.91 (0.03)
Chief complaint					
Bizarre	0.25*** (9.44)	0.25*** (8.90)	0.27*** (8.38)	0.27*** (7.81)	0.27** (7.54)
Depression	0.90 (0.14)	0.93 (0.07)	0.94 (0.05)	0.95 (0.04)	0.97 (0.01)
Drug screen					
Cocaine	0.61‡ (3.09)	0.60‡ (3.35)	0.63 (2.63)	0.63 (2.49)	0.59‡ (3.09)
Opiates	0.43‡ (3.30)	0.43‡ (3.29)	0.45‡ (2.84)	0.48 (2.29)	0.55 (1.47)
Discharge place					
Institutional		1.32 (0.72)	1.29 (0.62)	1.56 (1.27)	1.66 (1.58)
Accessibility					
Driving time			0.55* (4.54)	0.52* (5.01)	0.51* (4.63)
Socioeconomic					
Disadvantage				1.05 (0.06)	0.89 (0.12)
Diversity				1.01 (0.00)	1.23 (0.31)
Hispanic				1.04 (0.09)	1.26 (1.27)
Crime				0.50‡ (3.20)	0.40* (4.45)
Mobility				1.01 (0.00)	0.77 (1.20)
Socioeconomic gradient					
Disadvantage gradient					1.16 (0.29)
Diversity gradient					0.78 (0.74)
Hispanic gradient					0.79 (1.96)
Crime gradient					1.45* (4.04)
Mobility gradient					1.26 (2.09)
Constant	0.49 (1.31)	0.47 (0.72)	0.82 (0.09)	0.76 (0.16)	1.09 (1.01)
Nagelkerke R^2	0.08	0.08	0.11	0.12	0.17
AUC	0.64***	0.64***	0.66***	0.68***	0.71***

Note: Values are odds ratios. Wald statistic reported in parentheses. $N = 294$. AUC = area under the receiver operating curve.
‡$p < 0.10$. *$p < 0.05$. **$p < 0.01$. ***$p < 0.005$.

Of these characteristics, only bizarre behavior demonstrated an effect with 95 percent confidence, where a patient with a chief complaint of bizarre behavior is only 25 percent as likely to attend outpatient treatment within one month following discharge compared to a patient without a chief complaint of bizarre behavior. It is also notable that both cocaine and opiate use are also associated with a reduced likelihood of treatment continuity, although the confidence level for these two variables is 92 percent and 93 percent, respectively.

Model 2 indicates that patients discharged to an institutional setting, such as a shelter or recovery house, are no more likely or unlikely than other patients to attend treatment within one month of hospital discharge. Model 3 demonstrates that the longer it takes to drive to the program from the patient's discharge location, the less likely the patient is to attend treatment within one month of hospital discharge. This was a par-

ticularly strong effect, with each extra ten minutes of driving time reducing the likelihood of attendance by approximately half.

Model 4 adds each of the five neighborhood socioeconomic variables to the model. None is significantly related to treatment continuity at a confidence level of 95 percent or higher, although crime is significant at the 93 percent confidence level. Patients discharged to higher crime neighborhoods are less likely to attend treatment within a month of hospital discharge. Model 5 adds the socioeconomic gradient variables. In this model, both the magnitude of crime in the patient's discharge location and the change in crime from the discharge location to the program location are significant at greater than 95 percent confidence. Patients who are discharged to high-crime locations are less likely to attend treatment within one month of hospital discharge, as are patients who have to travel from a relatively lower

Table 6. The influence of driving time and crime gradient on treatment continuity as moderated by ethnicity

Independent variable	Model 1	Model 2	Model 3	Model 4	Model 5	Model 6	Model 7
Bizarre	0.32* (6.77)	0.31** (6.99)	0.32** (6.77)	0.31** (6.96)	0.32* (6.56)	0.34* (6.24)	0.31** (6.91)
Driving time	0.43*** (8.85)	0.44*** (7.74)	0.42*** (8.83)	0.43*** (8.18)	0.42*** (8.87)	0.42*** (8.66)	0.43*** (8.73)
Crime	0.37* (6.29)	0.38* (5.60)	0.37* (6.31)	0.38* (5.82)	0.32** (7.43)	0.34** (6.94)	0.37* (6.40)
Crime gradient	1.40* (4.34)	1.39* (4.25)	1.4* (4.34)	1.39* (4.12)	1.38‡ (3.74)	1.42* (4.30)	1.40* (4.29)
African American		1.17 (0.32)			1.10 (0.11)		
Hispanic			1.03 (0.01)			1.00 (0.00)	
White				0.78 (0.34)			0.87 (0.13)
Driving time × African American		2.25 (1.92)					
Driving time × Hispanic			1.35 (0.22)				
Driving time × White				0.16* (4.00)			
Crime gradient × African American					1.84‡ (3.43)		
Crime gradient × Hispanic						0.48* (4.02)	
Crime gradient × White							0.98 (0.00)
Constant	1.17 (0.39)	0.62*** (9.10)	0.63*** (8.50)	0.63*** (8.78)	1.05 (0.04)	1.03 (0.01)	1.03 (0.02)
Nagelkerke R^2	0.11	0.12	0.11	0.13	0.13	0.13	0.11

Note: Values are odds ratios. Wald statistic reported in parentheses. $N = 294$.
‡$p < 0.10$. *$p < 0.05$. **$p < 0.01$. ***$p < 0.005$.

crime intensity neighborhood to a higher crime intensity neighborhood to attend treatment.

Moderation by ethnicity was observed for the relationship between both driving time and crime gradient with treatment continuity (Table 6) but not for the association between crime and treatment continuity (not shown in the table for the sake of brevity). Regarding driving time, the influence of travel accessibility on treatment continuity is stronger for whites versus other ethnicities (Model 4). Concerning the relationship of crime gradient with treatment continuity, African Americans and whites are less likely to attend treatment within a month of discharge when traveling from a lower to a higher crime area for treatment compared to Hispanics, where the opposite relationship is observed—traveling to a higher crime area is associated with a higher likelihood of treatment continuity (Model 6).

Discussion

Perhaps the most interesting finding from the study is that, with the exception of bizarre behavior as the chief complaint, the most influential factors found to predict attendance at the first postdischarge outpatient appointment for dually diagnosed individuals are geographic in nature. Clearly, travel time has a substantial impact on treatment continuity. Although it is to be expected that having to take more time to travel to an appointment would decrease the likelihood of attendance, we note that this is often not taken into account

when referrals to treatment programs are made. Our study indicates that travel time should be taken explicitly into account in discharge planning to encourage continuing treatment engagement.

The finding that living in a high-crime area suppresses the likelihood of treatment continuity suggests that the socioeconomic conditions of a patient's home neighborhood do indeed influence treatment compliance. We speculate that in this study crime could be considered an indicator of a lack of neighborhood social control that facilitates deviant behaviors, such as illicit drug use, and thus there might be less peer influence toward treatment. The presence of crime might also be associated with greater availability of illicit drugs, consequently facilitating continued drug addiction and a decrease in motivation for seeking treatment.

We also found that traveling from a low- to a high-crime neighborhood for treatment also suppressed treatment continuity. Such a finding suggests that patients are sensitive to the neighborhood environment of their treatment program. We are not aware of any other studies that have incorporated the concept of socioeconomic gradients in modeling barriers to accessibility for mental health or addiction treatment, and our results suggest that this construct could be an important barrier to treatment for this population. Interestingly, other socioeconomic characteristics, such as concentrated disadvantage and residential mobility, were not found to influence treatment continuity. We note that our sample resides in relatively high concentrated disadvantage and high-crime neighborhoods as compared to Philadelphia as a whole. Some of these disadvantaged neighborhoods also have high concentrations of crime, but some do not. Our findings suggest that for this population, it is the presence of crime (and its associated social and environmental properties), not simply disadvantage or poverty, that influences health behavior regarding treatment attendance.

The fact that the driving time and crime gradient variables are moderated by ethnicity suggests that these contextual influences on treatment continuity operate differently for different population subgroups. We speculate that the greater influence of driving time for whites compared to the other ethnic groups reflects the strong pattern of racial and ethnic segregation in Philadelphia, where whites tend to live in several neighborhoods at the periphery of the city. African Americans, and especially Hispanics, tend to be concentrated in neighborhoods in the interior of the city. Indeed, the mean and standard deviation driving time for each ethnic group indicates that whites tend to live farther from their treatment program and have greater variation in driving times compared to other ethnic groups. Thus, it makes sense that their treatment attendance would be more affected by the time it takes to travel to their appointment.

The moderation of the influence of the crime gradient variable on treatment continuity by ethnicity is more difficult to interpret. The overall effect of the crime gradient is as expected—traveling to a high-crime area acts to suppress treatment attendance—so it is curious that this appears to not be the case for Hispanics. It is possible that for Hispanics, who are concentrated primarily in one region of the city, the crime gradient plays much less of a role than the impact associated with traveling outside of the Hispanic neighborhood due to language and cultural barriers. We note that of the three ethnic groups, African Americans are actually the group that both resides and attends treatment within the very highest crime areas. So even though African Americans are discharged to high-crime areas, they must then travel to even higher crime areas for treatment, and this appears to have a particularly negative impact on treatment continuity for African Americans.

Conclusion

We acknowledge several limitations to this study. First, we are limited to our sample of patients discharged from a single hospital; thus, we can only speculate as to how our results might be generalized to other populations of dually diagnosed patients or other mental health patients participating in community-based treatment more generally. Second, we suspect that driving time might not adequately capture travel time as experienced by patients traveling to their treatment program. Considering the poverty and high unemployment of this population, it could be that many patients who do attend treatment do not drive to their appointment but rather walk or take public transportation. Additionally, some programs might provide transportation, which would obviously affect the influence of travel accessibility on treatment continuity (Whetten et al. 2006). As noted earlier, we did compare a travel time measure using public transportation but we did not find this variable to be as predictive as driving time. Future research, however, should seek to incorporate more accurate measures of how patients actually travel to their treatment programs and inquire about commute times. In addition, the study was retrospective and relied on data extracted from hospital records with unknown reliability.

Despite these limitations, this research contributes to the growing recognition that geographic characteristics are an important consideration for the treatment of drug dependence and mental health disorders. Traditionally, the biopsychosocial model employed in conceptualizing the evaluation and treatment of psychopathology has focused on individual-level issues of the patient. Relevant patient information is largely restricted to medical data related to the underlying disease process and its response to treatment intervention, pharmacologic and nonpharmacologic, and psychosocial factors such as emotional stressors, family supports, and financial matters. Rarely, if ever, are geographic variables considered, if even elicited by the discharge planning team. Evidence from this study suggests that geographic characteristics might be a key factor in determining successful community-based treatment for dually diagnosed individuals and thus have important implications for mental health professionals involved with discharge planning and those establishing outpatient treatment facilities. Our results suggest that treatment providers, as well as those conducting treatment outcome studies, need to pay more attention to the role of geographic mechanisms, given their potential impact on adherence to follow-up outpatient care.

References

American Psychiatric Association. 2000. *Diagnostic and statistical manual of mental disorders*. 4th ed., text revision. Washington, DC: American Psychiatric Association.

Apparicio, P., V. Petkevitch, and M. Charron. 2008. Segregation Analyzer: A C#.Net application for calculating residential segregation indices. *Cybergeo: European Journal of Geography* 414:27.

Bradizza, C. M., P. R. Stasiewicz, and N. D. Paas. 2006. Relapse to alcohol and drug use among individuals diagnosed with co-occurring mental health and substance use disorders: A review. *Clinical Psychology Review* 26:162–78.

Carpiano, R. M. 2006. Towards a neighborhood resource-based theory of social capital for health: Can Bourdieu and sociology help? *Social Science & Medicine* 62 (1): 165–75.

Cummins, S., S. Curtis, A. V. Diez-Roux, and S. MacIntyre. 2007. Understanding and representing "place" in health research: A relational approach. *Social Science and Medicine* 65:1825–38.

Curtis, S., A. Copeland, J. Fagg, P. Congolton, M. Almog, and J. Fitzpatrick. 2006. The ecological relationship between deprivation, social isolation and rates of hospital admission for acute psychiatric care: A comparison of London and New York City. *Health and Place* 12:19–37.

Galea, S., S. Rudenstine, and D. Vlahov. 2005. Drug use, misuse and the urban environment. *Drug and Alcohol Review* 24:127–36.

Jacobson, J. O., P. Robinson, and R. N. Buthenthal. 2007. A multilevel decomposition approach to estimate the role of program location and neighborhood disadvantage in racial disparities in alcohol treatment completion. *Social Science in Medicine* 64:462–76.

Latkin, C., and A. Curry. 2003. Stressful neighborhoods and depression: A prospective study of the impact of neighborhood disorder. *Journal of Health and Social Behavior* 44:34–44.

Leventhal, T., and J. Brooks-Gunn. 2003. Moving to opportunity: An experimental study of neighborhood effects on mental health. *American Journal of Public Health* 93:1576–82.

Macintyre, S. 2007. Deprivation amplification revisited; or, is it always true that poorer places have poorer access to resources for healthy diets and physical activity? *The International Journal of Behavioral Nutrition and Physical Activity* 4 (32). doi: 10.1186/1479-5868-4-32

Mair, C., A. V. Diez Roux, and S. Galea. 2008. Are neighborhood characteristics associated with depressive symptoms? A review of evidence. *Epidemiology and Community Health* 62:940–46.

Mason, M. J., J. Mennis, D. J. Coatsworth, T. Valente, F. Lawrence, and P. Pate. 2009. The relationship of place to substance use and perceptions of risk and safety in urban adolescents. *Journal of Environmental Psychology* 29 (4): 485–92.

Mason, M., T. Valente, J. D. Coatsworth, J. Mennis, F. Lawrence, and P. Zelenak. 2010. Place-based social network quality and correlates of substance use among urban adolescents. *Journal of Adolescence* 33 (3): 419–27.

McLafferty, S., 2008. Placing substance abuse. In *Geography and drug addiction*, ed. Y. Thomas, D. Richardson, and I. Cheung, 1–16. Berlin: Springer.

McLellan, A. T., D. C. Lewis, C. P. O'Brien, and H. D. Kleber. 2000. Drug dependence, a chronic medical illness: Implications for treatment, insurance, and outcomes evaluation. *Journal of the American Medical Association* 284:1689–95.

Mennis, J., and M. J. Mason. 2011. People, places, and adolescent substance use: Integrating activity space and social network data for analyzing health behavior. *Annals of the Association of American Geographers* 101 (2): 272–91.

Mennis, J., and M. J. Mason. 2012. Social and geographic contexts of adolescent substance use: The moderating effects of age and gender. *Social Networks* 34 (1): 150–57.

Rankin, B. H., and J. M. Quane. 2002. Social contexts and urban adolescent outcomes: The interrelated effects of neighborhoods, families, and peers on African-American youth. *Social Problems* 49:79–100.

Sampson, R. J. 2003. Neighborhood-level context and health: Lessons from sociology. In *Neighborhoods and health*, ed. L. Berman, 132–46. New York: Oxford University Press.

Snedker, K. A., and J. R. Herting. 2008. The spatial context of adolescent alcohol use. In *Geography and drug addiction*, ed. Y. Thomas, D. Richardson, and I. Cheung, 43–64. Berlin: Springer.

Stahler, G., S. Mazzella, J. Mennis, S. Chakravorty, G. Rengert, and R. Spiga. 2007. The effect of individual, program, and neighborhood variables on continuity of treatment among dually diagnosed individuals. *Drug and Alcohol Dependence* 87:54–62.

Stahler, G., J. Mennis, R. Cotlar, and D. Baron. 2009. The influence of the neighborhood environment on treatment continuity and rehospitalization for dually diagnosed patients discharged from acute inpatient care. *The American Journal of Psychiatry* 166 (11): 1258–68.

Stockdale, S. E., K. B. Wells, L. Tang, T. R. Belin, L. Zhang, and C. D. Sherbourne. 2007. The importance of social context: Neighborhood stressors, stress-buffering mechanisms, and alcohol, drug, and mental health disorders. *Social Science & Medicine* 65:1867–81.

Subramanian, S. V. 2010. Multilevel modeling. In *Handbook of behavioral medicine: Methods and applications*, ed. A. Steptoe, 881–93. New York: Springer Media.

Thomas, Y., D. Richardson, and I. Cheung, eds. 2008. *Geography and drug addiction*. Berlin: Springer.

United Nations Office on Drugs and Crime–World Health Organization. 2008. Principles of drug dependence treatment. Discussion paper. Vienna, Austria: United Nations Office on Drugs and Crime–World Health Organization. http://www.unodc.org/documents/drug-treatment/UNODC-WHO-Principles-of-Drug-Dependence-Treatment-March08.pdf (last accessed 7 February 2012).

Wang, P. S., S. Aguilar-Gaxiola, J. Alonso, M. C. Angermeyer, G. Borges, E. J. Bromet, R. Bruffaerts, et al. 2007. Use of mental health services for anxiety, mood, and substance disorders in 17 countries in the WHO world mental health surveys. *Lancet* 370:841–50.

Whetten, R., K. Whetten, B. W. Pence, S. Reif, C. Conover, and S. Bouis. 2006. Does distance affect utilization of substance abuse and mental health services in the presence of transportation services? *AIDS Care* 18 (Suppl. 1): S27–S34.

Wilson, W. J. 1987. *The truly disadvantaged: The inner-city, the underclass, and public policy.* Chicago: University of Chicago Press.

Winstanley, E. L., D. M. Steinwachs, M. E. Ensminger, C. A. Latkin, M. L. Stitzer, and Y. Olsen. 2008. The association of self-reported neighborhood disorganization and social capital with adolescent alcohol and drug use, dependence, and access to treatment. *Drug and Alcohol Dependence* 92:173–82.

World Health Organization World Mental Health Survey Consortium. 2004. Prevalence, severity, and unmet need for treatment of mental disorders in the World Health Organization World Mental Health Surveys. *Journal of the American Medical Association* 291:2581–90.

Xu, J., R. C. Rapp, J. Wang, and R. G. Carlson. 2008. The multidimensional structure of external barriers to substance abuse treatment and its invariance across gender, ethnicity, and age. *Substance Abuse* 29 (1): 43–54.

Measurement, Optimization, and Impact of Health Care Accessibility: A Methodological Review

Fahui Wang

Department of Geography & Anthropology, Louisiana State University

Despite spending more than any other nation on medical care per person, the United States ranks behind other industrialized nations in key health performance measures. A main cause is the deep disparities in access to care and health outcomes. Federal programs such as the designations of Medically Underserved Areas/Populations and Health Professional Shortage Areas are designed to boost the number of health professionals serving these areas and to help alleviate the access problem. Their effectiveness relies first and foremost on an accurate measure of accessibility so that resources can be allocated to truly needy areas. Various measures of accessibility need to be integrated into one framework for comparison and evaluation. Optimization methods can be used to improve the distribution and supply of health care providers to maximize service coverage, minimize travel needs of patients, limit the number of facilities, and maximize health or access equality. Inequality in health care access comes at a personal and societal price, evidenced in disparities in health outcomes, including late-stage cancer diagnosis. This review surveys recent literature on the three named issues with emphasis on methodological advancements and implications for public policy.

尽管人均医疗花费超过其他任何国家，在关键的卫生性能指标上，美国还排在其它工业化国家之后。一个主要的原因是人们对保健和医疗成果的获取性存在很大的差距。诸如设立医疗服务欠缺地区/人口以及健康专业短缺地区的联邦项目，目的是提高这些地区的卫生服务业人士的数量，并帮助减轻看病难的问题。这些项目的有效性首先最主要依赖于对医保获取性的准确测量，使资源可以分配到真正需要的地区。需要把各种医保获取性的测量指标集成到一个框架中进行比较和评价。优化方法可用于改善医疗保健提供者的分配和供应，以最大限度地提高服务的覆盖范围，最大限度地减少患者的旅行需求，限制所需设施的数量，并最大限度地提高健康或机会平等。保健服务的不平等需个人和社会付出代价，以健康结果之间的差距为证据，包括晚期癌症的诊断。这篇评论调查最近的三期文献，这些文献强调方法的进步和公共政策的影响。*关键词：获取性的测量指标，保健服务，晚期癌症，优化。*

A pesar de que registre el mayor gasto por persona en atención médica, los Estados Unidos se ubican detrás de otras naciones industrializadas en términos de las medidas claves utilizadas sobre lo que se hace en salud. Una causa principal de esta situación tiene que ver con disparidades profundas de acceso a las instalaciones donde se presta este tipo de servicios. Programas federales como el de las designaciones de Áreas Médicamente Mal Servidas/Áreas de Escasez Profesional para Poblaciones y Salud se diseñan para estimular el número de profesionales de la salud que sirven estas áreas para ayudar a aliviar el problema de acceso. En primer término, la efectividad de estos programas depende sobre todo de una medida exacta de la accesibilidad, buscando que los recursos se asignen a las áreas verdaderamente necesitadas. Varias medidas de accesibilidad deben integrarse en un marco para comparación y evaluación. Se pueden utilizar métodos de optimización para mejorar la distribución y suministro de proveedores de cuidados de la salud para maximizar la cobertura del servicio, minimizar las necesidades de desplazamiento de los pacientes, limitar el número de instalaciones, y maximizar la igualdad en salud o en acceso. Las desigualdades de acceso a servicios de salud conllevan costos personales y sociales, que se evidencian en las disparidades de cómo son atendidos los problemas relacionados con este asunto, incluyendo el diagnóstico de cáncer en etapa tardía. En el presente estudio se exploró la literatura reciente sobre las tres cuestiones mencionadas, con énfasis en el desarrollo metodológico y en las implicaciones para las políticas públicas.

Despite spending more per capita on medical care than any other nation, the United States ranks behind other industrialized nations in key health performance measures (World Health Organization 2000). One major factor is the deep disparities in access to care and health outcomes. Maldistribution of the health care workforce leads to the "shortages amid surplus paradox" (Hart et al. 2002, 212). Disparities between races and between the haves and have-nots in health insurance lead to more than 100,000 excessive deaths each year (Physicians for Social Responsibility 2009). Enactment of the Patient Protection and Affordable Care Act will have enormous implications for the supply and distribution of health care providers and provide great opportunities for researchers, including geographers, on related issues.

The U.S. Department of Health and Human Services (DHHS) has implemented various programs including the designations of Medically Underserved Areas/Populations (MUA/P) and Health Professional Shortage Areas (HPSAs) for improving access to health care services for the underserved. The effectiveness of such programs relies on appropriate and accurate measures of accessibility so that resources can be allocated to the neediest areas. Recent advancements in this area have benefited from spatial analysis supported by geographic information systems (GIS) technologies. Many new methods have been developed to improve health care accessibility measures. These methods need to be integrated into one framework to reveal the connection among them and compare their advantages and weaknesses.

Most work uses optimization methods to site health care facilities to maximize service coverage, minimize travel needs of patients, limit the number of facilities, maximize health, or combine some of these goals. Equity in health and health care is widely accepted as an important goal of public policy. Among a diverse set of principles of equity, equal access to health care (for those in equal need) is considered the most appropriate principle for health care policymakers to pursue. Minimizing inequality in health care accessibility helps to identify the adjustments needed to close the gaps.

Inequality in health care access comes at a personal and societal price, evidenced in disparities in various health outcomes. Outcomes include differential rates in infant mortality and birth weight, vaccination, complications from preventive and common diseases, late-stage cancer diagnosis, and quality patient care and survival, among others. Cancer stage (based on tumor size and invasion) at the time of diagnosis plays a critical role in determining the prognosis of patients. This article uses the risk factors of late-stage cancer diagnosis as an example to examine the relationship between health access and outcomes.

This review surveys recent literature on these three closely linked issues with emphasis on methodological advancements. Due to limited space, only representative or the most recent literature is cited on an issue of discussion. The article attempts to synthesize related methods in existing work and identify room for improvements for future studies.

Health Care Accessibility Measures Synthesized

Accessibility refers to the relative ease by which services, here health care, can be reached from a given location. This article focuses on place accessibility, different from, although built on, the work on individual accessibility (Kwan 1998, 1999). Accessibility measures need to account for both spatial and nonspatial factors (Khan 1992). *Spatial access* emphasizes the importance of spatial separation between supply (i.e., health care providers) and demand (i.e., population) and how they are connected in space (Joseph and Phillips 1984) and thus is a classic issue for location analysis well suited for GIS to address. Nonspatial factors include many demographic and socioeconomic variables such as social class, income, age, sex, race, and so on, which also interact with spatial access (Meade and Earickson 2000, 389).

Spatial access is determined by where you are. A simple method measures spatial accessibility by the supply–demand match ratio in an area. For example, the DHHS (2008, 11236) uses a minimum population-to-physician ratio of 3,000:1 within a "rational service area" as a basic indicator for defining HPSAs. This neither reveals the detailed spatial variations within an area unit (e.g., a county or a subcounty area) nor accounts for interaction between population and physicians across areas, however.

Among others, the gravity-based accessibility model considers interaction between supply and demand located in different areas and has been applied in studying health care access (Joseph and Bantock 1982) and other areas, such as job access (Shen 1998). Accessibility at location i (A_i) is written as

$$A_i = \sum_{j=1}^{n} \left[S_j d_{ij}^{-\beta} / \left(\sum_{k=1}^{m} P_k d_{kj}^{-\beta} \right) \right] \quad (1)$$

where P_k is population at location k, S_j is the capacity of the health care provider (e.g., number of doctors or hospital beds) at location j, d is the distance or travel time between them, β is the travel friction coefficient, and n and m are the total numbers of physician locations and population locations, respectively. It is essentially the ratio of supply (S) to demand (P), each of which is

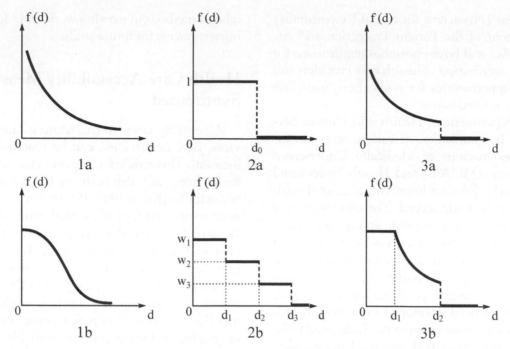

Figure 1. Conceptualizing distance decay in patient–physician interactions: (1a) gravity function, (1b) Gaussian function, (2a) binary discrete, (2b) multiple discrete, (3a) kernel density, and (3b) three-zone hybrid.

discounted by a distance factor. Although conceptually advanced, the model is not intuitive or transparent to public health professionals. Moreover, its distance friction parameter β requires additional data and work to define and might be region specific (Huff 2000).

Luo and Wang (2003) developed the two-step floating catchment area (2SFCA) method to measure spatial access. In the first step, define the catchment of physician location j as an area composed of all population locations (k) within a threshold travel time (d_0) from j and compute the physician-to-population ratio (R_j) within the catchment area as $R_j = S_j / \sum_{k \in \{d_{kj} \leq d_0\}} P_k$. In the second step, for each population location i search all physician locations (j) within the threshold travel time (d_0) from i and sum up the ratios R_j at these locations:

$$A_i = \sum_{j \in \{d_{ij} \leq d_0\}} R_j = \sum_{j \in \{d_{ij} \leq d_0\}} \left(S_j / \sum_{k \in \{d_{kj} \leq d_0\}} P_k \right) \quad (2)$$

Again, the model in Equation 2 is basically a ratio between supply (S) and demand (P), which interact with each other only within a catchment area (e.g., thirty-minute driving time). The method is easy to implement in GIS.

Since its inception, the 2SFCA method has been used in a number of studies measuring health care ac-

cessibility. Despite its relative popularity, the method's major limitation is its dichotomous approach that defines a doctor inside a catchment as accessible and one outside the catchment as inaccessible. Several studies have attempted to improve it. A kernel density function (Guagliardo 2004) or a Gaussian function (Dai 2010) have been proposed to model the distance decay effect (i.e., a continuously gradual decay within a threshold distance and no effect beyond). The catchment radius might also vary by provider types or neighborhood types (Yang, Goerge, and Mullner 2006). Weights can be assigned to different travel time zones to account for the distance decay effects within each catchment area (Luo and Qi 2009). McGrail and Humphreys (2009) proposed a constant weight within ten minutes, a zero weight beyond sixty minutes, and a weight of gradual decay between. The aforementioned methods have different assumptions for conceptualizing distance decay in patient–physician interactions, as illustrated in Figure 1. By generalizing the distance decay effect as a term $f(d)$, we can synthesize all measures of spatial accessibility in a model similar to the models in Equations 1 and 2:

$$A_i = \sum_{j=1}^{n} \left[S_j f(d_{ij}) / \left(\sum_{k=1}^{m} P_k f(dk_j) \right) \right] \quad (3)$$

where $f(d)$ can be a continuous function (Cases 1a and 1b in Figure 1), a discrete variable (Cases 2a and 2b), or a hybrid of the two (Cases 3a and 3b).

The main debate centers on what is (are) the reasonable catchment area size(s) of physician services and which is the best function to capture the distance decay behavior in physician visits. Any debate over the best function or the right size for catchment areas cannot be settled without analyzing real-world health care utilization behavior. This pursuit is related to a long-standing interest of geographers in *activity space*, defined as the local areas within which an individual moves or travels on a regular basis (Gesler and Albert 2000). For example, Sherman et al. (2005) used various GIS-based methods to define activity space for health care access by rural residents in North Carolina, and Arcury et al. (2005) demonstrated that geographic and spatial behavior factors (including location, availability of a driver's license, and transportation modes) play important roles in rural health care utilization. Such studies are useful to help define appropriate catchment areas (likely to be different in rural and urban areas) when the distance decay is treated as a discrete variable. To model the wide spectrum of propensity for physician visits corresponding to trip lengths, more rigorous modeling is needed to derive the best fitting analytical functions and related parameters.

We now turn our attention to nonspatial access, which captures how access is influenced by who you are. Nonspatial factors include a wide selection of demographic and socioeconomic variables that affect health care access. These variables include demographics (e.g., seniors, children, women of childbearing ages), socioeconomic status (e.g., poverty, female-headed households, homeownership, and median income), housing conditions (e.g., crowdedness, basic amenities), and linguistic barriers and education. A major challenge is integrating these variables. Some suggest standardizing the variables and then combining them to produce a composite score (e.g., Field 2000), but many variables are correlated and thus contain duplicate information. Wang and Luo (2005) used factor analysis to consolidate these variables into fewer independent factors. The DHHS (2008, 11263–69) also proposed using factor analysis to help design the weights to aggregate the variables together in HPSA designations.

Nonspatial factors also interact with spatial access. For example, transit-dependent residents might be considered a nonspatial issue because of their age, medical condition, or lack of economic means, but they also tend to travel longer times to health care providers, thus affecting their spatial access (e.g., Lovett et al. 2002; Martin, Jordan, and Roderick 2008). Various population groups might also have different levels of health care needs and travel behavior (Morrill and Kelley, 1970). The Agency for Healthcare Research and Quality (2010, 177) identified seven priority populations (racial and ethnic minorities, low-income groups, women, children, older adults, residents of rural areas, and individuals with disabilities or special health care needs) for high health care needs. McGrail and Humphreys (2009, 420) used principal component analysis to consolidate seven sociodemographic variables into one summary score of health needs. This allows nonspatial factors to be used to adjust the definition of demand in spatial accessibility measures, providing one way of integrating spatial access and nonspatial factors in a unified accessibility measure.

The preceding review provides a glimpse of the diversity and complexity of issues on health care accessibility measures. Naturally, the methodology gets more advanced and complicated as more issues arise, but the increasing complexity also hinders its implementation and adoption. An equally important and tall task is to develop simplified and transparent proxy measures that require limited data but capture variation patterns of actual health care utilization across space and demographic groups consistent with otherwise more sophisticated methods.

Optimization Models in Health Care Accessibility

Several models in classic location–allocation problems (e.g., Church 1999) can benefit the study of planning for health care facilities (Table 1). The *p median problem* seeks to locate a given number of facilities among a set of candidate sites so that the total travel distance or time between demands and supply facilities is minimized. For example, Wang (2006, 203–11) used a case study in Cleveland to demonstrate its implementation in ArcGIS (Environment System Research Institute, Inc. 2012) to allocate health clinics to serve clients in the most efficient way (i.e., minimizing total distance or time). The *location set covering problem* (LSCP) minimizes the number of facilities needed to cover all demand within a critical distance or time. For example, Shavandi and Mahlooji (2008) employed a queuing theory in a fuzzy framework to solve an LSCP in allocating health care facilities at different levels in Iran. The *maximum covering location problem* (MCLP) maximizes

Table 1. Optimization models of health care facilities

Model	Objective	Constraints	Health care application example
P median problem	Minimize total distance/time	Locate p facilities; cover all demands (optional: demand must be within a specified distance/time)	Wang (2006, 203–11)
Location set covering problem (LSCP)	Minimize the number of facilities	Cover all demands	Shavandi and Mahlooji (2008)
Maximum covering location problem (MCLP)	Maximize coverage	Locate p facilities; cover demand if within a specified distance/time (optional: demand not covered must be within a second larger distance/time)	Pacheco and Casado (2005)
Center model	Minimizes the maximum distance	Locate p facilities; cover all demands	N/A
Equity model	Minimize inequality in accessibility	Adjust supplies in p facilities but maintain the same total supply	Wang and Tang (2010)

the demand covered within a desired distance or time threshold by locating p facilities. For example, Pacheco and Casado (2005) used a hybrid heuristic algorithm to solve an MCLP in allocating health care resources in Burgos, Spain. The *center model* identifies a location arrangement for p facilities that minimizes the maximum distance to cover all clients. Additional constraints can be added to these models, and multiobjective models can also be constructed by combining the objectives of the models.

The preceding models emphasize various objectives such as minimal travel, minimal resources, maximal coverage, or a combination of them (i.e., multiobjective) that focuses on either the supply or demand side of health care service delivery. Some recent work accounts for both sides, particularly their match ratios considered in spatial accessibility measures. For example, Perry and Gesler (2000) used a target ratio of health personnel versus population and a maximum travel distance as criteria to adjust health personnel distribution to improve overall access. Zhang, Berman, and Verter (2009) developed an optimal health care location model with a unique feature of accounting for distance decay in probability of using a health care facility. Gu, Wang, and McGregor (2010) used a biobjective model to identify optimal locations for health care facilities that maximize total coverage of population as well as their total accessibility. None of these studies has equity as an objective, however. Equity in health and health care can be defined as equal access to health care, equal utilization of health care services, or equal (equitable) health outcomes, among others (e.g., Culyer and Wagstaff 1993). Most agree that equal access is the most appropriate principle of equity from a public health policy perspective (Oliver and Mossialos 2004, 656). Some (e.g., Hemenway 1982) have argued that

health maximization is the most justifiable objective. Quantifying the conversion from health care provision to outcome remains a challenge, however.

Wang and Tang (2010) developed an equity model with an objective of minimizing inequality in accessibility. Given an accessibility measure as defined in any of the Equations 1 through 3, it is known that the weighted mean of accessibility is equal to the ratio of total supply to total demand in a study area (Shen 1998), denoted by a constant a. The objective is to minimize the variance (i.e., least squares) of accessibility index A_i across all population locations by redistributing the total amount of supply S among health care facilities, written as:

$$\min = \sum_{i=1}^{m} P_i (A_i - a)^2$$

In this objective function, accessibility gaps $(A_i - a)^2$ are weighted by corresponding population P_i. One constraint is the total supply such as:

$$S_1 + S_2 + \dots + S_n = S$$

Additional constraints (e.g., threshold service population, as in Gu, Wang, and McGregor 2010) can be added.

This formulation fits a quadratic programming (QP), where the objective function is a quadratic function of variables S_i subject to linear constraints (here just one) on these variables (Nocedal and Wright 2006). There are various free and open-source programs to solve QP problems (Gould and Toint 2012). Note that the problem does not have a trivial solution such as distributing physicians proportionally to population in all areas because of complexity of interaction between health care supplies and demands across area units. In rare cases, the optimization yields equal accessibility across all population locations. Comparing the existing physician distribution to the "optimal" pattern that maximizes access equality, one could identify areas with severe shortage of services and adjustment needed toward maximal access equality.

Association of Health Care Access Inequality with Late-Stage Cancer Diagnosis

Inequality in health care access leads to disparities in various health outcomes, including differential rates in infant mortality and birth weight, vaccination, and complications from preventive and common diseases, among others. Cancer is a leading cause of death in the United States, second to heart disease (Centers for Disease Control and Prevention 2010). Cancer stage (based on tumor size and invasion) at the time of diagnosis plays a critical role in determining the patient's prognosis. This section uses late-stage cancer diagnosis as an example to illustrate the impact of health care access inequality. There is an enormously rich body of literature on analyzing various risk factors of late-stage cancer (Wang, Luo, and McLaffferty 2010). In summary, the risk factors include (1) spatial access to both cancer screening facilities and primary care physicians and (2) nonspatial factors at both the individual and neighborhood levels (see Table 2). For example, poorer spatial access to mammography is associated

Table 2. Regression models for analyzing late-stage cancer risks

Variables	OLS	Poisson	Multilevel logit
Late-stage cancer rate	Y		
Number of late-stage cancer cases		Y	
Number of all cancer cases		Offset	
Individual cancer cases (1 = late-stage, 0 = otherwise)			Y
Individual cancer patient sociodemographic attributes			X
Neighborhood demographic and socioeconomic factors	X	X	X
Neighborhood urban–rural classification	X	X	X
Spatial access to primary care	X	X	X
Spatial access to cancer screening	X	X	X

Note: Y indicates the dependent variable, and X indicates an independent variable in a model. OLS = ordinary least squares.

with its lower utilization and consequently higher late-stage breast cancer rate (Menck and Mills 2001). Similar effects are found for spatial access to primary care physicians (Wang et al. 2008) and nonspatial factors (McLafferty and Wang 2009). The accessibility measures discussed previously will help to improve the definitions of the preceding risk factors. This section focuses on methodological issues of examining the association of these factors with late-stage cancer diagnosis.

For data in large analysis units such as state and county, it is appropriate to use ordinary least squares (OLS) regression, where the dependent variable is late-stage cancer rate (i.e., ratio of number of late-stage cancer cases to total cancer cases) and independent variables are the aforementioned risk factors. For cancer data in small areas such as ZIP code area and census tract with a small number of cancer counts, late-stage cancer rate is sensitive to missing data and other data errors, has a high variance, and is less reliable, commonly known as the *small population problem*. Several spatial strategies have been proposed to mitigate the problem. Conceptually similar to moving averages that smooth observations over a longer time interval, spatial smoothing computes the average late-stage cancer rates around each area using a larger spatial window. Spatial smoothing methods include the floating catchment area method, kernel density estimation (Wang 2006, 36–38), empirical Bayes estimation (Clayton and Kaldor 1987), locally weighted average (Shi et al. 2007) and adaptive spatial filtering (Tiwari and Rushton 2004). Another geographic approach is regionalization, which groups small areas to form larger geographic areas with more reliable late-stage cancer rates. Some earlier methods emphasized either homogeneity of attributes within the new areas (e.g., Haining, Wises, and Blake 1994) or spatial proximity between the areas to be grouped (e.g., Lam and Liu 1996). Recent research aims to develop GIS-based automated methods that take into account both spatial contiguity and attribute homogeneity within the derived areas (e.g., Guo 2008; Mu and Wang 2008). When similar areas are merged, it mitigates the spatial autocorrelation problem commonly observed in data of geographic areas and simplifies subsequent regression analysis.

There are also regression methods that are suitable for analysis of small-area cancer data (see Table 2). One is Poisson regression (e.g., Wang et al. 2008), where the dependent variable is the number of late-stage cancer cases and the total number of cancer cases serves as an offset variable. A spatial Poisson regression model

(Best, Ickstadt, and Wolpert 2000) is needed to account for spatial autocorrelation in the data. Another method is logit regression to model the risk of individual cancer cases being late stage, where the dependent variable is binary (0, 1). A multilevel logistic model is needed to examine the effects of both individual- and neighborhood-level risk factors (e.g., McLafferty and Wang 2009). Similarly, more advanced logit models are needed to control for spatial autocorrelation (Griffith 2004).

Summary

This article reviews recent methodological advancements in three issues related to inequality in health care accessibility: measurement, optimization, and impact. Various methods have been proposed to measure health care accessibility, accounting for both spatial and nonspatial factors. Various measures of spatial accessibility differ in ways of conceptualizing the distance decay effect as a continuous function, a discrete variable, or a hybrid of the two. The selection of an appropriate model needs to be based on analysis of real-world health care utilization behavior. Nonspatial factors include a wide selection of demographic and socioeconomic variables, which can be consolidated into a few independent factors by factor analysis. The increasing complexity of accessibility models hinders its implementation and adoption by public health professionals and calls for the development of simplified and transparent proxy measures.

The classic location–allocation problems have been widely used in planning for health care facilities. There is a lack of operation research on equity in health care access. An equity model explicitly formulates the objective function as minimizing inequality in accessibility across demand locations and can be solved by quadratic programming. Results from the model identify adjustments needed for maximizing access equality.

The final issue is assessing the impact of accessibility disparities on an important health indicator, late-stage cancer diagnosis. OLS regression can be used to examine risk factors in influencing the variation of late-stage cancer rate across large areas. Poisson and logit regressions are often used for analysis of cancer data in small areas. Some geographic strategies, such as spatial smoothing and regionalization methods, have also been proposed to mitigate the small population problem.

Acknowledgments

I would like to acknowledge support by the National Institutes of Health (#1R01CA140319-01A1, Ming Wen as PI), and the National Natural Science Foundation of China (No. 40928001). I appreciate valuable inputs from Sara McLafferty of University of Illinois at Urbana-Champaign, Wei Luo of Northern Illinois University, and Imam Xierali and Robert L. Phillips, Jr., of the Robert Graham Center. Comments by three anonymous reviewers and Mei-Po Kwan helped me prepare the final version.

References

Agency for Healthcare Research and Quality. 2010. *National healthcare disparities report 2009*. AHRQ Publication No. 10-0004. http://www.ahrq.gov/qual/qrdr09.htm (last accessed 15 October 2010).

Arcury, T. A., W. M. Gesler, J. S. Preisser, J. Sherman, and J. Perin. 2005. The effects of geography and spatial behavior on healthcare utilization among the residents of a rural region. *Health Services Research* 40:135–55.

Best, N. G., K. Ickstadt, and R. L. Wolpert. 2000. Spatial Poisson regression for health and exposure data measured at disparate resolutions. *Journal of the American Statistical Association* 95:1076–88.

Centers for Disease Control and Prevention. 2010. Deaths and mortality. Atlanta, GA: Centers for Disease Control and Prevention. http://www.cdc.gov/nchs/fastats/deaths.htm (last accessed 8 February 2012).

Church, R. L. 1999. Location modelling and GIS. In *Geographical information systems*, ed. P. A. Longley, M. F. Goodchild, D. J. Maguire, and D. W. Rhind, 293–303. New York: Wiley.

Clayton, D., and J. Kaldor. 1987. Empirical Bayes estimates of age-standardized relative risks for use in disease mapping. *Biometrics* 43:671–81.

Culyer, A. J., and A. Wagstaff. 1993. Equity and equality in health and healthcare. *Journal of Health Economics* 12:431–57.

Dai, D. 2010. Black residential segregation, disparities in spatial access to health care facilities, and late-stage breast cancer diagnosis in metropolitan Detroit. *Health & Place* 16:1038–52.

Department of Health and Human Services. 2008. Designation of medically underserved populations and Health Professional Shortage Areas: Proposed rule. http://bhpr.hrsa.gov/shortage/proposedrule/frn.htm (last accessed 15 October 2010).

Environment Systems Research Institute, Inc. 2012. ArcGIS Network Analyst. http://www.esri.com/software/arcgis/extensions/networkanalyst/ (last accessed 8 February 2012).

Field, K. 2000. Measuring the need for primary healthcare: An index of relative disadvantage. *Applied Geography* 20:305–32.

Gesler, W. M., and D. P. Albert. 2000. How spatial analysis can be used in medical geography. In *Spatial analysis, GIS and remote sensing applications in the health sciences,* ed. D. P. Albert, W. M. Gesler, and B. Levergood, 11–38. Chelsea, MI: Ann Arbor Press.

Gould, N., and P. Toint. 2012. A quadratic programming page. http://www.numerical.rl.ac.uk/qp/qp.html (last accessed 8 February 2012).

Griffith, D. A. 2004. A spatial filtering specification for the auto-logistic model. *Environment & Planning A* 36:1791–1811.

Gu, W., X. Wang, and S. E. McGregor. 2010. Optimization of preventive healthcare facility locations. *International Journal of Health Geographics* 9:17.

Guagliardo, M. F. 2004. Spatial accessibility of primary care: Concepts, methods and challenges. *International Journal of Health Geography* 3:3.

Guo, D. 2008. Regionalization with dynamically constrained agglomerative clustering and partitioning (REDCAP). *International Journal of Geographical Information Science* 22:801–23.

Haining, R., S. Wises, and M. Blake. 1994. Constructing regions for small area analysis: Material deprivation and colorectal cancer. *Journal of Public Health Medicine* 16:429–38.

Hart, L. G., E. Salsberg, D. M. Phillips, and D. M. Lishner. 2002. Rural health care providers in the United States. *Journal of Rural Health* 18:211–32.

Hemenway, D. 1982. The optimal location of doctors. *The New England Journal of Medicine* 306:397–401.

Huff, D. L. 2000. Don't misuse the Huff model in GIS. *Business Geographies* 8:12.

Joseph, A. E., and P. R. Bantock. 1982. Measuring potential physical accessibility to general practitioners in rural areas: A method and case study. *Social Science and Medicine* 16:85–90.

Joseph, A. E., and D. Phillips. 1984. *Accessibility and utilization—Geographical perspectives on healthcare delivery.* New York: Harper & Row.

Khan, A. A. 1992. An integrated approach to measuring potential spatial access to healthcare services. *Socioeconomic Planning Science* 26:275–87.

Kwan, M.-P. 1998. Space–time and integral measures of individual accessibility: A comparative analysis using a point-based framework. *Geographical Analysis* 30:191–216.

———. 1999. Gender and individual access to urban opportunities: A study using space–time measures. *The Professional Geographer* 51:210–27.

Lam, N. S.-N., and K. Liu. 1996. Use of space-filling curves in generating a national rural sampling frame for HIV-AIDS research. *Professional Geographer* 48:321–32.

Lovett, A., R. Haynes, G. Sunnenberg, and S. Gale. 2002. Car travel time and accessibility by bus to general practitioner services: A study using patient registers and GIS. *Social Science & Medicine* 55:97–111.

Luo, W., and Y. Qi. 2009. An enhanced two-step floating catchment area (E2SFCA) method for measuring spatial accessibility to primary care physicians. *Health and Place* 15:1100–1107.

Luo, W., and F. Wang. 2003. Measure of spatial accessibility to healthcare in a GIS environment: Synthesis and a case study in the Chicago region. *Environmental and Planning B* 30:865–84.

Martin, D., H. Jordan, and P. Roderick. 2008. Taking the bus: Incorporating public transport timetable data into healthcare accessibility modelling. *Environment and Planning A* 40:2510–25.

McGrail, M. R., and J. S. Humphreys. 2009. A new index of access to primary care services in rural areas. *Australian and New Zealand Journal of Public Health* 33:418–23.

McLafferty, S., and F. Wang. 2009. Rural reversal? Risk of late-stage cancer across the rural–urban continuum in Illinois. *Cancer* 115:2755–64.

Meade, S. M., and R. J. Earickson. 2000. *Medical geography.* 2nd ed. New York: Guilford.

Menck, H. R., and P. K. Mills. 2001. The influence of urbanization, age, ethnicity, and income on the early diagnosis of breast carcinoma: Opportunity for screening improvement. *Cancer* 92:1299–1304.

Morrill, R. L., and M. Kelley. 1970. The simulation of hospital use and the estimation of locational efficiency. *Geographical Analysis* 2:283–300.

Mu, L., and F. Wang. 2008. A scale-space clustering method: Mitigating the effect of scale in the analysis of zone-based data. *Annals of the Association of American Geographers* 98:85–101.

Nocedal, J., and S. J. Wright. 2006. *Numerical optimization.* 2nd ed. Berlin: Springer-Verlag.

Oliver, A., and E. Mossialos. 2004. Equity of access to healthcare: Outlining the foundations for action. *Journal of Epidemiology and Community Health* 58:655–58.

Pacheco, J. A., and S. Casado. 2005. Solving two location models with few facilities by using a hybrid heuristic: A real health resources case. *Computers & Operations Research* 32:3075–91.

Perry, B., and W. Gesler. 2000. Physical access to primary healthcare in Andean Bolivia. *Social Science & Medicine* 50:1177–88.

Physicians for Social Responsibility. 2009. Physicians for Social Responsibility policy on health care reform in the United States. http://www.psr.org/social-justice/healthcare-reform.html (last accessed 8 February 2012).

Shavandi, H., and H. Mahlooji. 2008. Fuzzy hierarchical queuing models for the location set covering problem in congested systems. *Scientia Iranica* 15:378–88.

Shen, Q. 1998. Location characteristics of inner-city neighborhoods and employment accessibility of low-income workers. *Environment and Planning B* 25:345–65.

Sherman, J. E., J. Spencer, J. S. Preisser, W. M. Gesler, and T. A. Arcury. 2005. A suite of methods for representing activity space in a healthcare accessibility study. *International Journal of Health Geographics* 4:24.

Shi, X., E. Duell, E. Demidenko, T. Onega, B. Wilson, and D. Hoftiezer. 2007. A polygon-based locally-weighted-average method for smoothing disease rates of small units. *Epidemiology* 18:523–28.

Tiwari, C., and G. Rushton. 2004. Using spatially adaptive filters to map late stage colorectal cancer incidence in Iowa. In *Developments in spatial data handling*, ed. P. Fisher, 665–76. New York: Springer-Verlag.

Wang, F. 2006. *Quantitative methods and applications in GIS.* Boca Raton, FL: CRC Press.

Wang, F., and W. Luo. 2005. Assessing spatial and non-spatial factors in healthcare access in Illinois: Towards an integrated approach to defining Health Professional Shortage Areas. *Health and Place* 11:131–46.

Wang, F., L. Luo, and S. McLafferty. 2010. Healthcare access, socioeconomic factors and late-stage cancer diagnosis: An exploratory spatial analysis and public policy implication. *International Journal of Public Policy* 5:237–58.

Wang, F., S. McLafferty, V. Escamilla, and L. Luo. 2008. Late-stage breast cancer diagnosis and healthcare access in Illinois. *The Professional Geographer* 60:54–69.

Wang, F., and Q. Tang. 2010. Towards equal accessibility to services: A quadratic programming approach. Paper presented at the 18th International Conference on GeoInformatics, Beijing, China.

World Health Organization. 2000. *The world health report 2000: Health systems: Improving performance.* http://www.who.int/whr/2000/en/index.html (last accessed 8 February 2012).

Yang, D., R. Goerge, and R. Mullner. 2006. Comparing GIS-based methods of measuring spatial accessibility to health services. *Journal of Medical Systems* 30:23–32.

Zhang, Y., O. Berman, and V. Verter. 2009. Incorporating congestion in preventive healthcare facility network design. *European Journal of Operational Research* 198:922–35.

Spatial Heterogeneity in Cancer Control Planning and Cancer Screening Behavior

Lee R. Mobley,* Tzy-Mey Kuo,† Matthew Urato,* Sujha Subramanian,* Lisa Watson,* and Luc Anselin‡

*RTI International
†Lineberger Comprehensive Cancer Center, University of North Carolina at Chapel Hill
‡School of Geographical Sciences, Arizona State University

Each state is autonomous in its comprehensive cancer control (CCC) program, and considerable heterogeneity exists in the program plans, but researchers often focus on the concept of nationally representative data and pool observations across states using regression analysis to come up with average effects when interpreting results. Due to considerable state autonomy and heterogeneity in various dimensions—including culture, politics, historical precedent, regulatory environment, and CCC efforts—it is important to examine states separately and to use geographic analysis to translate findings in place and time. We used 100 percent population data for Medicare-insured persons aged sixty-five or older and examined predictors of breast cancer (BC) and colorectal cancer (CRC) screening from 2001 to 2005. Examining BC and CRC screening behavior separately in each state, we performed 100 multilevel regressions. We summarize the state-specific findings of racial disparities in screening for either cancer in a single bivariate map of the fifty states, producing a separate map for African American and for Hispanic disparities in each state relative to whites. The maps serve to spatially translate the voluminous regression findings regarding statistically significant disparities between whites and minorities in cancer screening within states. Qualitative comparisons can be made of the states' disparity environments or for a state against a national benchmark using the bivariate maps. We find that African Americans in Michigan and Hispanics in New Jersey are significantly more likely than whites to utilize CRC screening and that Hispanics in six states are significantly and persistently more likely to utilize mammography than whites. We stress the importance of spatial translation research for informing and evaluating CCC activities within states and over time.

虽然各州的综合癌症控制计划（CCC）是自治性的，并且在各州的工作计划中存在相当大的差异性，但研究者往往侧重于全国代表性的数据概念和跨州使用回归分析的池观察，目的是在解释结果时产生出平均效果。由于在包括文化，政治，历史先例，监管环境，和 CCC 的努力等各个方面，存在相当多的州自主性和差异性，单独地对各州进行审查和使用地理分析，以把结果转化到具体的时间和地点是重要的。我们使用 100% 的年龄六十五以上的政府医疗保健补贴制度的参保人的数据，并且研究了从 2001 年到 2005 年筛选出的乳腺癌（BC）和大肠癌（CRC）的预测因子。我们进行了 100 多级的回归分析，以分别检验在每个州的 BC 和 CRC 筛查行为。我们总结了在五十个州的二元地图里两种癌症筛查中的种族差异，产生了一幅单独的非裔和西班牙裔美国人相对于白人在各州的差距地图。这些地图可在空间上转换浩大的，有关各州白人和少数民族之间在癌症筛查上的，统计差别的回归结果。使用这些二元地图，可对州的差距环境，或将一个州针对整个国家的基准进行定性比较。我们发现在密歇根州的非裔美国人和新泽西州的拉丁裔美国人和白人相比更可能利用 CRC 筛查，有六个州的拉丁裔美国人比白人更显著和持续地利用乳房 X 光检查。我们强调转换研究在表示和评估各州内的和随时间推移的 CCC 活动的重要性。关键词：二元映射图，全面的癌症控制，地域差异，空间差异性，空间翻译。

Cada estado es autónomo en su programa amplio de control del cáncer (CCC) y existe una considerable heterogeneidad en los planes del programa, pero a menudo en donde los investigadores centran su atención es en el concepto de datos nacionalmente representativos y en el banco de observaciones a través de los estados, usando análisis de regresión para obtener efectos promedio cuando se trate de interpretar resultados. Debido a la considerable autonomía estatal y la heterogeneidad existente en varias dimensiones—incluyendo cultura, política, precedente histórico, ambiente regulador y esfuerzos del CCC—es importante examinar los estados separadamente y usar análisis geográfico para traducir los hallazgos en lugar y tiempo. Utilizamos en cien por ciento datos de población referidos a personas aseguradas con Medicare, con edades de 65 años o más, y examinamos los vaticinadores de cáncer del seno (BC) y cáncer colorrectal (CRC) entre 2001 y 2005. Al examinar el comportamiento de las exploraciones de BC y CRC de manera separada para cada estado, efectuamos 100 regresiones de nivel múltiple. Resumimos los hallazgos específicos de disparidades raciales por estado al efectuar los chequeos para cada tipo de cáncer en un mapa bivariado de 50 estados, generando un mapa separado para las disparidades afroamericanas e hispanas en relación con las de los blancos de cada estado. Los mapas sirven para traducir espacialmente los voluminosos hallazgos de regresión considerando las disparidades estadísticamente significativas en los chequeos por cáncer entre las minorías dentro de los estados. Se pueden hacer comparaciones cualitativas de los entornos de disparidad de los estados o por un estado contra el punto de referencia nacional utilizando los mapas bivariados. Encontramos que los afroamericanos en Michigan y los hispanos en Nueva Jersey tienen mayor probabilidad que los blancos de utilizar chequeos para CCR, y que las hispanas en seis estados significativamente tienen más propensión y persistencia en utilizar la mamografía que las blancas. Enfatizamos la importancia de la traducción espacial de la investigación para informar y evaluar las actividades de CCC dentro de los estados y a través del tiempo.

C ancer is a leading cause of death in the United States, second to heart disease (Heron et al. 2009; Edwards et al. 2010). Among all cancers, breast cancer (BC) and colorectal cancer (CRC) can be detected in early stages through effective screening methods. The five-year survival rate for BC and CRC exceeds 90 percent if detected at an early stage and exceeds 70 percent at a regional stage (American Cancer Society [ACS] 2010). BC and CRC screening rates are low, however (about 50 percent for BC and CRC in 2008; ACS 2011), and the mortality rates remain high.

In 2009, BC was the most common cancer in women, and CRC was the third most common cancer in both men and women (Jemal et al. 2009). Both of these cancers are more common among people aged sixty-five or older. Specifically, BC incidence is five times greater and CRC incidence is fifteen times greater among this age group than among younger populations (National Cancer Institute 2007). Cancer morbidity and mortality in the older population become even more salient as the population size and life expectancy of older persons continue to increase (Hetzel and Smith 2001; Administration on Aging 2006). Thus, it is imperative to understand factors associated with BC and CRC screening behavior in the older population to promote early detection and effective reduction in cancer morbidity and mortality.

Disparities in cancer screening, morbidity, and mortality are well documented across different races or ethnicities, levels of socioeconomic status or acculturation, insurance coverage or type, and geographic location (ACS 2010; Naishadham et al. 2011). Recent evidence from the Behavioral Risk Factor Surveillance System regarding CRC screening, coupled with the National Program of Cancer Registries data, suggests that states with higher prevalence of CRC screening have lower mortality from CRC (Richardson et al. 2011). Elimination of disparities in cancer screening and outcomes has been a particular focus of comprehensive cancer control (CCC) efforts (Coughlin et al. 2006). Therefore, we focus this article on disparities in screening utilization among whites and minorities.

Several studies have demonstrated that BC or CRC screening utilization varies across states (Nelson et al. 2003; Cooper and Koroukian 2004; Schneider et al. 2009; Mobley et al. 2010). Mobley et al. (2010) reported considerable state-level variation in CRC screening rates (sigmoidoscopy or colonoscopy), ranging from 34 percent in New Mexico to 44 percent in Maryland, using 100 percent Medicare fee-for-service claims data for people aged sixty-five or older from 2001 through 2005.

This article focuses on racial or ethnic disparities in two types of cancer screening (BC, CRC) and how these vary across states. The article's major contribution is in the spatial translation area; that is, we use mapping of state-specific population estimates reflecting racial or ethnic disparities in two types of cancer screening to convey the results of 100 independently estimated multilevel models all in one graphic. This efficient means of summarizing the vast amount of information allows CCC planners to easily assess how the screening behaviors of older minorities in their state compare to older whites and how their observed disparity environments compare to a national average and to findings in other states. We stress the statistical significance of differences within states, which is the most valid comparison for disparities research. Plotting the results from all states together in one map allows for qualitative assessments of how the disparity environments differ across states. Including the national average effect estimate from a pooled model in the map legend offers a benchmark that illustrates which states are driving national estimates.

This sort of information on place-specific variation could be useful in CCC planning and interventions, under new guidelines that encourage communities to collaborate with others and to pay attention to place-based differences. More specifically, federal preventive health policy has moved toward place-based initiatives, as exemplified by a 2009 White House memo (Office of Management and Budget 2009). The memo promotes place-conscious planning and place-based policies and programming at the interagency level to increase the impact of public investments. To do this, planners need a way to reliably compare results across different communities and state cancer control environments. Title IV of the Affordable Care Act of 2010 supports specific programs tailored to improving population health and integrated, place-based efforts to improve the well-being of persons and communities. The law provides substantial funding for public health and requires that the Centers for Disease Control and Prevention convene an independent Community Preventive Services Task Force to review interventions, including consideration of place-specific social, economic, and physical environments that can affect health and disease (Mueller et al. 2011).

Conceptual Model

Understanding health disparities in cancer screening requires a priori modeling of the sources of these disparities in a geospatial context and requires an approach that explicates the social ecology of the health behaviors. As such, states should be modeled as separate systems, which allows covariates (i.e., race or ethnicity relative to whites) to have unique effects on the outcome variables (i.e., probability of BC or CRC screening) across states. Independent modeling of each state's population in multilevel models, with person-level and community-level covariates, reveals the net effects of local area interactions between people and their environments for each state. Thus, we estimate multilevel models to generate population-level effect estimates that are allowed to vary across states, reflecting their local population and socioecological differences.

The conceptual model that we use here (Figure 1) draws from the literature and describes spatial interaction among people and characteristics of their contextual environments along the pathways to health care utilization (Mobley, Kuo, and Andrews 2008; Mobley et al. 2010). This is a hybrid model that incorporates the behavioral model of utilization (Aday and Andersen 1974) and spatial interactions in health care access and utilization (Khan and Bhardwaj 1994). In the model, each of the U.S. states represents a unique health care environment with unique and decentral-

ized cancer control programs that are state specific in funding and implementation. State health care environments are governed by local and regional politics, social systems, market-level forces that determine supply factors, and community or neighborhood-level forces that determine social factors. Individuals exhibit predisposing, enabling, and need characteristics, which interact with the forces in the broader system.

This conceptual model directs us to study state populations as separate systems and to examine racial or ethnic disparities in BC or CRC screening among each state's population in contexts that are relevant for each state. The separate, state-specific analyses are crucial for truly understanding disparities, because racial or ethnic disparities observed in studies conducted at the national level are due in part to geographic differences rather than to differences among racial or ethnic groups within geographic areas (Chandra and Skinner 2003). Because our goal is to disentangle true disparities within states from geographic ones across the nation, we focus on individual states in separate analyses.

Methods

Study Population

Our study population is the entire Medicare fee-for-service (FFS) population aged sixty-five or older and residing in the fifty U.S. states from 2001 through 2005. We used the traditional definition of Medicare FFS coverage (persons with both Parts A and B coverage). We defined an FFS Medicare cohort of persons aged sixty-five or older in 2001 and followed them for several years to assess whether they used screening for BC (in 2001–2003 and 2003–2005) or CRC (in 2001–2005). We used annual data from 100 percent of Medicare claims to record any endoscopy or mammography use by persons over these intervals. Persons included in the cohort must remain alive during the entire period, maintain coverage of Medicare Parts A and B, and remain living in the same state. In BC multilevel statistical models we included only females, whereas in CRC models we included both males and females. Table 1 provides sample statistics by state, where the number of cohort observations per state is reported.

Statistical Analysis

We used the generalized estimating equations (GEE) form of multilevel models in the analysis to obtain robust population-level effect estimates for each state

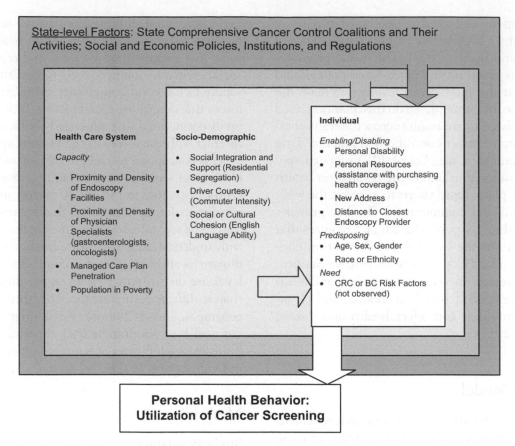

Figure 1. Socioecological model of factors impacting cancer screening.

from individual-level data with binary response (Liang and Zeger 1986; Horton and Lipsitz 1999; Hardin and Hilbe 2003; Gelman and Hill 2007). We separately and independently estimated cancer screening models for each state and each cancer site. All models included the same set of multilevel predictors (Table 2).

The neighborhood-community level is defined as the primary care service area (PCSA; Goodman et al. 2003), and the political system level is defined as the county. PCSAs are smaller and more numerous than counties (see Table 1) and might better represent local neighborhood conditions (Mobley, Kuo, and Andrews 2008). To demonstrate the deviation of state-specific estimates from a national benchmark, we needed a national estimate, which we derived by estimating a national-level model that pooled all observations across states to obtain national-level population estimates.

Statistical Significance

We used extracts from 100 percent Medicare FFS population data (not survey sample data), yielding the maximal sample size possible for each state. For each state, we tested whether the effect estimate of a minority group coefficient was significantly different (higher or lower) from zero and then displayed those that were higher (positive) versus lower (negative) versus no effect (zero) for each state.

False-positive results (statistically significant effect estimates that occur by chance and reflect Type 1 statistical errors) are common in the medical literature (Sainani 2009). We recognize that sampling error among small groups of Hispanics or African Americans in some states with small minority populations can create the appearance of variation in screening rates, even when none exists. In some states, the number of these individuals is relatively low and thus the power to detect a significantly different effect compared to whites might be low; it is therefore always possible that some findings are due to chance. To tightly control the probability of Type 1 errors, we used a 1 percent significance level for the tests.

False-positive results can also be hidden when researchers make multiple comparisons and do not properly account for the reduction in power that

Table 1. Cohort size in sample population, number of PCSAs and counties in each state, and U.S. totals

State	Number of persons in population cohort, CRC	Number of persons in population cohort, BC	Number of PCSAs	Number of counties	State	Number of persons in population cohort, CRC	Number of persons in population cohort, BC	Number of PCSAs	Number of counties
Alabama	319,335	238,401	144	67	Montana	79,539	53,217	71	56
Alaska	22,585	14,251	24	27	Nebraska	146,001	102,360	121	93
Arizona	223,305	155,084	74	15	Nevada	75,709	50,983	30	17
Arkansas	224,275	158,998	149	75	New Hampshire	95,298	66,428	46	10
California	1,126,335	781,832	338	58	New Jersey	567,836	415,796	139	21
Colorado	156,466	108,329	96	63	New Mexico	100,328	68,114	61	33
Connecticut	245,186	177,375	71	8	New York	1,040,451	780,421	324	62
Delaware	64,072	44,321	12	3	North Carolina	587,505	433,969	207	100
Florida	1,139,258	815,741	167	67	North Dakota	62,867	42,927	71	53
Georgia	464,828	338,081	169	159	Ohio	783,948	562,887	254	88
Hawaii	56,573	37,307	23	5	Oklahoma	248,870	177,463	156	77
Idaho	82,703	56,394	57	44	Oregon	151,816	104,809	78	36
Illinois	818,437	586,674	258	102	Pennsylvania	819,431	614,838	296	67
Indiana	471,278	335,827	172	92	Rhode Island	50,326	39,362	14	5
Iowa	274,939	197,168	225	99	South Carolina	308,796	222,581	110	46
Kansas	211,602	148,392	162	105	South Dakota	72,116	49,227	95	66
Kentucky	307,484	217,727	145	120	Tennessee	395,590	296,521	145	95
Louisiana	244,130	184,005	112	64	Texas	1,118,495	793,778	414	254
Maine	121,387	83,940	91	16	Utah	113,066	78,565	54	29
Maryland	346,573	248,298	62	24	Vermont	50,631	34,904	49	14
Massachusetts	362,711	270,039	107	14	Virginia	492,814	347,962	170	128
Michigan	765,461	533,627	191	83	Washington	314,345	214,054	119	39
Minnesota	314,019	225,862	176	87	West Virginia	162,307	114,791	123	55
Mississippi	211,398	155,479	141	82	Wisconsin	412,030	301,704	173	72
Missouri	387,278	278,696	213	115	Wyoming	37,384	24,669	41	23
Total United States						17,249,117	12,384,178	6,740	3,133

Note: BC = breast cancer; CRC = colorectal cancer; PCSA = primary care service area.

accompanies this practice. In our disparities research, we conducted several simultaneous tests to determine whether there were differences between various minority subgroups and whites (unless all minorities are grouped together in the model). Our model includes five subgroups (African American, Hispanic, Asian, Native American, and other) that we compared to whites in a series of tests for significance of coefficient estimates. To tightly control for Type 1 error, we accounted for this multiple-testing aspect of our model (Sainani 2009) using a Bonferroni correction. The Bonferroni correction

is a very conservative approach because it represents a worst case scenario where all of the tests being conducted are assumed to be completely independent (which is not likely to be the case for these subgroup comparisons). The Bonferroni correction is simple to conduct and to understand, and even if it is not the best approach, using a more stringent significance level helps weed out spurious results from states with small minority populations. Thus, we used the 1 percent level of significance in translating our findings via the maps, which reflects at least a 5 percent overall level of

Table 2. Variables included in multilevel regression

Factors in conceptual model	Variables in regression model
Individual	
Enabling/disabling	Moved to a new zip code within state, 2001–2005
	Months with state assistance to purchase Part B insurance
	Distance (miles) to closest screening facility
Predisposing	Age in 2001
	Gender (only in CRC models and using female as the reference group)
	Race or ethnicity (five groups relative to whites)
Sociodemographic factors (at PCSA level)	
Social integration and support	Residential segregation (isolation) index
Stressor, driver courtesy	Proportion of the workforce commuting $\geq$ 60 minutes to work
Social or cultural cohesion	Proportion of the people aged 65 or older who speak little or no English
Health system factors (at county area level)	
Capacity	Average number of screening facilities per 1,000 population aged 65 or older
	Number of oncologists per 1,000 population aged 65 or older
	Medicare managed care plan penetration
	Proportion of people living below the federal poverty level
	Proportion of the county population in rural tracts

Note: CRC = colorectal cancer; PCSA = primary care service area.

significance according to the Bonferroni correction approach ($0.05 = 0.01 \times 5$). Thus, the significance level for our tests is somewhere between 1 percent and 5 percent. As described next, we make only qualitative comparisons across states and between states and the nation by mapping the state-specific results to demonstrate differences across states in the state-specific population dynamics.

Spatial Translation of Findings from 100 Multilevel Regressions

After estimation, we compiled results for the race or ethnicity effect estimates for each state and cancer site. Each state's effect estimates for each minority group (African American and Hispanic) were classi-fied as follows: statistically significantly different from zero (at the 1 percent level of significance) and (a) with a positive value, or (b) with a negative value, or (c) not significantly different from zero. Using these, we then translated the disparities findings using three maps (Figures 2–4). We used a spatial join method in Arc-View 10.0 GIS software that combined the effect estimates for each cancer site and state, resulting in nine possible bivariate classes for the joined estimates. For example, one of these classes is defined as follows: a positive effect for minority group relative to whites for BC screening (horizontal legend) and also a positive effect for minority group relative to whites for CRC screening (vertical legend). All states meeting both of these criteria are classified in that cell of the map legend and colored light blue. No such states are evident for the African American disparities (Figure 2); however, for Hispanic disparities, one state (New Jersey) shows higher screening for both cancers among Hispanics relative to whites (Figure 3). For benchmarking, we display the national effect estimate for the covariate of interest in the map legend to help viewers understand which states are dominating national statistics.

This innovative spatial translation of GEE modeling results allows findings from 100 separate estimating equations to be compared visually across all states in one single graphic. This demonstrates the power of mapping to aid in visualizing vast quantities of information. For example, in Figures 2 and 3, we display disparities between two minority groups (African Americans and Hispanics) and whites for each state. In some states, we see a reversal of what we might expect based on national statistics. That is, minorities seem to display an advantage over whites in terms of screening utilization. We posit that these sorts of reverse disparities reflect successful state-specific initiatives designed to reduce disparities. In Figure 4, we display the Hispanic disparity estimate relative to whites for BC screening over two time intervals to investigate whether disparity outcomes for BC screening are changing over time. These sorts of temporal comparisons might also be useful in evaluating CCC efforts.

In summary, observed differences in map colors across individual states reflect different state environments and CCC efforts and other place-specific differences across states. Multistate patches of the same color reflect similar disparity environments in particular regions, reflecting geographic disparities. For example, patches of red color across multiple states suggest regional disparities in both of the dimensions displayed in the map.

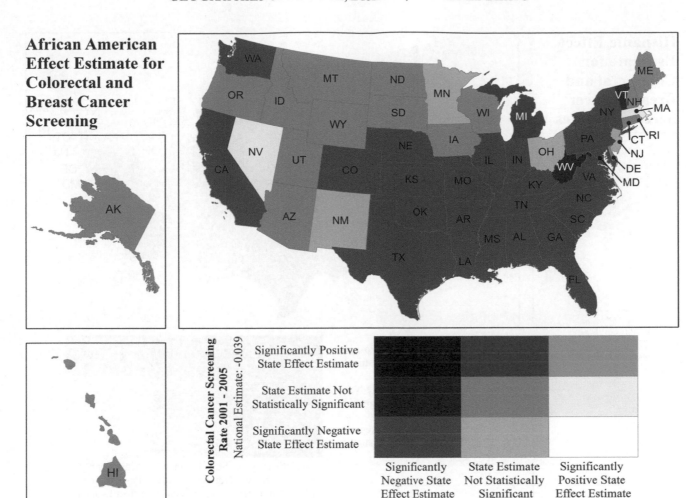

African American Effect Estimate for Colorectal and Breast Cancer Screening

Colorectal Cancer Screening Rate 2001 - 2005
National Estimate: -0.039

Significantly Positive State Effect Estimate

State Estimate Not Statistically Significant

Significantly Negative State Effect Estimate

| Significantly Negative State Effect Estimate | State Estimate Not Statistically Significant | Significantly Positive State Effect Estimate |

Breast Cancer Screening Rate 2003 - 2005
National Estimate: -0.032

Figure 2. Positive/negative effect estimates from 100 multilevel regressions: African American (relative to whites) colorectal or breast cancer screening. (Color figure available online.)

Findings

Disparities Estimates from Multilevel Models

Figure 2 displays effect estimates from our GEE multilevel models, focusing on the disparity between African Americans and whites in BC screening during 2003 to 2005 or CRC screening during 2001 to 2005. As shown in Figure 2, in Michigan, African American women have no significant differences in the probability of BC screening relative to whites ("state estimate not statistically significant" category), whereas African Americans have a significantly higher probability of CRC screening than whites ("significantly positive state effect estimate" category), resulting in a dark blue color classification. Michigan is the only state exhibiting this dichotomy. In seventeen states, African Americans have no significant differences in either type

of screening relative to whites, whereas in the majority of states, African Americans have a significantly lower probability of BC and CRC screening than whites. This explains the national statistic often reported that African Americans are less likely to be screened for BC or CRC than whites. Our results, however, suggest that in Michigan, this disparity in CRC screening has been reversed for African Americans.

Figure 3 displays effect estimates from the Hispanic indicator variable, relative to whites, on probability of cancer screening. It is constructed in the same way as Figure 2. In New Jersey, Hispanics have a higher probability of both BC and CRC screening than whites (light blue color, "significantly positive state effect estimate" in both dimensions). In twenty-one states, Hispanics have no significantly different probability of BC or CRC screening than whites (gray color). In the majority of states, Hispanics have significantly lower probability of

233

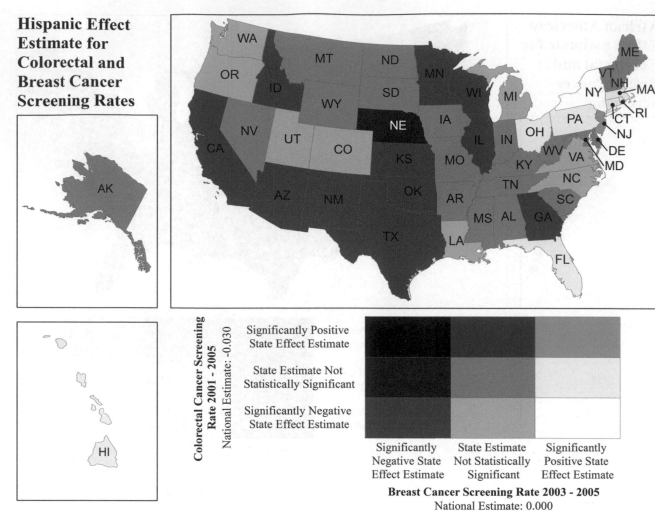

Hispanic Effect Estimate for Colorectal and Breast Cancer Screening Rates

Colorectal Cancer Screening Rate 2001 - 2005
National Estimate: -0.030

Significantly Positive State Effect Estimate

State Estimate Not Statistically Significant

Significantly Negative State Effect Estimate

Significantly Negative State Effect Estimate

State Estimate Not Statistically Significant

Significantly Positive State Effect Estimate

Breast Cancer Screening Rate 2003 - 2005
National Estimate: 0.000

Figure 3. Positive/negative effect estimates from 100 multilevel regressions: Hispanic (relative to whites) colorectal or breast cancer screening. (Color figure available online.)

BC and CRC screening than whites (red color). This explains the national statistic often reported that Hispanics are less likely to be screened for BC or CRC than whites. Our results, however, suggest that in New Jersey, this disparity in BC and CRC screening has been reversed for Hispanics.

Figure 4 displays effect estimates from the Hispanic indicator, relative to whites, on probability of BC screening over time, using bivariate mapping to compare estimates from an early period (2001–2003) and late period (2003–2005). The national estimate finds no statistically significant effect of being Hispanic on the probability of BC screening in either period, suggesting no disparity in BC screening for Hispanics relative to whites in either period. The state-specific analysis demonstrates that what seems to be the case nationally does not necessarily hold across the states. Although the majority of states exhibit no disparities, the map

shows that in six states (New York, Pennsylvania, Massachusetts, Connecticut, New Jersey, and Florida), Hispanic women were more likely than whites to have BC screening in both periods. Conversely, in eleven states, Hispanics were significantly less likely than whites to receive BC screening in both periods (red). Seven states showed changes in disparities over time, with disparities increasing over time in only one state (Idaho) and decreasing in the remainder (Ohio, Hawaii, Rhode Island, Michigan, Oregon, and Tennessee). No states experienced a complete reversal (black) from negative to positive disparities over time (relative to whites).

Discussion

In this study, we used a socioecological, spatial interaction framework to guide the multilevel statistical

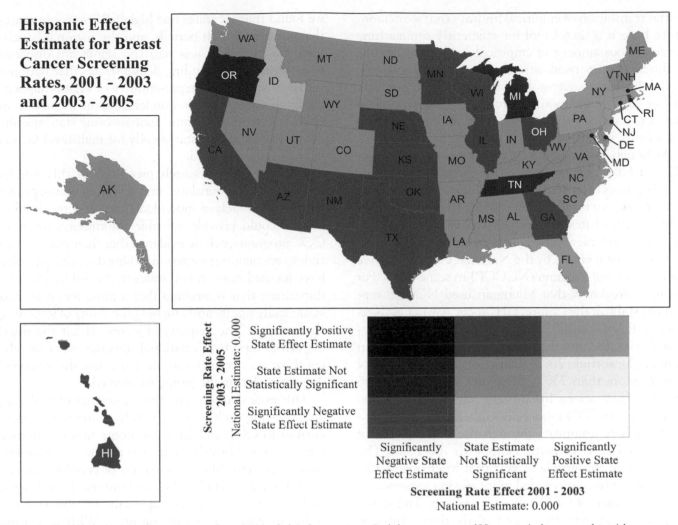

Hispanic Effect Estimate for Breast Cancer Screening Rates, 2001 - 2003 and 2003 - 2005

Figure 4. Positive/negative effect estimates from 100 multilevel regressions: Stability over time of Hispanic (relative to whites) breast cancer screening. (Color figure available online.)

modeling of minority disparities in BC and CRC screening. The study sample is based on 100 percent FFS Medicare data and thus is fully representative of this population in all states, providing large samples that allow for robust characterization of disparities between minorities and whites in most states. No other source of BC or CRC screening data exists to date that can be used for this purpose. Sample data, such as the Behavioral Risk Factor Surveillance System, cover a broader range of adults but sparsely represent population subgroups within states. In the future, with standardized medical records and greater national health security from increased availability of competitive health insurance, population-based analyses are expected to become more prevalent. This article demonstrates the advances in disparities research that are possible with availability of population health (vs. survey sample) data.

We used innovative spatial translation to present the study findings in a series of bivariate maps, which are useful for making comparisons of state-specific disparity environments across states and for revealing patterns in these disparity environments that reflect broader regional geographic disparities. Each state is treated as a separate environment, and each state's estimates describe behaviors in that state's FFS Medicare population. Mapping the findings for each state together in a single map of the United States enables comparisons of the disparity dynamics across the state environments, showing geographic differences in the disparities between minorities and whites. We provide a national estimate, derived from a pooled model, to use as a benchmark to demonstrate which states are driving the national estimate. We demonstrate that many states deviate from the national estimate, calling into question the utility of such a statistic for CCC efforts. The spatial translation of empirical findings that we demonstrate here is a useful tool for efficiently summarizing a tremendous amount of empirical evidence, allowing qualitative comparisons across states and over time. Observed differences across states can be used to motivate further CCC evaluation research that seeks to explain why some states exhibit disparities among minorities and whites, whereas others do not.

Although the majority of the states showed lower BC and CRC screening rates for African Americans and Hispanics than whites, consistent with the national benchmark, our findings demonstrate that minority disparities with whites are reversed in some states, such as Michigan and New Jersey. It is possible that this reflects successful efforts by the National Comprehensive Cancer Control Program (NCCCP) in some states. For example, we know that Michigan used NCCCP support to establish the Colorectal Cancer Awareness Network (CRAN) in 2002 with a mission to raise awareness about CRC and the need for screening (Michigan Cancer Consortium 2002). During its first year, CRAN grew to more than 230 participants, representing 145 unique organizations from every region of the state. New efforts in 2004 more fully embedded CRAN into Michigan's communities through the development of regional CRANs in partnership with the ACS. The findings from this article suggest that Michigan's CCC efforts might have succeeded in increasing awareness of the importance of CRC cancer screening. The state's African Americans were more likely than whites to utilize CRC screening, suggesting that the efforts to promote screening at the church and community levels reaped positive rewards for this minority group.

The finding that Hispanics in New Jersey are significantly more likely than whites to receive both types of screening is puzzling, as we found no description of CCC efforts that could help explain this phenomenon (New Jersey Comprehensive Cancer Control Plan 2002). Our examination of census data from the year 2000 show that Hispanics made up 13 percent of New Jersey's population; 33 percent of Hispanics in New Jersey were from Puerto Rico, and another 16 percent were from South America, with only 9 percent from Mexico and 7 percent from Central America. Perhaps Hispanics in New Jersey are on average wealthier or better educated; perhaps they are more firmly established over several generations and have greater social cohesion and support than Hispanics in other areas. Further research is needed to better understand this reverse disparity.

When looking at the change in BC screening rates over time for Hispanics relative to whites (see Figure 4),

we found that six states had higher BC screening rates than whites in both periods, and the national benchmark for Hispanics was zero (no significant disparity from whites). This finding demonstrates that national statistics are not very representative of what is happening in all states. To better understand the impacts from CCC efforts in each state, analysis using state-specific data and controlling statistically for multilevel factors in each state is needed.

The geospatial research presented in this article, which spatially translates findings from state-specific multilevel models of individuals' cancer screening behavior, could provide valuable information for state CCC programs seeking evidence that their program activities are making progress in closing disparity gaps. We have focused here on two states that exhibited reverse disparities; that is, findings that a minority group had statistically significantly higher probability of screening than whites. The majority of states exhibit the usual disparities reported in national statistics, whereas others show no disparities at all and a few show reverse disparities during the period we studied.

This work demonstrates the importance of analyzing and assessing each state separately. States are separate entities in CCC and have autonomy to set insurance regulation and health promotion policies. In addition, states represent different mixtures of peoples and cultures, have different baseline and historical conditions, and are so heterogeneous that it calls into question the utility of national average statistics. With increasing availability of population data and computing capabilities, we advocate use of population data and a focus on individual states, so that meaningful interventions to improve cancer screening behavior can be implemented and evaluated.

Acknowledgments

This work was supported by a National Cancer Institute grant (1R01CA126858) and an American Recovery and Reinvestment Act supplement to it. This article was completed while Tzy-Mey Kuo was at RTI International; she is now at the University of North Carolina at Chapel Hill. The content is solely the responsibility of the authors and does not necessarily represent the official views of RTI International, Arizona State University, the University of North Carolina at Chapel Hill, the National Cancer Institute, or the National Institutes of Health.

References

Aday, L. A., and R. Andersen. 1974. A framework for the study of access to medical care. *Health Services Research* 9:208–20.

Administration on Aging. 2006. *Federal interagency forum on aging-related statistics: Older Americans update 2006: Key indicators of well-being.* Washington, DC: U.S. Government Printing Office. http://www.aoa.gov/agingstatsdotnet/Main_Site/Data/2006_Documents/OA_2006.pdf (last accessed 10 July 2011).

Affordable Care Act. 2010. The Affordable Care Act of 2010. http://www.healthcare.gov/law/introduction/index.html (last accessed 10 July 2011).

American Cancer Society. 2010. *Cancer facts and figures 2010.* Atlanta, GA: American Cancer Society. http://www.cancer.org/acs/groups/content/@epidemiologysurveilance/documents/document/acspc-026238.pdf (last accessed 10 July 2011).

———. 2011. *Cancer prevention & early detection facts & figures 2011.* Atlanta, GA: American Cancer Society. http://www.cancer.org/Research/CancerFactsFigures/CancerPreventionEarlyDetectionFactsFigures/ACSPC-029459 (last accessed 10 July 2011).

ArcView, version 10.0 [software]. Redlands, CA: ESRI.

Chandra, A., and J. Skinner. 2003. Geography and racial health disparities. NBER Working Paper No. W9513, National Bureau of Economic Research, Cambridge, MA.

Cooper, G. S., and S. M. Koroukian. 2004. Geographic variation among Medicare beneficiaries in the use of colorectal carcinoma screening procedures. *American Journal of Gastroenterology* 99:1544–50.

Coughlin, S. S., M. E. Costanza, M. E. Fernandez, K. Glanz, J. W. Lee, S. A. Smith, L. Stroud, I. Tessaro, J. M. Westfall, J. L. Weissfeld, and D. S. Blumenthal. 2006. CDC-funded intervention research aimed at promoting colorectal cancer screening in communities. *Cancer* 107 (5 Suppl.): 1196–1204.

Edwards, B. K., E. Ward, B. A. Kohler, C. Eheman, A. G. Zauber, R. Anderson, A. Jemal, et al. 2010. Annual report to the nation on the status of cancer, 1975–2006, featuring colorectal cancer trends and impact of interventions (risk factors, screening, and treatment) to reduce future rates. *Cancer* 116 (3): 544–73.

Gelman, A., and J. Hill. 2007. *Data analysis using regression and multilevel/hierarchical models.* New York: Cambridge University Press.

Goodman, D. C., S. S. Mick, D. Bott, T. Stukel, C. H. Chang, N. Marth, J. Poage, and H. J. Carretta. 2003. Primary care service areas: A new tool for the evaluation of primary care services. *Health Services Research* 38:287–309.

Hardin, J., and J. Hilbe. 2003. *Generalized estimating equations.* New York: Chapman & Hall/CRC.

Heron, M., D. L. Hoyert, S. L. Murphy, X. Jiaquan, K. D. Kochanek, and B. Tejada-Vera. 2009. Deaths: Final data for 2006. *National Vital Statistics Report* 57 (14). Hyattsville, MD: National Center for Health Statistics. http://www.cdc.gov/nchs/data/nvsr/nvsr57/nvsr57_14.pdf (last accessed 10 July 2011).

Hetzel, L., and A. Smith. 2001. The 65 years and over population: 2000. Census 2000 brief C2KBR/01-10, U.S. Census Bureau, Washington, DC.

Horton, N., and S. Lipsitz. 1999. Review of software to fit generalized estimation equation regression models. *The American Statistician* 53:60–169.

Jemal, A., R. Siegel, E. Ward, Y. Hao, J. Xu, and M. Thun. 2009. Cancer statistics, 2009. *CA: A Cancer Journal for Clinicians* 59:225–49.

Khan, A. A., and S. M. Bhardwaj. 1994. Access to health care: A conceptual framework and its relevance to health care planning. *Evaluation & the Health Professions* 17:60–76.

Liang, K., and S. Zeger. 1986. Longitudinal data analysis using generalized linear models. *Biometrika* 73:13–22.

Michigan Cancer Consortium. 2002. Statewide Colorectal Cancer Awareness Network, 2002–2004. http://www.michigancancer.org/PDFs/ColorectalCancerProjects/StatewideColoCaAwareNetwork-2002–2004/SummaryDocument.pdf (last accessed 10 July 2011).

Mobley, L. R., T. Kuo, and L. S. Andrews. 2008. How sensitive are multilevel regression findings to defined area of context? A case study of mammography use in California. *Medical Care Research and Review* 65:315–37.

Mobley, L., T. Kuo, M. Urato, and S. Subramanian. 2010. Community contextual predictors of endoscopic colorectal cancer screening in the USA: Spatial multilevel regression analysis. *International Journal of Health Geographics* 9:44.

Mueller, K., C. MacKinney, M. Gutierrez, and J. Richgels. 2011. Place based policies and public health: The road to healthy rural people and places. Policy paper, Rural Policy Research Institute, Columbia, MO. http://www.rupri.org/Forms/HHSPanels_Integration_March2011.pdf (last accessed 10 July 2011).

Naishadham, D., I. Lansdorp-Vogelaar, R. Siegel, V. Cokkinides, and A. Jemal. 2011. State disparities in colorectal cancer mortality patterns in the United States. *Cancer Epidemiology, Biomarkers & Prevention* 20 (7): 1296–1302.

National Cancer Institute. 2007. NCI SEER cancer statistics review 1975–2007. http://seer.cancer.gov/csr/1975_2007/browse_csr.php?section=1&page=sect_01_table.25.html (last accessed 10 July 2011).

Nelson, D. E., J. Bolen, S. Marcus, H. E. Wells, and H. Meissner. 2003. Cancer screening estimates for U.S. metropolitan areas. *American Journal of Preventive Medicine* 24 (4): 301–309.

New Jersey Comprehensive Cancer Control Plan. 2002. New Jersey Comprehensive Cancer Control Plan, 2002. http://www.nj.gov/health/ccp/ccc_plan.htm (last accessed 10 July 2011).

Office of Management and Budget. August 2009. Memorandum: Developing effective place-based policies for the FY 2011 budget. http://www.whitehouse.gov/omb/assets/memoranda_fy2009/m09-28.pdf (last accessed 10 March 2010).

Richardson, L., E. Tai, S. Rim, D. Joseph, and M. Plescia. 2011. Vital signs: Colorectal cancer screening, incidence, and mortality—United States, 2002–2010. *Morbidity and Mortality Weekly Report* 60 (26): 884–89.

http://www.cdc.gov/mmwr/preview/mmwrhtml/mm6026 a4.htm?s_cid=mm6026a4_w (last accessed 10 July 2011).

Sainani, K. 2009. Statistically speaking: The problem of multiple testing. *American Academy of Physical Medicine and Rehabilitation* 1:1098–1103.

Schneider, K. L., K. L. Lapane, M. A. Clark, and W. Rakowski. 2009. Using small-area estimation to describe county-level disparities in mammography. *Preventing Chronic Disease* 6 (4). http://www.cdc.gov/pcd/ issues/2009/oct/08_0210.htm (last accessed 10 July 2011).

Spatial Access and Local Demand for Major Cancer Care Facilities in the United States

Xun Shi,* Jennifer Alford-Teaster,† Tracy Onega,‡ and Dongmei Wang*

*Department of Geography, Dartmouth College
†New England Center for Emergency Preparedness, Dartmouth Medical School
‡Department of Community & Family Medicine, Dartmouth Medical School

The Cancer Centers designated by the National Cancer Institute (NCI Centers) and academic medical centers (AMCs) form the "backbone" of the cancer care system in the United States. We conducted a nationwide analysis and generated a high-resolution map detailing spatial variation in the potentially unfulfilled demand for these facilities. A local demand value incorporates spatial access to the facilities and the number of local potential patients. The spatial access was estimated using the two-step floating catchment area method, taking into account both travel time and facility capacity. The travel time was measured using service-area rings created around each facility based on road networks. The facility capacity was measured as the ratio between the bed count of the facility and the number of potential patients in its three-hour catchment. The number of local potential patients was estimated from local demography and standard cancer rates. The demographic information is a combination of LandScan data and U.S. Census data, and the cancer rates are from the Surveillance Epidemiology and End Results. The final demand map shows distinctive patterns in the western and eastern halves of the contiguous United States. The demand in the east is spatially continuous but relatively low, whereas in the west it is sporadic but tends to have high values. We also examined the inherent relationships between several methods for measuring spatial access and found that the differences between them are technical rather than conceptual, which sets a theoretical basis for selecting and adapting those methods.

由国家癌症研究所（NCI 中心）和学术医疗中心（AMCs）指定的癌症研究中心，形成了美国癌症保健系统的"支柱。"我们总结了一个全国范围的分析，并生成一个高分辨率的地图，详细说明对这些设施的潜在的，未实现的需求的空间差异。本土的需求值结合考虑对设施的空间访问度和当地潜在的患者人数。该空间访问度是通过使用两步浮动集水区法，同时考虑到旅行时间和设施的容量估计出来的。通过测量基于道路网络的设施周围的服务环区，得出旅行时间。通过计算三个小时的集水区里这些设施的床位数，和潜在的患者人数之间的比例，来测量设施容量。从当地人口和标准的癌症发病率，估计出本地潜在患者人数。人口信息是综合 LandScan 数据和美国人口普查数据得来，癌症发病率是从流行病学和对最终结果的监测得来。最终的需求地图显示了美国本土的西部和东部的独特模式。东部的需求在空间上是连续的，但是相对较低，而在西部，它是零散但往往具有较高值的。我们还研究了测量空间访问度的几种方法之间的内在关系，并发现它们之间的差异是技术上的，而不是概念上的，这给选择和适应这些方法设置了一个理论依据。关键词：癌症，浮动的集水区，保健，核密度估计，空间的访问。

Los Centros del Cáncer, designados por el Instituto Nacional de Cáncer (Centros NCI) y los centros médicos académicos (AMCs), constituyen la "columna vertebral" del sistema de atención por cáncer en los Estados Unidos. En el artículo llevamos a cabo un análisis de alcance nacional y generamos un mapa de alta resolución en el que se detalla la variación espacial en la demanda potencialmente no satisfecha de estas facilidades. Un valor de demanda local incorpora el acceso espacial a los servicios y el número potencial de pacientes locales. El acceso espacial fue calculado mediante el uso del método de área de demarcación flotante de dos etapas, tomando en consideración tanto el tiempo de viaje como la capacidad de la instalación. El tiempo de viaje se midió utilizando los anillos de área de servicio creados alrededor de cada instalación con base en la red de carreteras. La capacidad de la instalación se midió como la razón entre el número de camas de la instalación y el número de pacientes potenciales en su área de demarcación de tres horas. El número de pacientes potenciales locales se estimó a partir de la demografía local y las tasas estándar de incidencia de cáncer. La información demográfica es una combinación de datos LandScan y datos del Censo de los EE.UU., en tanto que las tasas de cáncer son las de Vigilancia Epidemiológica y Resultados Finales. El mapa final de demanda muestra patrones distintivos en las mitades occidental y oriental de los Estados Unidos contiguos. La demanda en el este es espacialmente continua aunque relativamente baja, mientras que en el oeste es esporádica pero con tendencia a acusar valores altos. Examinamos también las relaciones inherentes entre varios métodos para medir el acceso espacial y encontramos que las diferencias entre ellos son más técnicas que conceptuales, lo cual provee una base teórica para seleccionar y adaptar tales métodos.

Access to health care services received unprecedented attention within the context of the recent health care reform in the United States. Two different notions of accessibility have been distinguished: *place accessibility*, primarily determined by physical separation, and *individual accessibility*, involving adjustments to that separation by individual characteristics (Kwan 1998, 1999). This article focuses on place accessibility, which is also referred to as *geographic access* or *spatial access* in the literature (e.g., Guagliardo 2004; Onega et al. 2008; F. Wang et al. 2008).

Spatial access can be further analyzed into *accessibility* and *availability* (Joseph and Phillips 1984; Guagliardo 2004; F. Wang and Luo 2005), which essentially correspond to the relative and integral measures of access in the literature (Morris, Dumble, and Wigan 1979; Kwan et al. 2003), respectively. *Accessibility* refers to the impedance in patients' travel to services, and *availability* can be understood as the number of facilities from which the patient can choose (Guagliardo 2004). Guagliardo (2004) stressed that the two should be considered simultaneously.

As cancer is now the leading cause of death in the United States for those aged younger than eighty-five years (Twombly 2005), access to cancer care has become a focal issue in both research and policymaking (Onega et al. 2008). For highly specialized services as found in cancer care, which are typically characterized by limited geographic distribution, spatial access might be particularly important to patients' utilization of the services (Onega et al. 2008). Evidence shows that associations between travel time to cancer care services and risk of advanced cancer (e.g., Gumpertz et al. 2006), utilization of certain therapy (e.g., Celaya et al. 2006), and enrollment in clinical trials (Avis et al. 2006).

In the United States, the National Cancer Institute (NCI) designates a number of facilities as the NCI Cancer Centers, which "are characterized by scientific excellence and the capability to integrate a diversity of research approaches to focus on the problem of cancer" (National Cancer Institute 2010). The association between NCI Center attendance and reduction in cancer mortality has been established (Onega et al. 2010a,

2010b). In addition to the NCI Centers, other high-volume hospitals, especially academic medical centers (AMCs), defined as either independent or integrated with medical schools and being members of the Council of Teaching Hospitals, also play a major role in providing cancer care services (Birkmeyer et al. 1999; Onega et al. 2008, 2010a, 2010b). The NCI Centers and AMCs (the two have considerable overlap) form the "backbone" of the cancer care system in the United States.

Onega, Duell, Shi, Demidenko, and Goodman (2009) reported an 11 percent decreased likelihood of attendance to the NCI Centers for every ten minutes of additional travel time, along with considerable disparities in terms of geographic region, race and ethnicity, rurality, and types of cancers (Onega et al. 2008; Onega, Duell, Shi, Demidenko, and Goodman 2009; Onega, Duell, Shi, Demidenko, Gottlieb, et al. 2009; Onega et al. 2010a, 2010b). In these works, however, spatial access is measured simply as the travel time from the patient's location to the nearest facility, with the patient's location represented by the population weighted centroid of the ZIP code polygon. Although commonly used in health care studies, this measurement does not fully capture availability. In addition, the centroid of the ZIP code polygon (however defined) can be too coarse a representation of a patient's location, especially in rural areas where such polygons tend to be large.

A more sophisticated method for evaluating spatial access is the two-step floating catchment area (2SFCA) procedure. Its general idea is to first (step 1) calculate the service that a facility can provide to its catchment and then (step 2) for each individual location sum up the services from all the facilities whose catchments cover that location. In 2SFCA, accessibility is represented by the travel impedance in the calculation of the service from a facility to a given location, and availability is measured when services from the accessible facilities are summed up at a location. Luo and Wang (2003) first applied this method to a study of primary health care in the Chicago region and have since been followed by others (F. Wang and Luo 2005; Yang, Goerge, and Mullner 2006; L. Wang 2007; Cervigni et al. 2008; F. Wang et al. 2008). Luo and Wang (2003) proved that 2SFCA is a special case of the gravity model (e.g., Hansen 1959; F. Wang and Minor 2002). They also noticed the connection between 2SFCA and the kernel density estimation (KDE) that was later applied to health care studies by Guagliardo et al. (2004) and McLafferty and Grady (2004). In the next section, we discuss the inherent relationships between these methods under different names. All of these studies, however, are at the regional level and thus might suffer from the *edge effect*; that is, the catchment of a facility might unreasonably stop at the region boundary. Also, most of these studies use highly aggregated and thus low-precision data at the ZIP code or census tract level.

This study goes beyond spatial access estimation and intends to quantify spatial variation in the potential demand for major cancer care facilities. We define the local potential demand as a combination of the local spatial access and the number of local potential cancer patients, which is inspired by a study of McLafferty and Grady (2004) on special prenatal care services. Our study is nationwide and includes all of the NCI Centers and AMCs in the contiguous United States ($N = 290$), which significantly reduces the edge effect. In addition, we maximized the utilization of the publicly available data and worked at the pixel rather than the ZIP code or census tract level.

Two-Step Floating Catchment Area, Gravity Model, and Kernel Density Estimation

In addition to Luo and Wang's (2003) deduction that 2SFCA is a special case of the gravity model, we demonstrate that the gravity model itself belongs to the more general KDE over inhomogeneous backgrounds. Within the context of this study, 2SFCA can be represented as follows (adapted from Luo and Qi 2009):

$$\text{Step 1}: \quad R_k = \frac{S_k}{\sum_{j \in (t_{k,j} < T)} p_j W_{k,j}} \tag{1}$$

where R_k is a measurement of potential service intensity of facility k, originating from the classic doctor–patient ratio; S_k is k's service capacity; p_j is the number of potential cancer patients at location j; $W_{k,j}$ is a weight determined by $t_{k,j}$, the travel impedance between k and j; T is a threshold that defines k's catchment; and $j \in (t_{k,j} < T)$ indicates that only the patients within k's catchment will be counted.

$$\text{Step 2}: \quad A_i = \sum_{k \in (t_{i,k} < T)} R_k W_{i,k} \tag{2}$$

where A_i is the spatial access at location i; $W_{i,k}$ is a weight determined by $t_{i,k}$, the travel impedance between i and k; and $k \in (t_{i,k} < T)$ indicates that only the services from those facilities accessible to i will be summed up.

The gravity model is a combination of Equations 1 and 2 (adapted from Luo and Qi 2009):

$$A_i = \sum_{k \in (t_{i,k} < T)} \frac{S_k W_{i,k}}{\sum_{j \in (t_{k,j} < T)} p_j W_{k,j}} \tag{3}$$

It should be noted that using T as a threshold makes Equation 3 a simplified version of the general gravity model. In the general model, T is a parameter in the functions for determining $W_{i,k}$ and $W_{k,j}$.

The gravity model can then be fitted into the framework proposed by Shi (2009, 2010) for sorting methods of performing KDE over an inhomogeneous background. The framework includes four KDE estimators, which are combinations of two types of bandwidths (fixed vs. adaptive) and two types of adjustments (supply-side vs. demand-side, termed as case-side vs. site-side in Shi's work). The adjustments, performed by the denominators in Equations 1 and 3, are for addressing an inhomogeneous background. Within the context of this study, a *supply-side* approach sets the kernel around cancer care facilities and performs the adjustment on the facility side; that is, it counts patients around the facilities. Alternatively, the kernel can also be set around patients' locations and such a *demand-side* method counts patients around their locations. Adjustments on different sides could generate quite different results, which has been noted by Luo and Wang (2003). Essentially, the difference between the supply-side and demand-side methods can be represented as

$$\underset{\text{(supply — side)}}{\sum \frac{f(S)}{\sum g(p)}} \quad \text{vs.} \quad \underset{\text{(demand — side)}}{\frac{\sum f(S)}{\sum g(p)}} \tag{4}$$

where S indicates supplies and p indicates demands, and $f(.)$ and $g(.)$ are general function forms. In terms of Shi's framework, the gravity model is a supply-side-fixed-bandwidth estimator. T in Equations 1 through 3 is eventually the bandwidth in KDE. It is fixed, as being a specified constant threshold. As a comparison, the (one-step) floating catchment area (FCA) method used by Luo (2004) is a demand-side-fixed-bandwidth estimator.

From the KDE perspective, a fundamental assumption of 2SFCA becomes clear: Number of nearby facilities and distance to a facility can compensate each other. In other words, availability and accessibility are mutually compensating. Based on 2SFCA's density nature, we can also infer that (1) different functions for determining distance decay weight (i.e., function for $W_{i,k}$ in Equations 2 and 3) will not significantly impact the spatial pattern of the estimates, which is a well-accepted conclusion in the KDE study (Silverman 1986); and (2) bandwidth T in Equations 1, 2, and 3 is a critical parameter (Silverman 1986; Shi 2010).

The revelation of the inherent relationships between the methods under different names leads to the understanding that their differences are technical rather than conceptual, which in turn might lead to more informed selection, better design, and further development of those methods. For example, different selections of the bandwidth, including its magnitude and type, could result in different methods that fit into diverse situations and problems.

Estimation of Potential Local Demand for Major Cancer Care Facilities

The local potential demand estimated in this study incorporates two primary factors: spatial access to cancer care facilities and spatial distribution of potential patients. We assume that the poorer the spatial access and the more patients at a location, the higher the demand for the service, which can be represented as follows:

$$D_i = p_i / A_i \qquad (5)$$

where D_i is the potential demand at location i, and p_i and A_i are the number of patients and the spatial access at i, respectively. This definition is simplistic, but it can be a preliminary step toward a deeper understanding and reasonable estimation of this quantity. Also, this definition measures the potentially unfulfilled demand (i.e., insufficiency) under existing facilities, which might serve the purpose of evaluating the locations, coverage, and spatial distribution of existing facilities and planned new facilities. In this study, A_i in Equation 5 was estimated using 2SFCA:

$$D_i = p_i \bigg/ \sum_{k \in (t_{i,k} < T)} \frac{S_k W_{i,k}}{\sum_{j \in (t_{k,j} < T)} p_j W_{k,j}} \qquad (6)$$

Estimation of Number of Potential Patients at a Location (p)

This estimation integrated three data sets, namely, LandScan, census, and national, standardized cancer rates, and generated a raster layer. The LandScan and census data were used to derive the population in each age, sex, and race and ethnicity category in a cell of the raster. The number of potential patients in each category in a cell was then calculated by multiplying the population in that category by the corresponding cancer rate. The final estimate is the sum of patients from each category.

LandScan, generated by the Oak Ridge National Laboratory through sophisticated modeling (Bhaduri et al. 2002), is raster data of population distribution throughout the world. In this study, we used LandScan 2005, the most updated version when this research began. After projection to North America Albers Equal Area, the resolution of the data set is 843 m, and this resolution was applied to all the other data layers and the resulting maps. The main reason for using LandScan was to obtain higher spatial resolution in rural areas where census blocks tend to be large. LandScan only has the total number of people for each cell, however, and we had to obtain the demographic information from the census data. Specifically, for each LandScan cell we calculated proportion of the people in an age, sex, and race and ethnicity category to the total population based on the data of the census blocks that intersect with that cell. If a polygon only partially fell within the cell, its population was split based on its area within the cell. The estimate of the number of people in an age, sex, and race and ethnicity category in a cell was then calculated as the product of the total number of people from the LandScan and the corresponding proportion from the census data.

The cancer prevalence rate for each age, sex, and race and ethnicity category came from the Estimated Cancer Prevalence Table published by Surveillance Epidemiology and End Results (SEER) Cancer Registry (NCI 2001). The rates in the table are the national averages. We understand that there is considerable geographic variation in cancer rates, but we were not able to find prevalence data with higher resolution that cover the entire contiguous United States.

Estimation of Facility's Service Capacity (S)

We used bed counts to represent a facility's capacity. Although the bed counts of the entire medical center or hospital might not be an accurate representation of its capability for cancer care, to our knowledge the two have a close relationship and bed counts are the best data available to us that are relatively consistent across facilities. Data of the number of staffed hospital beds were obtained from the American Hospital Directory (2011) and Web sites of individual hospitals.

Estimation of the Weights Determined by Travel Impedance (W)

$W_{k,j}$ in Equation 1 is the weight of the patients at location j in the calculation of the service-to-patient ratio. In this study, we assigned 1 to $W_{k,j}$, because a facility is supposed to serve all potential patients in its catchment and a patient farther from the facility should not be considered less important, as long as the facility is within an accessible distance. We used a three-hour travel time to define the catchment of a facility (i.e., $T = 3$ hours) and created the catchment using the Service Area function in ArcGIS, version 10 (ESRI, Redlands, CA). Three hours of travel time (one way) is likely to be the limit for a patient to travel to and from the facility (six hours) and obtain services in one day. The sum of patients within each catchment (i.e., $\sum_{j \in (t_{k,j} < T)} p_j$ in Equation 1) was calculated by sending the service area polygons and the patient layer described earlier into the Zonal Statistics tool in ArcGIS.

$W_{i,k}$ in Equation 2 represents the accessibility of facility k to the patients at location i, and should vary along with the travel impedance. For its estimation, we created six *service area rings* around each facility using thirty minutes as the interval. For example, the innermost ring defines the area in which travel to the facility takes 0 to 30 minutes, and the outermost ring defines the area in which travel to the facility takes 2.5 to 3 hours. The values for $W_{i,k}$ were calculated using a Gaussian function:

$$W_X = e^{(\frac{X-1}{7})^2 \ln(0.01)} \quad (X = 1, 2, \ldots, 6) \quad (7)$$

where W_X is the weight for ring X and e is the natural base (2.71828 ...), and we numbered the service area rings from inner to outer as 1, 2, 3, ... 6. Essentially, Equation 7 is the kernel function in KDE, and thus various choices for this function should have little effect on the resulting pattern (Silverman 1986). We proved this by our empirical testing with different functions and parameter values. We chose a Gaussian function because it is widely used to define bell-shaped curves. The specific parameter values in Equation 7 are for making the

Table 1. A lookup table for converting travel time to weight using a Gaussian function

Travel time	Weight
0–30 Minutes	1.00
30–60 Minutes	.91
60–90 Minutes	.69
90–120 Minutes	.43
120–150 Minutes	.22
150–180 Minutes	.10
> 180 Minutes	.015 (subjectively specified)

innermost ring have weight 1 and the outermost ring 0.1. The weights of all six service area rings are listed in Table 1.

Calculation of Final Demand Value

The raw values from Equation 6 vary from 0 to more than 41 million. To facilitate visualization and interpretation, we applied logarithmic transformation to the raw values:

$$D'_i = \log_{10} \left(p_i \bigg/ \sum_{k \in (t_{i,k} < T)} \frac{S_k W_{i,k}}{\sum_{j \in (t_{k,j} < T)} p_j W_{k,j}} \right) \quad (8)$$

Results and Discussion

Figures 1, 2, and 3 are maps of potential cancer patients (i.e., the numerator in Equation 6), the local spatial access to the major cancer care facilities (i.e., the denominator in Equation 6), and the local potential demand for the service of the facilities (i.e., the output from Equation 8), respectively. These maps show two distinctive patterns in the contiguous United States, and the boundary between them runs through those states in the middle of the continent, including North Dakota, South Dakota, Nebraska, Kansas, Oklahoma, and Texas, almost perfectly dividing the contiguous United States into two halves. In the rest of this section, we refer them as the western half and the eastern half.

Figure 1 shows that the distribution of potential cancer patients is largely continuous and the values are relatively high in the eastern half, whereas in the western half the patients are generally scattered and sparse. Concentrations are around major cities across the continental United States. Figure 2 shows that the eastern half

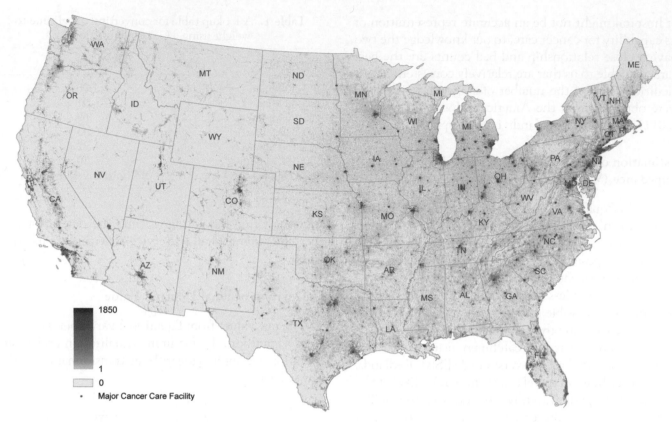

Figure 1. Spatial distribution of potential cancer patients in the contiguous United States. (Color figure available online.)

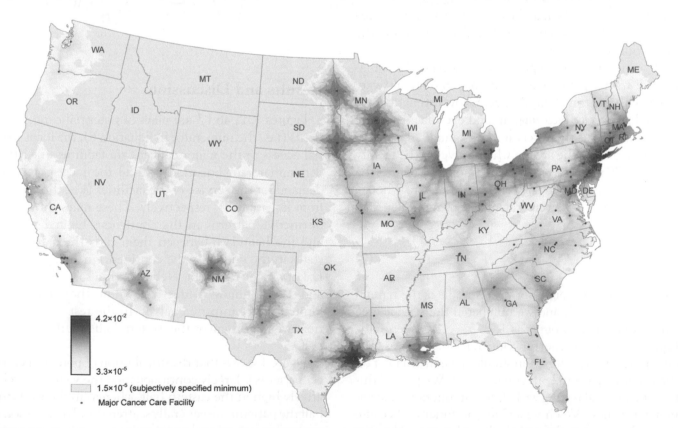

Figure 2. Spatial access to major cancer care facilities in the contiguous United States. (Color figure available online.)

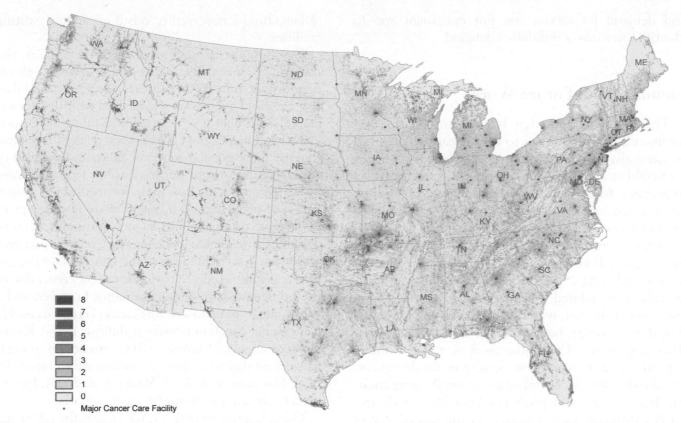

Figure 3. Demand for major cancer care facilities in the contiguous United States (logarithmically transformed value). (Color figure available online.)

is almost completely covered by the services, with only a few "islands," whereas in the western half the covered areas become "islands." Nevada, Idaho, Montana, and Wyoming, and dominant proportions of North Dakota, South Dakota, and Nebraska have almost no coverage at all.

Although Figure 3 was created based on Figures 1 and 2, it reveals new information. The pattern in Figure 3 largely follows that in Figure 1, because whether Figure 3 has a zero or nonzero value at a location is determined by its corresponding value in Figure 1. Although in Figure 3 the eastern half is also featured by continuity, the demand value is generally low. In the western half, except the West Coast, although the demand is sporadic, the values tend to be high, likely resulting from sporadic population and scarce services. The sporadic high demand pattern is most typical in Montana, Idaho, Wyoming, Nevada, Utah, Colorado, Arizona, and New Mexico and can also be observed in the northern parts of Minnesota, Wisconsin, Michigan, and Maine. The most visible high-demand area (due to its size and position in the map), however, is in the

eastern half and is located at the contact of Kansas, Missouri, Arkansas, and Oklahoma, corresponding to the biggest uncovered "island" in the eastern half in Figure 2. Several other such isolated, uncovered areas in the eastern half also host high-demand patches, including the southern tip of Texas; the contact of Alabama, Georgia, and Florida; and an area in southwestern West Virginia. In the small "island" in upper New York, the area between Oklahoma and Arkansas, and northeastern Maine, however, there are no continuous distributions of pixels with high demand values, a result of the small number of patients. This mismatch between lack of service and large continuous high-demand areas is far more typical in the western half. In those largely empty western areas in Figure 2, there are no corresponding vast high-demand patches in Figure 3. In this sense, Figure 3 offers a more precise characterization of the spatial distribution of the demand for major cancer care facilities and therefore is more useful in evaluating the existing facilities and planning new facilities. In fact, it is our key objective in this study to reveal the areas where service coverage and demand for service are not congruent and to identify locations of unfulfilled demand.

Summary and Future Work

The methodologies that have been implemented for assessing spatial access to health care services can be generally classified into two categories: the nearest neighbor approach, which only considers the travel impedance to the nearest (however defined) facility, and the density approach, which covers both availability and accessibility. Between the two, the density approach is more sophisticated in terms of incorporating more information. The density methods under different names, including 2SFCA, the gravity model, and KDE, are inherently related. The first two are essentially special cases of the last, which leads to the understanding that the differences between them are technical rather than conceptual. This understanding makes a fundamental assumption of these density methods explicit: Availability and accessibility are mutually compensating. It also leads to the prediction that different choices for the distance decay function in the second step of 2SFCA will not significantly affect the general pattern in the estimation result, whereas the travel impedance threshold is a critical parameter.

In this study, we use 2SFCA to assess the local spatial access to major cancer care facilities in the contiguous United States, including the NCI Centers and AMCs. Based on that assessment, we further estimated the local demand for those facilities. The nationwide analysis minimizes the edge effect that is common in research of this nature. The process was designed to take advantage of the best available data. The final demand map in raster format has a resolution of 843 m and shows distinctive patterns in the western and eastern halves of the contiguous United States. Generally, the demand in the eastern half is spatially continuous but has relatively low values, whereas in the western half it is sporadic but the values tend to be high.

A series of further analyses can be proposed; for example, evaluating the discrepancies in the demand at state, regional, and other spatial scales; characterizing demands in different places (e.g., urban vs. rural); and detecting the associations between demand and demographic features, particularly race and ethnicity and socioeconomic status. It is also possible to use a similar approach to evaluate the optimality of a candidate location for a new facility and examine competing scenarios of launching a new facility versus expanding existing facilities.

For the most important parameter in 2SFCA, the travel impedance threshold, we simply specified a three-hour driving time for all facilities. We understand that the threshold might vary for different facilities; for example, some high-reputation facilities might have much larger catchments; travel by foot and public transportation are more popular in urban areas; and disabled or elderly patients might not have access to private cars. More important, in this study we focused on spatial access or place accessibility and made no attempt to incorporate nonspatial factors such as insurance status, cultural barriers, patient preferences, referral patterns, and so on. Two individuals at the same location might not have same access to the same cancer center due to their different space–time constraints, however, and a person might not find a nearby cancer center accessible if his or her constraints make it difficult (M. P. Kwan, personal e-mail, 21 January 2011). Studies on individual accessibility to health care service are still sporadic (e.g., Sherman et al. 2005; Wang et al. 2008), but we perceive it as a direction with great potential.

Other limitations of this study include the following:

1. In the census data, the attributes for Hispanic and white are not mutually exclusive, which could result in an overestimation of white and Hispanic patient populations.
2. The national-level cancer prevalence rates conceal the geographic variation in such rates.
3. Number of beds of a facility might not be an accurate representation of the facility's capacity for cancer care.
4. LandScan data attempt to represent the average population at a location over a twenty-four-hour period (Bhaduri et al. 2002), which is different from traditional estimates of residential population and the impact of this difference on the analysis result is still unknown.

Due to these limitations, the results of this particular study might not meet the quality for detailed planning, and the article is more of a demonstration of methodology.

Acknowledgment

This work was supported by the National Institutes of Health (Grant P20 R018787).

References

American Hospital Directory. 2011. Hospital statistics by state. http://www.ahd.com/state_statistics.html (last accessed 10 February 2012).

Avis, N. E., K. W. Smith, C. L. Link, G. N. Hortobagyi, and E. Rivera. 2006. Factors associated with participation in breast cancer treatment clinical trials. *Journal of Clinic Oncology* 24:1860–67.

Bhaduri, B., E. Bright, P. Coleman, and J. Dobson. 2002. LandScan: Locating people is what matters. *Geoinformatics* 5 (2): 34–37.

Birkmeyer, J. D., S. R. Finlayson, A. N. Tosteson, S. M. Sharp, A. L. Warshaw, and E. S. Fisher. 1999. Effect of hospital volume on in-hospital mortality with pancreaticoduodenectomy. *Surgery* 125:250–56.

Celaya, M. O., J. R. Reese, J. J. Gibson, B. L. Riddle, and E. R. Greenberg. 2006. Travel distance and season of diagnosis affect treatment choices for women with early-stage breast cancer in a predominantly rural population (United States). *Cancer Causes and Control* 17:851–56.

Cervigni, F., Y. Suzuki, T. Ishii, and A. Hata. 2008. Spatial accessibility to pediatric services. *Journal of Community Health* 33:444–48.

Guagliardo, M. F. 2004. Spatial accessibility of primary care: Concepts, methods and challenges. *International Journal of Health Geographics* 3:3.

Guagliardo, M. F., C. R. Ronzio, I. Cheung, E. Chacko, and J. G. Joseph. 2004. Physician accessibility: An urban case study of pediatric providers. *Health and Place* 10:273–83.

Gumpertz, M. L., L. W. Pickle, B. A. Miller, and B. S. Bell. 2006. Geographic patterns of advanced breast cancer in Los Angeles: Associations with biological and sociodemographic factors (United States). *Cancer Causes and Control* 17:325–39.

Hansen, W. G. 1959. How accessibility shapes land use. *Journal of the American Institute of Planners* 25:73–76.

Joseph, A. E., and D. R. Phillips. 1984. *Accessibility and utilization—Geographical perspectives on healthcare delivery.* New York: Harper & Row.

Kwan, M. P. 1998. Space–time and integral measures of individual accessibility: A comparative analysis using a point-based framework. *Geographical Analysis* 30:191–217.

———. 1999. Gender and individual access to urban opportunities: A study using space–time measures. *The Professional Geographer* 51:211–27.

Kwan, M. P., A. T. Murray, E. O. Morton, and T. Michael. 2003. Recent advances in accessibility research: Representation, methodology and applications. *Journal of Geographical Systems* 5:129–38.

Luo, W. 2004. Using a GIS-based floating catchment method to assess areas with shortage of physicians. *Health and Place* 10:1–11.

Luo, W., and Y. Qi. 2009. An enhanced two-step floating catchment area (E2SFCA) method for measuring spatial accessibility to primary care physicians. *Health and Place* 15:1100–07.

Luo, W., and F. Wang. 2003. Measures of spatial accessibility to healthcare in a GIS environment: Synthesis and a case study in Chicago region. *Environment and Planning B: Planning and Design* 30:865–84.

McLafferty, S., and S. Grady. 2004. Prenatal care need and access: A GIS analysis. *Journal of Medical Systems* 28 (3): 321–33.

Morris, J. M., P. L. Dumble, and M. R. Wigan. 1979. Accessibility indicators for transport planning. *Transportation Research* 13A:91–109.

National Cancer Institute. 2001. SEER cancer statistics review 1975–2001. http://seer.cancer.gov/csr/1975_2001/results_single/sect_01_table.18_2pgs.pdf (last accessed 24 November 2010).

———. 2010. Cancer centers. http://cancercenters.cancer.gov/cancer_centers/index.html (last accessed 24 November 2010).

Onega, T., E. J. Duell, X. Shi, E. Demidenko, and D. Goodman. 2009. Determinants of NCI Cancer Center attendance in Medicare patients with lung, breast, colorectal, or prostate cancer. *Journal of General Internal Medicine* 24 (2): 205–10.

———. 2010a. Influence of place of residence in access to specialized cancer care for African Americans. *The Journal of Rural Health* 26 (1): 12–19.

———. 2010b. Race versus place of service in mortality among Medicare beneficiaries with cancer. *Cancer* 116 (11): 2698–2706.

Onega, T., E. J. Duell, X. Shi, E. Demidenko, D. Gottlieb, and D. Goodman. 2009. Influence of NCI-Cancer Center attendance on mortality in lung, breast, colorectal, and prostate cancer patients. *Medical Care Research and Review* 66 (5): 542–60.

Onega, T., E. J. Duell, X. Shi, D. Wang, E. Demidenko, and D. Goodman. 2008. Geographic access to cancer care in the U. S. *Cancer* 112 (4): 909–18.

Sherman, J. E., J. Spencer, J. S. Preisser, W. M. Gesler, and T. A. Arcury. 2005. A suite of methods for representing activity space in a healthcare accessibility study. *International Journal of Health Geographics* 4:24.

Shi, X. 2009. A geocomputational process for characterizing the spatial pattern of lung cancer incidence in New Hampshire. *Annals of the Association of American Geographers* 99 (3): 521–33.

———. 2010. Selection of bandwidth type and adjustment side in kernel density estimation over inhomogeneous backgrounds. *International Journal of Geographical Information Science* 24 (5): 643–60.

Silverman, B. W. 1986. *Density estimation for statistics and data analysis.* Boca Raton, FL: Chapman & Hall/CRC.

Twombly, R. 2005. Cancer surpasses heart disease as leading cause of death for all but the very elderly. *Journal of National Cancer Institute* 97:330–31.

Wang, F., and W. Luo. 2005. Assessing spatial and nonspatial factors for healthcare access: Towards an integrated approach to defining health professional shortage areas. *Health and Place* 11:131–46.

Wang, F., S. McLafferty, V. Escamilla, and L. Luo. 2008. Late-stage breast cancer diagnosis and healthcare access in Illinois. *The Professional Geographer* 60 (1): 54–69.

Wang, F., and W. W. Minor. 2002. Where the jobs are: Employment access and crime patterns in Cleveland. *Annals of the Association of American Geographers* 92:435–50.

Wang, L. 2007. Immigration, ethnicity, and accessibility to culturally diverse family physicians. *Health and Place* 13:656–71.

Yang, D., R. Goerge, and R. Mullner. 2006. Comparing GIS-based methods of measuring spatial accessibility to health services. *Journal of Medical Systems* 30 (1): 23–32.

Patterns of Patient Registration with Primary Health Care in the UK National Health Service

Daniel J. Lewis* and Paul A. Longley†

*Department of Geography, Queen Mary, University of London
†Department of Geography, University College London

The UK National Health Service (NHS) is a long-established universal provider of health care. Most primary care is delivered by general practitioner (GP)-run health centers (surgeries) that, subject to proposed policy changes, are increasingly central to the welfare geographies of the NHS. This article develops an analysis of a unique and hitherto underexploited data set, comparing the observed pattern of patient registrations at GP surgeries with an optimum geographic pattern in the London borough of Southwark. In addition to evaluation of the level of geographic order that arises in a locally administered, centralized system of health care provision, we also use a new and innovative ethnicity classification tool to assess the ethnic dimensions to deviations from the normative arrangement. These results are considered in light of current and recent initiatives regarding patient choice in the United Kingdom.

英国国家卫生服务（NHS）是一个历史悠久的医保通用供应商。大多初级保健服务是由全科医生（GP）运行的健康中心（手术）提供，它们随着政策的变化，已经日渐成为了 NHS 的福利地域的中心。本文开发了一个独特的，前所未有的数据集分析，把观察到的 GP 手术病人的登记模式与伦敦南沃克市镇的最佳地理格局进行比较。除了评估在一个本地管理的卫生保健服务的集中式系统中所产生的地理秩序水平，我们也使用一个新的和创造性的种族分类工具，评估种族尺度离规范安排的偏差。并鉴于当前和近期的英国病人的选择来考虑这些结果。关键词：种族分类，国民健康服务，病人的选择，初级保健。

El Servicio Nacional de Salud del Reino Unido (NHS) es un proveedor universal de servicios de salud de vieja data. La mayor parte del cuidado primario es proporcionado por centros de salud (consultorios) manejados por practicantes generales (GP) que, sujetos a los cambios de política que puedan proponerse, son crecientemente centrales en las geografías del bienestar del NHS. Este artículo desarrolla un análisis de un conjunto de datos único y hasta ahora poco explotado, comparando el patrón de registro de pacientes observado en los consultorios GP con un patrón geográfico óptimo en el barrio londinense de Southwark. Adicionalmente a la evaluación del nivel de orden geográfico que resulta de un sistema centralizado para la provisión de servicios de salud, localmente administrado, utilizamos también una nueva e innovadora herramienta de clasificación de la etnicidad, para evaluar las dimensiones étnicas de desviaciones del esquema normativo. Estos resultados se consideran a la luz de iniciativas actuales y recientes en lo que toca a la selección del paciente en el Reino Unido.

Reform of primary health care is often viewed as politically necessary in many health care systems, as in the Obama administration's initiatives to extend eligibility and coverage but also more widely in systems affected by neoliberal government agendas for transparency and cost savings in public finance. Ongoing debate in the UK National Health Service (NHS) centers on the role of general practitioners (GPs) not only as providers of primary care but also as procurers of secondary (hospital) care. The organization and management of local delivery of health care through the system of (privately owned) GP surgeries, whose staff are usually the first point of patient contact in the UK system, is key to proposed health care reforms. Prospective patients have the right to register with a single GP surgery and receive health care that is free at the point of delivery. Reforms envisage the devolution of budgets behind commissioning of services to GP-led consortia and competition between public and private providers in providing care in a more market-oriented way, akin to the operation of health maintenance organizations (HMOs) in the United States. Central government recommendations (Department of Health [DH] 2010) to give patients both a wider and freer choice in accessing health care, if implemented, will have profound geographic implications on primary care, with similarly far-reaching consequences for hospital care.

The NHS health improvement agenda is increasingly primary-care led as well as primary-care focused (Moon and North 2000; DH 2005) and is seen as providing better value for money than relying on expensive hospital services. Concurrently, the articulation of choice has shifted from a doctor–patient dialogue to one more firmly centered on the rights of the patient (DH 2006, 2008a). Reforms set out in the radical 2010 white paper "Equity and Excellence: Liberating the NHS" have stated that any patient should be able to choose or change their GP surgery without being "limited to one that is nearest to your home" (DH 2010).

This policy shift significantly loosens the geographic basis for organization of NHS services, which has always hitherto advocated that patients should register with geographically proximate GP surgeries, in the interests of efficient and effective provision of primary health care. Catchment areas, initially ad hoc, later formalized (Martin and Williams 1991), although never fully implemented in practice, formed a basis for service delivery. In densely populated urban areas, however, NHS concerns have focused on the needs of the elderly, disabled, and young families for whom GP surgeries should, for example, be "within walking distance for mothers with prams" (Ministry of Health 1962; Sumner 1971). Morrill, Earickson, and Rees (1970) noted that in U.S. city centers, where several health centers might serve a small, densely populated area, patients are less likely to use their nearest physician. This trade-off in choice and accessibility by patients will become an increasingly interesting facet of provision of primary health care in the United Kingdom should stated NHS reform proceed. Given the centrality of this premise to provision, there is something of a dearth of literature on the geographies of patient registration in UK cities (but see, inter alia, Knox 1978; Joseph and Phillips 1984; Wilkin, Metcalfe, and Leavey 1987).

This article uses a case study of the London borough of Southwark to consider how patient registration behavior deviates from a normative geographic arrangement. A complete listing of patient registration provided by Southwark Primary Care Trust at the patient level represents a privileged view of patient–GP surgery interaction, wherein, given the varied ethnic composition of the study area, we focus on indicators of patient ethnicity to analyze variation. We provide a contemporary insight into the pattern of choice of GP surgery by patients, in which the behaviors manifest in existing spatial patterns of registration add weight to the suggestion that choosing a GP surgery is not an innovation to be imposed but a preexisting condition of primary care in the urban context.

Equity and Choice of GP in the NHS

The NHS has a duty to provide a universal service to the UK population, underpinned by the notion of equity; that is, "a just distribution justly arrived at" (Harvey 1973, 16) defined for health care by Asthana and Gibson (2008) as "equal opportunities of access to healthcare for equal needs" (4). We focus on the spatial equity of access to the health care system, namely, "the question of who benefits and why in the provision of urban services and facilities" (Talen and Anselin 1998, 596). The trade-offs between promoting choice and equity are numerous and contested, and we do not propose to cover them here, beyond the obvious apparent view of Dixon and Le Grand (2006) that "[g]iving choice to individuals and groups who previously had none . . . will extend to all a privilege that was previously confined to those who could afford private healthcare" (166).

Corrigan (2005) contended that choice has been a factor in primary care ever since the creation of the NHS in 1948, consistent with the stated intention of ensuring equal quality of care wherever it is sought (see also Moon and North 2000). The uniform nature of GP services, and lack of information on differences where they do exist, however, ensured that patient choice of GP surgery was principally driven by location. Exworthy and Peckham (2006) and Greener (2007) noted that it is unusual for patients to be willing to travel beyond local services, and a series of studies of the NHS in the late 1980s and 1990s similarly identify location as key to choice of GP surgery (Salisbury 1989; Billinghurst and Whitfield 1993; Gandhi et al. 1997). In the urban context, however, available local services can include numerous GP surgeries. In the case of Southwark, for example, many catchment areas defined by GP surgeries overlap each other, overlap the artificial boundaries of the borough, might not correspond to the areas where many of their patients live, or might simply be so extensive as to effectively stipulate no catchment at all.

Patients living within a catchment area can expect to be provided with GP services, including home visits if required, assuming that they are registered to that GP surgery. If the surgery is operating at capacity, however, the list might be closed to new registrants, irrespective

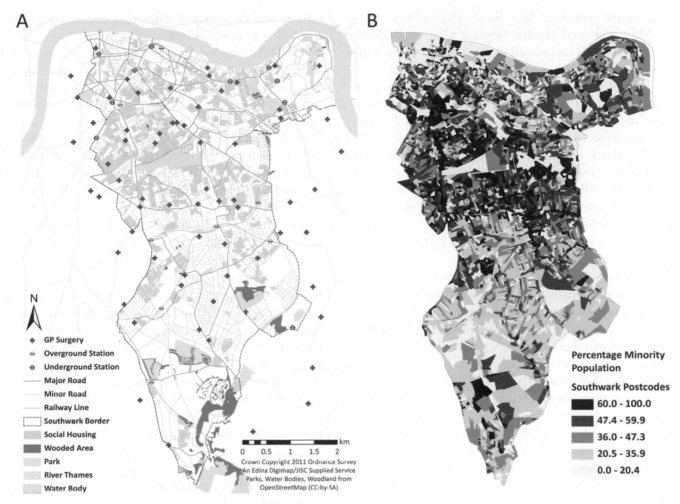

Figure 1. Location map of (A) Southwark and (B) proportions of ethnic minority patients by postcode. (Color figure available online.)

of their residential locations, and the GPs will not be obliged to provide health care services to additional patients. The basis for defining a catchment area comes from the need to regulate the workload of the GPs at any given surgery, and in practice catchments are agreed upon by negotiation between the GP and the relevant local NHS body. The NHS Constitution (DH 2009) deems registration with a local NHS GP surgery a "right," but managing catchment areas and service quality is challenging, with a recent NHS survey reporting that "members of some black and ethnic minority groups, commuters, and people living in more deprived areas are more likely to report dissatisfaction with services. These groups want greater control of how and when they access primary care" (DH 2008b). Reforms, in their present state, would seek to abolish catchment areas.

Framing Evidence for Choice in Southwark

The borough of Southwark (Figure 1A) is an Inner London local authority some 30 km² in extent, home to almost 300,000 people. It is ranked the twenty-fifth (of 326) most deprived local authority in England (Index of Multiple Deprivation 2010: an index of hardship[1]). Over 30 percent of Southwark's residents are members of ethnic minorities (Figure 1B), and the local authority is landlord to the largest stock of social housing in London (Southwark Council 2010). Affluent enclaves are also to be found in Southwark, though, particularly along the River Thames and to the south. There are forty-seven GP surgeries in Southwark, providing employment for about 200 general practitioners.

Southwark patient data were drawn from the May 2009 National Health Service Central Register (NHSCR), which records all patient registrations with a GP surgery in the United Kingdom. The extract identifies the full names and addresses of every Southwark resident registered with any GP surgery. These unique patient data were obtained following successful application for ethical approval to the appropriate NHS research ethics committee. Difficulties in handling patient residential mobility mean that the number of records in the NHSCR is likely to be an overestimate of the actual size of GP lists. The study population for Southwark includes about 325,000 people using forty-seven GP surgeries in Southwark and 127 GP surgeries in its environs.

The motivation for this research is to investigate the deviations between the observed and normative patterns of registration, particularly with respect to patient and GP ethnicity. We do this by considering the network distance between each patient and his or her GP of registration, characterizing patients as traveling an additional distance if they use a GP surgery that is further away than their nearest, assumed to be the normative choice. Each GP surgery and patient residence is address geo-referenced using Ordnance Survey MasterMap Address Layer 2 (a database of residential and commercial building locations), and distances are computed on the Ordnance Survey MasterMap Integrated Transport Network.

The classification of ethnicity is often inherently subjective and error prone in medical records, and our chosen approach was to classify patients using the Onomap classification, which indicates the likely cultural, ethnic, or linguistic origin of each individual patient based on their forename and surname pairings (Mateos, Webber, and Longley 2007; Mateos, Longley, and O'Sullivan 2011; Petersen et al. 2011). This adds considerable value and consistency to the NHSCR birthplace records, which are error prone and might fail to identify second-generation members of ethnic groups. Coding the ethnicity of individual patients makes it possible for the first time to undertake a nonecological study; ethnic variation in GP registration is an important, but underresearched, facet of primary care provision, and coding patient ethnicity by their name presents an innovative approach to resolving this. Lakha, Gorman, and Mateos (2011), in a validation study based on Scottish public health data, suggested that "Onomap offers an effective methodology for identifying population groups in both health-related and educational datasets, categorizing populations into a variety of ethnic groups" (1).

The ethnicity-based segmentation of Southwark GP registration data makes it possible to see which groups are more likely and which are less likely to register with GP surgeries that are close to their residence. These behaviors are likely influenced by numerous factors (Joseph and Phillips 1984; Hays, Kearns, and Moran 1990) reflecting differences in age (Hopkins et al. 1967; Ahmad, Kernohan, and Baker 1991), sex (Salisbury 1989), social class (Goddard and Smith 2001), wealth (Knox and Pacione 1980), and other locational factors (Bullen, Moon, and Jones 1996). As such, observed behaviors might reflect clear-cut patient preferences but equally might simply reveal the effect of patients conforming to, or constrained by, other aspects of the system. Hawthorne and Kwan (2012) have suggested that a qualitative understanding of the humanistic factors regarding access to health care is important, because patients can impose an added perceived distance when faced with low-quality health provision. Knowledge of the local patterns of patient registration could be an important precursor to uncovering these subjective drivers to choice, with evidence for demographic variations providing a basis for qualitative enquiry.

Patterns of Registration with GP Surgeries in Southwark

The proximity rank of a patient's GP surgery of registration offers the possibility of a baseline measure of spatial efficiency, in which patients not using their nearest GP surgery (assuming unconstrained capacities) introduce spatial inefficiencies. The cumulative distribution of patient proximity ranks for registration with Southwark GP surgeries is shown in Figure 2.

Approximately 40 percent of Southwark residents use their nearest GP surgery, and 80 percent of residents use one of their nearest six GP surgeries. This suggests that a relatively large number of patients are either willing to or required to make small trade-offs in accessibility against other considerations in registering with a particular GP surgery. Table 1 demonstrates the geographic inefficiencies consequent on patients not using their nearest GP surgery, measured as additional distance traveled.

Roughly 197,000 patients each travel an average additional 790 m (around ten minutes of walking time) to use a GP surgery other than their nearest, with the

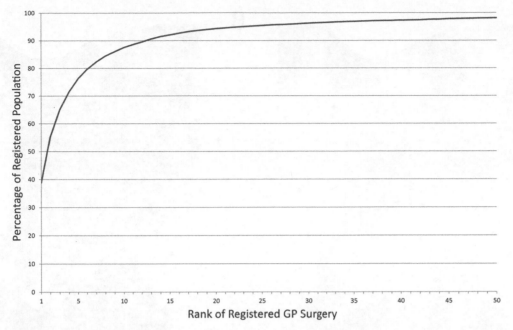

Figure 2. Cumulative percentage of general practitioner (GP) surgery usage by rank.

median additional distance being approximately 479 m. The preponderance of short additional distances provides good evidence for patients exercising choice in some form. The geography of additional distance traveled is not itself geographically random: Figure 3 presents two smoothed representations of the proportion of people using their nearest GP surgery.

Figure 3A makes apparent that patients living in areas that lie at borough boundaries (e.g., the River Thames to the north) or in areas that have a lower density of GP surgeries (e.g., the south) are more likely to use their nearest GP surgery. At a finer level of granularity, site-specific effects become more apparent, with the lowest percentages of patients traveling extra distances in the immediate vicinities of GP surgeries. In the more service-rich areas, the islands of higher registration surrounding each surgery are less intense, with smaller spatial extent than the less service-rich areas. The nature of these distance decay effects is not identical across the borough as a whole, though, suggesting that there might be additional factors behind the observed pattern of patient registration behaviors.

Southwark's multiethnic character provides the motivation for investigating variability in additional distances traveled. Table 2 shows the mean and median additional distances traveled by patients of different

Table 1. Additional distance traveled to general practitioner surgery by rank

Rank order	No. of patients	Additional total distance traveled (km)	Additional mean distance traveled (m)	Additional median distance traveled (m)
1 (40%)	128,137	0	0	0
2	53,494	11,226	210	151
3 (66%)	32,828	11,546	352	293
4	20,529	10,269	500	440
5	15,412	8,721	566	550
6 (80%)	11,205	7,659	684	641
7	8,547	6,544	766	749
8	6,476	5,463	844	799
9	5,608	6,790	1,211	975
10	4,492	4,787	1,066	953
11	3,152	3,454	1,096	1,004
12 (90%)	3,126	3,897	1,247	1,119
≥ 13	32,258	75,389	2,337	1,865

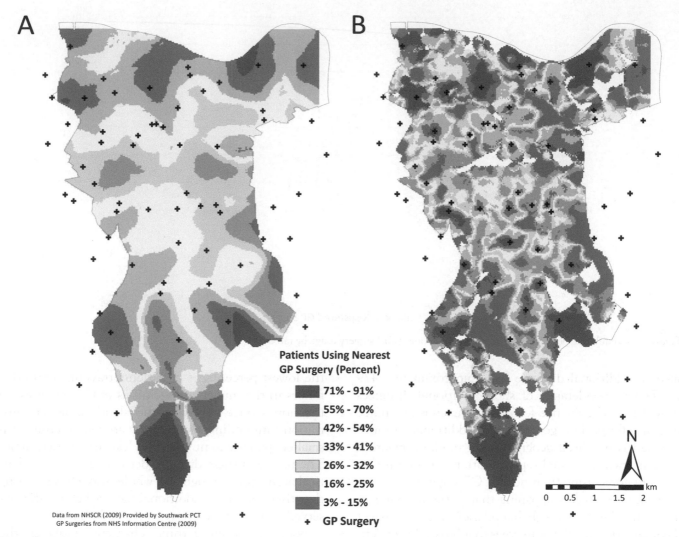

Figure 3. Percentage of patients using their nearest general practitioner (GP) surgery. Gaussian kernel smoothing for (A) 500 m and (B) 100 m. (Color figure available online.)

Table 2. Distance to nearest general practitioner surgery and general practitioner surgery of registration by ethnic group in Southwark (Ratio Registered : Nearest)

Patient ethnicity	No. of patients	Mean distance to nearest GP (m)	Mean distance to registered GP (m)	Median distance to nearest GP (m)	Median distance to registered GP (m)
African	35,091	489.2	1,138.7 (2.3)	467.3	770.1 (1.6)
British	166,058	515.6	979.0 (1.9)	491.8	773.8 (1.6)
East Asian	9,451	513.3	1,001.2 (2.0)	492.5	720.3 (1.5)
Eastern European	7,182	505.1	912.2 (1.8)	489.3	698.7 (1.4)
European	45,944	510.4	948.6 (1.9)	490.5	736.2 (1.5)
Hispanic	11,470	480.4	852.9 (1.8)	465.3	688.8 (1.5)
Muslim	31,263	484.5	958.5 (2.0)	464.0	737.9 (1.6)
South Asian	6,012	542.6	1,075.7 (2.0)	504.8	736.5 (1.5)
Other	12,793	501.8	973.7 (1.9)	485.0	723.7 (1.5)

Note: GP = general practitioner.

ethnic groups, classified using the Onomap[2] system. It is apparent that African patients travel further on average to use a GP surgery than any other group, although the median distances traveled by all ethnic groups are broadly similar. A chi-square test (following Hays, Kearns, and Moran, 1990: $\chi^2 = 2,504$) of whether patients of different ethnicities exhibit similar registration behaviors with respect to the rank of the GP surgery they use shows significant difference between groups at the greater than 1 percent level. Differences in registration patterns are most apparent among the African group, within which only 35 percent of patients use their nearest GP surgery (compared to 40 percent for the population as a whole). Conversely, the Eastern European, European, Hispanic, and East and South Asian groups show a more marked tendency to use their nearest GP surgery (44, 41, 43, 43, and 45 percent, respectively).

This pattern can be formally tested with a logit model in which the likelihood of a patient registering with his or her nearest GP surgery is related to demographic characteristics (age, sex, ethnicity), distance to his or her nearest surgery, the characteristics of provision at the nearest surgery (the number of full-time GPs and their ethnicity), and an estimate of the number of competing destinations available to the patient. The idea of competing destinations stems from Fotheringham's (1983) work; however, he initially used the concept in the formulation of spatial interaction models, whereas it is used here as an indicator of available choices. In this way, as the number of possible destinations (GP surgeries) varies for any given patient, the likelihood of registration with the nearest GP surgery can also vary.

In the model, a patient's distance to the nearest GP surgery is defined by the road network distance from the patient's residence. The patient's age and sex are extracted from the patient register and the ethnicity is coded using Onomap, as is each GP's ethnicity. The number of full-time GPs per surgery is calculated by taking the number of registered GPs and subtracting a value of 0.5 for each reported part-time GP, which was considered to best represent a patient's impression of the number of GPs at a practice. Further, the proportion of GPs in each GP surgery belonging to an ethnic minority is calculated with reference to the Onomap-derived ethnicity of each GP. Finally, the choice set of competing destinations (local GP surgeries) for each patient is calculated by estimating a service area for each of Southwark's GP surgeries based on the observed distribution of patients who are registered with it. The num-

ber of service areas that a patient falls within provides a measure of that patient's local opportunity to access a GP surgery. Service areas are created using Gibin, Longley, and Atkinson's (2007) method of enclosing a given percentage of the density distribution of patients registered with a GP surgery. The procedure for creating a service area is as follows:

1. Estimate the continuous density surface of the pattern of patient registrations with a GP surgery using kernel density estimation (KDE). This has many desirable properties (de Smith, Goodchild, and Longley 2009) including, in this application, the maintenance of confidentiality of individual patient records.
2. Each cell in the resultant raster has a volume, which can be expressed as a proportion of the whole raster.
3. The raster cells are sorted from high to low volume. Cells are then recoded to value 1 up until the desired cumulative percentage of volume (Shortt et al. [2005] noted that 75 percent is popular). The remaining cells are coded to zero.
4. The raster cells are resorted into their original order, and a percentage volume contour (PVC) is drawn to bound the extent of the cells coded 1 in the binary raster. This can create several distinct polygons if the service area created is multinucleated.

Following Shortt et al. (2005), the 75 percent PVCs were calculated and taken to characterize a principal service area for each Southwark GP surgery. As the service areas reflect the pattern of patient registration, rather than an arbitrary distance or potential statistic, they can be thought of as a spatial representation of the patient community using a GP surgery. As such, any patient who falls within a service area can be considered as likely to have local neighbors who use the service, designating that GP surgery as lying within the choice set of accessible GP surgery choices for that patient. Patients who fall within more than one service area thus have more scope to exercise choice of GP surgery. The data set is only complete for Southwark, meaning that service areas are not calculated for non-Southwark GP surgeries. The use of the 75 percent PVC generally mitigates edge effects, in view of the strong geographical basis for registration and the resultant tightly defined service areas. The overlaps due to congruent service areas are shown in Figure 4.

Some characteristics of the service areas are shown in Figure 5: The clustering of Muslim and African

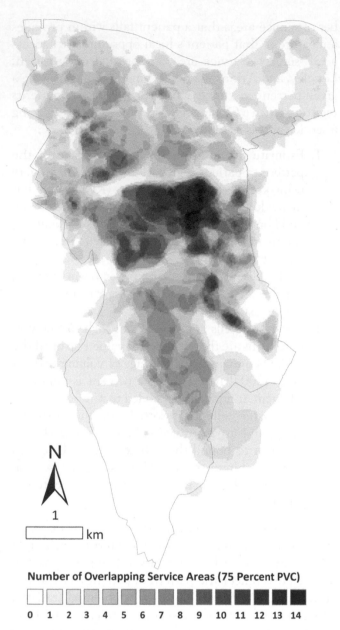

N

1
km

Number of Overlapping Service Areas (75 Percent PVC)

0 1 2 3 4 5 6 7 8 9 10 11 12 13 14

Figure 4. Overlapping service areas in Southwark, derived from the 75 percent percentage volume contour (PVC) for each general practitioner surgery.

patients with respect to GP surgery locations is shown in Figure 5A, in which South Asian patients are the least well served. The complexity of this relationship is demonstrated in Figure 5B, which shows the cumulative proportion of patients in each group that fall within the principal service area of their chosen GP surgery. Muslim and African patients are more likely to reside within the principal service area of their chosen GP surgery than their South Asian or white British counterparts, but this only becomes evident when the number of congruent service areas is very high (>10); below this, African and Muslim patients are consider-

ably less likely to reside within the service area of their GP surgery of registration.

Table 3 documents a model of patient registration behaviors with respect to their nearest GP surgery, calculated for all patients, excluding patients under sixteen years old, whose registrations are constrained by parental registrations. In general, the further away a patient lives from his or her nearest GP surgery, the less likely he or she is to use it; similarly, the greater the number of service areas belonging to other GP surgeries that a patient's residence falls within, the less likely he or she is to use the nearest GP surgery. This supports evidence of patients making small distance-based trade-offs in accessing a local GP surgery that might not necessarily be the nearest. The larger the nearest GP surgery is (in terms of number of GPs), however, the more likely patients are to use it, providing evidence for the growing utility and attractiveness of consolidated health centers over more traditional surgeries staffed by only one or two GPs. In terms of differentiation by ethnicity, the modeled results are to be expected: African and Muslim patients are less likely to use their nearest GP surgery than British patients and vice versa for East Asian, European, and Hispanic patients at the 1 percent level of significance and South Asian patients at the 5 percent level. The ethnicity of the GPs at each surgery is also an important factor: As the proportion of ethnic minority GPs (African, Muslim, East and South Asian, and other) at the nearest GP surgery increases, the likelihood of a patient using that surgery decreases. This result is not consistent over all ethnic groups, however; there is an interaction effect that suggests that as the percentage of minority GPs at the nearest GP surgery increases, the likelihood that a non-British patient will use his or her nearest GP surgery increases compared to the British group. This is particularly true for African and Muslim patients, as well as Hispanic, unclassified, and Eastern European patients, although there is no difference between British and East Asian groups, whereas the relationship is only significant at the 10 percent level for the European and South Asian groups. This suggests that the ethnicity of GPs at particular surgeries can play a role in patient registration behaviors, particularly among African and Muslim patients.

Discussion and Conclusion

This analysis of patient registrations in Southwark demonstrates that different ethnic groups of the population, classified using patient forenames and

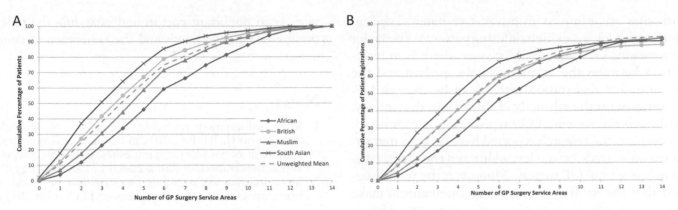

Figure 5. Cumulative proportions of (A) of patients resident within the principal service areas of any general practitioner (GP) surgery and (B) principal service area residents who are registered with a general practitioner serving the area. Both differentiated according to the ethnic groups of African, Muslim, British, and South Asian.

Table 3. Logit regression results testing patient usage of their nearest GP surgery in Southwark

Variable	Coefficient	SE	z	P > |z|
Distance to nearest GP surgery (km)	−1.68000	0.018900	−88.87	0.000
Number of accessible GP surgeries	−0.20130	0.002841	−70.84	0.000
Number of GPs at nearest surgery	0.11321	0.001724	65.66	0.000
Percentage of minority GPs at nearest surgery	−0.00478	0.000196	−24.38	0.000
Patient sex (female)	0.00919	0.008767	1.05	0.295
Age of patient[a]				
16–24	−0.22569	0.014061	−16.05	0.000
35–44	−0.16663	0.011965	−13.93	0.000
45–54	−0.35146	0.013871	−25.34	0.000
55–64	−0.39594	0.017489	−22.64	0.000
65–74	−0.46513	0.021669	−21.47	0.000
75+	−0.46167	0.023489	−19.65	0.000
Ethnicity of patient[b]				
African	−0.69489	0.031875	−21.80	0.000
East Asian	0.20510	0.047313	4.34	0.000
Eastern European	0.05819	0.055112	1.06	0.291
European	0.06931	0.024081	2.88	0.004
Hispanic	−0.11843	0.045139	−2.62	0.009
Muslim	−0.41616	0.031656	−13.15	0.000
South Asian	0.13051	0.058795	2.22	0.026
Unclassified	0.00882	0.044964	0.20	0.845
Interaction between percentage of minority GPs at nearest surgery and ethnicity of patient				
African interaction	0.011145	0.000489	22.78	0.000
East Asian interaction	−0.000220	0.000776	−0.28	0.777
Eastern European interaction	0.002594	0.000854	3.04	0.002
European interaction	0.000641	0.000386	1.66	0.097
Hispanic interaction	0.004685	0.000709	6.61	0.000
Muslim interaction	0.006092	0.000500	12.18	0.000
South Asian interaction	0.001658	0.000963	1.72	0.085
Unclassified interaction	0.002255	0.000708	3.19	0.001
Constant	0.721757	0.021294	33.89	0.000
	Number of observations = 239,525		Log likelihood = −151,077.29	
	LR $\chi^2(27)$ = 21,101.01		Prob > χ^2 = 0.0000	

Note: GP = general practitioner; LR = likelihood ratio.
[a]Base category for age is 25–34 years old.
[b]Base category for ethnicity is British.

surnames according to an innovative name classification methodology, exhibit differing patterns of behavior in accessing GP surgeries. Moreover, there is an interaction between the ethnicity of the patient and the likelihood of registration with his or her nearest GP, contingent on the ethnicity of the GPs in that surgery. Although the system of GP registration in Southwark is complex, the research reported here suggests that patients often trade off modest additional travel distances to access a local GP surgery. Moreover, the results suggest that the willingness, or the requirement, to make these trade-offs is more common among the African and Muslim populations and is likely to be connected with the characteristics of the GP surgery, particularly the GPs themselves. Such trade-offs are highly spatially contingent, however; all groups have a higher likelihood of using their nearest GP surgery the closer they live to it. This reflects the role of the GP surgery as a place that provides local services in a way that tries to serve the population as a whole (i.e., spatial equity). This analysis benefits from the individual level at which it is conducted, deriving door-to-door network distances and spatial referencing at the household level. Further work investigating the characteristics of the GP surgeries that might be driving differential registration behaviors could help develop delivery of health care in the United Kingdom within the local community remit specified by the NHS, consistent with the mantra of improving patient choice.

It is also evident that ethnicity is only one of a number of factors driving patient registration behaviors, albeit an important one in the context of Southwark and one that predates the most recent proposals on promoting patient choice in NHS primary care. Opening up choice in the ways suggested by UK NHS reforms might weaken the effect of distance on the patterning of registrations with a GP surgery and confer greater importance on the characteristics of the patients and the services they seek. In this respect, it is important to understand the preconditions of registration as a benchmark to assessing whether reform manages to effectively maintain levels of equity or whether there is a discernable polarization of patients and services. This is set to become an issue of increasing importance if NHS reform creates a significantly increased role for GPs in the procurement and delivery of secondary health care services. Such insights might also be gained from analyzing reforms to other health care systems: By analogy, has the U.S. health care reform really made accessing health care any fairer on the ground? Similarly, and more broadly, how people use public (and private) services is an important part in understanding the functioning of neighborhoods and the coherence of communities.

Notes

1. More information regarding the index of multiple deprivation (IMD) for 2010 can be found at http://www.communities.gov.uk/communities/research/indicesdeprivation/deprivation10/ (last accessed 17 October 2011).
2. More information regarding Onomap can be found at http://www.onomap.org (last accessed 17 October 2011).

References

Ahmad, W., E. Kernohan, and M. Baker. 1991. Patients' choice of general practitioner: Importance of patients' and doctors' sex and ethnicity. *British Journal of General Practice* 41:330–31.

Asthana, S., and A. Gibson. 2008. Health care equity, health equity & resource allocation: Towards a normative approach to achieving the core principles of the NHS. *Radical Statistics* 96: 4–28.

Billinghurst, B., and M. Whitfield. 1993. Why do patients change their general practitioner? A postal questionnaire study of patients in Avon. *The British Journal of General Practice* 43:336–38.

Bullen, N., G. Moon, and K. Jones. 1996. Defining localities for health planning: A GIS approach. *Social Science & Medicine* 42 (6): 801–16.

Corrigan, P. 2005. *Registering choice: How primary care should change to meet patient needs.* London: Social Market Foundation. http://www.smf.co.uk/assets/files/publications/Choice%20&%20Health.pdf (last accessed 17 October 2011).

Department of Health. 2005. *Choosing health: Making healthy choices easier.* London: TSO.

———. 2006. *Our health, our care, our say: A new direction for community services.* London: TSO.

———. 2008a. *High quality care for all.* London: TSO.

———. 2008b. *Impact assessment of NHS next stage review proposals for primary and community care.* http://www.ialibrary.berr.gov.uk/uploaded/IA%20NHS%20Next%20Stage%20Review%20Proposals%20for%20primary&Community%20CarevF.doc (last accessed 17 October 2011).

———. 2009. *NHS constitution.* http://www.nhs.uk/choice-intheNHS/Rightsandpledges/NHSConstitution/Pages/Overview.aspx (last accessed 17 October 2011).

———. 2010. *Equity and excellence: Liberating the NHS.* London: TSO.

De Smith, M., M. Goodchild, and P. Longley. 2009. *Geospatial analysis.* 3rd ed. Leicester, UK: Matador.

Dixon, A., and J. Le Grand. 2006. Is greater patient choice consistent with equity? The case of the English NHS. *Journal of Health Services Research and Policy* 11:162–66.

Exworthy, M., and S. Peckham. 2006. Access, choice and travel: Implications for health policy. *Social Policy and Administration* 40 (3): 267–87.

Fotheringham, A. 1983. A new set of spatial-interaction models. *Environment and Planning A* 15 (1): 15–36.

Gandhi, I., J. Parle, S. Greenfield, and S. Gould. 1997. A qualitative investigation into why patients change their GPs. *Family Practice* 14 (1): 49–57.

Gibin, M., P. Longley, and P. Atkinson. 2007. *Kernel density estimation and percent volume contours in general practice catchment area analysis in urban areas.* Paper presented at GISRUK 2007, National University of Maynooth, Maynooth, Ireland.

Goddard, M., and P. Smith. 2001. Equity of access to health care services: Theory and evidence from the UK. *Social Science & Medicine* 53 (9): 1149–62.

Greener, I. 2007. Are the assumptions underlying patient choice realistic?: A review of the evidence. *British Medical Bulletin* 83:249–58.

Harvey, D. 1973. *Social justice and the city.* London: Edward Arnold.

Hawthorne, T., and M.-P. Kwan. 2012. Using GIS and perceived distance to understand the unequal geographies of healthcare in lower-income urban neighbourhoods. *The Geographical Journal* 178 (1): 18–30.

Hays, S., R. Kearns, and W. Moran. 1990. Spatial patterns of attendance at general practitioner services. *Social Science and Medicine* 31 (7): 773–81.

Hopkins, E., A. Pye, M. Solomon, and S. Solomon. 1967. A study of patients' choice of doctor in an urban practice. *Journal of the Royal College of General Practitioners* 14:282–88.

Joseph, A., and D. Phillips. 1984. *Accessibility and utilization: Geographical perspectives on health care delivery.* New York: Harper & Row.

Knox, P. 1978. The intraurban ecology of primary medical care: Patterns of accessibility and their policy implications. *Environment and Planning A* 10:415–35.

Knox, P., and M. Pacione. 1980. Locational behaviour, place preferences and the inverse care law in the distribution of primary medical care. *Geoforum* 11 (1): 43–55.

Lakha, F., D. Gorman, and P. Mateos. 2011. Name analysis to classify populations by ethnicity in public health: Validation of Onomap in Scotland. *Public Health* 125 (10): 688–96.

Martin, D., and H. Williams. 1991. Market-area analysis and accessibility to primary health-care centres. *Environment and Planning A* 24:1009–19.

Mateos, P., P. Longley, and D. O'Sullivan. 2011 Ethnicity and population structure in personal naming networks. *PLoS ONE* 6 (9): e22943.

Mateos, P., R. Webber, and P. Longley. 2007. The cultural, ethnic and linguistic classification of populations and neighbourhoods using personal names. *CASA Working Paper* No. 116. http://www.casa.ucl.ac.uk / publications / workingPaperDetail.asp?ID=116 (last accessed 17 October 2011).

Ministry of Health. 1962. *Local authority building note, local health authority clinics.* London: Her Majesty's Stationery Office.

Moon, G., and N. North. 2000. *Policy and place: General medical practice in the UK.* London: Macmillan.

Morrill, R., J. Earickson, and P. Rees. 1970. Factors influencing distances traveled to hospitals. *Economic Geography* 46 (2): 161–71.

Petersen, J., P. Longley, M. Gibin, P. Mateos, and P. Atkinson. 2011. Names based classification of accident and emergency department users. *Health & Place* 17 (5): 1162–69.

Salisbury, C. 1989. How do people choose their doctor? *British Medical Journal* 299:608–10.

Shortt, N., A. Moore, M. Coombes, and C. Wymer. 2005. Defining regions for locality health care planning: A multidimensional approach. *Social Science & Medicine* 60 (12): 2715–27.

Southwark Council. 2010. Housing. http://www.southwark.gov.uk/site/scripts/documents.php?categoryID=100007 (last accessed 17 October 2011).

Sumner, G. 1971. Trends in the location of primary medical care in Britain: Some social implications. *Antipode* 3:46–53.

Talen, E., and L. Anselin. 1998. Assessing spatial equity: An evaluation of measures of accessibility to public playgrounds. *Environment and Planning A* 30:595–613.

Wilkin, D., D. Metcalfe, and R. Leavey. 1987. *Anatomy of urban general practice.* London and New York: Routledge.

(Un)Healthy[1] Men, Masculinities, and the Geographies of Health

Deborah Thien* and Vincent J. Del Casino Jr.†

*Department of Geography, California State University, Long Beach
†School of Geography and Development, University of Arizona

Being men and being healthy seem to be contradictory sociospatial states. Although research on the interrelationships between gender and health is strongly represented in geography, and masculinity has been examined, geographical perspectives examining the contradictory spatialities of men's health are lacking. This article addresses this absence by working through a feminist and relational framework to examine how sociospatial forces linking gender, health, and emotion intertwine in the process of being (un)healthy men. We argue that any representation of men's health as situated within a singular narrative of hegemonic masculinity is refuted by tracing the multiple processes of how gender, health, and emotion intersect to define (un)healthy men's bodies and spaces. To flesh out the conceptual argument, we employ two illustrative case studies: (1) a set of narratives of living with HIV from gay and bisexual men in the United States and (2) a set of veterans' responses to a posttraumatic stress disorder program in Canada. These examples demonstrate men's fraught practices of their masculinities in relation to health and illustrate how variegated sociospatial practices of hegemonic masculinity affect men's health, men's affective relationships with support systems for health, and the contexts within which men's health takes place. This article offers a modest beginning to the inclusion of men in health geography and to an extended conceptual terrain for geographies of health encouraging the rethinking of linkages between health and gender and gender and emotion.

成为男人和身体健康，似乎是矛盾的社会空间分异状态。虽然性别和健康之间的相互关系在地理学中研究得很多，阳刚之气已经被研究过，但是研究男性健康的矛盾空间性的地理观点还很缺乏。本文针对这一现况，通过一个女权主义者和关系的框架，审查社会空间分异的力量是如何在成为（不）健康男性的过程中，与性别，健康和情感交织连接。我们认为，在霸权阳刚之气的奇特叙述范围内的男性健康的任何代表性言论，都可以通过跟踪交错的性别，健康，情感等多个进程来定义（不）健康男性的身体和空间而被驳斥。为了充实概念上的参数，我们使用了两个典型案例研究：（1）一套来自美国同性恋和双性恋男子的艾滋病毒感染者的生活叙述，和（2）一套加拿大退伍军人对创伤后应激障碍方案的反应。这些例子显示了这些人的与健康相关的男子气概的实践，也说明霸权阳刚之气的杂色的社会空间分异的做法是如何影响男性的健康，男人健康支持系统的情感关系，和男性健康发生的背景。本文提供了一个把男性列入健康地理研究的温和开端，并通过鼓励对健康和性别，以及性别和情感之间联系的重新考虑，拓展了健康地理的概念地域。关键词：情感，艾滋病毒/艾滋病，阳刚之气，创伤后应激障碍，性特性。

Ser hombres, y ser hombres saludables, parecieran ser estados socio-espaciales contradictorios. Aunque la investigación sobre las interrelaciones entre género y salud se encuentra muy bien representada en geografía, y también algo los estudios sobre masculinidad, todavía son pobres las perspectivas geográficas para examinar las espacialidades contradictorias de la salud de los varones. Este artículo aboca esa limitación trabajando por medio de un marco feminista y relacional con el fin de explorar cómo interactúan las fuerzas socio-espaciales que ligan género, salud y emoción en el proceso por el cual los hombres llegan a estar saludables o no. Lo que argüimos es que cualquier representación que sitúe la salud de los hombres dentro de una narrativa singular de masculinidad hegemónica queda refutada al trazar los múltiples procesos de cómo el género, la salud y emoción se intersectan para definir lo que son cuerpos y espacios saludables o no. Para darle cuerpo al argumento conceptual, empleamos dos estudios de caso ilustrativos: (1) un conjunto de narrativas de hombres gay y bisexuales sobre vivir afectados con VIH en los Estados Unidos y (2) un conjunto de las respuestas de veteranos a un programa sobre trastorno por estrés postraumático en Canadá. Estos ejemplos demuestran las tensas prácticas de sus masculinidades en los hombres en relación con la salud e ilustran cómo las variadas prácticas socio-espaciales de masculinidad hegemónica afectan su salud, sus relaciones afectivas con los sistemas de apoyo para la salud y los contextos dentro de los cuales se fragua la salud masculina. Este artículo es un modesto aporte al comienzo de la incorporación de los varones en la geografía de la salud y a un terreno conceptual ampliado para las geografías de la salud, promoviendo una nueva forma de pensar sobre los vínculos que se dan entre salud y género, y entre género y emoción.

Being men and being healthy seem to be contradictory sociospatial states. For example, rejecting health care, minimizing or dismissing health needs, and engaging in (socially sanctioned) risky behaviors are ways in which some men have historically demonstrated manliness while compromising their health (Courtenay 2000). Although research on the interrelationships between gender and health is strongly represented in geography, and masculinity has been examined, geographical perspectives examining the contradictory spatialities of men's health are lacking. Health geographers have yet to interrogate men's overall (un)healthiness, their health behaviors, experiences, and outcomes, including how sociospatial practices of hegemonic masculinities affect men's health, men's spatial and affective relationships with and in support systems for health, and the contexts within which men's health takes place. This article contributes toward a more robust health geography for men by drawing on work in health and feminist as well as social and emotional geographies to investigate the intertwined sociospatial forces and embodied experiences linking health, emotion, and masculinity. We argue that any representation of men's health as situated within a singular narrative of hegemonic masculinity is refuted by tracing the multiple processes of how gender, health, and emotion intersect to define (un)healthy men's bodies and spaces.

To flesh out the conceptual argument, we employ two illustrative case studies: (1) a set of narratives of living with HIV from self-identified gay and bisexual men in the United States and (2) a set of responses from veterans with posttraumatic stress disorder (PTSD) to a treatment program in Canada. These examples demonstrate men's fraught practices of their masculinities in relation to health. The juxtaposition of these two cases, one that focuses on gay and bisexual men and one that concentrates on veterans' emotional well-being in a militarized heterosexual domain, illustrate how "being men" is normalized through feminized emotional and caring practices within divergent contexts of health care and health promotion. Both examples highlight contradictions in how men perform their health as male and masculine bodies and demonstrate how presumptive performances of masculinity construct the (un)healthy male subject. We see an opportunity for geographers to interrogate how the practices of masculinity shape and are shaped by (un)healthy men's embodiment; that is, their "lived spaces" (Moss 2008) and "embodied social practices" (Moss and Dyck 1999, 392) in relation to shifting gendered and emotional contexts. This article advances an agenda for the study of men's health within the context of health geography and argues that there is more work to be done on masculinity and health, spaces of (masculine) caring and support, and the contexts within which men's health takes place.

Gender and Health, Masculinity and Emotion

Within geography there is a strong tradition of studying both the interrelationships between gender and health (e.g., Moss and Dyck 2002; Dyck and Dossa 2007) and masculinity as a geographic subject (Berg and Longhurst 2003a, 2003b; Van Hoven and Hörschelmann 2005; Hopkins and Noble 2009). In this section, we take inspiration not only from these previously referenced feminist, social, and cultural geographers, who have demonstrated the ways in which relations and identities are organized in uneven geographies, but also from feminist and critical race theorists who emphasize the constitutive relations between social identities and the differential effects and affects of interleaved processes of dominance and oppression (e.g., Ahmed 2000). We argue for a feminist and relational interrogation of the interrelationship between men's health and masculinity within the context of health geography (Dyck 2011), challenging the notion that "'gender and health' . . . [is] synonymous with 'women's health'" (Courtenay 2000, 1386). We conclude, as Parr (2004, 251) did, that there is space within health geography for "an explicit focus on (un)healthy masculinities . . . as opposed to an approach in which men's health status is an unmarked norm." We suggest three areas for more explicit development: (1) the interrelated gendered contexts within which men's health takes place, (2) men's affective relationships with differently spatialized support systems for health, and (3) the ways in which the variegated sociospatial practices of hegemonic masculinities affect men's health.

To begin, then, we want to argue that although gender continues to be a contested concept, we understand gender as embodiments and discourses taking place within continuously emerging and relational contingencies, a process where "differences are determined at the level of encounter" (Ahmed 2000, 145). This relational stance emphasizes the mobile coconstituencies of spaces, subjectivities, and sociospatial processes of power and inequality as embodied processes through which genders take place. In the area of geography, gender, and health, Moss and Dyck have employed this relational understanding of gender while exploring the material and discursive female body (Moss and Dyck 1999, 2002), investigating how women and their health can be understood to embody processes of marginalization (Moss and Dyck 2000), how women construct healthy spaces for their families via everyday activities of food preparation and religious practices (Dyck and Dossa 2007), and how diagnostic categories and ill (female) bodies mutually constitute one another (Moss 2008). This work provides a close reading of women's lives and embodied experiences in the context of health and place, making space for investigating the contours of women's lives and examining how gendered social relations are deeply embedded in health contexts. In particular, this research conceptualizes embodiment as "lived spaces"; that is, "spaces . . . where the *specifics* of any one body or sets of bodies are momentarily *fixed* as bodies, replete with identities, subjectivities, and power as well as senses, content, and expression" (Moss 2008, 161, emphasis in original). Although this work does not explore male bodies, per se, it does frame a discussion of how men, as materially and discursively coconstituted relational bodies, shape, intervene, affirm, and challenge presumptive categories of male (un)healthiness.

Within geography, Berg and Longhurst (2003a, 2003b), Van Hoven and Hörschelmann (2005), and Hopkins and Noble (2009) suggested how we can interrogate the gendered processes through which hegemonic masculinities, "as ways of living in everyday local circumstances" (Connell and Messerschmidt 2005, 838), remain "in crisis" (Connell 1995). This definition of masculinity (as "in crisis") strongly informs geographic work, although as Berg and Longhurst (2003b) noted, Connell "does not explicitly acknowledge [masculinity's] geographic context" (352). Berg and Longhurst continued, however: "given the importance of contexts, relationships, and practices in both the (re)construction of masculinity and the way that we come to understand the meanings of the term, it should be clear that masculinity is both temporally and geographically contingent" (352). Hopkins and Noble (2009, 814) further proposed that Connell and Messerschmidt's work is valuable for geographers as it suggests the interrelated spatialities that are necessary for temporarily establishing masculinities through practices of "being men." It is not surprising, then, that in *Spaces of Masculinities*, Van Hoven and

Hörschelmann (2005) also argued: "A focus on the relational formation of male identities and masculine spaces [has been] long overdue in both feminist and gender-oriented geographical work" (5). Their edited collection draws on the wider tradition in geography, which is to examine "[m]asculinities . . . [as] highly contingent, unstable" (Berg and Longhurst 2003b, 352), by emphasizing masculinities in transition, as well as the relationship of masculinities to cultural change, violence, embodiment, and sexuality.

Although there is limited geographic research investigating the intersections of men, masculinity, and health, there is an important literature examining the (un)healthy geographies of men with HIV and AIDS (e.g., Wilton 1996; Brown 1997; Raimondo 2005; Sothern and Dyck 2009) and the intersections of drug use with sexualized and gendered identities (Del Casino 2007a). A few geographers have also explored the emotional complexities of masculinity and health more generally (Valentine 1999; Aitken 2009; Thien 2009). Brown's (1997) work is particularly illustrative here, as he engages with the important question of citizenship and its relation to the healthy political subject. Building on this question, Del Casino (2010, 197) further argues in the case of HIV prevention campaigns in Los Angeles that a "discourse of responsibility firmly reorganizes the (dis)eased body: it is not enough to have to deal with your own personal health, you must now be responsible for everyone else as well." The (un)healthy male subject in this case is only made healthy socially by taking on an individual responsibility that is simultaneously a community responsibility—the individual subject remains the center of the epidemic and the center of the response to reduced HIV transmission.

For her part, Valentine investigated "what it means to be a man" by interrogating the links among the body, masculinities, and disability, via a case study of Paul, a former miner with spinal cord injuries. Drawing from Butler's theorization of gender performance and Connell's work on masculinities, Valentine argued that everyday spatialities shape hegemonic performances of being a man, such that masculinity is (re)negotiated through spatially and emotionally nuanced encounters, reconstituted in and through (ill) health. Aitken's (2009) work on sensitively detailing the "work of fathers as an emotional practice" (xi) has also broken new ground in geographies of masculinities and health by expanding understandings of gendered emotional geographies. Aitken emphasized the process of *fathering* in contrast with the institution of *fatherhood* (2–3). In so doing, he "affirms the ability of fathers to inhabit bodies and spaces in diverse ways" (124) by engaging with men's diverse, embodied, felt experiences of fathering. Finally, Thien (2009) examined masculinity and emotional health via a consideration of normatively militaristic and masculine veteran spaces popularly associated with stereotypically "manly" attitudes. She concluded that these spaces unexpectedly function to offer significant space for male feeling in a manner normatively accorded to a feminized repertoire of emotionality (that of community, sharing of emotions, and emotionality through community).

Building on this body of work, wherein feminist and health geographers are listening to people talk about what is making them (un)healthy, we see exciting possibilities for further interrogation of the interrelationships between men's (un)health and masculinity within the context of health geography (Parr 2004). We argue that health is both intimately interwoven with expressions of gender, as evidenced by work on masculinity and health (e.g., Courtenay 2000) and in the scant literature on the geographies of men's health (Del Casino 2007a), and inextricably tied to emotional well-being (Thien 2009). Indeed, the field of emotional geographies with its broad critique of the masculinist ideal of an "emotion-free or emotionally controlled human subject" (Smith et al. 2009, 7) demands that we think through how the gendering of health and health care is simultaneously about the processes of emotion and emotionally constituted subjects (Ahmed 2004; Bondi, Davidson, and Smith 2005; Lipman 2006). In the gendered, heteronormative arena of health, men are often portrayed as (un)healthiest when not expressing emotion, unless expressing it in titularly masculine ways, such as appropriately controlled aggressive anger (O'Brien, Hunt, and Hart 2005). Turning our attention to emotionality, its presence or its lack, is one important way in which to understand the relational experiences of men, masculinity, health, and emotion (Thien 2011); that is, the ways in which (un)healthy men feel their way, in place (Craddock and Brown 2009; Parr and Davidson 2009). Within the health geography of emotional well-being, the performance of masculinity remains a vital if underexamined subject. As a result, we believe that emotional geographies of health become "another dimension through which to examine the articulation of power through a discourse of gender" (Sothern and Dyck 2009, 235).

We propose, then, that gender and health research within health geography has the considerable conceptual tools to address (un)healthy masculinities but not yet the substantive body of work to demonstrate this. Drawing together the emergent works that address (un)healthy masculinities, we suggest employing a fem-

inist and relational conception of gender to avoid reiterating a singular version of masculinity that neutralizes a more complex understanding of the geographies of men's health, emotions, and bodies (Dyck 2011); instead, we can consider how "bodies, gestures and turns of phrase" are continually (re)shaped through our (emotional) relations with others (Ahmed 2004, 166). Such nuanced understanding of the simultaneously embodied and discursive relations continuously taking place is vital in the context of health, where "health risks associated with men's gender or masculinity have remained largely unproblematic and taken for granted" (Courtenay 2000, 1387). Building on this call, then, we briefly examine two illustrative case studies so that we might begin to engage the complex relations men face as they try to act as (un)healthy subjects.

Being Men, Being Responsible, Being Healthful: HIV/AIDS and Masculinities

My name is Jason. I'm HIV *positive* and I *protect* with *condoms*.

—HIVStopsWithMe.org video (last accessed 1 November 2010)

In opening up the first page of the HIVStops WithMe.org Web site in the fall of 2010,[2] one would have been confronted by seventeen separate introductions from people living with HIV. Each person, dressed in black, spoke his or her name and then filled in the blanks to the following sentence "I'm HIV _____ and I _____ with _____" (see Figure 1). All of them are HIV positive. As each person spoke, his or her words were transcribed on a billboard behind him or her. Set against a clear blue sky with white puffy clouds, each person, in black neutralizing attire, was front and center in the campaign against the spread of HIV. On entering the Web site, readers were directed to this campaign's national history—within the United States—and its 2010 location in three states—Alaska, New York, and Virginia—and within New York, two cities—Buffalo and New York City. Each state linked the reader to a site with biographies of the seventeen spokesmodels—eleven men and six women—for the campaign. Of the men, ten were self-identified gay or bisexual. This campaign grounds its work in the geographies of each place, as it has long relied on local demographics to identify faces that are appropriate to a wider community.[3] Like the larger campaign, the biographies on the Web site offer a confessional style, as each person outs himself or herself as to how he or she contracted HIV and identified this reality to the

Figure 1. One image from the rotating stories illustrated on the HIVStopsWithMe.org Web site (2010) entry window. Source: HIV Stops with Me, http://www.hivstopswithme.org (last accessed November 2010). (Color figure available online.)

people around him or her. Their words have an affirmative tone, as they profess faith in their ability to stem the tide of HIV.[4] Their neutrality stands as a powerful metaphor of rebirth, signaling the commitment to a healthier body, one that is more than individual; it is a socially responsible body—responsible for the communities that each body is intended to represent—even as the responsibility remains firmly grounded in the autonomous individual subject.

This campaign is now in its second decade and represents a wider turn in HIV prevention efforts, which has focused on "prevention with positives" (Fisher, Smith, and Lenz 2010; see also Gordon, Stall, and Cheever 2004). Although there have been both affirmation and criticism of these campaigns (see Davis 2008 for further discussion), we briefly interrogate HIVStops WithMe.org for the ways in which it constructs masculinity, health, and masculine responsibility in the context of the HIV epidemic, particularly for self-identified gay and bisexual men (compare Brewis and Jack 2010). Despite the campaign's inclusive framework, the often problematic placing of male bodies in the context of certain core dualities—sexual and nonsexual, positive and negative, appropriate and inappropriate actors—might fail to fully appreciate the complexities of risk and prevention, disclosure and nondisclosure, as well as serostatus (as HIV positive or negative or, for that matter, viral load free) that

can limit the efficacy of such outreach efforts (cf. Armendinger 2009; see also Del Casino 2007b). The campaign assumes that a masculine homonormativity (cf. Casey 2007) can be established, denying the multiplicity of masculinities performed in gay spaces—for example, drag queens, bears, leathermen, tops and bottoms (Hennen 2008)—and centering individual responsibility for health and well-being in the context of one's (hegmonically masculine) obligation to the wider community. The hegemonic construction of this male subject is only made possible through the reiterative and emotional relationalities that emerge in the context of the campaign. Put another way, the responsible male subject emerges because this campaign grounds that responsibility in an emotional engagement with the wider community. Their individual responsibility is thus riddled with the tensions that emerge around their own masculine individuality and the need to be a (feminized) caring subject.

In the majority of the narratives at HIVStops WithMe.org, these representatives of a healthier self offer core lessons about HIV prevention and a restructured and rebuilt masculinity. The result is a new man—a healthier, safer, and responsible citizen. In the words of Waylon:

> After thinking about things some more, I finally realized that I *needed to start over from square one.* I moved back home to Anchorage once again, this time without any

264

sort of timeline in place. All I knew was that I needed to basically pretend like I was eighteen again and figure out how to live life as a responsible adult. I'm proud to say that this was the right thing to do. It's almost three years later, and I'm happier and healthier than I've ever been. I'm a full-time college student (studying music, of course) and I'm in the longest relationship I've ever been in. (emphasis added)

The lessons are projected backward in time as well. Billy offers how his infection history helped reconstitute him as a better man.

About two weeks after I found out [I was HIV positive], I was able to and needed to call the three guys I had been with in the past year. . . . The last call was to my friend in Portland. When I told him I was positive he replied, "That makes both of us now." Not knowing what he was talking about, I asked him, and he said, "Remember I told you I was positive in June?" It was a boldface lie . . . I called him on his lie and he hung up on me. Months went by before he returned a call or e-mail. He admitted I was right and profusely apologized. If nothing else, I needed to hear that. Had he told me he was positive, I still would have had sex with him, but we would have been safe. *Knowledge is power.* (emphasis added)

Even in the context of prison, HIV-positive men find recovery in their solitude. Isolation can produce self-reflection and self-awareness and, in the end, individual responsibility, as Jason suggests.

But I can honestly say, spending those four months in solitary confinement helped me. It built up my love for myself and helped me grasp a better knowledge of the disease. In the hole, I started to take my life back little by little. Schooling myself about my condition, reading articles, and gaining confidence to love me.

Within the context of this campaign, a more complex geography of men's health emerges. First, a "good" HIV-positive male gay subject is one who can get in touch with himself and ground his health in his own personal responsibilities. Being (un)healthy demands self-reflection and solitude—both material and emotional—and it requires suffering as well as rebirth. Second, a responsible HIV-positive gay man picks up the phone, connects with those in his own sociosexual network, and uses his own knowledge to protect not only himself but others. These narratives emphasize the value of constructing a responsible subject, a citizen protector who cares for both self and others—his community space becomes part of his individual responsibility. Third, this new healthy male subject is, in many ways, a feminized subject, as the men must of-

fer how they will nurture themselves and others in the fight against the further spread of HIV. In this context, they are asked to be responsible caretakers—not just appropriate men but appropriate gay men. The intersection of these subjects' masculinities and homosexualities further mark their bodies as reborn. Finally, like the Million Man March, which asked black men to take responsibility for their community as men, and other disparate men's movements (Poling and Kirkley 2000), the HIVStopsWithMe.org campaign asks HIV-positive gay and bisexual men to "man up," be responsible for, and take care of those around them—a healthier individual self ultimately presages a healthier community. As they stand there in their all-black attire, using the same messaging, the campaign reiterates their uniform masculinity (Butler 1999) and grounds their health identities in their HIV disease—they are no longer men or gay men, they are HIV-positive gay men. It is in admitting this new identity position achieved through feminized tactics of emotional knowing ("the confidence to love me") and community caretaking, alongside hegemonically masculine claims for responsibility and autonomy, that this campaign creates a new (un)healthy male body.

Putting Masculinity Back Together Again (With Feeling): Combating Posttraumatic Stress Disorder

Like HIV and AIDS, PTSD is a health issue as profoundly devastating as it is difficult to reconcile with normative ideas of masculinity. Indeed, the Vietnam veteran experience on which the diagnostic category of PTSD was founded (Thomas 2004) has been characterized as the "failure of masculinity" (Karner 1996). Resulting treatments have been animated by a desire to remasculinize not only veterans but also society at large (in the case of Vietnam, the United States as a whole; Jeffords 1989). For sufferers, PTSD is experienced as both spatially confounding and anxiety producing, deeply affecting health and masculinity. The pathological reexperiencing of trauma in PTSD is characterized by a vertigo-like placing of the sufferer into another time and place (American Psychiatric Association 2000). This loss of a locatable self together with difficult-to-locate feelings of anxiety present an emotional, even existentialist, challenge: a "falling away of the self" that painfully impinges "on the model of masculine self-containment" (Callard 2006, 886).[5] The institutional anchor is also cast loose: Men return from service feeling out of place, isolated, and alone with a "combined felt sense of allegiance to and betrayal by" their military command (Westwood et al. 2008, 297–98). Struggles to resolve tensions between disordered (feminized) feelings and the desired hegemonic masculinity of military subjectivity have shaped a complex social–spatial and emotionally laden health experience for many veterans with PTSD.

The Canadian Forces Transition Program for Peacekeepers and Veterans (the VTP) was developed in 2005 at the University of British Columbia, in Vancouver, by Dr. Marv Westwood. It is designed to provide a therapeutic space for soldiers dealing with psychological injuries as they transition from war zone to home. The program works by "normalizing the experience of expected trauma stressors to help soldiers better understand their military experience and its impact on their lives," offering "skills/strategies of how to cope with the effects of the trauma related stress," and, lastly, "enacting repair to help restore and reintegrate the parts of the self that have become fragmented thus facilitating their readjustment to civilian life" (Westwood 2009).

Soldiers' Stories (Legion Version) is a media presentation[6] detailing soldiers' experience of the VTP. Based on conversations with nine VTP participants (Dennis, Mike, Robert, Bill, Vytas, George, Doug, Paul, and Tony) in British Columbia, Canada, *Soldiers' Stories* alternates between narratives as the men tell their stories of coming home from military service:

> George: I did my tour in '91, I had a stroke and we got shot at and you know we seen a lot of death and destruction but we never talked about it when we got back, so [nods affirmatively, eyes are welling with tears].
> Bill: It was a very, very bad time for me. I considered it to be spiraling into a very, very deep depression the only way out that I could see was, uh, was by taking my own life.
> Paul: [he cannot speak—he swallows hard]

In sharp contrast to the DVD's opening shots of heroism in action and a rallying voice-over about bravery, our first glimpse of these soldiers is in painfully close-up shots of men in distress filmed against a somber gray background (see Figure 2). Unlike the upbeat framing of confessionals against a blue sky background on the HIV/AIDS Web site, the testimonials offered in *Soldiers' Stories* begin with unconcealed despair: Death, depression, and suicidal behavior are raised in the first minute; the men fight tears, sigh, shake their heads, and are overcome to the point of being unable to testify. Similar to the HIV/AIDS campaign, however, the eventual lesson in this instructional video is that the

Figure 2. Still of George, participant in the Veterans' Transition Program. Source: *Soldiers' Stories (Legion Version).* (Color figure available online.)

(male) self can be restructured, rebuilt, and reborn. In the course of the video, these men describe their transition from the spaces of military action shown in the opening scenes, through their feelings of broken-down selves and homes on return, and finally to their achievement of re-formed health and well-being via the transformational space of the VTP. The collective story that emerges is one of sociospatial transformation of health and well-being for (male) veterans with PTSD.

Again, we see a complex geography of masculinity and health emerge. First, the men describe how connecting to their emotions and to one another enables them to transform suffering into bringing out the best in their soldier selves:

> George: I've had more joy with my kids in the past two months than I've had in the last ten years, it's amazing to see what other soldiers, working with other soldiers you know [in the VTP], once again brings the best out of ya, just allow ya to tell your emotions and never once feel you're being judged for being uh, being a soldier.
> Doug: You know there's a lot of guys that you know out there in the same boat.
> Paul: I came to this group and now I have a connection again [nodding]. Um, I can call any of these guys anytime day or night and I can tell them what I'm feeling and it's the same that goes for them as well [sighs].

Part of the therapeutic intent of the VTP is to recognize and normalize these emotional expressions in situ, to be able, as George says, to "tell your emotions" and still "be a soldier" and to know, as Doug says, you are in "the same boat"; that is, to be emotionally knowing and to engage in health-seeking behavior through their bonds with other soldiers. Emotional expression is explicitly encouraged (Westwood and McLean 2007).

Second, although their experiences speak to a larger discourse of assigning responsibility for health to the autonomous, rational, and male subject—arguably, a shifting of responsibility for war-related injury to the soldier/citizen instead of the state (Terry 2009, 206)—the participants shape a "therapeutic citizenship" in subversive relation to this edict of responsibility (Nguyen 2010). Similar to the sense of duty reported by volunteers in early HIV/AIDS testing (see Nguyen 2010), soldiers are encouraged to participate in PTSD recovery for the good of their fellow soldiers. Their shared "boat" is shared feeling in the spaces of military culture, where "being a solider" is a normative masculine touchstone paradoxically characterized by caring. As both our case studies show, this question of an active (i.e., healthy) citizenship, wherein being well becomes a moral imperative, is an area that needs further examination in health geographies.

Third, through the processes of emotional knowing and caring, men are reborn and their personal geographies are reconfigured:

> Robert: When I finished telling my story, through the course of ongoing counseling with the group and individually in some cases, I felt, uh, like a new man, I was reconnecting with the person I was before. [Clears throat] I improved my personal relationship with my wife, with my family, I was much more open about my feelings and about my future.
> Bill: I got to find out a lot about myself and traits that I had buried um, very deep within myself for a very long period of time [eyes are filled with tears]. And, they're pretty cool some of them [the barest of laughs].
> Paul: My kids like to cuddle with me now, my little guy tells me that he loves me [takes a big breath, emotion-filled and the camera cuts away].

As Robert expresses, he is a "new man," connected to who he was before but not the same; Bill has experienced a profound sense of self rediscovery; and Paul, after experiencing disconnection with his young son, is newly experiencing himself through the process of fathering (see Aitken 2009). Reintegrating and restoring the (masculine) self is achieved through (feminized) emotional expression in a nurturing, witnessing space. Finally, then, norms of masculinity are preserved (i.e., the brotherhood made manifest) to safely express the self in the embodied, emotional terms demonstrated in

Soldiers' Stories: "Group members report that participation in these [VTP] groups helps foster both pride and a reconnection to 'the brotherhood,' which is, for them, the best part of ever being a soldier" (Westwood 2009).

In these ways, masculinity, health, and emotion intersect to redefine (un)healthy men's bodies and spaces in the context of militarization and PTSD. The program, where participants are invited to reenact the narrative of their traumatic experience and to go "beyond language to express the self through action, movement, emotion, and reflection" (Westwood 2009), has proven highly successful. The VTP illustrates that emotional and feminized expression proves integral to the restoration and reintegration of fragmented male selves, paradoxically forming them anew into (un)healthy soldiers, citizens, and men.

Men, Masculinities, and the Emotional Geographies of Health

What emerges from our consideration of the geographies of masculinity and health? Both sets of narratives demonstrate that men's health becomes spatially constituted in the individual subject, an autonomous, rational being who is socially networked but individually contained. For the autonomous subject, good health is framed by his abilities to reform his self-identity through specific strategies for management of embodiment, emotionality, and well-being. Yet, in both contexts, men's health remains an individual trial of rebirth and conquest over a lack of health, which is simultaneously a lack of manliness. The paradoxically feminized conquest ritual that these men perform in the management of their HIV status or PTSD therapy (re)constitutes a safe, bounded masculinity. Rather than performances of the "dangerous" or "unruly" queer subject—the drag queen, for example (Butler 1993)—or the "dangerous" and "out-of-control" PTSD sufferer, we find the recovered subject and an appropriate male citizen. Indeed, within the HIVStopsWithMe.org campaign, transgendered and other queer subjects are erased through the subjects' all-black clothing[7] (Halberstam 2005), whereas in the context of PTSD, masculinity remains sutured to the soldiers' identities as militarized men, the presumptively heterosexual brotherhood. These health programs thus stand to obscure the complexity of embodied masculinities in their efforts to enhance health, effectively remasculinizing male subjects whose lives are more complex than these campaigns could hope (or want) to represent.

Disclosure is also a key aspect of the achievement of health and well-being for these men—gay men with HIV/AIDS out themselves as positive in a climate of silencing (Del Casino 2010) and veterans with PTSD reveal their feelings for a "witnessing group of others" (Westwood 2009) whose cultural norms are the hypermasculine regulatory gender practices of the military (but note Atherton's [2009] work on the domestication of the soldier). Davidson and Henderson (2010) argued that disclosure (and its other, concealment) is a series of "complex and selective strategies of information and identity management" (155). Here, we are especially interested in disclosure as a form of agency—paradoxically a feminized tactic for the recovery of masculinity. The emotionality of the experience of men's health in both case studies suggests that men can somehow manage their health through the public presentation of their own (dis)ease, that the outward expression of "knowledge as power" is equivalent to being healthy. The normalizing of therapy as part of Western popular culture and, more specifically, the increasing sense that disclosure is therapeutic is certainly at work here. For the military man to disclose the out-of-control emotional nature of traumatic experience is thus an action simultaneously in control of the self and profoundly transgressive, leading to particular configurations of (un)healthy masculinities. Although homonormative social networks and spaces for gay men encourage (some) emotional expression, silence has characterized feelings about HIV and AIDS in a climate of fear and mistrust. This is particularly true in the complex world of the politics related to HIV serostatus, sex, and identity (Del Casino 2007b).

Putting together these two examples provides a thought-provoking pairing. In each case, specific performances of masculinity are rendered normalized within their respective contexts of health and well-being. Both examples offer lessons of how men can and should perform their health and male/masculine bodies appropriately within their particular context and each reiterates how men's engagement with norms of masculinity work to shape (un)healthy men. Going forward, there is a broader concern demonstrated here: Men's health is not only about the management of their responsibilities as political citizens but as biological citizens (Rose 2006) with all the attendant emotional geographies. Performing one's identity as an out gay HIV-positive man is equivalent to extending the self into the community and into the role of appropriate social actor (Del Casino 2009). In the context of the PTSD treatment program, the citizenship of the militarized normatively heterosexual man is grounded in his renewal as a productive male body, ready for battle and for civilian life.

Within both health domains there remains no accountability for the wider sociospatial processes—cultural politics or political economies—that put men in spaces of risk. There is an apolitical power to these campaigns that seek to define an (un)safe, (un)healthy, unitary masculine subject while failing to contextualize the geographies of violence and prejudice that are culpable in the risks related to HIV or PTSD. Courtenay (2000) argued, "Masculinity is continually contested, [and so] it must be renegotiated in each context that a man encounters" (1393). What is paradoxical in our case studies is that (un)healthy men renegotiate their masculinity through their (feminized) practices of health. Our two illustrative examples have thus offered the opportunity to more closely scrutinize our key concerns, namely, the ways in which masculinities shape men's lived experiences of health, the (emotional) ways in which men use support networks for health, and the diverse and paradoxical contexts in which their health takes place.

Acknowledgments

We are indebted to the anonymous reviewers and Mei-Po Kwan for their close and insightful readings of our original manuscript. These thoughtful critiques allowed us to considerably improve our resulting article. We thank the Canadian Institutes for Health Research for supporting the PTSD research via Open Doors/Closed Ranks: Locating Mental Health after the Asylum (Grant #MOP-84510); the men who participated in *Soldiers' Stories*; and Dr. Marv Westwood, Officer Joanne Henderson, and Sharel Fraser for their insights into PTSD and the Royal Canadian Legion. And, we send our appreciation out to those who are working in the HIV prevention community, particularly Lee Kochems, whose work and experience helped shape some of the ideas about HIV prevention and identity formation found in this article. We also thank our colleagues in the Department of Geography at California State University, Long Beach, where much of this work was first discussed.

Notes

1. We use this parenthetical to draw attention to the mutual constitution of healthy and unhealthy masculinities. It is impossible to simply disentangle and equate one form of masculinity, for example, with one form of healthy or

unhealthy practice, behavior, identity, or subjectivity. As we discuss, men's "healthiness" is paradoxically and problematically often founded in the active dismissal of health concerns for men. Thus, men's health is simultaneously its lack, as there can be no health without some level of unhealth.

2. It is important to note that since this analysis was developed, the Web site has been updated and changed. Although some of the "spokesmodels" who spoke in 2010 are still there, the design and the aesthetic of the site have been significantly adapted. These adaptations include the use of a distinct color for each participant as well as a more significant library of videos. The narrative style of many of the stories remains consistent with the analysis found in this article.

3. One example of the campaign for Long Beach from 2006 can be found at the following YouTube link: http://www.youtube.com/watch?v=PY9fp7cR_5I&feature =relmfu (last accessed 1 April 2012). In this narrative, the spokesmodels discuss how they can protect others. This leads one model to suggest that not disclosing one's status helps you "take care of the next person."

4. The religiosity of this metaphor is not lost on the authors, as the confessional, which is also used in other prevention programs, such as Alcoholics Anonymous, is a common trope underwriting such prevention programs and efforts. See a further discussion of what Poling and Kirkley (2000) called "phallic spirituality" in their eponymous analysis of Promise Keepers, the Million Man March, and Sex Panic.

5. Compare Callard (2006) and Davidson (2003), whose different work on agoraphobia nonetheless dovetails around the assertion of a devastating loss of a spatial sense of self.

6. The presentation was funded by the Royal Canadian Legion, Canada's long-established organization for veterans (see Thien 2009).

7. There are traces of that alterity, at least surficially, in the image of Billy, who wears a headband.

References

Ahmed, S. 2000. *Strange encounters: Embodied others in postcoloniality*. London and New York: Routledge.

———. 2004. *The cultural politics of emotion*. London and New York: Routledge.

Aitken, S. C. 2009. *The awkward spaces of fathering*. Farnham, UK: Ashgate.

American Psychiatric Association. 2000. *Diagnostic and statistical manual of mental disorders*. 4th ed., text revision. Washington, DC: American Psychiatric Association.

Armendinger, B. 2009. (Un)touchability: Disclosure and the ethics of loss. *Journal of Medical Humanities* 30:173–82.

Atherton, S. 2009. Domesticating military masculinities: Home, performance and the negotiation of identity. *Social & Cultural Geography* 10 (8): 821–36.

Berg, L., and R. Longhurst. 2003a. A bibliography of geography and masculinities. Acme Journal. http://www.acmejournal.org/MascBib.pdf (last accessed 25 May 2012).

———. 2003b. Placing masculinities and geography. *Gender, Place & Culture: A Journal of Feminist Geography* 10 (4): 351–60.

Bondi, L., J. Davidson, and M. Smith. 2005. Introduction: Geography's emotional turn. In *Emotional geographies*, ed. J. Davidson, L. Bondi, and M. Smith, 1–16. Aldershot, UK: Ashgate.

Brewis, J., and G. Jack. 2010. Consuming chavs: The ambiguous politics of gay chavinism. *Sociology* 44 (2): 251–68.

Brown, M. 1997. *Replacing citizenship: AIDS activism and radical democracy*. New York: Guilford.

Butler, J. 1993. *Bodies that matter: On the discursive limits of "sex."* London and New York: Routledge.

———. 1999. *Gender trouble: Feminism and the subversion of identity*. 10th anniversary ed. London and New York: Routledge.

Callard, F. 2006. "The sensation of infinite vastness"; or, the emergence of agoraphobia in the late 19th century. *Environment & Planning D: Society & Space* 24 (6): 873–89.

Casey, M. 2007. The queer unwanted and their undesirable "otherness." In *Geographies of sexualities: Theory practice, and politics*, ed. K. Browne, J. Lim, and G. Brown, 125–35. Surrey, UK: Ashgate.

Connell, R. W. 1995. *Masculinities*. Los Angeles and Berkeley: University of California Press.

Connell, R. W., and J. W. Messerschmidt. 2005. Hegemonic masculinity: Rethinking the concept. *Gender Society* 19 (6): 829–59.

Courtenay, W. H. 2000. Constructions of masculinity and their influence on men's well-being: A theory of gender and health. *Social Science & Medicine* 50 (10): 1385–1401.

Craddock, S., and T. Brown. 2009. Representing the un/healthy body. In *A companion to health and medical geography*, ed. T. Brown, S. McLafferty, and G. Moon, 301–21. Oxford, UK: Wiley-Blackwell.

Davidson, J. 2003. "Putting on a face": Sartre, Goffman, and agoraphobic anxiety in a social space. *Environment and Planning D: Society and Space* 21 (1): 107–22.

Davidson, J., and V. L. Henderson. 2010. "Coming out" on the spectrum: Autism, identity and disclosure. *Social & Cultural Geography* 11 (2): 155–70.

Davis, M. 2008. The "loss of community" and other problems for sexual citizenship in recent HIV prevention. *Sociology of Health and Illness* 30 (2): 182–96.

Del Casino, V. J., Jr. 2007a. Flaccid theory and the geographies of sexual health in the age of Viagra. *Health & Place* 13 (4): 904–11.

———. 2007b. Health/sexuality/geography. In *Geographies of sexualities: Theory, practice, and politics*, ed. K. Browne, G. Brown, and J. Lim, 39–52. Aldershot, UK: Ashgate.

———. 2009. *Social geography: A critical introduction*. Oxford, UK: Wiley-Blackwell.

———. 2010. Living with and experiencing disease. In *A companion to health and medical geography*, ed. T. Brown, S. McLafferty, and G. Moon, 188–204. London: Blackwell.

Dyck, I. 2011. Embodied life. In *A companion to social geography*, ed. M. T. V. J. Del Casino, Jr., R. Panelli, and P. Cloke, 346–61. Oxford, UK: Wiley-Blackwell.

Dyck, I., and P. Dossa. 2007. Place, health and home: Gender and migration in the constitution of healthy space. *Health & Place* 13 (3): 691–701.

Fisher, J. D., L. R. Smith, and E. M. Lenz. 2010. Secondary prevention of HIV in the United States: Past, current, and future perspectives. *Journal of Acquired Immune Deficiency Syndromes* 55:S106–15.

Gordon, C. M., R. Stall, and L. W. Cheever. 2004. Prevention intervention with persons living with HIV/AIDS. *Journal of Acquired Immune Deficiency Syndromes* 37 (Suppl. 2): S53–57.

Halberstam, J. 2005. Shame and white gay masculinity. *Social Text* 23 (3–4): 219–33.

Hennen, P. 2008. *Faeries, bears, and leathermen: Men in community queering the masculine.* Chicago: University of Chicago Press.

Hopkins, P., and G. Noble. 2009. Masculinities in place: Situated identities, relations and intersectionality. *Social & Cultural Geography* 10 (8): 811–19.

Jeffords, S. 1989. *The remasculinization of America: Gender and the Vietnam War.* Bloomington: Indiana University Press.

Karner, T. 1996. Fathers, sons, and Vietnam: Masculinity and betrayal in the life narratives of Vietnam veterans with post traumatic stress disorder. *American Studies* 37 (1): 63–94.

Lipman, C. 2006. The emotional self. *Cultural Geographies* 13 (4): 617–24.

Moss, P. 2008. Edging embodiment and embodying categories: Reading bodies marked with myalgic encephalomyelitis as a contested illness. In *Contesting illness: Processes and practices,* ed. P. Moss and K. A. Teghtsoonian, 158–80. Toronto: University of Toronto Press.

Moss, P., and I. Dyck. 1999. Body, corporeal space, and legitimating chronic illness: Women diagnosed with M.E. *Antipode* 31 (4): 372–97.

———. 2000. Material bodies precariously positioned: Women embodying chronic illness in the workplace. In *Geographies of women's health: Place, diversity and difference,* ed. I. Dyck, N. D. Lewis, and S. McLafferty, 231–47. London and New York: Routledge.

———. 2002. *Women, body, illness: Space and identity in the everyday lives of women with chronic illness.* Lanham, MD: Rowman & Littlefield.

Nguyen, V.-K. 2010. *The republic of therapy: triage and sovereignty in West Africa's time of AIDS.* Durham, NC: Duke University Press.

O'Brien, R., K. Hunt, and G. Hart. 2005. "It's caveman stuff, but that is to a certain extent how guys still operate": Men's accounts of masculinity and help seeking. *Social Science & Medicine* 61 (3): 503–16.

Parr, H. 2004. Medical geography: Critical medical and health geography? *Progress in Human Geography* 28 (2): 246–57.

Parr, H., and J. Davidson. 2009. Mental and emotional health. In *A companion to health and medical geography,* 258–77. Oxford, UK: Wiley-Blackwell.

Poling, J. N., and E. A. Kirkley. 2000. Phallic spirituality: Masculinities in Promise Keepers, the Million Man March and Sex Panic. *Theology and Sexuality* 12: 9–25.

Raimondo, M. 2005. "AIDS capital of the world": Representing race, sex and space in Belle Glade, Florida. *Gender Place and Culture* 12 (1): 53–70.

Rose, N. 2006. *The politics of life itself: Biomedicine, power, and subjectivity in the twenty-first century.* Princeton, NJ: Princeton University Press.

Smith, M., J. Davidson, L. Cameron, and L. Bondi. 2009. Introduction: Geography and emotion—Emerging constellations. In *Emotion, place and culture,* ed. M. Smith, J. Davidson, L. Cameron, and L. Bondi, 1–18. Farnham, UK: Ashgate.

Sothern, M., and I. Dyck. 2009. "... A penis is not needed in order to pee": Sex and gender in health geography. In *A companion to health and medical geography,* 224–41. Oxford, UK: Wiley-Blackwell.

Terry, J. 2009. Significant injury: War, medicine, and empire in Claudia's case. *Women's Studies Quarterly* 37 (1 & 2): 200–25.

Thien, D. 2009. Death and bingo? The Royal Canadian Legion's unexpected spaces of emotion. In *Emotion, place and culture,* ed. M. Smith, J. Davidson, L. Cameron, and L. Bondi, 207–25. Farnham, UK: Ashgate.

———. 2011. Emotional life. In *A companion to social geography,* ed. V. J. Del Casino Jr., M. E. Thomas, P. Cloke, and R. Panelli, 309–25. Malden, MA: Wiley-Blackwell.

Thomas, S. P. 2004. From the editor—The debate about posttraumatic stress disorder and some thoughts about 9/11. *Issues in Mental Health Nursing* 25 (3): 223–25.

Valentine, G. 1999. What it means to be a man: The body, masculinities, disability. In *Mind and body spaces: Geographies of illness, impairment and disability,* ed. R. Butler and H. Parr, 163–75. London: Routledge.

Van Hoven, B., and K. Hörschelmann. 2005. *Spaces of masculinities.* London and New York: Routledge.

Westwood, M. 2009. The Veterans' Transition Program—Therapeutic enactment in action. *Educational Insights* 13 (2). http://www.ccfi.educ.ubc.ca/publication/insights/v13n02/articles/westwood/index.html (last accessed 25 May 2012).

Westwood, M., T. G. Black, S. Kammhuber, A. C. McFarlane, N. Arthur, and P. Pedersen. 2008. Case incident 18: The transition from veteran life to the civilian world. In *Case incidents in counseling for international transitions,* 297–311. Alexandria, VA: American Counseling Association.

Westwood, M. J., and H. B. McLean. 2007. Traumatic memories and life review: Individual and group approaches. In *Transformational reminiscence: Life story work,* ed. J. A. Kunz and F. G. Soltys, 181–96. New York: Springer.

Wilton, R. 1996. Diminishing worlds: HIV/AIDS and the geography of everyday life. *Health and Place* 2: 1–17.

Therapeutic Imaginaries in the Caribbean: Competing Approaches to HIV/AIDS Policy in Cuba and Belize

Cynthia Pope

Department of Geography, Central Connecticut State University

In this article, I put forward a therapeutic imaginaries framework, developed from previous geographic work on therapeutic landscapes. In particular, I briefly trace the history of HIV and HIV policy in these countries, resulting from field work I have conducted in Havana since 1997 and in Belize since 2005. I highlight how therapeutic imaginaries are created and experienced through governmental AIDS policies, and how these strategies in Cuba and Belize influence individuals' perceptions of the salubriousness of these countries' natural and built landscapes. These case studies demonstrate how countries in the same region can develop health care policies that represent different biomedical and sociocultural outcomes. For example, Cuba's policies result in the lowest HIV/AIDS prevalence rates in the Caribbean, whereas Belize experiences the second highest rates. On one hand, the Cuban government espouses comprehensive and centralized health care as a political goal. On the other hand, Belize has a health care system characterized by a decentralized knowledge base and a reliance on Cuban medical personnel. I argue that geopolitics, gender dynamics, economic philosophies, and cultural norms intertwine to create differing disease outcomes in these countries. In both cases I emphasize the roles of HIV policies in influencing perceptions (from individual to international) about whether a particular landscape is healthy or diseased. These perceptions inform the relationship among therapeutic landscapes, therapeutic narratives, and therapeutic imaginaries. Although the causes and outcomes of risk, from geopolitical to individual behavior, vary in each context, the importance of places, filtered through these different scales, remains constant.

在本文中，我提出了一个治疗的想像框架，该框架产生于以前的有关治疗景观的地理工作。特别的是，我简要地追述这些国家的艾滋病毒和艾滋病政策的历史，其结果来自于我从 1997 年以来在哈瓦那，以及从 2005 年来在伯利兹所进行的实地考察工作。我强调治疗想像是如何创建并通过政府的艾滋病政策来贯彻的，以及这些在古巴和伯利兹实施的策略是如何影响个人对这些国家的自然和建筑景观的健康看法的。这些案例研究表明，在同一地区的国家是如何可能产生代表不同的生物医学和社会文化成果的卫生保健政策。例如，古巴的政策虽然导致了加勒比地区的最低的艾滋病毒/艾滋病患病率，但是伯利兹却经历了第二高的比率。一方面，古巴政府奉行全面而集中的卫生保健，并把其作为一项政治目标。另一方面，伯利兹的卫生保健系统以分散的知识基础和对古巴医务人员的依赖为特点。我认为，地缘政治，性别动态，经济哲学，和文化规范相互交织，在这些国家创造出了不同的疾病结果。在这两个例子中，我强调艾滋病防治政策的作用，它影响对某一个特别的景观健康与否的看法（从个人到国际）。这些看法告知治疗景观，治疗叙述，和治疗想像之间的关系。虽然从地缘政治到个人行为，风险的原因和结果在各种情况下有所不同，在不同的尺度下，地方的重要性保持不变。*关键词：伯利兹，古巴，卫生政策，艾滋病毒/艾滋病，治疗的想像。*

En este artículo propongo un marco de imaginarios terapéuticos, desarrollado a partir de trabajo geográfico anterior sobre paisajes terapéuticos. En particular, trazo brevemente la historia del HIV y de las políticas sobre HIV en estos países, resultantes del trabajo de campo que he llevado a cabo en La Habana desde 1997 y en Belice desde 2005. Destaco la manera como los imaginarios terapéuticos se crean y experimentan a través de las políticas gubernamentales sobre SIDA, y cómo estas estrategias influyen en Cuba y Belice las percepciones de las personas sobre el grado de salubridad de los paisajes naturales y construidos de estos países. Estos estudios de casos demuestran cómo países pertenecientes a una misma región pueden desarrollar políticas de cuidado de la salud que representan diferentes resultados biomédicos y socioculturales. En Cuba, por ejemplo, sus políticas han dado por resultado las tasas de prevalencia del HIV/SIDA màs bajas del Caribe, en tanto que Belice experimenta la segunda de las tasas màs altas. Por una parte, el gobierno cubano propugna con la atención de la salud centralizada y amplia un objetivo político; y por la otra, Belice tiene un sistema de atención de la salud caracterizado por una base de conocimiento descentralizada y dependencia del servicio de personal médico cubano. Mi argumento es que la geopolítica, la dinámica de género, filosofías económicas y normas culturales se entrelazan para crear diferentes resultados en el control de enfermedades en estos países. En ambos casos yo destaco el papel de las políticas de HIV para influir las percepciones (de la individual a la internacional) sobre si un paisaje particular es saludable o deletéreo. Estas percepciones informan la relación entre paisajes terapéuticos, narrativas terapéuticas e imaginarios terapéuticos. Aunque las causas y consecuencias de riesgo, de comportamiento geopolítico e individual, varían en cada contexto, la importancia de los lugares, filtrados por estas diferentes escalas, permanece constante.

This article develops the concept of therapeutic imaginaries, drawing on ideas put forward in psychology (Flaskas 2009), nursing (Long 2008), sociology (Dawney 2011), and the geographic literature (Lea 2008). I highlight how therapeutic imaginaries are created and experienced through AIDS policies and how governmental HIV strategies in Cuba and Belize influence individuals' perceptions of the salubriousness of these countries' landscapes. I compare these two countries because of their medical relationship and also because of their differing health philosophies and statistics, despite their proximity. Whereas Cuba has the Western Hemisphere's lowest HIV/AIDS adult prevalence rate (0.01 percent), Belize has the second highest rate (2.3–2.7 percent; UNAIDS and World Health Organization [WHO] 2010a, 2010b). On one hand, Cuba espouses the ideal of universal health care. On the other hand, Belize's health care system is characterized by a patchwork of private and public health facilities and a decentralized knowledge base. Although the causes and outcomes of disease risk vary in both countries, the importance of place, filtered through these various scales, remains constant.

Social scientists have increasingly highlighted the cultural, structural, and environmental influences on HIV risk behaviors (Friedman, Cooper, and Osborne 2009; Pope, White, and Malow 2009; Latkin et al. 2010), in essence advancing geographies of AIDS "beyond epidemiology" (Kalipeni et al. 2003). Indeed, geographers present ever more nuanced ways of interpreting how places constitute, as well as contain, social relations and physical resources (Jones and Moon 1993; Kearns 1997; Faubion 2009; Myers and Kearns 2009).

The therapeutic imaginaries framework I am developing here builds on the concept of therapeutic landscapes, a cornerstone of critical health geography research (Williams 1998, 2007; Gesler 1992; Kearns and Barnett 2000; Wilson 2003; Smyth 2005). In these landscapes, therapeutic narratives and metaphors cross boundaries and reach across many scales (Kearns 1997). This article's therapeutic imaginaries viewpoint goes beyond addressing the physical landscapes in Cuba and Belize. I highlight how health philosophies can contribute to the mythology of these places and their reputations (e.g., in the Cuban case) or can be obfuscated by the mythology and history of these places (e.g., in the Belizean case). Indeed, as Andrews (2004, 309) stated, "Despite the range of possible ways of defining structures, the 'norm' in therapeutic geographies of physical landscapes is ultimately restrictive." Conradson (2005, 338) also acknowledged the therapeutic landscape "as something that emerges through a complex set of transactions between a person and their broader socio-environmental setting." As such, "interactions between an individual and a landscape are understood as being complex and multifaceted, emerging out of particular embodied encounters but also subject to later interpretations" (338). Thus, the landscape transcends the physical to the symbolic, emphasizing the relationship between self and society as filtered through the lens of healing landscapes. Cummins et al. (2007) also offered a relational approach to health and place. This view of place differs from a more conventional view in a number of critical ways, including acknowledgment of multiscaled definitions of place, the importance of sociorelational distance, the mobility of individuals between places, changes in paths of access to certain resources over space and time, and the importance of social power relations. Cummins et al. thus argued for a "concentration on the *processes* and *interactions* [sic] occurring between people and places and over time which may be important for health" (1832).

These geographers' studies are critical to progress a therapeutic imaginaries framework that has an eye toward sorting out biosocial relationships between individuals and the state. I argue that connections among geopolitics, gender dynamics, economic philosophies, history, and cultural networks intertwine to create differing disease outcomes and prevention strategies in Cuba and Belize. In turn, whether outsiders and nationals perceive the built and natural landscapes as healthy, diseased, or (more likely) a combination of the two is based on lived experiences, political discourses, and local economies.

This article stems from my fieldwork in Cuba since 1997 and in Belize since 2005. I have conducted more than 600 surveys, formal interviews, informal conversations, and life histories in Cuba with medical tourists, sanatorium workers, AIDS organization organizers, women living with HIV, commercial sex workers, caregivers, and family doctors discussing women's risk for HIV in Cuba. In Belize, I have conducted several studies that span from oral narratives of women's experience with the health care system to a recent survey of approximately 200 individuals addressing perceptions of ethnicity, risky spaces, and HIV stigma.

Contextualizing Therapeutic Narratives: HIV Discourses in Cuba and Belize

The Cuban Context

Cuba has the lowest adult HIV prevalence rate in the Americas (0.01 percent), the lowest prevalence among women in the Caribbean (19 percent of people living with HIV as compared to 43 percent regionally), universal and free antiretroviral (ARV) medications, and free HIV tests (Gorry 2008; UNAIDS and WHO 2010b). The low HIV rate results from several factors: education about sexually transmitted infections (STIs), access to primary care, and culturally appropriate disease control policies (Aragonés et al. 2011).

The health of Cuba's revolutionary government is embodied in the strength of its medical system, a system that is then reflected in the health of individual Cubans. Since the triumph of the 1959 revolution, medical provision has been an integral part of the Cuban sociopolitical landscape. Health care is guaranteed by the Cuban constitution, and persons living with HIV are guaranteed adequate medical care, employment, and social security (Ministerio de Salud Pública 2007). In the Cuban model, HIV prevention and interventions have a top-down approach that reaches from the medical structures through educational structures, regional hospitals, neighborhood clinics, and finally into the spaces of the home.

The comprehensive health program, including sending doctors abroad to volunteer, contributes to Cuba's self-proscribed reputation as a "world medical power" (Feinsilver 1989). Although meeting the population's basic needs is the primary raison d'être for the extraordinary medical effort, the accumulation of capital, both symbolic and financial, has played an important role in Cuba's health policies (Feinsilver 2008). The need for political legitimacy at home and abroad and a desire for international prestige and influence further Cuba's "medical diplomacy" as a foreign policy tenet (Huish and Spiegel 2008; Kirk 2009).

Cuban doctors abroad have trained more than 50,000 medical students in developing countries (Reed 2007). Cuba's economic straits and the need for convertible currency have made further development of international services a necessity. The number of medical "volunteers" jumped from about 5,000 in 2003 to more than 25,000 in 2005, with 75 percent in Venezuela, their services provided in exchange for inexpensive oil (De Vos et al. 2008). In addition, Cuba's Latin American Medical School, established in 1999, educates students of more than 101 ethnicities from twenty-nine countries, including eighty-five students from the United States (Medical Education Cooperation with Cuba [MEDICC] 2010). Cuba's earnings from foreigners seeking care in Havana and the export of Cuban doctors amounted to US$2.3 billion in 2007—28 percent of total export receipts and net capital payments (Feinsilver 2008).

Cuba's geopolitical maneuvers have come at a cost; the first case of HIV was diagnosed in 1985 in a soldier returning from Angola, and the second case was in his wife. Cuban health officials had tested most of the adult population by the end of 1990, and thus the country has the most complete population-based countrywide serologic information in the world (Pérez et al. 2004). Currently, all returning volunteers are tested for HIV after their medical "tours of duty."

Cuba's medical system exemplifies a geographic approach to health and social monitoring. The 1984 introduction of the Family Doctor Program placed physicians in each neighborhood to determine the health picture for that catchment area (Gorry 2008). The pattern extends (and reflects) the Committees for the Defense of the Revolution charged with neighborhood vigilance and creating social programs. This family doctor delivery of health care has led to extraordinary statistics, including a low infant mortality rate, long life expectancy, and low rates of infectious disease. That success comes at the expense of personal privacy, however. For example, the health system reaches into the intimate space of the home; doctors make house calls and can be quite influential in getting patients tested for HIV. Indeed, all pregnant women are legally required to be tested (and receive medication if HIV-positive), ensuring that no children are born with the virus, according to an interview with a public health researcher, who is currently serving in Venezuela (R. González Cruz, personal interview 2007).

Those considered at highest risk in both Belize and Cuba are men who have sex with men (MSM), heterosexual women, and incarcerated individuals (Jaramillo and Gough 2006; Gorry 2008). I highlight MSMs and women in this section on Cuba because they demonstrate that the therapeutic imaginary, whereby political leaders frame Cuba as a medical haven, is much more nuanced than typically expressed. Gay men have been persecuted and prosecuted by Cuban revolutionary officials and labeled as psychologically and morally diseased (Leiner 1994). AIDS policies, though, have brought the MSM community to the forefront of health discourse in a more positive way than in other Caribbean countries. They actually have been able to convert what could have been a stigmatizing lens on them in this epidemic to becoming one of the primary forces in educating the population about HIV, thereby reducing negative stereotypes. For example, in 1996 the gay community was instrumental in creating the first *Memorias* AIDS quilt, which became a traveling symbol of HIV, and education about HIV, sexuality, and gender norms reached from Havana into rural villages across the country (Aragonés et al. 2011). The education efforts have created innovative gateways for information, including training hairdressers and practitioners of Afro-Cuban religions as HIV educators.

The fact that women are still at risk merits attention, however, as women have gained political and legal power but continue to be vulnerable to HIV in the space of the home. Most women are infected through their male partners; unequal power relationships have yet to be completely erased by the revolution (Romero and Echevarría 2011). As such, an important element of the therapeutic imaginary in Cuba is the perception that gender equality has been reached through legislation (Lichtenstein et al. 2005). Cuba is often imagined, and promotes itself, as a unique socialist nation with progressive gender and racial equity laws; however, domestic violence is still prevalent, mainly affecting women (Pope 2005). One of the opinions that younger generations often express is that the revolution has become an imagined space and concept, perhaps what Andrews (2004) termed "a place of the mind." Moreover, Cuban politicians and citizens often speak of "the Revolution" not as an event but as an organism capable of making decisions and informing culture, politics, and health.

Thus, the Revolution created one of the most controversial therapeutic places in Cuba—the countrywide AIDS sanatoria system. The first sanatorium, outside Havana, opened in 1986 and was directed by the military. The goal was to separate citizens diagnosed with HIV from the rest of the population. The Ministry of Public Health (MINSAP) took over by 1993 when outpatient care was allowed. In 1995 Cuba began to manufacture HIV test kits, and ARV medication was universally accessible by the early 2000s (UNAIDS and WHO 2010b). Many outsiders still judge the Cuban program as an anachronism in the civil libertarian climate of the late twentieth century (Hoffman 2004). Scheper-Hughes (1993, 966) echoed international public health sentiment that Cuba represented a "nightmare of hyper-vigilant medical police and of over-observed and over-disciplined bodies: A Foucauldian nightmare of medical 'discipline' verging on 'punishment.'" One of the outcomes of medicine-as-panopticon is that individuals might believe that the state itself is responsible for keeping HIV at bay, rather than engaging in individual risk aversion (Pope 2005).

The largest sanatorium in the country recently has been renamed Havana's Comprehensive Care Center for People with HIV/AIDS. The sanatorium policy shifted as more information about the virus was discovered, education campaigns diffused across the country, and ARVs were developed domestically. The change represents a shift in discourse, marking a new period in HIV management. Despite the power of MINSAP, individual experts and health workers have created spaces to navigate and shape the therapeutic narratives. In one instance, government-funded GPSIDA[1] has used revolutionary health dogma, and originally the space of the AIDS sanatorium, to resist mandatory confinement and challenge traditional stigmas (Pérez Avila 2008). One female health professional who contracted the virus while volunteering in Ethiopia notes the change in the Havana Sanatorium since she arrived in the 1980s:

> When I got here it was still a secret, obligatory place. But I never felt isolated or imprisoned or humiliated. . . . Also, the concept of the sanatorium has changed—now it's like a recuperative center, where you go when something extraordinary happens. . . . Welfare cases are different—they'll live here if they have nowhere else to go. (Gorry 2008, 28)

She stays at the sanatorium because "there are all kinds of germs out there. At least here I'm somewhat protected" (Gorry 2008, 28). Note the juxtaposition where this hospice, although portrayed as a violation of civil rights in international public health circles, is now reconceptualized as a safe space that protects residents from outsiders, a sentiment that was commonly expressed to me in interviews from the late 1990s.

Therapeutic narratives also resonate in the state's promotion of the island as a medical tourism destination (Reed 2007). The entire nation, as encapsulated and reified in its biomedical system, has thus become a built therapeutic landscape. It was because of the therapeutic landscape, political achievements, and the need for hard currency that Fidel Castro even admitted to the national assembly in the 1990s that, "We can say that they are highly educated *jineteras* [sex workers] and quite healthy, because we are in a country with the lowest number of AIDS cases. . . . Therefore there is truly no tourism healthier than Cuba's" (Paternostro 2000).

The Belizean Context

One of the most compelling reasons to highlight Belize is the relative dearth of academic literature on the country. As such, a relational perspective is useful here. Approximately 150 Cuban doctors, nurses, and technicians work in Belize and perform a critical function in the Belizean health care system, according to the Cuban Embassy. Between 2004 and 2008, Cuban doctors saw at least 2,000 Belizeans ("Cuban Ambassador" 2008). The therapeutic ideal is imported to Belize from other countries as well. In fact, in one of the main islands off the coast, Caye Caulker, the health care office is staffed only by a Cuban doctor and a Guyanese nurse.

HIV was first diagnosed in Belize in 1987. As in Cuba, the virus is assumed to have entered from abroad (in this case from Honduras), and Belize's location, economic policies, and gender norms have led to increasing transmission rates (National AIDS Commission [NAC] 2010). Belize is more reliant on international funding sources and foreign volunteers than Cuba, and the majority of its HIV program funding comes from the Global Fund (NAC 2010).

Three reasons why so little few exist on HIV are (1) Belizeans' reluctance to get tested, (2) the patchwork of private and public doctors whereby private doctors are not required to report HIV diagnoses, and (3) legal status. My participant observations at a testing center and interviews with medical officials have demonstrated that the close-knit communities and stigma surrounding HIV diagnosis influence individuals to travel to Belize City for anonymous testing. Thus, positive test results skew the geographic pattern of HIV prevalence; it appears that a much higher rate of HIV exists on the coast where the larger testing centers are located (Pope and Shoultz 2010). This stands in contrast to the Cuban system, where health care provision is inherently geographic and based on place of residence.

The Belizean education system, with its combination of public and private schools, is ingrained in its colonial past. No standardized curriculum about HIV exists, and because religious groups (including U.S. evangelical groups) administer and finance many schools, children often receive the messages that HIV is a moral disease and that abstinence is more important than using condoms (DeRose et al. 2010). Thus, many adolescents in Belize are wary of using condoms and getting tested for HIV and other STIs. Additionally, condoms are relatively expensive.[2] The Ministry of Health in Belize does not carry as much legal strength as that in Cuba. Although individuals working at the Ministry of Health and the NAC have a strong intent to educate, prevent, and diagnose HIV, recent interviews conducted with individuals living with HIV/AIDS show that the majority ($n = 45$ out of 50) did not know that Belize had an HIV policy, often because these participants could not read English.

Like the influential Cuban HIV prevention group, GPSIDA, several nongovernmental groups exist in Belize to educate the public, create safer places for individuals with HIV, and decrease stigma for gays and lesbians. In Belize sodomy is still illegal, which means that very few people who are gay feel empowered enough, either within personal relationships or in the public sphere, to engage in open dialogue about a stigmatizing subject (E. Castellanos, personal interview 2010). Although stigma against MSMs still exists in Cuba, it has tapered in recent years due to GPSIDA's work and the current proposal to legalize gay marriage. From personal observations and conversations, it appears that Belizean advocates still struggle to create safe places and spaces for MSMs and to create ways to combat gender norms that put women at risk. These gender norms include women's political and economic disadvantages in society at large and are particularly salient in minority communities, such as among Mayans. One such issue that increases women's risk for HIV and STIs is the fear that questioning a man's sexual experiences outside a relationship can lead to domestic violence (McClusky 2001).

Another important distinction between the two systems is the role of ethnicity in HIV discourse. In Belize, ethnicity has been one axis from which to interpret the geography of disease (Jaramillo and Gough 2006; Pope and Shoultz 2010; Pope forthcoming). Andrewin and

Chien (2008) found that Cuban doctors, perhaps ironically, held more discriminatory attitudes against people with HIV than doctors from other countries. Cuban data on ethnicity and health are not disseminated to international audiences. By erasing race and ethnicity, health inequality is symbolically erased in the therapeutic imaginary. Given the quantity of data collected by and disseminated to the medical community in Cuba, it seems the only reason for this not being a central analytical concept is that it does not dovetail with national sociopolitical goals.

Cuba's methods to monitor HIV, whether testing through the family doctor program or the sanatoria, cannot be duplicated in most countries, including Belize. In fact, in e-mail discussions I was privy to in 2011 with the Belize Ministry of Health, the idea to build an AIDS hospice in the countryside was considered discriminatory, as it would not target citizens with other diseases. Belize promotes its economy through the therapeutic landscape but in a much different sense from Cuba. The therapeutic landscape in Belize is a primarily rural escape meant to entice tourists to the Caribbean's "sun, sex, and sand" (Kempadoo 2004). Once the surface of paradise is scratched, however, traditional gender norms, economic marginalization, and regional linkages that lead to sex tourism create a topography of HIV and STI risk reminiscent of other developing countries in the region (Pattullo 2005). Economic desperation leads to male and female sex work and an increase in drug trafficking, and Belize's location allows men to cross into Mexico and Guatemala to buy sex, often without condoms (E. Castellanos, personal interview 2010). Conversely, Central American women migrate to Belize to sell sex because Belize's economy is stronger than those of its neighbors. Its geopolitical position, in terms of its reliance on international tourism and on foreign doctors to support its health care system, highlights that the image of Belize as a healing landscape might be more imagined than real.

Conclusion: From Therapeutic Landscapes to Therapeutic Imaginaries

These two countries are at historical crossroads. In Cuba, President Raul Castro recently announced that 500,000 health care workers will be leaving the state payroll. In Belize, a medical system that pits privatized care against public care seems to be intensifying. How will neoliberal economic changes affect the medical industry and health care provision in Cuba? Will universal access be guaranteed? Will Cuba continue to send doctors to Belize or will individuals be able to enter into private practice?

These two countries are "sites of pilgrimage," as Cuba attracts medical and political tourists and Belize attracts holiday goers and ecotourists. Increasingly, at both sites health has become a marketed commodity. Both of these examples represent a critical view of medical policy as an extension of history, geopolitics, marketing, health care discourses, and economic structures, all of which are filtered through multiscaled lenses to produce different outcomes and therapeutic narratives. This article shows how environments are marketed and that traditional notions of healing landscapes can be expanded and contextualized in unique ways. Cuba's therapeutic environment is a built one, originating in social justice movements and resulting in political capital. Thus, some citizens perceive the former sanatorium space as healthy and protective. Officials have marketed and transformed the island from a colonial space to one that serves a therapeutic purpose in the region, despite its limited civil rights. In Belize a natural, healing landscape might actually represent a therapeutic paradox; its preindustrial landscape might be promoted by the government as a restorative sojourn for tourists, but this obfuscates the disease patterns that reify traditional medical literature characterizing the tropics as diseased, or at least unhygienic.

Health care systems cannot be disentangled from political philosophies, nor can an individual's place in the political body be disentangled from medical philosophies. I argue for a critical approach to Cuba and Belize as two examples of divergent therapeutic imaginaries informed by the images, politics, and economic systems that, on the Cuban side, result from a unique revolutionary rupture with a colonial past and, on the Belizean side, still contend with colonial legacies of poverty and discrimination. I hope to have demonstrated throughout this article that these Caribbean therapeutic landscapes are worthy of being analyzed through the nuanced lens of therapeutic imaginaries. This article recognizes a crossover between the ideas of therapeutic landscapes and therapeutic imaginaries and among natural, built, and symbolic landscapes. As such, this framework can be extended to other regions providing a useful analytical lens to filter how perceptions of place interact with medical data and natural and built landscapes to create unique perceptions of healthy and diseased places.

Acknowledgment

The author would like to thank the anonymous reviewers for their valuable insights, suggestions, and knowledge about the article's subject matter.

Notes

1. Translated as AIDS Prevention Group, formed in the late 1980s in the Havana AIDS Sanatorium.
2. Based on fieldwork, the price of a condom in Cuba is currently about US$0.01, whereas in Belize the price is closer to US$1.00.

References

Andrewin, A., and L.-Y. Chien. 2008. Stigmatization of patients with HIV/AIDS among doctors and nurses in Belize. *AIDS Patient Care and STDs* 22 (11): 897–906.

Andrews, G. 2004. (Re)thinking the dynamics between healthcare and place: Therapeutic geographies in treatment and care practices. *Area* 36 (3): 307–18.

Aragonés, C., J. R. Campos, O. Nogueira, and J. Pérez. 2011. Raising HIV/AIDS awareness through Cuba's *Memorias* project. *MEDICC Review* 13 (2): 38–42.

Conradson, D. 2005. Landscape, care and the relational self: Therapeutic encounters in southern England. *Health and Place* 11 (4): 337–48.

Cuban ambassador concludes tour of duty in Belize. 2008. LoveFM.com: News and Power Music from Belize 7 August 2008. http://www.lovefm.com/ndisplay.php?nid=8400 (last accessed 1 December 2011).

Cummins, S., S. Curtis, A. V. Diez-Roux, and S. Macintyre. 2007. Understanding and representing "place" in health research: A relational approach. *Social Science & Medicine* 65:1825–38.

Dawney, L. 2011. Social imaginaries and therapeutic self-work: The ethics of the embodied imagination. *The Sociological Review* 59 (3): 535–52.

DeRose, K. P, D. E. Kanonse, D. P. Kennedy, K. Patel, A. Taylor, K. J. Leuschner, and H. Martinez. 2010. *The role of faith-based organizations in HIV prevention and care in Central America*. Santa Monica, CA: Rand Corporation.

De Vos, P., W. de Ceukelaire, M. Bonet, and P. Van der Stuyft. 2008. Cuba's health system: Challenges ahead. *Health Policy and Planning* 23 (4): 288–90.

Faubion, T. 2009. Multiplicity of meaning: Living with HIV/AIDS. In *HIV/AIDS: Global frontiers in prevention/intervention*, ed. C. Pope, R. White, and R. Malow, 451–58. London and New York: Routledge.

Feinsilver, J. M. 1989. Cuba as a "world medical power": The politics of symbolism. *Latin American Research Review* 24 (2): 1–34.

———. 2008. Oil-for-doctors: Cuban medical diplomacy gets a little help from a Venezuelan friend. *Nueva Sociedad* 216.

Flaskas, C. 2009. The therapist's imagination of self in relation to clients: Beginning ideas on the flexibility of empathic imagination. *Australian & New Zealand Journal of Family Therapy* 30 (3): 147–59.

Friedman, S., H. L. F. Cooper, and A. H. Osborne. 2009. Structural and social contexts of HIV risk among African Americans. *American Journal of Public Health* 99 (6): 1002–1008.

Gesler, W. M. 1992. Therapeutic landscapes: Medical issues in light of the new cultural geography. *Social Science and Medicine* 34 (7): 735–46.

Gorry, C. 2008. *Cuba's HIV/AIDS strategy: An integrated, rights-based approach*. Havana, Cuba: OXFAM.

Hoffman, S. Z. 2004. HIV/AIDS in Cuba: A model for care or an ethical dilemma? *African Health Sciences* 4 (3): 208–209.

Huish, R., and J. Spiegel. 2008. Integrating health and human security into foreign policy: Cuba's surprising success. *The International Journal of Cuban Studies* 1 (1): 1–13.

Jaramillo, R., and E. Gough. 2006. *Belize national specialist, Dangriga composite policy index—2006*. Belize City, Belize: National AIDS Committee.

Jones, K., and G. Moon. 1993. Medical geography: Taking space seriously. *Progress in Human Geography* 17:515–24.

Kalipeni, E., S. Craddock, J. Oppong, and J. Ghosh, eds. 2003. *HIV and AIDS in Africa: Beyond epidemiology*. New York: Wiley-Blackwell.

Kearns, R. 1997. Narrative and metaphor in health geographies. *Progress in Human Geography* 21 (2): 269–77.

Kearns, R., and J. R. Barnett. 2000. "Happy meals" in the Starship Enterprise: Interpreting a moral geography of health care consumption. *Health & Place* 6 (2): 81–93.

Kempadoo, K. 2004. *Sexing the Caribbean: Gender, race, and sexual labor*. London and New York: Routledge.

Kirk, J. M. 2009. Cuban medical internationalism and its role in Cuban foreign policy. *Diplomacy and Statecraft* 20:275–90.

Latkin, C. A., S. J. Kuramoto, M. A. Davey-Rothwell, and K. E. Tobin. 2010. Social norms, social networks, and HIV risk behavior among injection drug users. *AIDS Behavior* 14 (5): 1169–81.

Lea, J. 2008. Retreating to nature: Rethinking "therapeutic landscapes." *Area* 40 (1): 90–98.

Leiner, M. 1994. *Sexual politics in Cuba: Machismo, homosexuality, and AIDS*. Boulder, CO: Westview Press.

Lichtenstein, B., B. A. Abboud, S. L. Brodsky, and A. Oakes. 2005. Mujeres de Carácter: The strong women of Cuba. Paper presented at the meeting of the American Sociological Association, Philadelphia, PA.

Long, P. G. 2008. *Therapeutic imagination: A key feature of nursing care in forensic mental health care settings*. Newcastle, UK: University of Newcastle, School of Nursing and Midwifery.

McClusky, L. J. 2001. *Here, our culture is hard: Stories of domestic violence from a Mayan community in Belize*. Austin: University of Texas Press.

Medical Education Cooperation with Cuba. 2010. Cuba & the global health workforce: Health professionals abroad. (last accessed 1 December 2011).

Ministerio de Salud Pública [Ministry of Public Health]. 2007. *Anuario Estadística de Salud 2006* [Health Statistics Yearbook 2006]. Havana, Cuba: Ministerio de Salud Pública.

Myers, J., and R. Kearns. 2009. Feelings, bodies, places: New directions for geographies of HIV/AIDS. In *HIV/AIDS: Global frontiers in HIV prevention/intervention*, ed. C.

Pope, R. White, and R. Malow, 501–10. London and New York: Routledge.

National AIDS Commission. 2010. *UNGASS country progress report: Belize*. Geneva, Switzerland: UNAIDS.

Paternostro, S. 2000. Communism versus prostitution: Sexual revolution. *New Republic Online* 10 and 17 July. http://www.cubanet.org/CNews/y00/jun00/30e17.htm (last accessed 1 December 2011).

Pattullo, P. 2005. *Last resorts: The cost of tourism in the Caribbean*. New York: Monthly Review Press.

Pérez, J., D. Pérez, I. González, M. Diaz Jidy, M. Orta, C. Aragonés, J. Joanes, et al. 2004. *Approaches to the management of HIV/AIDS in Cuba: Case study*. Geneva, Switzerland: World Health Organization.

Pérez Avila, J. 2008. *SIDA: Confesiones a un médico* [AIDS: Confessions to a doctor]. Havana, Cuba: Casa Editora Abril.

Pope, C. 2005. The political economy of desire: Gendered construction of sex work in Havana, Cuba. *Journal of International Women's Studies* 6 (2): 99–118.

———. Forthcoming. Geographies of HIV and marginalization: A case study of HIV/AIDS risk and Mayan communities in western Belize. In *Ecologies and politics of health*, ed. B. King and K. Crews. London and New York: Routledge.

Pope, C., R. White, and R. Malow. 2009. Global convergences: Emerging issues in HIV risk, prevention, and treatment. In *HIV/AIDS: Global frontiers in prevention/intervention*, ed. C. Pope, R. White, and R. Malow, 1–10. London and New York: Routledge.

Pope, C. K., and G. Shoultz. 2010. An interdisciplinary approach to HIV/AIDS stigma and discrimination in Belize: The roles of geography and ethnicity. *GeoJournal*. DOI:10.1007/s10708-010-9360-z

Reed, G. 2007. From the source: Cuba's health minister on "SICKO" and more. http://www.medicc.org/cubahealthreports/chr-article.php?&a=1032 (last accessed 1 December 2011).

Romero, M., and D. Echevarría. 2011. *Convergencia de géneros* [Convergence of genders]. Havana, Cuba: University of Havana Press.

Scheper-Hughes, N. 1993. AIDS, public health, and human rights in Cuba. *Lancet* 342 (8877): 965–68.

Smyth, F. 2005. Medical geography: Therapeutic places, spaces and networks. *Progress in Human Geography* 29:488–95.

UNAIDS and World Health Organization. 2010a. *Epidemiological factsheets on HIV and AIDS: Belize*. http://www.unaids.org/en/dataanalysis/epidemiology/epidemiologicalfactsheets/#d.en.52699 (last accessed 1 December 2011).

———. 2010b. *Epidemiological factsheets on HIV and AIDS: Cuba*. http://www.unaids.org/en/dataanalysis/epidemiology/epidemiologicalfactsheets/#d.en.52699 (last accessed 1 December 2011).

Williams, A. 1998. Therapeutic landscapes in holistic medicine. *Social Science & Medicine* 46 (9): 1193–1203.

———, ed. 2007. *Therapeutic landscapes*. Surrey, UK: Ashgate.

Wilson, K. 2003. Therapeutic landscapes and First Nations peoples: An exploration of culture, health and place. *Health & Place* 9:83–93.

Producing Contaminated Citizens: Toward a Nature–Society Geography of Health and Well-Being

Farhana Sultana

Department of Geography, Syracuse University

A nature–society geography approach to health and well-being demonstrates that socioecological parameters, in addition to economic and political factors, are critical to explaining outcomes of health crises. In expounding on this multifaceted understanding of health and well-being in the context of development, I draw on research on chronic arsenic poisoning and water contamination in rural Bangladesh. A public health crisis has arisen from naturally-occurring arsenic poisoning millions of people who drink, cook, and irrigate with arsenic-laced groundwater pumped up by tubewells, where the very sources that were promoted to bring health are now bringing illness, hardship, and death. In examining the interlinked ways that arsenic and water come to influence well-being and illness, I pay particular attention to social stigma and the production of contaminated citizens. By engaging the insights from nature–society geographies of health and feminist geographies of well-being in contributing to scholarship in geographies of health, the article highlights that the experiences of health and well-being are complex and evolving in instances where slow poisoning is simultaneously an outcome of development endeavors and environmental factors.

健康和福祉的自然—社会地理方法表明，除经济和政治以外，社会生态学参数，是解释健康危机结果的至关重要的因素。我在阐述发展的背景下，多方面了解健康和福祉这一问题，研究孟加拉国农村的慢性砷中毒和水污染情况。因使用泵管井泵出的含砷地下水喝，做饭，和灌溉而自然地发生在百万人身上的中毒现象，已经导致了一场公共卫生危机。这一水源原本被认为会带来健康，现在带来的是疾病，困苦，和死亡。通过审议影响福祉和疾病的砷和水相互关联的方式，我特别关注社会耻辱和污染公民的产生。通过洞察自然与社会健康地理和促进健康地理学术发展的女权地理的福祉，文章强调，健康和福祉的经验是复杂的，是在慢性中毒同时作为发展的努力和环境因素的一种结果的大情况下发展的。*关键词：砷，健康，自然—社会，耻辱，福祉。*

Abordar el tema de la salud y el bienestar con el enfoque geográfico expresivo de la relación naturaleza–sociedad demuestra que los parámetros socioecológicos, además de los factores económicos y políticos, son cruciales para explicar lo que sobreviene de las crisis de la salud. Para una mayor elaboración de esta manera multifacética de entender la salud y el bienestar en el contexto del desarrollo, me baso en investigaciones sobre envenenamiento crónico con arsénico y aguas contaminadas en el espacio rural de Bangladesh. Ha surgido una crisis sanitaria por el envenenamiento de origen natural entre millones de personas que beben, cocinan y riegan con agua cargada de arsénico, la cual es bombeada a la superficie a través de pozos entubados, donde las propias fuentes que se abrieron para traer salud están ahora aportando enfermedad, sufrimiento y muerte. Al examinar los entrelazamientos por medio de los cuales el arsénico y el agua llegan a influir bienestar y enfermedad, pongo particular atención al estigma social y a la producción de ciudadanos contaminados. Al buscar las luces de las geografías de la salud inspiradas en la relación naturaleza–sociedad y las geografías feministas del bienestar para contribuir de manera académica específica a las geografías de la salud, el artículo destaca que las experiencias de salud y bienestar son complejas y evolucionan en instancias en las que el envenenamiento lento puede ser simultáneamente un resultado de propósitos de desarrollo y de factores ambientales.

Health geographers have significantly contributed to debates about health, place, and well-being in a variety of contexts globally (for overviews, see Gatrell and Elliot 2009; Brown, McLafferty, and Moon 2010; Kearns and Collins 2010). Given the linkage among health, well-being, and overall social development, it is critical to look at the multifaceted ways that various health concerns affect people's everyday lives and opportunities to be healthy (Kearns and Andrews 2010). Critical scholars of nature–society geography have been enriching existing literatures in recent years by more forcefully engaging with environmental and social systems simultaneously. Mayer (1996, 2000), Mansfield (2008, 2011), and King (2010) have pointed to fruitful avenues of research that engage nature–society geography, especially insights from political ecology, highlighting the salience of broader political economy and the environment to explanations of health and well-being. Such perspectives have also shifted the dominant focus from infectious diseases to exploring various aspects of environmental change and impacts on human health, such that health is understood as inherently a nature–society issue. Critical political ecology of noninfectious health is thus an emerging body of scholarship that is contributing to such debates (e.g., McGee 1999; Eyles and Elliott 2001; Richmond et al. 2005; Sultana 2006, 2007b; Hanchette 2008; Biehler and Simon 2011). To this end, political ecologies of health are being contextualized more broadly within development interventions, where environment–development contradictions animate the trajectories of public health debates in the global south, as this article demonstrates.

Engaging such insights with those from feminist geography provides further nuanced ways to analyze health and well-being. Scholarship in gender and health has specifically emphasized the ways that well-being is closely tied to notions and experiences of gender (Dyck, Lewis, and McLafferty 2001; Moss and Dyck 2002). Feminist geographers have richly debated and elucidated the ways that embodied notions of well-being are important to explaining the everyday ramifications of chronic illnesses to life and livelihood (Chant and McIlwaine 2009; Del Casino 2010), as well as the ways that gendered embodiments are constructed, challenged, and experienced in everyday spaces and practices (Longhurst 2001; Sothern and Dyck 2010; Sultana 2011). Informed by such insights, I analyze chronic arsenic poisoning from contaminated drinking water in Bangladesh. I demonstrate the importance of investigating the complex intersections of social processes, environmental change, and embodied well-being to shed greater light on the nature–society geographies of health and well-being.

Arsenic Poisoning: The Making and Unmaking of a Public Health Success Story

In 2000, the World Health Organization (WHO) declared that arsenic poisoning of nearly 30 million people in Bangladesh was the "largest mass poisoning of a population in history" (Smith, Lingas, and Rahman 2000, 1; WHO 2000). Such a claim alluded to, and brought international attention to, the severity of the problem of drinking water contamination by naturally occurring arsenic. People had been drinking, cooking, and irrigating with arsenic-contaminated water for years, in response to the promotion of using groundwater from aquifers in previous decades through development policies aimed to provide "safe" groundwater for human use. As groundwater was already contaminated with arsenic unknown to people at first, however, the usage of groundwater for human consumption continued for many years and exposed unsuspecting societies to slow poisoning from arsenic. As the situation was increasingly becoming dire, scholars in various fields started to undertake research on the arsenic crisis in Bangladesh, predominantly from public health, epidemiological, geological, and policy perspectives but also from critical social and geographical perspectives (for greater detail, see Ahmed and Ahmed 2002; Ahmed 2003; Paul 2006; Sultana 2006, 2007a, 2007b, 2009b; Atkins, Hassan, and Dunn 2007).

The tragic irony of the crisis arises from the fact that carcinogenic arsenic occurs naturally in the sediments of the delta aquifers, thereby making arsenic enter socioecological worlds through abstraction of groundwater by tubewells that are used for domestic and irrigation purposes. Tubewells were heavily promoted by the state and development institutions in the 1970s and 1980s as part and parcel of development planning in safe water provision in much of South Asia, to reduce waterborne diseases (e.g., cholera, diarrhea, etc.) that led to high infant mortality and morbidity rates (Briscoe 1978). The goal of development endeavors was to reduce overall incidences of illnesses from polluted water consumption through the large-scale introduction of tubewells that would pump up "safe" groundwater (as aquifers were largely free of pathogens). Groundwater was deemed a safe alternative and one that was available in abundance in the delta (and annually recharged by the monsoonal climate). This was heralded as a public health success story, and it was touted around the world that rural populations had switched over to drinking safe groundwater within only a few years.

Yet the presence of arsenic was not tested for in the 10 million tubewells that were installed by government, nongovernmental organization (NGO), and private funding. This situation enabled millions of people to consume groundwater with the belief that they were drinking safe water. Although this water was largely free of living contaminants, it was laced with the nonliving contaminant of carcinogenic arsenic. Because it is impossible to detect whether water is arsenic-free from taste, color, or smell, people continued to drink deadly water. Scientific testing is needed to identify whether trace amounts of arsenic exist in the water or not and whether the allowable limits have been exceeded. Chronic exposure to small amounts of arsenic over years causes arsenicosis (arsenic poisoning), which leads to various health problems, such as skin spots (melanosis, keratosis), cancer, organ failure, and ultimately death. Given the time frame of about five to fifteen years for chronic arsenic poisoning to manifest physical symptoms, it was not until growing numbers of people showed signs of arsenic poisoning in the late 1990s that development planners and policymakers started to come to grips with the rapidly escalating problem. Estimates of 30 to 35 million people consuming arsenic-laced water quickly came to underscore the enormity of the crisis, and the statistics of people not only exposed to drinking contaminated water but also falling ill were expected to rise over time. The rising number of people manifesting signs of arsenic-related illnesses prompted a large-scale awareness and information-sharing campaign throughout the country, heavily funded by the same international donor institutions that were responsible for promoting the tubewells in the first place.

Despite the awareness campaigns and testing of water quality, millions of people have continued to drink arsenic-contaminated water, largely due to lack of viable safe alternatives. The uncertainty about the amount of arsenic in the aquifer, and subsequently in drinking water, confounded early attempts to convince people that their tubewell water was no longer safe. Skepticism and concerns about the quality of alternative water sources kept many households from switching away from groundwater on which they had come to depend. The risk substitution involved weighing very difficult options for each household: either consuming pathogen-contaminated water and falling ill immediately (especially if insufficient resources existed to obtain fuel wood to boil the water or treat it with some other method) versus taking a chance of risking long-term illness from arsenic but having almost no immediate ill health. Such decisions became more complicated when officials tested the water in a national water screening program in 2001 and 2002 and identified contaminated tubewells with different colors: safe tubewells' spouts were painted green and unsafe tubewells' spouts were painted red (British Geological Survey/Department of Public Health Engineering 2001). Red-painted tubewells had arsenic in concentrations above 50 ppb (parts per billion), whereas green-painted tubewells had concentrations below 50 ppb; such guidelines followed the standards set by the Bangladesh government, although the WHO has more stringent standards of 10 ppb for allowable amounts of arsenic in drinking water. Due to the heterogeneous nature of arsenic deposits in the aquifer sediments, and the different rates of release into groundwater, tubewells within close proximity could display very different levels of arsenic contamination in the water pumped out. Thus, safe or unsafe tubewells have to be tested and identified as such.

Although color-coding was supposed to present quick visual markers of safe and unsafe water sources to enable people to identify and obtain safe water, the conflicts over scarce safe water sources (i.e., green-painted tubewells) made social access to safe water a challenge in fulfilling everyday water needs (Sultana 2009b, 2011). Some households benefited from having safe green tubewells, whereas others faced increasing challenges of owning an unsafe red tubewell and having to negotiate their access to safe water (Sultana, 2007b). Mass awareness campaigns attempted to communicate basic information about the sources of arsenic and the contamination of tubewells, as well as the symptoms and causes of arsenicosis, but information alone was insufficient to enable people to switch to safer water sources. Safe alternative water sources that are physically, socially, and financially viable were not available in most places, posing severe constraints on those who knew about arsenic but could not find viable alternatives, as well as exacerbating a particularly gendered burden of fetching safe water because women are tasked with fetching drinking water daily for households (Sultana 2009b). Although many households switched to safe sources as best they could (often by investing in a considerably more expensive, deeper tubewell that accessed the deeper ancient aquifer that was largely free of arsenic), millions of households continued to consume arsenic-laced water because they did not have access to safe water sources or funds to invest in a deeper well. As a result, a public health success story of safe water consumption in the country (with statistics of 97 percent of the population having access to safe potable water in the late 1990s being heralded by various development institutions and the government) quickly turned into a public health nightmare (where millions of people were identified to be at risk of consuming arsenic-contaminated drinking water). Within a short time, by the early 2000s, increasing numbers of tubewells were identified as contaminated, and more people were found to be ill with arsenicosis and related health problems. Cancer rates in Bangladesh are expected to rise over time as more patients are identified to have been exposed to chronic arsenic poisoning.

Producing Contaminated Citizens

Contaminated water's impacts are largely felt in the multiple arenas of health and well-being, social stigma and ostracization, and socioeconomic burdens. The social implications of chronic water poisoning and illness manifest in many aspects of everyday life and livelihood. Although physical illnesses are widely prevalent and documented, this is further compounded by emotional stress and incidences of depression (see also Brinkel, Khan, and Kraemer 2009). The ways that arsenic comes to affect both physical and emotional well-being needs much greater attention from health geographers as well as practitioners in the field, as their focus has primarily been on identifying and addressing bodily symptoms and disease burdens of arsenic poisoning and related complications (e.g., cancer of the kidney, heart, and liver). Although these are critically important, there are complex ways that the well-being of entire families is affected by having arsenicosis patients in the home, as well as from living with fear and uncertainty, dealing with rejection from society, and coping with the multifaceted lived experiences of ostracism and stigmatization.

Most of the areas with arsenic contamination have been targeted by development programs that focused on awareness and mitigation endeavors (such as trialing of water filtration systems or community-based water projects), but one aspect that has lagged behind is identification of arsenicosis patients, providing adequate health care, and addressing the social outcomes of arsenicosis for patients, their households, and their communities. There exist significant gaps in the awareness about and understanding of arsenic, its implications and transmission, ways to deal with different symptoms of arsenicosis, and health management options (Mosler, Blochliger, and Inauen 2010). Among rural communities that are grappling with the uncertainties of water poisoning, there appears to be considerable misperception and confusion about what arsenic does, how it affects the body, how it can be treated at different stages, and how to avoid misdiagnosis (Rosenboom 2004). Time is a challenging factor here, as manifestations of arsenicosis can take years, making awareness

campaigns more effective in reducing health impacts if people have been exposed to consuming contaminated water for a shorter time and have alternative safe water sources to which they can switch over. Many people have some general knowledge about skin spots and rashes, as these skin-level symptoms are often most visible in the early onset of poisoning, but most people are generally unaware of or confused about other symptoms (especially those that can lead to various health complications over time). Those who have seen arsenicosis patients or were afflicted themselves are more aware of the health issues involved and more keen about accessing health care (e.g., ameliorative supplements, ointments, medicines, and more aggressive treatment if needed). The prevalence of arsenicosis patients in different areas varies considerably, however, so not everyone in arsenic-affected areas has seen arsenicosis symptoms. Given differences in individual physiologies and exposure levels, there has been a wide variance in arsenicosis occurrence rates across areas with similar levels of arsenic contamination of water and contradictory statistics of morbidity or mortality. To what extent people know exactly in what ways they can help themselves in dealing with the health impacts of arsenic is still debatable. A large proportion of the population rely on information from second- or third-hand sources, with high rates of illiteracy that reduce the effectiveness of written information. There also persist superstitions about the symptoms of arsenicosis (e.g., fear of the condition being contagious). In many cases, the medical costs, especially for people with advanced stages of arsenicosis, are prohibitive for many households, which can also influence patients' abilities to obtain medical assistance in the first place. Furthermore, faith in traditional healers can compound problems when arsenic poisoning goes misdiagnosed or untreated for longer periods of time.

Many people do not think that arsenicosis is a contagious disease or understand that it resulted from drinking contaminated water, and few are knowledgeable about the stages of arsenicosis. Some people perceived arsenic water to be deadly and were more fearful, but there were also skeptical people who did not believe arsenic was a problem, especially if they were drinking water from a contaminated tubewell but had not developed any symptoms yet. Because the predominant way that arsenic has been described in awareness programs in the vernacular is *beesh* (lethal venomous poison), confusion and skepticism arose when no immediate deaths resulted from consuming poisonous water. High levels of variation exist in understanding and acceptance of information on chronic arsenic poisoning.

Those who were more scared were convinced that they should avoid contaminated water, but often this did not stop them from consuming contaminated water if there were no other safe alternatives in their village (Sultana 2006). This creates considerable anxiety and frustration among households who want to obtain safe water but cannot. The sense of despair, grief, and anger in some instances is linked to feeling cheated by development programs that promised safe water but inadvertently ended up poisoning so many people. Thus, poisoned citizens were produced through intersections of geology, development planning, and social realities.

The degree of arsenicosis determines what affect it has on the body and what interventions might be effective. If caught early on, symptoms of arsenicosis can be ameliorated with ceasing further intake of arsenic-laced water (and food) as well as increasing the consumption of nutritional supplements. More advanced cases of poisoning often require aggressive medical treatment to deal with the various health complications. Messages informing people to consume arsenic-free water and more nutritious food to combat arsenic's effects are likely to be useful to those who can afford to do so, however. It is more challenging for poorer households, where individuals are generally malnourished to begin with and have access to even fewer resources for nutritional food or medical treatment. Class is compounded by gender inequalities, where women are often further marginalized in having their health concerns identified, heeded, or addressed. As scholars have widely noted regarding the gender disparity in access to health care globally (Kabeer 1994; Dyck, Lewis, and McLafferty 2001; Curtis 2004), women in rural areas of Bangladesh are less likely to be able to afford and obtain medical attention for health manifestations of arsenic poisoning (Nasreen 2003). Access to adequate health care is a problem throughout rural areas, due to lack of sufficient health care facilities and doctors, as well as the distances and costs involved in accessing health care. This often results in households underplaying illnesses, and often women are denied health care due to their lack of voice, financial resources, or chaperone to accompany them to medical facilities. Many women are also reluctant to be identified as arsenicosis patients and thus become marked socially as someone with arsenicosis, because the stigma of being ill is often a greater emotional stressor. It has been noted by medical fieldworkers that women often cover their bodies even more to hide visible signs of arsenicosis when diagnostic teams come to a village. The overall fear of arsenicosis is explained by not only the bodily health and illness factors but also from social stigmatization associated with it and the gendered dynamics of stigma, as I discuss in the next section.

Stigma and Well-Being in Poisoned Waterscapes

Scholars have posited that stigma is a social process that affects not only health but also people's sense of self, well-being, and place (Das 1997). Stigma is faced by individual arsenicosis patients, their families, and even entire areas that have large numbers of arsenic-contaminated tubewells. This is largely due to earlier beliefs that arsenicosis was contagious but also due to opinions that association with ill people is generally a bad idea. Socially constructed norms of who is valued or devalued and stigmatized have thus further complicated the well-being and suffering of people in arsenic-affected areas. Across villages and households, general discomfort exists in associating or socializing with people who have fallen ill with arsenicosis and related medical conditions. General ostracism and marginalization of afflicted families and patients occurs in both subtle and overt ways. People with arsenicosis are often denied work, terminated from their jobs, or treated as social pariahs. Such social outcomes result in exacerbating the general ill-being that people felt from their bodily afflictions and living in contaminated environments. Although most people did not feel that they deserved such treatment (because it was not a contagious illness), the subtle ways that socially constructed notions of acceptance, value, and stigma operate can complicate clear-cut analysis of the situation.

Many people who are afflicted or have arsenicosis victims in their family find that the wider public does not always understand that they are not contagious and that it is difficult to change perceptions. Only the very aware or more educated persons believe that arsenicosis would not be a problem in general socializing, but there is still reluctance by the majority to fraternize with afflicted patients. A substantial minority of the people are openly willing to shun arsenicosis sufferers, highlighting the broader societal problems faced by those living with the condition. This reflects that there are awareness and acceptance gaps in rural societies where arsenic is acute. Overall, both men and women agree that social acceptance and integration are major issues for arsenicosis victims in their community, but there is greater reluctance to associate with a female arsenicosis patient than a male one, as ill women are often shunned in

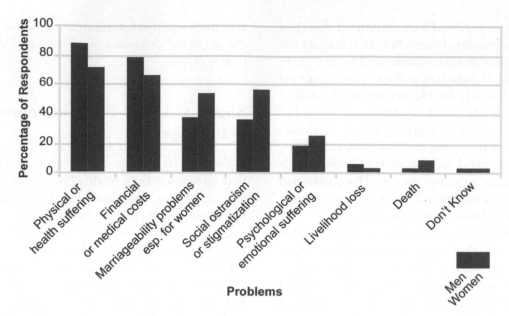

Figure 1. Perceptions of problems facing arsenicosis patients. Source of data: Author. (Color figure available online.)

general. There is a general sense that women are agents of bad luck, and an ill one would be a curse to the family. Although entire households and areas might be stigmatized as outsiders, the situation is particularly difficult for afflicted women and girls in any household.

Figure 1 demonstrates the gender differences in the perceptions about arsenicosis and the problems faced by afflicted people in a survey of 232 people across eighteen villages.[1] Both men and women identified that physical and bodily health suffering was the primary problem. This was followed by financial costs incurred from both medical expenses and costs of trying to obtain safe water (whether purchasing a deeper tubewell or cost-sharing in a water project; see Sultana 2009a). A higher percentage of men, compared to women, identified these two issues as the top two critical problems. The next two items are largely social (social stigmatization and marriageability), where higher percentages of women compared to men deemed the issues to be significant. Nearly 53 percent of the women, compared to 34 percent of the men, identified the biggest social problem to be marriageability issues for women and general social ostracism, stigmatization, and rejection of ill women. Many mothers were worried about the prospect of not finding husbands for their daughters, and younger women were worried about whether they would be valued or desired as wives if they were showing symptoms of arsenicosis. The psychological stress and emotional angst experienced by women who were already ill were followed by the anxiety felt by those who were not ill yet but concerned about what would happen in the fu-

ture. The superstitions that prevailed about arsenicosis, and the stigma against people who were ill, had resulted in divorces, abandonments, and spinsterhood for many women. Although some men felt social stigma as well, it was predominantly women who experienced it disproportionately and were increasingly concerned about it. The complex intersections of stigma, well-being, and social relations are thus evident.

In terms of how the situation played out across socioeconomic class categories, the nature of the problem is starker. The relational nature of class means that the relations that reproduce inequalities also maintain class differences and poverty. In such relations, it is seen that water and arsenic have come to play an important role in the ways that people are variously marginalized or impoverished. One of the most noticeable outcomes of arsenic poisoning has been the ways that ill health and subsequent treatment costs have dramatically affected those with uncertain or limited access to medical resources. It has also resulted in the loss of livelihood from inability to work, as well as from ostracism and stigmatization that resulted in difficulty in finding employment or keeping existing jobs. The less common pathway that impoverishment had come to affect households is from arsenicosis deaths, especially among earning members of the household. Social implications of arsenicosis manifested across a range of issues that iteratively combined to produce illness and ill-being. As a result, the outcomes of arsenic poisoning affected multiple arenas of everyday life and jeopardized the well-being of individuals, households, and entire communities, where a hazardous environment interacted with society in producing differentiated levels of risk, contaminated bodies, and opportunities to be healthy.

Conclusion

Health and well-being in developing contexts are imbricated with a host of issues, but a nature–society geography approach highlights complex interactions of ecological and geological systems with social systems. Feminist geography insights, as well as those from critical geographies of health, further illuminate the ways that health, well-being, stigma, and illness are lived and experienced by gendered bodies and communities in hazardous environments. As demonstrated in this article, the ways that people cope with and respond to environmental risks such as water poisoning are complicated by broader social processes, histories, and policies of development interventions, which are further compounded by gender and class differentiations and environmental heterogeneity in the groundwater and local geology that produced an acute but uneven crisis. Geological factors, social processes, and power relations in uneven hazardscapes thus intersected to produce contaminated citizens. These intersections of social dimensions of environmental risks, development processes, and nature further underscore the complexities involved in assessing well-being and health in a developing context. By engaging insights from nature–society geographies of health and feminist geographies of well-being in contributing to existing scholarship in geographies of health, the article highlighted that the experiences of health and well-being are complex and evolving in instances where slow poisoning is simultaneously an outcome of development endeavors and environmental factors and that attention to complex socioecological relations is vital to explanations of well-being and health.

Acknowledgments

I am very grateful to the anonymous reviewers and Mei-Po Kwan for excellent feedback. All errors remain mine.

Note

1. Lack of space prevents me from elaborating on the details of the study, but further information is available in Sultana (2007a).

References

Ahmed, M. 2003. *Arsenic contamination: Bangladesh perspective.* Dhaka, Bangladesh: ITN-Bangladesh.

Ahmed, M., and C. Ahmed. 2002. *Arsenic mitigation in Bangladesh.* Dhaka, Bangladesh: Local Government Division, Government of Bangladesh.

Atkins, P., M. Hassan, and C. Dunn. 2007. Environmental irony: Summoning death in Bangladesh. *Environment and Planning A* 39:2699–2714.

Biehler, D., and G. Simon. 2011. The great indoors: Research frontiers on indoor environments as active political ecological spaces. *Progress in Human Geography* 35 (2): 172–92.

Brinkel, J., M. Khan, and A. Kraemer. 2009. A systematic review of arsenic exposure and its social and mental health effects with special reference to Bangladesh. *International Journal of Environmental Research and Public Health* 6 (5): 1609–19.

Briscoe, J. 1978. The role of water supply in improving health in poor countries (with special reference to Bangladesh). *American Journal of Clinical Nutrition* 31.2100–13.

British Geological Survey/Department of Public Health Engineering. 2001. Arsenic contamination of groundwater in Bangladesh. Final BGS Technical Report WC/00/19, British Geological Survey, Keyworth, UK.

Brown, T., S. McLafferty, and G. Moon. 2010. *A companion to health and medical geography.* Malden, MA: Wiley-Blackwell.

Chant, S., and C. McIlwaine. 2009. *Geographies of development in the 21st century: An introduction to the global south.* Cheltenham, UK: Edward Elgar.

Curtis, S. 2004. *Health and inequality: Geographical perspectives.* London: Sage.

Das, V. 1997. Language and body: Transactions in the construction of pain. In *Social suffering,* ed. A. Kleinman, V. Das, and M. Lock, 67–92. Berkeley: University of California Press.

Del Casino, V. 2010. Living with and experiencing (dis)ease. In *A companion to health and medical geography,* ed. T. Brown, S. McLafferty, and G. Moon, 188–204. Chichester, UK: Wiley-Blackwell.

Dyck, I., N. Lewis, and S. McLafferty. 2001. *Geographies of women's health.* London and New York: Routledge.

Eyles, J., and S. Elliott. 2001. Global environmental change and human health. *Canadian Geographer* 45:99–104.

Gatrell, A., and S. Elliot 2009. *Geographies of health: An introduction.* 2nd ed. Chichester, UK: Blackwell.

Hanchette, C. 2008. The political ecology of lead poisoning in eastern North Carolina. *Health and Place* 14 (2): 209–16.

Kabeer, N. 1994. *Reversed realities: Gender hierarchies in development thought.* London: Verso.

Kearns, R., and G. Andrews. 2010. Geographies of wellbeing. In *The Sage handbook of social geographies,* ed. S. Smith, S. Marston, R. Pain, and J. P. Jones III, 309–28. London: Sage.

Kearns, R., and D. Collins. 2010. Health geography. In *A companion to health and medical geography,* ed. T. Brown,

S. McLafferty, and G. Moon, 15–32. Malden, MA: Wiley-Blackwell.

King, B. 2010. Political ecologies of health. *Progress in Human Geography* 34 (1): 38–55.

Longhurst, R. 2001. *Bodies: Exploring fluid boundaries*. London and New York: Routledge.

Mansfield, B. 2008. Health as a nature–society question. *Environment and Planning A* 40:1015–19.

———. 2011. Is fish health food or poison? Farmed fish and the material production of un/healthy nature. *Antipode* 43:413–34.

Mayer, J. 1996. The political ecology of disease as one new focus for medical geography. *Progress in Human Geography* 20 (4): 441–56.

———. 2000. Geography ecology and emerging infectious diseases. *Social Science and Medicine* 50 (7–8): 937–52.

McGee, T. 1999. Private responses and individual action: Community responses to chronic environmental lead contamination. *Environment and Behavior* 31 (1): 66–83.

Mosler, H., O. Blochliger, and J. Inauen. 2010. Personal, social, and situational factors influencing the consumption of drinking water from arsenic-safe deep tubewells in Bangladesh. *Journal of Environmental Management* 91 (6): 1316–23.

Moss, P., and I. Dyck. 2002. *Women, body, illness: Space and identity in the everyday lives of women with chronic illness*. Lanham, MD: Rowman and Littlefield.

Nasreen, M. 2003. Social impacts of arsenicosis. In *Arsenic contamination: Bangladesh perspective*, ed. M. Ahmed, 340–53. Dhaka, Bangladesh: ITN-Bangladesh.

Paul, B. 2006. Health seeking behavior of people with arsenicosis in rural Bangladesh. *World Health and Population* 8 (4): 16–33.

Richmond, C., S. Elliott, R. Matthews, and B. Elliott. 2005. The political ecology of health: Perceptions of environment, economy, health and wellbeing among Namgis First Nation. *Health and Place* 11:349–65.

Rosenboom, J. 2004. *Not just red or green: An analysis of arsenic data from 15 Upazilas in Bangladesh*. Dhaka, Bangladesh: Arsenic Policy Support Unit, Government of Bangladesh.

Smith, A., E. Lingas, and M. Rahman. 2000. Contamination of drinking water by arsenic in Bangladesh: A public health emergency. *Bulletin of the World Health Organization* 78:1093–103.

Sothern, M., and I. Dyck 2010. Sex and gender in health geography. In *The Sage handbook of social geographies*, ed. S. Smith, S. Marston, R. Pain, and J. P. Jones III, 224–41. London: Sage.

Sultana, F. 2006. Gendered waters, poisoned wells: Political ecology of the arsenic crisis in Bangladesh. In *Fluid bonds: Views on gender and water*, ed. K. Lahiri-Dutt, 362–86. Kolkata, India: Stree Publishers.

———. 2007a. Suffering for water, suffering from water: Political ecologies of arsenic, water and development in Bangladesh. Unpublished PhD dissertation, Department of Geography, University of Minnesota, Minneapolis, MN.

———. 2007b. Water, water everywhere but not a drop to drink: Pani politics (water politics) in rural Bangladesh. *International Feminist Journal of Politics* 9 (4): 1–9.

———. 2009a. Community and participation in water resources management: Gendering and naturing development debates from Bangladesh. *Transactions of the Institute of British Geographers* 34 (3): 346–63.

———. 2009b. Fluid lives: Subjectivities, water and gender in rural Bangladesh. *Gender, Place, and Culture* 16 (4): 427–44.

———. 2011. Suffering for water, suffering from water: Emotional geographies of resource access, control and conflict. *Geoforum* 42 (2): 163–72.

World Health Organization (WHO). 2000. *Towards an assessment of the socioeconomic impact of arsenic poisoning in Bangladesh*. Geneva, Switzerland: World Health Organization.

"We Pray at the Church in the Day and Visit the Sangomas at Night": Health Discourses and Traditional Medicine in Rural South Africa

Brian King

Department of Geography, The Pennsylvania State University

Research within geography and cognate disciplines has worked to demonstrate the significant impacts of human disease on social and ecological systems. Although human disease fundamentally reshapes demographic patterns and regional and national economies, scholarly and policy research has tended to concentrate at the macroscale, thereby reducing attention to local-level dynamics that directly influence health decision making. This absence is notable given the invocation by various governmental agencies of the importance of traditional cultural practices, including the employ of traditional medicine, in responding to illness. South Africa's particular experience is representative of this, with national and provincial governmental agencies continuing to advocate traditional medicine in managing human health. Yet understandings of disease within South Africa remain deeply contested and expose underlying tensions about how health decision making is shaped by varied perceptions of illness and treatment options. This article draws on research that began in 2000 to analyze perceptions of health and the use of traditional medicine within rural areas. I work to uncover the divergent, and often conflicting, views on traditional medicine, and examine how they intersect with sociocultural systems that mediate health decision making. The article concludes that future geographic research on human health needs to engage with the social and cultural systems that contribute in shaping health perceptions and decision-making in various settings.

地理和同源学科的研究曾经已证明人类疾病对社会和生态系统所产生的重大影响。虽然人类疾病从根本上重塑人口结构以及区域和国家经济，学术和政策研究往往集中在宏观尺度，从而减少对直接影响健康决策的地方一级的动态关注。这种缺乏是值得注意的，鉴于各政府机构所调用的传统文化习俗的重要性，包括使用传统医药应对疾病。南非的特别经验是其中一个代表，国家和省级政府机构继续在人类的健康管理中提倡传统医药。然而，在南非对疾病的认识，仍然有很深的争议，揭示出有关公共卫生决策的底层的紧张局势是如何由各种疾病和治疗方案的看法所塑造的。本文借鉴 2000 年开始的研究来分析健康的看法和传统医药在农村的使用。我努力发掘传统医学上的意见分歧和经常冲突，并研究它们如何与调解卫生决策的社会文化系统相交错。文章得出结论认为，未来人类健康的地理研究，需要考虑有助于塑造健康的看法，和考虑在不同环境中的社会和文化系统决策。*关键词：健康，政治生态，传统的精神医生，南非，传统医药。*

La investigación en geografía y disciplinas afines ha servido para demostrar los impactos significativos de la enfermedad humana sobre los sistemas sociales y ecológicos. Aunque la enfermedad humana fundamentalmente reconfigura los patrones demográficos y las economías regionales y nacionales, la investigación académica y de políticas ha tendido a concentrarse en la macroescala, reduciendo así la atención hacia la dinámica de nivel local que influye directamente en la toma de decisiones sobre salud. Tal ausencia es notable si se tiene en cuenta la invocación originada en varias agencias gubernamentales sobre la importancia de las prácticas culturales tradicionales, incluyendo el empleo de la medicina tradicional para responder a las dolencias humanas. Una experiencia particular representativa de esto es la de Sudáfrica, donde las agencias gubernamentales nacionales y provinciales siguen defendiendo el uso de la medicina tradicional para el manejo de la salud humana. Con todo, las maneras de entender la enfermedad en Sudáfrica siguen siendo profundamente debatidas y ponen de manifiesto tensiones ocultas sobre cómo la toma de decisiones sobre salud se configura a través de las variadas percepciones que existen sobre las enfermedades y las opciones de tratamiento. Este artículo se basa en una investigación que se inició en el año 2000 para analizar las percepciones de la salud y el uso de la medicina tradicional en áreas rurales. Mi trabajo se orienta a descubrir los puntos de vista divergentes y a veces en conflicto de la medicina tradicional, y a examinar la manera como aquellos se intersectan con los sistemas socioculturales que median la toma de decisiones sobre la salud. El artículo concluye que la futura investigación geográfica sobre salud humana necesita involucrarse con los sistemas sociales y culturales que contribuyen a moldear las percepciones de la salud y la toma de decisiones en varios entornos.

The expansion of human disease, including HIV/AIDS, in sub-Saharan Africa has had significant impacts on both social and ecological systems (Drimie 2003; Barnett and Whiteside 2006; International Union for Conservation of Nature and Natural Resources 2010). Although human disease fundamentally reshapes demographic patterns and regional and national economies, scholarly and policy research has tended to concentrate at the macroscale, thereby reducing attention to local-level dynamics that directly influence health decision making. In addition to the local, and contextually specific, impacts of human disease, less attention has been directed toward understanding how health and prevention programs are themselves shaped by sociopolitical and cultural systems that extend beyond the geographies of disease. This is particularly needed given that perceptions of disease, and hence the opportunities for healthy decision making, are themselves often deeply contested. South Africa's experience with HIV/AIDS is a clear example of this, with the country having endured years of deeply rooted conflicts over how the disease was understood and best managed. In fact, the national government received international condemnation for questioning the links between HIV and AIDS and the efficacy of antiretroviral medications (Jones 2005; Fassin 2007). Within the competing disease discourses at the time were regular invocations of traditional medicine as a viable strategy for managing illness. The previous Health Minister, Dr. Manto Tshabalala-Msimang, regularly asserted the value of diet and traditional medicine as preventatives for HIV/AIDS and also drafted the 2008 policy that proposed further integrating traditional medicine into the national health care scheme (Department of Health 2008).

The intention of this article is to demonstrate how geographic scholarship, particularly from within the fields of health geography and political ecology, can contribute to emerging understandings of health–environment interactions and health decision making. This article draws on long-term research in South Africa that began in 2000 focusing on changing livelihood patterns following the 1994 democratic elections, shifting institutional systems shaping resource access, and the impacts of conservation and development processes within rural areas. I concentrate here on the findings from a structured household survey and series of semistructured interviews that detail community perceptions of medical options, including the use of traditional medicine and the collection and use of medicinal plants. Although rural residents have a range of options for medical treatment, health decision making is shaped not only by constraints to access, which have been well documented in the literature (McIntyre and Gilson 2002; Schneider et al. 2006), but also by historical processes and sociocultural factors that influence perceptions of illness and discourses of health. This article details how perceptions of health and traditional medicine are differentially understood and how these views intersect and are shaped by local context, cultural practices, and historical geographies. While rural populations utilize traditional medicine with some regularity, governmental agencies and other stakeholders have reified the use of traditional medicine in particular ways that do not neatly align with local views and practices. As such, this article works to uncover the "subaltern health narratives" that conflict with national discourses to identify policy failures that might stem from misunderstandings of local practices and knowledge systems (King 2010, 50).

In the first section of the article, I provide a review of the research and policy literatures on traditional medicine within South Africa. This is followed by an overview of the case study and methodology, which involves research completed in northeast South Africa that examines processes of livelihood change and political and spatial transitions following the 1994 democratic elections. The case study research reported in this article concentrates on the findings on the collection and use of medicinal plants and perceptions and decision making of medical treatment options. These findings are discussed in detail in two sections, which outline general patterns of medicinal plant collection by residents of the Mzinti community and the divergent views on the efficacy of traditional healers. The article concludes by discussing the benefits of this study for emerging research on human health from within geography, with particular emphasis on the subfields of political ecology and health geography.

Traditional Medicine in South Africa

The use of traditional medicine within South Africa, and other countries within sub-Saharan Africa, is relatively common, with individuals and families either collecting medicinal plants themselves or visiting with traditional healers to receive treatment (Kale 1995; Flint 2008; Peltzer 2009). Bhat and Jacobs (1995) surveyed the use of traditional medicine in the Eastern Cape Province and find that traditional doctors, herbalists, herb sellers, tribal priests, and local people record medicinal benefits from twenty-six plants. In a later study from the same region, Dlisani and Bhat (1999) found twenty-seven plants identified at the species level that are used for the primary health care of mothers and children. Other work similarly details the significance of medicinal plant collection to household economy and livelihood production (Twine et al. 2003; Makhado et al. 2009). In addition to the collection of medicinal plants, there is a growing literature on the reliance on traditional healers in both urban and rural areas (Nattrass 2005; Peltzer and Mngqundaniso 2008; Peltzer 2009). South Africa has multiple types of traditional healers, including *sangomas*, *inyangas*, and *umthandazis* (faith healers). Within the study region *inyangas* and *sangomas* are two types of healers that are seen by local residents as generally equivalent, with the exception being that *sangomas* have the power of communing with ancestral spirits. Consequently, *sangomas* can receive additional training and are viewed as more powerful by some community members. *Umthandazis* might be affiliated with one of the African Christian churches and use the Bible, prayer sessions, and other approaches for treatment. In a widely cited piece, Kale (1995) reported as many as 200,000 traditional healers in the country on whom 80 percent of black South Africans depend for certain forms of treatment. Several studies indicated a preference for traditional medicine for specific purposes, such as improving personal well-being or spiritual needs (Cocks and Møller 2002; Ross 2008). Liverpool et al. (2004) argued that because traditional medicine tends to cost less than other medical care it is routinely sought out, reporting that 75 percent of inpatients at a hospital in Johannesburg and a health care center in Soweto have used traditional healers. Cook (2009, 264–65) noted that given the number of traditional healers, and presumption that many people seek them out as first caregivers, there is a "necessity to expedite registration, integration, and monitoring of traditional healers into the South African health care system."

Numerous studies, including several of those already cited, indicate a limited understanding of the factors shaping the use of traditional medicine and assert the need for more research, particularly given the expansion of HIV/AIDS. Given that networks of exchange exist between traditional healers in referring patients to one other, and also to local clinics or hospitals, the national government has pursued strategies to regulate traditional medicine. The South African Traditional Health Practitioners Act 35 was passed in 2004 prohibiting traditional healers from diagnosing or providing treatment to patients with HIV/AIDS, cancer, or other terminal illnesses. This was later found to be invalid by the Constitutional Court because of a lack of public hearings, but efforts continued, including those by the former Health Minister, Dr. Manto Tshabalala-Msimang, to formalize the role of traditional medicine within the national health care system. The Traditional Health Practitioners Act (No. 22 of 2007) established the Traditional Health Practitioners Council to provide oversight on traditional healers. This was followed by the Policy on African Traditional Medicine for South Africa that was drafted in 2008, stating that "the official recognition, empowerment, and institutionalization of African Traditional Medicine, and its incorporation and its utilisation within the National Health System, would be an important step towards delivery of cost effective and accessible clients based healthcare" (Department of Health 2008, 6). Building on previous efforts, the policy is designed to institutionalize and regulate traditional medicine, establish a pharmacopeia of medicinal products, and collaborate with other countries and the World Health Organization to exchange information intended to "harmonize policies and regulations according to international standards" (14). The Draft Policy is also notable for two reasons; first, for its advocacy of traditional medicine and stated assumptions about its widespread use within the country. Second, the language of the policy indicates a preference for an "evidence-based public health and epidemiological approach" (35) that, as noted by Bishop (2010), minimizes the importance of spirituality in how traditional medicine is understood and practiced. Given that the national government is advocating the use of traditional medicine, yet adhering to a formalized regulatory framework coupled with a specific discourse of traditional medicine, research on local perceptions and practices is much needed.

Case Study and Methodology

The majority of this research was completed from 2001 to 2002, employing a mix of qualitative and quantitative methods, including a structured household survey, participant observation within the community of

Mzinti, and semistructured interviews with residents and representatives from national and provincial conservation and development agencies. The Mzinti community is situated within Mpumalanga Province in northeast South Africa, in a region that comprised the KaNgwane bantustan during apartheid.[1] Follow-up fieldwork was completed in 2004 and 2006 with additional interviews completed in the Mzinti community and also with health care providers in Schoemansdal and the provincial capital of Nelspruit. This article relies primarily on data derived from fifty semistructured interviews that were completed with male and female household heads to analyze livelihood production systems, dependence on natural resources, collection of medicinal plants, and perceptions of health options within the region. These interviews focused on the use of traditional medicine and whether family members visited with sangomas for health treatment. Given the diversity of traditional healers in the area, this research specifically asked whether people visit with sangomas, as opposed to inyangas or umthandazis, for treatment. In some cases, however, people responded by discussing their views of inyangas, affirming Bishop's (2010) contention that there is fluidity within these categories. Interviews with community residents also provided insight into the varied options pursued for health care, including decision making regarding visiting the local clinic or nearby hospital and the sequence in which these particular options were pursued.

In addition to the semistructured interviews, this article reports on the findings from a structured survey of 478 randomly selected households that collected information on livelihood production patterns and reliance on economic and natural resources. The surveys were designed to collect data on household histories and demographic characteristics, household assets, natural resource collection strategies, and the collection and use of medicinal plants. I concentrate here on the reported reliance on traditional medicine as a source of treatment, in addition to the collection of medicinal plants from communal areas surrounding the community. Although the qualitative semistructured interviews provide insight into how Mzinti residents view traditional medicine, these interviews are limited in understanding generalized patterns within the community. As a result, this article works to integrate the findings from both qualitative and quantitative methods of data collection to generate a contextually rich understanding of health discourses and decision making. The following two sections outline the major findings, first discussing the collection and use of medicinal plants by Mzinti residents. This is followed by an analysis of the varied perceptions of traditional medicine within the region and how these views are shaped by social, political, and cultural processes unfolding throughout contemporary South Africa.

Medicinal Plant Collection and Use in Mzinti

According to the structured surveys, 6 percent of households report collecting medicinal plants and 26 percent indicate that a family member had visited a sangoma for treatment. It is worth noting that the structured surveys and semistructured interviews generated different percentages of reportage for collecting medicinal plants or visiting a traditional healer. As evidence of this, the semistructured interviews generated higher numbers, with roughly 18 percent of households collecting medicinal plants and 45 percent having a family member visiting with a sangoma at some point in time. These discrepancies are not surprising, because as I discuss in the next section, there are multiple social pressures to underreport the use of traditional medicine. As one of my research assistants explained after the completion of the household surveys, people were uncomfortable stating they visited traditional healers because "we pray at the church in the day and visit the sangomas at night." Given these circumstances, it should be clearly stated that the intention of this article is not to report the use of traditional medicine with precision; rather, I find the inconsistencies in reportage to be a valuable entry point into the varied ways in which health is understood within the Mzinti community, as well as the ways that health decision-making is shaped by social, political, and cultural factors.[2]

The collection of medicinal plants occurs at a number of points surrounding the community, although changing patterns of land cover are impacting the ability of collectors to locate plants. Additionally, communal areas in rural South Africa, which remain vital locations for the collection of a number of natural resources for livelihood production, exist at the intersection between overlapping and sometimes conflicting rules shaping access (King 2005, 2011). Following the 1994 democratic elections, new pieces of legislation and empowered national and provincial governmental agencies have been working to reshape land use patterns and resource access in rural South Africa. Within the Mzinti community, this includes the Mpumalanga Tourism and Parks Agency, which has jurisdiction in enforcing the 1998 Mpumalanga

Table 1. Select medicinal plant use within Mzinti

Medicinal plant (Western or local name)	Usage
Marula	Chest problems and headache
Guava	Coughing
Peaches	Stomach
Inyokane	Newborn child
Nunankulu	Stomach
Mixture of trees	Work problem and wounds
Rosaline	Luck, vomiting, and all diseases
Shongi	Knees, shivering, and muscles
Sihlangu	Stomach
Limphambo	Stomach
Matema	Stomach

Nature Conservation Act that placed new restrictions on natural resource collection, including the collection of medicinal plants from within communal areas. The Act obligates provincial governmental agencies to enforce these restrictions on resource access, challenging the historical power held by the Matsamo Tribal Authority, which has jurisdiction over Mzinti and other communities in the region. One of the semistructured interviews was completed with a *sangoma* who discussed at length the constraints in accessing medicinal plants within the area. As he explained:

> [The tribal authority] has a problem with us being *sangomas*. They are asking for a rule that you may not come from Mahlalela Tribal Authority to Matsamo Tribal Authority for the herbs. That is why I am here because I think I have the privilege when I go to the back of my place I just go to the bush to dig. You see this is Nkomazi West and the left side is Nkomazi East, so it differs in the soil. This one is sandy soil and that one there is clay so the red and black soil is more fertile. There are more medicines on that side than this side and the medicines that are on that side are not here and also the ones here are not on the other side.

Although 10 percent of households that collect traditional medicine report gathering plants monthly, the remainder collects only when necessary for treating specific illnesses. Plants are used for a number of purposes, and Table 1 reports on some of the more common uses by Mzinti residents.

The lack of medicine within the Mzinti clinic and nearby hospitals is also a factor contributing to the use of traditional medicine within the community. Household members noted that the clinic would recommend guava leaves and other traditional remedies for treating specific symptoms, particularly if there was an absence of medicine at the clinic. Although the percentage of households collecting medicinal plants is relatively small, an additional 7 percent of households report purchasing traditional medicine in markets either in Mzinti or in nearby villages. The average amount that households spend purchasing traditional medicine is roughly $4, although as discussed in the next section, the cost of visiting with traditional healers was noted as a barrier by some community members.

Health Discourses and Decision Making in Mzinti

In addition to collecting medicinal plants and purchasing traditional medicine at the market, many community members visit traditional healers. Twenty-six percent of surveyed households report having a family member visiting a *sangoma* for a variety of purposes, including headache, cancer, stroke, tuberculosis, stomach problems, and foot problems. One of the benefits of the semistructured interviews is that they assist in revealing several understudied elements shaping health decision making. First, Mzinti residents explained that some illnesses are "traditional" illnesses that require visiting *sangomas* for treatment. By comparison, it is believed that other illnesses do not fall into this category and hence the sick individuals decided to visit the clinic or hospital. As one resident explained, "Some diseases are better healed by the *sangoma*. We look at the sickness to know whether it is better to treat with the *sangoma* or by the doctor." Another community member indicated that he would go see a *sangoma* for a "traditional problem, like maybe if you are bewitched by a spirit." One resident explained that *muti* (traditional medicine[3]) can be used to be released from prison, "even if you are found guilty by the court." In a number of the semistructured interviews, community members indicated that *sangomas* would be approached to remedy a traditional problem that in some cases could not be resolved by visiting the clinic or hospital. These interviews also help detail the order in which residents make specific decisions about their potential treatment options. Although traditional medicine has been invoked by various governmental stakeholders as an invaluable component of the national health care scheme, many residents explained that they visit traditional healers only *after* seeking out assistance from clinics and hospitals. As one man explained, "When someone is sick there are those diseases that the doctors cannot cure and those that they can cure, so if they fail to cure, we go to the *sangoma*." Based on the semistructured interviews, it seems that

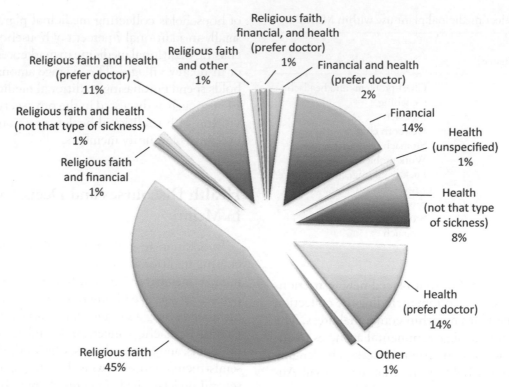

Figure 1. Reasons for not visiting traditional healers. (Color figure available online.)

the majority of residents approach traditional healers after first visiting the clinic or hospital for treatment.

Second, the research collected detailed information on the varied reasons for why traditional healers are not approached by certain community members. Figure 1 reports the percentages from the structured survey of 478 households. The different categories assigned to explain health decision making regarding traditional healers were religion, financial, other, and health, which was disaggregated to indicate "prefer doctor" and "not that type of sickness." The financial costs of visiting traditional healers were described as a constraint by some residents. One person indicated that he never visits a *sangoma* "because the *sangoma* needs money and we do not have money since I am not working, so it is better at the clinic because it is free." Another respondent discussed his anxiety about visiting a traditional healer and becoming sick, explaining that "the *sangomas* use a razor to cut you and they use the same razor with different people and therefore I am trying to avoid some diseases because AIDS can be transmitted in that way. In the hospital they use a new razor. That is why I prefer the hospital." The invocation of biomedicine in explaining health decision-making was notable in a number of the comments provided by community members. As an example of this, in discussing why she goes to the doctor before seeing a

traditional healer, one resident indicated, "The doctors have drips for a shortage of blood or water, and they are able to cure types of diseases that the *inyanga* cannot. So it is important to go to the doctors first." Yet another Mzinti resident indicated that she does not see a *sangoma* because "they are not professionals. Now we have professional doctors who are able to heal us well, even if there is something in your body. The problem with the *sangomas* is that they cannot see that." These statements reflect the acceptance by some residents that clinical medicine is preferable for treating disease and its symptoms. It is also notable the ways that biomedicine is infused with local discourses of health decision making, which seemed to predispose these individuals against seeking out traditional healers.

In addition to the financial constraints and negative perceptions on the efficacy of traditional medicine, there are sociocultural beliefs that are quite pervasive in shaping health decision making. Specifically, it was common for respondents to explain that their decision to not visit the *sangoma* was determined by their religious views. This is evidenced in Figure 1, which reports that 45 percent of respondents identify religious faith as the sole reason for not visiting a *sangoma*. Numerous respondents are conflicted about admitting visiting traditional healers, sharing that they are Christian and expected to believe in the power of prayer and Western science. One Mzinti resident indicated the decision to not visit with traditional healers because "we are a religious family. We believe in prayers and the divine." Another woman explained:

I am a Christian and we do not believe in that. We believe in God. I have seen that some of my family members where I am married believe too much in *sangomas*. They sometimes go to *sangomas* and I still remember after the death of my husband they brought most of the *sangomas* and then I was very, very sick. Then my sister came here and called the Christians and they came and held prayers and after that I was well. That is why I never turn my face towards the *sangomas*.

The number of households that report purchasing traditional medicine, taken together with the number that visit *sangomas* for treatment, suggests that a segment of the population relies on traditional medicine for a variety of purposes. This clearly places a demand on the natural resource base and requires ongoing negotiations between various systems to determine the institutions of access in the rural areas. Yet the collection and use of medicinal plants, in addition to visiting traditional healers for treatment, are specific decisions shaped at least in part by sociocultural factors that are deeply rooted and contested within rural communities such as Mzinti.

Conclusions

The central objective of this article was to draw on the findings on medicinal plant collection and reliance on traditional healers in Mzinti to show how perceptions of health and decision making are shaped by multiple processes unfolding within contemporary South Africa. Whereas external actors, including the national government, remain insistent on the value of traditional medicine in attending to disease, local actors have specific views on the types of illnesses warranting treatment from particular sources. The research informing this article suggests that many community members visit with a *sangoma* only after going to the clinic or hospital, which does not fit with some governmental representations of traditional medicine use. Additionally, people visit with traditional healers for a variety of purposes, including spiritual reasons or good luck, which suggests that the national government's policies on traditional medicine, and insistence on an epidemiological approach, do not fully engage with the diversity of health decision making in rural areas. This demonstrates that health decisions, which are deeply rooted and mediated by divergent perceptions of disease and health, are varied and do not align neatly with discourses generated by public health and governmental officials. Health decision making is fluid, pluralistic, and dynamic, and intersects with sociocultural processes in addition to individual biases and spiritual belief systems.

Although the findings from this work are specific to South Africa, I believe that the larger argument has value for geographic research on human health in two specific ways. First, this article shows that perceptions of disease, and by extension human health, are often multifaceted, historically and spatially situated, and conflicted. Research within health geography and political ecology has explored the intersections between health care and cultural and social geography (Gesler and Kearns 2002; Andrews and Evans 2008; King 2010) and challenged the hegemony of the biomedical model (Mansfield 2008). This article contributes to these studies by demonstrating the specific ways that sociocultural systems shape health perceptions and decision making. Second, this article shows the importance of engaging with subaltern health narratives that conflict with state discourses and policies that might be based on misunderstandings of local practices and knowledge systems. Although this article focuses on traditional medicine in South Africa, I believe it demonstrates the need for geographic scholarship to more critically engage with the ways that health perceptions and decision making are shaped by sociocultural systems in various settings.

Acknowledgments

The research for this article was supported through institutional affiliations with the Mpumalanga Tourism and Parks Agency and the Centre for Environmental Studies at the University of Pretoria. I am appreciative of the tireless efforts of my research assistants, Erens Ngubane, Cliff Shikwambane, and Wendy Khoza. My thanks to Jamie Shinn, who assisted with background research for this article, and also Kayla Yurco, who helped with Figure 1. I would also like to thank the editor and three anonymous reviewers for their generous comments and Kristina Bishop for her invaluable suggestions on a previous version. Finally, my sincere gratitude to the many residents of the Mzinti community who provided the time and insights that made this work possible.

Notes

1. KaNgwane was one of ten territories that were constructed by the apartheid government to enforce its ideology of separate development. The bantustans expanded on existing systems of segregation established under colonialism.
2. As further evidence of this, the structured surveys probed whether anyone in the household was sick or had died from HIV/AIDS. At the time of the survey in 2002, only one respondent indicated a household death from AIDS. Obviously, this was not an indication of the prevalence of the disease at that time but of its significant social and cultural stigmas, which have been documented in the literature (Campbell 2003; Posel, Kahn, and Walker 2007).
3. Muti (also muthi) can also include potions that can be used to inflict harm on another person. Much of the power of muti is ascribed to the belief in its effectiveness (Ashforth 2005).

References

Andrews, G. J., and J. Evans. 2008. Understanding the reproduction of health care: Towards geographies in health care work. *Progress in Human Geography* 32 (6): 759–80.

Ashforth, A. 2005. *Muthi,* medicine and witchcraft: Regulating "African science" in post-apartheid South Africa? *Social Dynamics* 31 (2): 211–42.

Barnett, T., and A. Whiteside. 2006. *AIDS in the twenty-first century: Disease and globalization.* 2nd ed. New York: Palgrave Macmillan.

Bhat, R. B., and T. V. Jacobs. 1995. Traditional herbal medicine in Transkei. *Journal of Ethnopharmacology* 48 (1): 7–12.

Bishop, K. 2010. *The nature of medicine in South Africa: The intersection of indigenous and biomedicine.* Unpublished PhD dissertation, University of Arizona, Tucson.

Campbell, C. 2003. *"Letting Them Die": Why HIV/AIDS intervention programmes fail.* Oxford, UK: James Currey.

Cocks, M., and V. Møller. 2002. Use of indigenous and indigenised medicines to enhance personal well-being: A South African case study. *Social Science & Medicine* 54 (3): 387–97.

Cook, C. T. 2009. Sangomas: Problem or solution for South Africa's health care system. *Journal of the National Medical Association* 101 (3): 261–65.

Department of Health. 2008. *Draft policy on African traditional medicine for South Africa.* Pretoria, South Africa: Government Gazette.

Dlisani, P. B., and R. B. Bhat. 1999. Traditional health practices in Transkei with special emphasis on maternal and child health. *Pharmaceutical Biology* 37 (1): 32–36.

Drimie, S. 2003. HIV/AIDS and land: Case studies from Kenya, Lesotho and South Africa. *Development Southern Africa* 20 (5): 647–58.

Fassin, D. 2007. *When bodies remember: Experiences and politics of AIDS in South Africa.* Berkeley: University of California Press.

Flint, K. E. 2008. *Healing traditions: African medicine, cultural exchange, and competition in South Africa, 1820–1948.* Athens: Ohio University Press.

Gesler, W. M., and R. A. Kearns. 2002. *Culture/place/health.* London and New York: Routledge.

International Union for Conservation of Nature and Natural Resources. 2010. *Interactions between HIV/AIDS and the environment: A review of the evidence and recommendations for next steps.* Nairobi, Kenya: IUCN-ESARO Publications Unit.

Jones, P. S. 2005. "A test of governance": Rights-based struggles and the politics of HIV/AIDS policy in South Africa. *Political Geography* 24 (4): 419–47.

Kale, R. 1995. Traditional healers in South Africa: A parallel health care system. *British Medical Journal* 310 (6988): 1182–85.

King, B. 2005. Spaces of change: Tribal authorities in the former KaNgwane homeland, South Africa. *Area* 37 (1): 64–72.

———. 2010. Political ecologies of health. *Progress in Human Geography* 34 (1): 38–55.

———. 2011. Spatialising livelihoods: Resource access and livelihood spaces in South Africa. *Transactions of the Institute of British Geographers* 36 (2): 297–313.

Liverpool, J., R. Alexander, M. Johnson, E. K. Ebba, S. Francis, and C. Liverpool. 2004. Western medicine and traditional healers: Partners in the fight against HIV/AIDS. *Journal of the National Medical Association* 96 (6): 822–25.

Makhado, R. A., G. P. von Maltitz, M. J. Potgieter, and D. C. J. Wessels. 2009. Contribution of woodland products to rural livelihoods in the northeast of Limpopo Province, South Africa. *South African Geographical Journal* 91 (1): 46–53.

Mansfield, B. 2008. Health as a nature–society question. *Environment and Planning A* 40 (5): 1015–19.

McIntyre, D., and L. Gilson. 2002. Putting equity in health back onto the social policy agenda: Experience from South Africa. *Social Science & Medicine* 54 (11): 1637–56.

Nattrass, N. 2005. Who consults sangomas in Khayelitsha? An exploratory quantitative analysis. *Social Dynamics* 31 (2): 161–82.

Peltzer, K. 2009. Utilization and practice of traditional/complementary/alternative medicine (TM/CAM) in South Africa. *African Journal of Traditional, Complementary, and Alternative Medicines* 6 (2): 175–85.

Peltzer, K., and N. Mngqundaniso. 2008. Patients consulting traditional health practitioners in the context of HIV/AIDS in urban areas in Kwazulu-Natal, South Africa. *African Journal of Traditional, Complementary, and Alternative Medicines* 5 (4): 370–79.

Posel, D., K. Kahn, and L. Walker. 2007. Living with death in a time of AIDS: A rural South African case study. *Scandinavian Journal of Public Health* 35 (69): 138–46.

Ross, E. 2008. Traditional healing in South Africa: Ethical implications for social work. *Social Work in Health Care* 46 (2): 15–33.

Schneider, H., D. Blaauw, L. Gilson, N. Chabikuli, and J. Goudge. 2006. Health systems and access to antiretroviral drugs for HIV in Southern Africa: Service delivery and human resources challenges. *Reproductive Health Matters* 14 (27): 12–23.

Twine, W., D. Moshe, T. Netshiluvhi, and V. Siphugu. 2003. Consumption and direct-use values of savanna bio-resources used by rural households in Mametja, a semi-arid area of Limpopo Province. *South African Journal of Science* 99:467–73.

Critical Interventions in Global Health: Governmentality, Risk, and Assemblage

Tim Brown,* Susan Craddock,† and Alan Ingram‡

*School of Geography, Queen Mary, University of London
†Institute for Global Studies and the Department of Gender, Women, and Sexuality Studies, University of Minnesota
‡Department of Geography, University College London

The rise of the term *global health* reflects a concern with rethinking the meaning of health in the context of globalization. As a field of practice, however, global health renders problems, populations, and spaces visible and amenable to intervention in differentiated ways. Whereas some problems are considered to be global, others are not. Some are considered to be matters of global security, whereas others lack this designation and remain in the realm of health or development. Attention is drawn to individual global health problems, even as their broader structural dimensions are often obscured. We suggest that a critical geographical approach to global health therefore entails reflexivity about the processes by which problems are constituted and addressed as issues of global health and identify three analytical approaches that offer complementary insights into them: governmentality, risk, and assemblage. We conclude by outlining some further issues for critically reflexive geographies of global health.

全球健康这一词的兴起，反映了在全球化的背景下，人们通过反思健康的意义而对健康的一种关注。然而，作为一个实践领域，全球卫生业呈现了可见的和可服从的，以不同方式干预的问题，人口和空间。有些问题被认为与全球安全事项有关，而其它一些问题则缺少这个称谓并保持在健康或发展的领域。本文提请关注个别的，甚至在其更广的结构尺寸往往模糊不清的情况下的全球健康问题。我们提出针对全球健康的一个重要的地理方法，对构成有关问题并被当作全球健康问题处理的过程进行反思，并确定三个互补观察的分析方法：治理性，风险，和组合。最后，我们概述了全球卫生批判性反思地理中的一些更进一步的问题。关键词：组合，全球化，治理，健康，风险。

El surgimiento de la expresión *salud global* refleja una preocupación por repensar el significado de la salud en el contexto de la globalización. Como campo de práctica, sin embargo, la salud global rinde problemas, poblaciones y espacios visibles y susceptibles de intervención de maneras diferentes. Si bien algunos problemas se aprecian como globales, otros no. Algunos son considerados materia de seguridad global, en tanto otros carecen de esta designación y permanecen dentro del mundo de la salud o del desarrollo. Se llama la atención sobre problemas globales de salud individuales, incluso si sus dimensiones estructurales más amplias a menudo son oscuras. Sugerimos que un enfoque geográfico crítico de la salud global requiere consecuentemente una reflexión acerca de los procesos con los que se constituyen los problemas y son abocados como asuntos de salud global, e identificamos tres enfoques analíticos que ofrecen comprensión complementaria para los mismos: gobernabilidad, riesgo y ensamblaje. Concluimos esquematizando algunas cuestiones adicionales para geografías críticamente reflexivas de la salud global.

"Health is global," or so we are encouraged to think.[1] The rise in prominence of the term *global health* has gone hand in hand with a multitude of academic analyses, think tank papers, policy initiatives, media reports, institutional innovations, and political controversies over the course of the last twenty years. The global health field has seen an influx of new actors, including activist networks, high-profile philanthropies, and new kinds of public–private partnerships, giving rise to a plethora of global health initiatives. All of this has coincided with a doubling of donor aid for health over the course of the last decade. This has triggered a rapid growth in academic global health centers and programs, many of which recognize that global health now goes beyond biomedicine, epidemiology, public health, and development to embrace matters of law, politics, history, economics, trade, diplomacy, and security.

Although geographers have long contributed to the understanding of health worldwide via medical geography, epidemiology, public health, and development studies, they are increasingly turning a critical eye toward its political, economic, cultural, and ethical dimensions, helping to shape a broader interdisciplinary field of critical global health studies. An important part of this agenda is a critical rethinking of how the global health field itself shapes the way global health is imagined, understood, and thus addressed. A key concern here is the differentiated manner in which particular problems, populations, and spaces are rendered visible and amenable to intervention. Although some problems (like emerging infectious diseases or tobacco-related disease) are considered to be global, others are not. Some (like influenza, HIV/AIDS, or biological weapons) are designated as matters of global security, whereas others (e.g., maternal health, diarrheal, and other so-called neglected diseases) remain in the realm of health or development. Attention is often drawn to individual problems (like epidemics or other crises) or solutions (e.g., vaccination or antiretroviral therapy for HIV and AIDS), whereas the broader structural dimensions and determinants of ill health, diminished life chances, and shortened lives (which derive from long-term and wide-scale political and economic processes) are often obscured (cf. Brown and Moon 2012).

We suggest that a critical geographical approach can contribute further to the process of revisioning global health by opening the processes whereby health is rendered global to reflexive scrutiny. Building on recent interventions by geographers and others (Braun 2007; Ali and Keil 2008; Hinchliffe and Bingham 2008; Lakoff and Collier 2008; Kearns and Reid-Henry 2009; Sparke 2009), we outline three analytical approaches that offer complementary insights into these processes: governmentality, risk, and assemblage. Each is relevant, because it draws attention to how global health problems and responses are not given but enabled, imagined, and performed via particular knowledges, rationalities, technologies, affects, and practices across a variety of sites, spaces, and relations.

Global Health Governmentality

A full account of today's global health field would need to explore multiple temporalities and spatialities, including imperial and colonial historical geographies, the formation of the world economy and patterns of state formation in terms of their mutual constitution with ecological and epidemiological processes, post-colonial problematizations of the Third World as requiring developmental intervention, the wreckage of postcommunist transitions, and the emergence of disputes over neoliberal globalization. It would also include a series of specific developments during the 1990s: the formation of the concept of emerging infectious diseases; alarm at epidemics spreading within and beyond the global south; efforts of policy entrepreneurs and activists on health, human rights, development, and peace issues; concern about bioweapons stocks, infrastructures, and expertise; and shifting concepts and practices of security. It would trace, too, the emergence of an ecosystem of global health governance, constituted by dozens of organizations and institutions concerned in some way about global health issues. With the term *governmentality* we signal additionally, first, the reflexive implication of particular knowledges, rationalities, and technologies in the constitution of global health (Larner and Walters 2004; Elbe 2009; Nguyen 2009; de Larrinaga and Doucet 2010; Ingram 2010). We wish to signal three further aspects of global health governmentality in particular.

The first is the manner in which the global health field deals with problems of space. Foucault's (2007) discussion of how the management of population and circulation comes to form the preeminent object and logic of modern government is particularly suggestive when it comes to global health. As Braun (2007) argued, in the case of biosecurity, what is apparent is an aspiration to govern the global biological that is enabled by a variety of surveillance and monitoring technologies and institutional arrangements. Furthermore, as Braun (2007), Ali and Keil (2008), and Hinchliffe and Bingham (2008) argued, the Euclidean geography of the nation-state system, on which global health governance had been based, is ill suited to the complex topologies of emerging infections, where one thing can morph into another and distinctions between the inside and the outside are blurred (see also Sarasin 2008). Efforts to secure the globe from emerging infections thus link cutting-edge surveillance, post-Westphalian sovereignty, and emergency alert and response (Fidler 2004; Elbe 2009).

Global health governmentality thus intersects with geopolitics in complex ways. Global health security systems, which are largely grounded in and funded from within the global north, reflect global north priorities, and render sovereignty provisional, start to look like a form of empire (see also King 2002; Weir and Mykhalovskiy 2006). Things might be changing, however. Controversies over global health security (Aldis

2008) have spurred the emergence of a discourse on global health diplomacy (Kickbusch, Silberschmidt, and Buss 2007). Although this marks a "rendering technical" (Li 2007) and thus potentially a depoliticization, it is evident that the emergence of global health diplomacy is also linked with the increasingly assertive role being played within global governance by countries like Brazil, India, South Africa, Indonesia, and Thailand, especially when it comes to issues such as intellectual property and access to medicines. What might be required now is less a critique of empire than attention to the emerging contours of a posthegemonic global health agenda.

Second, to confine our discussion of global health to such high-profile security issues, however, is to replicate a selective focus. Global health is a field where the "will to improve" (Li 2007) is at work much more widely. We recognize the broad-ranging nature of global health today, where governmental rationalities, technologies, and practices are deployed in relation not just to emerging infections but the trade, marketing, and consumption of tobacco; accidents and road safety; diet, lifestyle, and nutrition; and noncommunicable diseases (Herrick 2009; Elbe 2010). There is much to be gained, we suggest, from looking synthetically across such domains, to consider through what kinds of technologies and tactics particular issues are problematized and made visible, and how this relates to the production of particular kinds of spaces. International responses to HIV/AIDS (Smith and Siplon 2006; Grebe 2008; Pisani 2008) and tobacco (World Bank 1999; World Health Organization 2002) provide telling examples here, where scaled-up responses were leveraged on efforts to characterize, quantify, and render visible particular problems, populations, and spaces (see also Bowker and Starr 2000).

Third, we problematize the articulation between governmentality and neoliberalism. To put things starkly, the central concern is that the extension of infrastructures to manage circulation and improve the health and well-being of populations remains a thin response to deep-rooted problems of structural violence and inequality that emerge from global integration (Sparke 2009). Indeed, key global health commentators (Fidler 2008–2009; Schneider and Garrett 2009) have observed that the landmark global health initiatives (like the international response to HIV/AIDS) of the 2000s have been narrowly focused, short term, and unsustainable. Although recent interventions have attempted to force the issues of disparities in human resources for health, the social determinants of health, and health systems development back onto the global health agenda, the relationship between global health and the global economy is often only alluded to or addressed indirectly. Furthermore, many of the new private actors in global health are heavily invested in the current configuration of the global economy (Williams and Rushton 2011).

We suggest that these issues represent key points for investigation for critical geographies of global health. Before considering how to develop this further, we turn to risk as a key form of knowledge articulating global health practices.

Global Health, Global Risk

As noted, to understand global health, and to critically intervene in it, we need to ask questions about the kinds of knowledges, technologies, and tactics that are used to render it visible or to make it doable (Kickbusch, Silberschmidt, and Buss 2007). One possibility here would be to focus on specific devices—such as the disability adjusted life year—and on the ways in which they are used to fix the geographical contours of global health and around which specific assemblages are formed. A more extensive technology associated with global health discourse, however, is what Dean (1999) referred to as "epidemiological risk"; it is on this concept, which we suggest is an example of what Foucault referred to as a *dispositif* (see Foucault 1980, 194–95), that we focus here. Before doing so, we briefly discuss Beck's (1992) original articulation of the risk society thesis in which risk is defined as a "systematic way of dealing with the hazards and insecurities" that are the somewhat paradoxical result of scientific and technological advances associated with late modernity (21).

Beck's formulation contributed to the development of a new research paradigm in which risk was elevated to the status of a key analytical rubric. This is apparent in analyses of the heightened threats to health (and life) associated with globalization. As Brundtland (2001) commented, there are "two critically important forces shaping the world we live in: the revolution that is taking place in information and biotechnology, and the growing momentum of globalization. Both of these forces carry with them immense potential for good. But, as we are all aware, they carry risks." Brundtland went on to highlight the nature of these risks, noting that "[w]ith globalization, a single microbial sea washes all of humankind" and that there are "no health sanctuaries." She also pointed to the growing recognition that it is not only infectious diseases that are spread with globalization but heightened risks of heart disease, diabetes, and cancer.

Brundtland's general statements on the consequences of globalization can be associated with the emergence of particular interventions; for example, the antitobacco campaign mentioned earlier or others such as the global strategy against noncommunicable diseases and the Global Fund to Fight HIV/AIDS, Malaria, and tuberculosis. It is, then, important that we acknowledge the centrality of risk to current understandings of global health because ideas about its current and predicted future contours are in part shaped by this particular analytical frame (Brown and Bell 2008). We would argue, however, that it is equally important that we recognize that this conceptualization, which, as Dean (1999) suggested, assumes that *"real riskiness has increased"* (182, italics added), is not the only approach that we might take. Beck (2009) himself acknowledged this in his recent reworking of the risk society thesis. As he noted, *"[r]isks are social constructions and definitions based on corresponding relations of definition,"* and, as such, are open to dramatization, transformation, and even denial (30).

It is at this juncture that we turn to readings that have been informed by Foucault's work on governmentality (e.g., Castel 1991; Dean 1999). When thought of in these terms, risk, and specifically here epidemiological risk, emerges not so much as an objective reality of late modernity but as a dispositif. In making this connection, our goal is not to undervalue the material realities that shape people's lives and render them more or less exposed to health-related hazards. Rather, it is to acknowledge that the ways in which these realities are rendered visible, the ways in which they come to be known both now and in the future, is to a great extent shaped by the heterogeneous ensemble of elements that Foucault identified as being a part of the apparatus of a dispositif (Foucault 1977; see also Agamben 2009).

Clearly, then, we are referring here to a very different type of analytical frame than that elaborated by Beck in his initial conceptualization of the risk society thesis, and it is one that demands a focus on the constitution of things as risks and on the material consequences of this rendering within particular assemblages and in the context of particular forms of rule. As Dean (1999) argued, it is possible, when using governmentality as the lens through which to analyze risk, to "demonstrate that risk rationalities are not only multiple but heterogeneous and that practices for the government of risk are assembled from diverse elements and put together in different ways" (182). Two points are crucial here. First, we recognize that epidemiological risk is central

to current conceptualizations of global health and that it represents the calculative basis on which the health status of a population is determined or rendered visible (Brown and Bell 2008). Further, it is increasingly on the basis of these factors of risk, what Osborne (1997) referred to as "surrogate values" (186), that public health interventions are put into place.

This leads us to our second point. There is an acknowledgment in some critical accounts of global health discourse that the risks to health and well-being around which global health assemblages cohere reflect the priorities of the global north rather than those of the global south. Elbe provided an illustration of why this is important, when he noted that interventions to secure populations against HIV/AIDS were justified when they appeared to serve the national interests of those seeking security (those in the global north) rather than those being secured against (those in the global south; Elbe 2005, 2006). It is, however, not simply in relation to the risk posed by epidemics of infectious diseases that this question arises. As several commentators have acknowledged, despite recognizing the threat posed by the globalizing of lifestyle diseases and especially the double burden of disease that it poses to the world's most vulnerable populations, questions have been raised about the global commitment to tackle their causes (Beaglehole and Yach 2003; Yach et al. 2004; Marmot 2008).

Clearly, this suggests that we cannot interpret risk in terms of simple dichotomies; there is more than one at-risk population in operation here and they are affected by risk in different ways. This is an important point to acknowledge because risk is not a technology that remains static. It is adapted to or, in Foucault's terms, fabricates, organizes, and plans the milieu within which it is operationalized and the specific problematic on which it is brought to bear. A second question is that although epidemiological risk renders visible a particular present and future problematic (O'Malley 1996), there is a material as well as subjective facet to this rendering process. Put differently, not only are some individuals more prone to particular diseases or states of being but the subjects of risk discourses stake claims of identity, entitlement (biological citizenship), and voice on the basis of their status (Rose and Novas 2005; Rose 2007).

Assembling Global Health

The discussion so far leaves open the question of the epistemologies through which we might reflexively investigate, for example, the articulation between the constitution of risk and the global political economy (Farmer 2005; Global Health Watch 2005; Sparke 2009). Our use of the concept of assemblages is helpful here in highlighting the various actors and forces coming together within the milieu of late twentieth- and early twenty-first-century global capitalism and shaping the regulatory structures, social practices, and knowledge formations constituting global health. Our take on assemblage is purposefully syncretic, as we draw from a number of scholars who employ the concept (explicitly or not) to elicit the ethical complexities of new technologies and the "regimes of living" they help determine (Collier and Lakoff 2005; Ong and Collier 2005); highlight the productive frictions possible within the convergence of transnational, grassroots, and institutional forces (Tsing 2005); or track the unpredictable movements within networks (Latour 2005). To varying degrees for these scholars, tracing the interactions of highly diverse actors better elucidates the logic, contradictions, and negotiations within global processes; it also brings visibility to the role of nonhuman actors in changing the course of global practices. With few exceptions, however (cf. Nichter 2008), assemblages as a theoretical framework have been applied to examinations of environment, biosecurity, or technological deployment rather than in the investigation of global health. Often, attention to how complex networks interact also comes at the expense of retaining analytical sight of uneven power relations, making most global processes highly inequitable. We argue for a deployment of assemblages that addresses these gaps.

Returning to the differential actions of technologies of risk, although the milieu of global health is the globe itself, the spatial and political tensions between conditions of vulnerability (to infectious disease, obesity, smoking, cancer, etc.) and the rendering of securitization and governability take place within particular transnational circuits and regional locations. One practice helping to obscure these relations is the way the universal, as noted by Tsing (2005), has typically been produced within hegemonic regimes valorizing Western ideals and economies; it is a technique of obfuscation, as it works to suggest a commonality. Such hegemonic understandings of the universal or global commonly underlie high-level policy reports on globalization and health, placing in tension assumptions of universal practices of globalization with ascriptions of risk to particular populations. This tension then leads to what Lakoff and Collier (2008) called the "emergency modality of intervention" (17); that is, quick fixes aimed at shoring up, rather than extinguishing, risk.

An antidote to these problematic renderings is to look at the specificity of connections, as these reveal various interactions of multiple forces, technologies, agencies, actors, and power dynamics. As Ong and Collier (2005) suggested, assemblages are never "reducible to a single logic," and they do not "always involve new forms, but forms that are shifting, in formation, or at stake" (12). Every situation brings with it its own different set of connections that are unpredictable in both their constitution as well as their outcome (Tsing 2005), yet not every component of an assemblage plays an equally important role. Latour's (2005) notion of the intermediary versus the mediator is helpful here: The intermediary is "what transports meaning or force without transformation," whereas "mediators transform, translate, distort, and modify" (39).

Influenza vaccines are a good illustration of how a focus on assemblage elucidates aspects of global health. During the H1N1 pandemic, egg shortages thwarted rapid development of vaccine serum, forcing questions concerning why laboratories were still dependent on outdated technologies when new genetically based methods are known to be quicker. The answer in part lay in high liability risks and low profits, generating little incentive for pharmaceutical companies to develop new technologies. The areas of the world most at risk of inadequate vaccine supplies, however, are those that lack legal capacity to negotiate contracts with, or afford sizable fees to, the few pharmaceutical companies producing flu vaccines—fees governed in large part by global regulatory mechanisms such as the World Trade Organization's Trade Related Aspects of Intellectual Property (TRIPS). Within the context of regulatory regimes, legal negotiations, scientific parameters, pathogenic vicissitudes, and national financial capabilities, influenza vaccines become mediators changing the biopolitical terrain of influenza prevention and—in the event of a more virulent outbreak—survival (Craddock and Giles-Vernick 2010). As many have noted (cf. especially Hinchliffe and Bingham 2008), techniques of security actually end up rendering some populations more vulnerable, thereby ironically diminishing the efficacy of those techniques, no matter how robust. Yet these formations are in constant tension as well as emergence, as countries contest the terms of TRIPS, new organizations arise to respond to inequitable distribution, and philanthropic agencies underwrite the costs of better vaccine technologies.

Thus, embedded within the narratives as well as practices of intervention and security, yet belied in the declarations of universal vulnerability, is the fact that some people are profoundly more likely to be at risk of disease than others (Brown 2011). In other words, it is important to look not only at the ways in which practices of governability and security shape understandings of risk and their spatial contours: Equally trenchant is the recognition that these practices are a part of the broader network of global economic practices actually *creating* patterns of risk and disease burden.

Conclusion

The field of global health has emerged in large part as a response to the increased mobilities of populations, commodities, and pathogens associated with late twentieth- and early twenty-first-century globalization, yet the particular financial, regulatory, economic, and political structures determining these movements and the risks that result often remain in the margins of dominant global health imaginaries. A focus on mobilities but not the interrelations of their determinants consequently produces a turn toward securing countries and populations against a constant state of risk "emergence" (Cooper 2006). Whereas Cooper focused on the biological, we extend the analysis to include such risks as the global spread of obesity, cancer, or deaths from road accidents, as well as on the techniques utilized to secure against them. Too often, though, the technologies mobilized by governments, or even pharmaceutical companies, contain risk by "conducting conduct." Containment is simultaneously about making particular risks visible through increasingly robust technologies of surveillance, which in turn produce definitions of vulnerable populations and their socio-geographic loci of intervention. What these techniques are, how they operate, and the geopolitical terrain informing them constitute a field of inquiry requiring more attention from geographers.

Furthermore, these same technologies obfuscate as much as they elucidate. Although health risks are made visible through ever more sophisticated scientific, communications, and biotechnological capabilities, these cannot be construed as neutral. Not only are some issues made visible through practices of security and medical discourse (AIDS, SARS, H1N1) whereas others are not (vitamin deficiency, domestic violence), vulnerable populations are themselves redefined in sometimes problematic ways through the production of statistics or the mapping of epidemiology onto geopolitical and historical understandings of behavior. Further, at-risk populations are themselves not entirely absent from this rendering process. As Rose (2007) has indicated, human beings are increasingly coming to understand themselves in somatic terms and as such play a crucial role in the framing both of themselves and the issues that affect them as global health concerns. Going beyond such fixes requires critical contextualization of the kind that we have begun to outline here.

Note

1. Here we borrow the title of a recent UK government report (HM Government 2008).

References

Agamben, G. 2009. *What is an apparatus?* Stanford, CA: Stanford University Press.

Aldis, W. 2008. Health security as a public health concept: A critical analysis. *Health Policy and Planning* 23 (6): 369–75.

Ali, S. H., and R. Keil, eds. 2008. *Networked disease: Emerging infections in the global city.* Oxford, UK: Blackwell.

Beaglehole, R., and D. Yach. 2003. Globalisation and the prevention and control of non-communicable disease: The neglected chronic diseases of adults. *The Lancet* 362:903–908.

Beck, U. 1992. *Risk society: Towards a new modernity.* London: Sage.

———. 2009. *World at risk.* Cambridge, UK: Polity.

Bowker, G., and Starr, S. L. 2000. *Sorting things out: Classification and its consequences.* Cambridge: Massachusetts Institute of Technology Press.

Braun, B. 2007. Biopolitics and the molecularization of life. *Cultural Geographies* 13:6–28.

Brown, T. 2011. Vulnerability is universal: Considering the place of "security" and "vulnerability" within contemporary global health discourse. *Social Science and Medicine* 72 (3): 319–26.

Brown, T., and M. Bell. 2008. Imperial or postcolonial governance? Dissecting the genealogy of a global public health strategy. *Social Science and Medicine* 67:1571–79.

Brown, T., and G. Moon. 2012. Geography and global health. *The Geographical Journal* 178:13–17.

Brundtland, G.-H. 2001. *Globalization as a force for better health.* http://www.who.int/director-general/speeches/2001/index.html (last accessed 17 May 2010).

Castel, R. 1991. From dangerousness to risk. In *The Foucault effect: Studies in governmentality,* ed. G. Burchell, C. Gordon, and P. Miller, 281–98. Chicago: University of Chicago Press.

Collier, S., and A. Lakoff. 2005. On regimes of living. In *Global assemblages: Technology, politics, and ethics as anthropological problems,* ed. A. Ong and S. J. Collier, 22–39. Boston: Blackwell.

Cooper, M. 2006. Pre-empting emergence: The biological turn in the war on terror. *Theory, Culture, and Society* 23 (4): 113–35.

Craddock, S., and T. Giles-Vernick. 2010. Introduction. In *Influenza and public health: Learning from past pandemics*, ed. T. Giles-Vernick and S. Craddock, 1–21. London and New York: Taylor & Francis.

Dean, M. 1999. *Governmentality: Power and rule in modern society*. London: Sage.

de Larrinaga, M., and M. Doucet. 2010. *Security and global governmentality: Globalization, governance and the state*. London and New York: Routledge.

Elbe, S. 2005. AIDS, security, biopolitics. *International Relations* 19:403–19.

———. 2006. Should HIV/AIDS be securitized? The ethical dilemmas of linking HIV/AIDS and security. *International Studies Quarterly* 50:119–44.

———. 2009. *Virus alert: Security, governmentality and the AIDS pandemic*. New York: Columbia University Press.

———. 2010. *Security and global health: Toward the medicalization of insecurity*. Cambridge, UK: Polity.

Farmer, P. 2005. *Pathologies of power: Health, human rights and the new war on the poor*. Berkeley: University of California Press.

Fidler, D. 2004. SARS: Political pathology of the first post-Westphalian pathogen. *Journal of Law, Medicine and Ethics* 31:485–505.

———. 2008–2009. After the revolution: Global health politics in a time of economic crisis and threatening future trends. *Global Health Governance* 2 (2): 1–21. http://www.ghgj.org/Fidler_After%20the%20Revolution.pdf (last accessed 21 July 2011).

Foucault, M. 1977. *Language, counter-memory, practice: Selected interviews*. Ithaca, NY: Cornell University Press.

———. 1980. The confession of the flesh. In *Power/knowledge: Selected interviews and other writings 1972–1977*, ed. C. Gordon, 194–228. New York: Pantheon Books.

———. 2007. *Security, territory, population: Lectures at the Collège De France, 1977–1978*. Basingstoke, UK: Palgrave Macmillan.

Global Health Watch. 2005. *Global Health Watch 2005–2006*. London: Zed.

Grebe, E. 2008. Transnational networks of influence in South African AIDS treatment activism. AIDS 2031 Working Paper No. 5. http://www.aids2031.org/pdfs/workingpaper5-leadership-grebe.pdf (last accessed 9 December 2010).

Herrick, C. 2009. Shifting blame/selling health: Corporate social responsibility in the age of obesity. *Sociology of Health and Illness* 31:51–65.

Hinchliffe, S., and N. Bingham. 2008. Securing life: The emerging practices of biosecurity. *Environment and Planning A* 40:1534–51.

HM Government. 2008. *Health is global: A UK government strategy 2008–2013*. London: HM Government.

Ingram, A. 2010. Biosecurity and the international response to HIV/AIDS: Governmentality, globalisation and security. *Area* 42:293–301.

Kearns, G., and S. Reid-Henry. 2009. Vital geographies: Life, luck and the human condition. *Annals of the Association of American Geographers* 99:554–74.

Kickbusch, I., G. Silberschmidt, and P. Buss. 2007. Global health diplomacy: The need for new perspectives, strategic approaches and skills in global health. *Bulletin of the World Health Organization* 85:230–32.

King, N. 2002. Security, disease, commerce: Ideologies of postcolonial global health. *Social Studies of Science* 32:763–89.

Lakoff, A., and S. Collier, eds. 2008. *Biosecurity interventions: Global health and security in question*. New York: Columbia University Press.

Larner, W., and W. Walters, eds. 2004. *Global governmentality: Governing international spaces*. London and New York: Routledge.

Latour, B. 2005. *Reassembling the social: An introduction to actor-network-theory*. Oxford, UK: Oxford University Press.

Li, T. M. 2007. *The will to improve: Governmentality, development and the practice of politics*. London: Duke University Press.

Marmot, M. 2008. *Closing the gap in a generation*. Geneva, Switzerland: World Health Organization.

Nguyen, V.-K. 2009. Government-by-exception: Enrollment and experimentality in mass HIV treatment programmes in Africa. *Social Theory and Health* 7:196–217.

Nichter, M. 2008. *Global health: Why cultural perceptions, social representations, and biopolitics matter*. Tucson: University of Arizona Press.

O'Malley, P. 1996. Risk and responsibility. In *Foucault and political reason: Liberalism, neo-liberalism, and rationalities of government*, ed. A. Barry, T. Osborne, and N. S. Rose, 189–207. Chicago: University of Chicago Press.

Ong, A., and S. J. Collier, eds. 2005. *Global assemblages: Technology, politics, and ethics as anthropological problems*. Boston: Blackwell.

Osborne, T. 1997. Of health and statecraft. In *Foucault, health and medicine*, ed. A. Petersen and R. Bunton, 173–88. London and New York: Routledge.

Pisani, E. 2008. *The wisdom of whores: Bureaucrats, brothels and the business of AIDS*. London: Granta.

Rose, N. 2007. *The politics of life itself: Biomedicine, power, and subjectivity in the twenty-first century*. Princeton, NJ: Princeton University Press.

Rose, N., and C. Novas. 2005. Biological citizenship. In *Global assemblages: Technology, politics, and ethics as anthropological problems*, ed. A. Ong and S. J. Collier, 439–63. Boston: Blackwell.

Sarasin, P. 2008. Vapours, viruses, resistance(s): The trace of infection in the work of Michel Foucault. In S. H. Ali and R. Keil, eds. 2008. *Networked disease: Emerging infections in the global city*, 267–80. Oxford, UK: Blackwell.

Schneider, K., and L. Garrett. 2009. The end of the era of generosity? Global health amid economic crisis. *Philosophy, Ethics and Humanities in Medicine* 4:1.

Smith, R., and P. Siplon. 2006. *Drugs into bodies: Global AIDS treatment activism*. Westport, CT: Praeger.

Sparke, M. 2009. Unpacking economism and remapping the terrain of global health. In *Global health governance: Crisis, institutions and political economy*, ed. A. Kay and O. D. Williams, 131–59. Houndmills, UK: Palgrave Macmillan.

Tsing, A. 2005. *Friction: An ethnography of global connection*. Princeton, NJ: Princeton University Press.

Weir, L., and E. Mykhalovskiy. 2006. The geopolitics of global public health surveillance in the twenty-first century. In *Medicine at the border: Disease, globalization and security, 1850 to the present*, ed. A. Bashford, 240–63. Houndmills, UK: Palgrave Macmillan.

Williams, O., and S. Rushton, eds. 2011. *Partnerships and foundations in global health governance*. Houndmills, UK: Palgrave Macmillan.

World Bank. 1999. *Curbing the epidemic: Governments and the economics of tobacco control*. Washington, DC: International Bank for Reconstruction and Development.

World Health Organization. 2002. *The tobacco atlas*. Geneva, Switzerland: WHO.

Yach, D., C. Hawkes, C. Linn Gould, and K. Hofman. 2004. The global burden of chronic diseases: Overcoming impediments to prevention and control. *Journal of the American Medical Association* 291:2616–22.

Spatial Epidemiology of HIV Among Injection Drug Users in Tijuana, Mexico

Kimberly C. Brouwer,* Melanie L. Rusch,* John R. Weeks,[†] Remedios Lozada,[‡] Alicia Vera,* Carlos Magis-Rodríguez,[§] and Steffanie A. Strathdee*

*University of California San Diego School of Medicine
[†]Department of Geography, San Diego State University
[‡]Patronato Pro-COMUSIDA, and PrevenCasa, A.C., Tijuana, México
[§]Centro de Investigaciones en Infecciones de Transmisión Sexual, Programa de VIH/SIDA de la Ciudad de México

The northwest border city of Tijuana is Mexico's fifth largest and is experiencing burgeoning drug use and human immunodeficiency virus (HIV) epidemics. Because local geography influences disease risk, we explored the spatial distribution of HIV among injection drug users (IDUs). From 2006–2007, 1,056 IDUs were recruited using respondent-driven sampling and then followed for eighteen months. Participants underwent semiannual surveys, mapping, and testing for HIV, tuberculosis, and syphilis. Using average nearest neighbor and Getis-Ord Gi* statistics, locations where participants lived, worked, bought drugs, and injected drugs were compared with HIV status and environmental and behavioral factors. Median age was thirty-seven years; 85 percent were male. Females had higher HIV prevalence than males (10.2 percent vs. 3.4 percent; $p = 0.001$). HIV cases at baseline ($n = 47$) most strongly clustered by drug injection sites (Z score = -6.173, $p < 0.001$), with a 16-km^2 hotspot near the Mexico–U.S. border, encompassing the red-light district. Spatial correlates of HIV included syphilis infection, female gender, younger age, increased hours on the street per day, and higher number of injection partners. Almost all HIV seroconverters injected within a 2.5-block radius of each other immediately prior to seroconversion. Only history of syphilis infection and female gender were strongly associated with HIV in the area where incident cases injected. Directional trends suggested a largely static epidemic until July through December 2008, when HIV spread to the southeast, possibly related to intensified violence and policing that spiked in the latter half of 2008. Although clustering allows for targeting interventions, the dynamic nature of epidemics suggests the importance of mobile treatment and harm reduction programs.

西北边境城市蒂华纳是墨西哥的第五大城市，新兴药物的使用和人类免疫缺陷病毒 (HIV) 在该地流行。因为当地的地理环境影响疾病风险，我们探讨了 HIV 在注射吸毒者中 (IDUs) 的空间分布。从 2006 至 2007 年，我们采用应答驱动抽样和其后 18 个月的随访，招募了 1056 位注射吸毒者。对参与者进行了长达半年的 HIV，肺结核，梅毒等疾病的调查，测绘，和检测。使用平均近邻和 Getis-Ord 的 Gi 统计量，把参加者生活工作，买药，和注射药物的地点，与 HIV 状况，以及环境和行为因素相比较。他们年龄中位数为 37 岁，85% 为男性。女性的 HIV 感染率高于男性 (10.2% 比 3.4%，$p = 0.001$)。HIV 病例在基线组 ($n = 47$) 最强烈的聚集在药物注射集中的地区 (Z 值=-6.173，$p < 0.001$)，即墨西哥与美国边境近 16 平方公里的热点，包括红灯区。与 HIV 空间相关的因素包括梅毒感染，女性，更小的年龄，每天在街上呆更多的时间，和较多的注射伙伴。几乎所有 HIV 血清转换者在转换血清前，会在彼此的 2.5 个街区半径内注射毒品。仅有梅毒感染历史和女性性别因素，在发病病例注射的地区，紧密地与 HIV 联系在一起。定向趋势表明，直到 2008 年 7 月至 12 月，当 HIV 蔓延到东南部，才发生了一场大的静态疫情，这有可能与加剧的暴力和 2008 年下半年飙升的治安管辖有关。虽然疾病的聚集使我们能够实施有目标的干预措施，疫情的动态性质表明了移动地治疗和实施减少伤害的方案的重要性。关键词：人类免疫缺陷病毒，注射毒品的使用，当地的地理，空间流行病学。

La ciudad fronteriza de Tijuana, en la frontera noroeste de México, la quinta más grande del país, está padeciendo una tremenda epidemia de uso de drogas y de contagio con el virus de inmunodeficiencia humana (VIH). Debido a que la geografía local influye sobre el riesgo de contraer la enfermedad, exploramos la distribución espacial del VIH entre los usuarios de drogas inyectables (IDU, acrónimo inglés). Desde 2006–2007 se reclutaron para el estudio 1.056 IDUs utilizando un muestreo de base entre quienes respondieron, para ser seguidos luego durante dieciocho meses. Los participantes fueron sometidos a estudios de duración semestral, ubicación cartográfica y exámenes de laboratorio para VIH, tuberculosis y sífilis. Utilizando estadísticas del promedio de vecino más cercano y Geits-Ord GI*, las localizaciones donde los participantes vivían, trabajaban, compraban drogas y se las inyectaban, se compararon con el estatus del VIH y con factores ambientales y de comportamiento. La edad promedio fue de treinta y siete años; el 85 por ciento de los participantes fueron masculinos. Las mujeres registraron una más alta prevalencia del VIH que los varones (10.2 por ciento vs. 3.4 por ciento; $p = 0.001$). Los casos de VIH en el nivel base ($n = 47$) se agrupaban fuertemente alrededor de los sitios donde se producía la inyección de drogas (el puntaje de $Z = 6.173$, $p < 0.001$), con un punto crítico de 16-km2 cerca de la frontera México-EE.UU., el cual abarcaba todo el distrito de farol-rojo. Los correlatos espaciales del VIH incluyeron infección sifilítica, género femenino, edad más joven, mayor número de horas por día en la calle y un mayor número de compañeros de inyección. Casi todos los seropositivos con VIH se habían inyectado dentro de un radio de 2.5 manzanas de cada uno de ellos inmediatamente antes de la seroconversión. Solamente la historia de infección sifilítica y género femenino estuvieron fuertemente asociados con el VIH en el área donde ocurrieron casos incidentales de inyección. Las tendencias direccionales sugirieron una epidemia en gran medida estática hasta el tiempo transcurrido de julio a diciembre de 2008, cuando el VIH se difundió hacia el sudeste, algo posiblemente relacionado con la intensificación de la violencia y el incremento de la acción policial, que alcanzaron su pico en la segunda mitad del 2008. Aunque la aglomeración facilita la intervención claramente enfocada, la naturaleza dinámica de la epidemia sugiere la importancia de los tratamientos móviles y los programas de reducción del daño social.

Infectious disease rates are often reported at a state or national level, yet transmission operates at a much finer spatial scale. In the case of an infection such as the human immunodeficiency virus (HIV), which is influenced by social stigma and frequently concentrated in marginalized populations, exploring local geography could help to elucidate factors driving transmission. Spatial analysis can also assist in better allocation of scarce public health resources.

Injection drug use accounts for nearly 10 percent of HIV cases globally, with an estimated 20 percent of all injection drug users (IDUs) being infected (UNAIDS/World Health Organization 2010). There is growing recognition of the importance of "place," in terms of its physical, social, and geographical characteristics, on HIV risk (Rhodes et al. 1999). For IDUs, "unsafe places" associated with risky behaviors, such as needle sharing, include parks, alleys, abandoned buildings, and "shooting galleries" (where drugs can be obtained and injected with other IDUs; Small et al. 2007; Tempalski and McQuie 2009). Injecting outdoors might increase exposure to authorities, and fear of arrest or police harassment has been associated with needle sharing, injection in riskier environments, rushed injections, and unsafe syringe disposal (Koester 1994; Bluthenthal et al. 1997; Feldman and Biernacki 1998; Cooper et al. 2011). Social, structural, and economic disparities between neighborhoods and laws governing substance use and access to clean syringes also influence HIV transmission (Friedman, Perlis, and Des Jarlais 2001; Tempalski et al. 2007). Further, geographic proximity (which can be a correlate for social interactions) to social networks with high prevalence of infection or risky injection group norms might increase one's risk of HIV (Unger et al. 2006; Davey-Rothwell and Latkin 2007; Wylie, Shah, and Jolly 2007).

The HIV epidemic in Mexico is concentrated in at-risk groups, such as IDUs (Centro Nacional para la Prevencíon y el Control del VIH/SIDA [CENSIDA] 2010). Although drug use is not common in Mexico, Tijuana—a metropolitan area of 1.8 million adjacent to San Diego, California (Instituto Nacional de Estadística y Geografía [INEGI] 2010)—is situated on a major northbound drug trafficking route, with spillover use becoming entrenched in the urban environment (Bucardo et al. 2005; Brouwer, Case, et al. 2006). Tijuana is also on the busiest land border crossing in the world (see Weeks, Jankowski, and Stoler 2011). Greater exposure to U.S. media (music, TV, etc.) and interactions with those from the United States have been associated with higher drug use intentions and recent drug use among Tijuana adolescents (Becerra 2011). In 2003, it was estimated that approximately 6,000 IDUs attend shooting galleries in Tijuana, with the total IDU population (including those not frequenting shooting galleries) estimated to be 10,000 (Magis-Rodriguez et al. 2005).

HIV among IDUs can increase exponentially. In Jakarta, Indonesia, HIV prevalence in IDUs rose from 0 percent in 1997 to 48 percent in 2002 (Monitoring the AIDS Pandemic (MAP) Network 2004). We found crude HIV prevalence among Tijuana IDUs to be 1.9 percent in 2005 (Frost et al. 2006), 4.5 percent in 2007, and 10.2 percent among female IDUs in 2008 (Brouwer, Strathdee, et al. 2006; Strathdee, Lozada,

Ojeda, et al. 2008; Iniguez-Stevens et al. 2009), indicating the importance of vigilance to prevent a more generalized epidemic (Strathdee and Magis-Rodriguez 2008). In multivariate analysis, correlates of HIV infection among Tijuana IDUs included being female, syphilis titers suggestive of an active infection, increased numbers of injection partners, shorter time in Tijuana, and arrest for having track marks (Strathdee, Lozada, Pollini, et al. 2008). A subsequent analysis stratified by sex found additional associations, including younger age, lifetime syphilis infection, and longer time in Tijuana for females and history of deportation for males (Strathdee, Lozada, Ojeda, et al. 2008). In this analysis, we explore HIV among IDUs from a spatial perspective to better understand the distribution of cases over time and contextual factors associated with infection.

Methods

Study Population

From 2006 to 2007, 1,056 IDUs residing in Tijuana, Mexico, were enrolled in a study examining behavioral and contextual factors associated with HIV, syphilis, and tuberculosis infection, as described previously (Strathdee, Lozada, Pollini, et al. 2008). Participants were recruited through respondent-driven sampling (RDS) to achieve a more representative sample of this hard-to-reach population (Heckathorn 1997). Briefly, a group of "seeds" was selected based on diversity of neighborhoods, gender, and drug preferences and given three uniquely coded coupons to refer IDUs in their social networks, who were themselves given coupons to recruit three peers. Eligibility criteria included being eighteen years or older and a Spanish or English speaker, having injected drugs within the previous month, providing informed consent, and having no plans to move out of the city in the next eighteen months. The Ethics Board of the Tijuana General Hospital and Institutional Review Board of the University of California, San Diego, approved the study.

Data Collection

At baseline and every six months thereafter, staff from a municipal HIV/AIDS program (Patronato Pro-COMUSIDA A.C.) administered a quantitative survey to elicit information on demographic and economic factors, drug use practices, sexual behaviors, and HIV testing history. Additionally, participants were asked where they live (or most often sleep at night), earn money, and most commonly buy or use drugs. Mapping took place in an interactive manner with interviewers helping to narrow down locations from colonias or neighborhoods to more precise locations based on a discussion of landmarks and major crossroads. Data were digitized in ArcMap 9.3 (ESRI, Redlands, CA).

Participants were tested for HIV, tuberculosis, and syphilis, as described previously (Strathdee, Lozada, Ojeda, et al. 2008; Strathdee, Lozada, Pollini, et al. 2008). Pre- and posttest counseling were provided to all participants and those testing positive were referred to local public health care providers for free care, if indicated. Additionally, those with syphilis titers $\geq 1:8$ were given antibiotic treatment and counseling on risk behaviors.

Data Analysis

Using average nearest neighbor and Getis-Ord Gi* statistics and analytical extensions in ArcGIS version 9.3 software (ESRI), spatial distribution of where participants lived, worked, bought drugs, and injected drugs was compared with HIV status and in relation to contextual factors. The Getis-Ord Gi* analysis was based on a fixed Euclidean distance band of 3,000 m. This distance was derived from a previous analysis of this population indicating 3 km to be a natural break point in the data with respect to intraurban mobility (Brouwer et al. 2012). It is also the distance an average person walks in thirty to forty minutes (Knoblauch, Pietrucha, and Nitzburg 1996; Marx et al. 2000), which is how most of our participants travel. Results were corrected for multiple testing using the false discovery rate statistic (Benjamini and Hochberg 1995; Castro and Singer 2006).

Truncated spatial correlograms were constructed using Spatial Analysis in Macroecology v.4.0 (Rangel, Diniz-Filho, and Bini 2010) to estimate the spatial correlation between HIV and variables previously found to be associated with HIV in this population (Strathdee, Lozada, Ojeda, et al. 2008; Strathdee, Lozada, Pollini, et al. 2008). The effective number of degrees of freedom was estimated using Duttileul's estimator (Dutilleul 1993). The mean center and directional distribution (standard deviational ellipse) of HIV cases were calculated for each six-month time period of the study. Characteristics of colonias with more versus fewer than the expected numbers of HIV cases were compared based on year 2000 census data (INEGI 2000) or data from health service or other registries. In most cases, such analyses were performed using the Mann–Whitney U test to compare distributions across groups. Two-sided p values < 0.05 were considered statistically significant.

Results

Characteristics of the Study Population

Median age of participants was 37 years (interquartile range [IQR] = 31–42), with 85 percent being male and 69 percent single (unmarried, divorced, widowed, or separated). Two thirds (67 percent) were born outside of Baja California (the state where Tijuana is located) and 39 percent had come to Tijuana because of deportation from the United States. Although most (78 percent) had ever traveled to the United States, only 7 percent did so in the past year, with 6 percent accessing U.S. health services and 10 percent buying syringes. Nearly half (48 percent) reported ever injecting drugs with someone from the United States. Although 13.4 percent were homeless, the median amount of time spent on the street per 24 hours was 10 hours (IQR = 7–12). Participants reported having twenty friends who also inject drugs (IQR = 10–50), although the usual number of injection partners was two (IQR = 1–3). HIV prevalence was higher in females compared to males (10.2 percent vs. 3.4 percent, $p = 0.001$) and over 15 percent of participants had a history of syphilis infection, with half of these having titers suggestive of an active infection. Participants reported that they most commonly injected heroin (57.1 percent) or methamphetamine with heroin (39.6 percent), and 83 percent injected drugs at least daily. These demographics and drug behaviors did not change significantly over time, although the first and last time periods of the study deviated the most from mean values. Of the 1,056 participants, 943 (89 percent) had at least one follow-up interview, with 82 percent still represented at eighteen months. A total of twenty participants (1.9 percent) died between baseline and eighteen months of follow-up, of whom five were HIV-positive.

HIV Distribution

In comparing where one lives, works or earns money, and buys or injects drugs, HIV cases at baseline ($n = 47$) most strongly clustered by drug injection site, followed by residence (Table 1). There was no global clustering by work site. Drug use was unevenly spread throughout the city (Figure 1); certain injection locations had a higher than expected number of HIV-positive participants (Figure 1, red circles), with an ~16-km^2 hotspot

Table 1. Clustering of HIV-positive participants at baseline by activity location

Activity	Z-score	P-value
Live	−4.447	<0.001
Buy drugs	−2.113	0.035
Inject drugs	−6.173	<0.001
Work	−1.535	0.125

Note: $n = 47$. Average nearest neighbor.

near the Mexico–U.S. border encompassing the red-light district. Other neighborhoods, especially to the southeast, had fewer cases than expected (Figure 1, blue circles). Local clustering of HIV was not as extensive for IDU residences, work, or drug purchase sites; did not reveal new clusters; and tended to occur when injection and the other activity overlapped. In examining the distribution of HIV by sex, males most strongly clustered by injection site—with many traveling between their main injection site and where they lived or worked. For females, the four types of activities generally occurred in the same place or in very close proximity to each other (data not shown).

Incident HIV Cases

Eleven of the twelve who seroconverted during follow-up injected within 2.5 blocks (380 m) of the mean center of each other during the study visit immediately prior to seroconversion. After some initial changes in distribution (Figure 2A and B), directional trends suggested a largely static epidemic until July through December 2008, when HIV cases spread to the southeast (Figure 2). To explore whether these changes were due to movement of HIV-positive participants or a change in the underlying sample comprising a particular period, we measured the distance that each individual's injection sites shifted over consecutive follow-up visits.

The median distance moved was < 350 m for all time periods, with the exception of July through December 2008, with a median of zero for July through December 2007 and January through June 2008 (Figures 2D and 2E). For the time periods with the largest ellipses (Figures 2B and 2F), median distance injection sites moved was just 113.5 m (IQR = 72, 218) for July through December 2006 but 1,674 m (IQR = 609.5, 5343.5) for July through December 2008. Most (68 percent) of the 2006 period included new enrollees (not attending a follow-up visit), suggesting that the ellipse size might reflect an evolving study population as it became established through RDS. However, during the July through

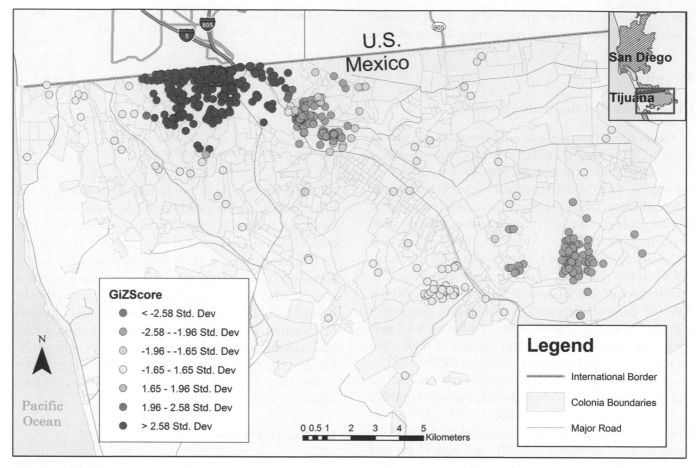

Figure 1. HIV hot and cold spots of injection sites among substance-using participants at baseline (N = 1,056) in Tijuana, Mexico. (Color figure available online.)

December 2008 time period there was a sudden shift in preferred injection sites of a number of individuals. The period at the close of the study (January through June 2009) does not suggest a continuation of the dispersion seen in July through December 2008; however, the follow-up period for the participants who had moved from the HIV core in the latter half of 2008 had ended by 2009, so their data are not included in this final period.

In comparing neighborhood characteristics, those making up the HIV hot spot had higher divorce and female-headed household rates, but lower home-ownership and health insurance rates (Table 3). Fewer residents in the hot spot area had an educational deficit. Although of marginal significance, population density

Associations with HIV Distribution

To better understand the distribution of HIV and incident cases, we explored a number of variables for spatial correlation with HIV. Those with significant or marginally significant correlations are listed in Table 2. RXR diagonal maps (data not shown) showed that only lifetime syphilis infection and gender showed strong correlation in the area where most incident cases occurred.

Table 2. Spatial correlations with HIV distribution

Characteristic	Pearson's r	p value
Age	−0.063	0.041
Hours on street per day	0.093	0.003
Born in Baja, California	−0.060	0.059
History of deportation	0.057	0.061
Female gender	0.119	<0.001
Number of injection partners (past 6 months)	0.075	0.014
History of syphilis infection	0.163	<0.001
Active syphilis infection (titer ≥ 1:8)	0.144	<0.001

Note: N = 1,056.

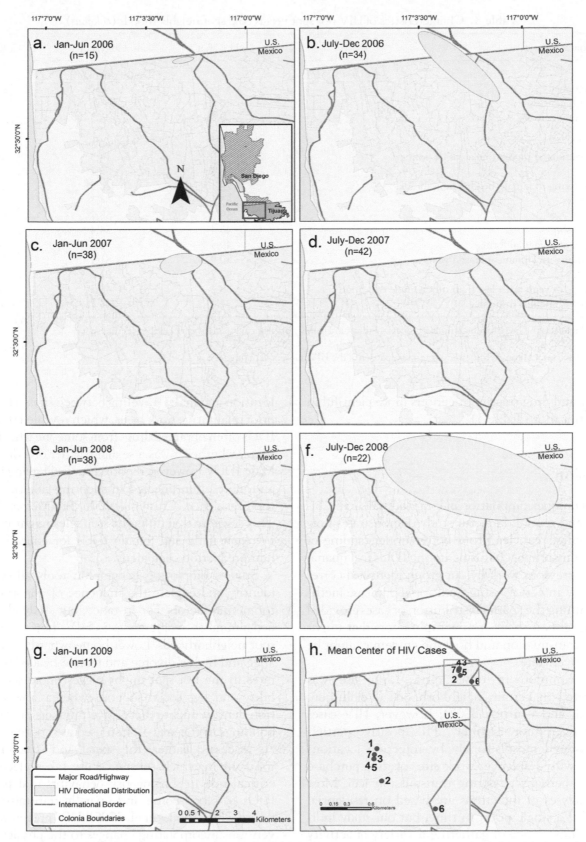

Figure 2. (A) and (B) represent January through June 2006 and July through December 2006, respectively. (C) and (D) are the respective six-month time periods for 2007, (E) and (F) represent 2008, and (G) represents January through June 2009. Red dots in (H) denote the mean center of HIV cases at each of the time periods indicated in the previous panels. The scale and legend (G) applies to all maps. (Color figure available online.)

Table 3. Characteristics of HIV hot spot versus cold spot neighborhoods (*colonias*)

Colonia characteristic	Hot spot	Cold spot 1[a]	Cold spot 2[a]
		Density per km^2	
Population density	5,452	9,501	10,451
Public health centers	0.43	0	0.18
Drug treatment centers	0.98	0*	0.18
		Median%	
Colonias considered high crime areas by police	50.0	14.3	50.0
Divorced	6.3	5.1*	4.5**
Population living in female-headed households	24.1	22.5*	19.5**
Catholic	85.1	84.3	78.6**
Employed	59.6	58.7	62.8
Home ownership	45.4	54.4**	71.4**
Born outside of Baja, California	49.5	50.7	56.5
Residents living in Tijuana at least 5 years	80.9	81.1	76.8**
Adult literacy	97.7	97.2	96.1*
Population > 15 years with less than ninth-grade education	38.2	48.2*	56.2**
Population with health insurance	37.4	44.7*	50.8**

[a]Cold spot 2 refers to the neighborhoods furthest to the southeast in Figure 1, whereas cold spot 1 is closer to the HIV hot spot.
*$p < 0.05$.
**$p < 0.01$, based on a Mann–Whitney U or t test comparison of the HIV hot spot vs. cold spot 1 or 2.

was lower and drug treatment centers more plentiful in the HIV hot spot (Table 3).

Discussion

From a unique compilation of longitudinal survey, biological, and spatial data, our study demonstrates how spatial analysis can lead to a better understanding of HIV transmission mechanisms among IDUs in Tijuana. Although previous work by our group identified correlates of HIV infection using traditional database methods, examining the spatial distribution of cases provided further insights into environmental factors that might influence transmission and how the epidemic might be changing over time.

As is common in other cities, drug use was concentrated in certain neighborhoods (Veldhuizen, Urbanoski, and Cairney 2007); however, HIV cases clustered even more strongly. HIV-positive participants clustered most strongly by injection locations compared with residence, work site, or drug purchase locations—perhaps reflecting transmission foci. Most spatial analyses of infections are based on residential addresses at a single point in time, but our study indicates the importance of gathering a variety of activity locations to improve understanding of disease transmission. We also found differences between activity locations by sex, with females tending to stay in one location and males traveling between sites. This reflects prior qualitative work by us, which revealed that female IDUs often obtained drugs from someone who delivered them, whether it was a sexual partner or a drug dealer. Male IDUs, however, reported several buying locations and more often injected in shooting galleries far from their residence (Cruz et al. 2006; Brouwer et al. 2012). This suggests that different strategies might be needed to reach male and female IDUs for educational and harm reduction campaigns.

Studies employing geographic tools allow one to identify and explore the influence of important environmental factors. For instance, our study also identified a number of differences in HIV "hot" versus "cold" spot neighborhoods. Lower homeownership and insurance and higher divorce and female-headed household rates in the hot spot might suggest more social instability and less social support, whereas lower population density might reflect large numbers of abandoned buildings in the area—often used as shooting galleries. As discussed earlier, the social and structural environment in which substance use takes place plays a critical role in shaping risk behaviors that predispose IDUs to blood-borne and sexually transmitted infections (STIs; Rhodes et al. 2005). HIV prevention interventions incorporating changes to the physical, social, economic, and legal environments to facilitate individual and community-level behavior change have had the most success (Rhodes et al. 2005).

The highly concentrated distribution of incident HIV cases suggests that targeting only a small area in Tijuana could have a large impact on incidence of HIV. Proximity to syringe exchange programs and drug treatment has been linked with increased use (Rockwell et al. 1999; Sarang, Rhodes, and Platt 2008; Cooper et al. 2011), but resource availability does not guarantee full access or utilization. Fear of police or being identified as an IDU, stigma, operating hours, and local social and political conditions can shape access (Bluthenthal et al. 1997; Tempalski et al. 2007; Simmonds and Coomber 2009; Tempalski and McQuie 2009). The border region is particularly complicated in that migratory and legal status might affect access and continuity of care (Moyer et al. 2008; Brouwer et al. 2009; Volkmann et al. 2011). This speaks to the importance of cross-border HIV health policies.

Clustering of incident cases also facilitated identification of structural factors possibly associated with transmission. For instance, female sex and syphilis infection were the primary spatial associations with HIV where incident cases occurred, and this area overlaps with the Zona Roja (red-light district) of Tijuana. Previous work by us and others indicates considerable overlap between drug use and the sex trade, with HIV prevalence among female sex workers who are also IDUs in Tijuana surpassing 12 percent (Strathdee, Philbin, et al. 2008). Because the Zona Roja is a major attractor of sexual tourism, drawing clients from the United States, Asia, and elsewhere, the potential for further spread of HIV in the region is also a concern.

Injection locations of HIV-positive participants remained fairly static until July through December 2008, when HIV cases shifted to the southeast. The reasons for this are unclear but might relate to changes in the local environment, such as economic conditions, drug availability, and so on. In recent years, there has been a tremendous increase in the number of murders in Tijuana (Negron 2008; Welch 2009), which peaked at 843 in the latter half of 2008—up 2.5-fold from 2007 (Welch 2009). At the same time, military presence increased, creating an uneasy relationship between local police and federal forces. Anecdotal reports from participants suggested that this led to a number of police "sweeps" through the canal area near the Mexico–U.S. border, where many participants resided or injected drugs. Due to an end in the follow-up periods of a number of participants, it is uncertain whether the directional trend seen in the latter half of 2008 continued or whether HIV cases who had shifted injection sites eventually moved back to the HIV hot spot. If they did

remain outside of the main HIV core, one of the unintended consequences of enforcement activities might have been displacement of HIV cases to other parts of the city, facilitating the penetration of HIV into new social networks. Displaced persons have been shown to be particularly vulnerable to HIV (Friedman, Rossi, and Braine 2009; Kim et al. 2009; Marshall et al. 2009). Likewise, political–economic events and transitions are theorized to create risk environments that might facilitate drug use and HIV outbreaks (Friedman, Rossi, and Braine 2009). Because the HIV risk environment in northern Mexico–U.S. border cities is constantly shifting due to drug-related violence, migration, economic factors, and policing, their individual and joint impact on transmission of HIV and STIs warrants continued surveillance.

This analysis was limited in that it focused on the locations where a subject "most often" participated in an activity in the past six months. Such data might fail to capture exposures of participants who often move or inject in a variety of locations. Although RDS seeds were selected to promote geographic diversity, more socially isolated IDUs might be underrepresented and network bottlenecks might have affected the spatial patterns observed. Our fairly large sample ($N = 1,056$) compared to the estimated number of IDUs in the city (10,000) might have tempered this, however. Another possible limitation is that not all participants were enrolled at the same time, so their follow-up period might represent exposure to different environmental circumstances. Further, not every six-month time period in the longitudinal analysis reflects the same participants, due to rolling enrollment as well as loss to follow-up or death. In particular, the first and last time periods were based on fewer participants due to study initiation and subsequent closure. Despite these limitations, the stability of the HIV epidemic for most of the study suggests that such changes did not greatly affect our results.

Our study highlighted spatial and environmental factors related to the IDU HIV epidemic in Tijuana, Mexico. Spatial clustering of HIV cases by injection site, but less so by other activity locations, suggests the importance of collecting the most pertinent location data possible when exploring disease distribution. Our study also indicates the importance of collecting longitudinal data and exploring spatial data by sex. The dynamic nature of this epidemic suggests the need for intensified prevention efforts involving community outreach, mobile treatment, and harm reduction programs. A high correlation with syphilis infection suggests that treatment of STIs, in addition to efforts to reduce drug-related harm, is needed. Additional studies are necessary to determine the consequences and causes of the dispersion of HIV cases in the latter half of 2008 and whether this trend has continued.

Acknowledgments

This study was supported by grants K01DA020364 and R01DA019829 of the U.S. National Institute on Drug Abuse (NIH).

References

ArcMap, Version 9.3. Redlands, CA: ESRI.

Becerra, D. 2011. The relationship between pre-migration acculturation and substance use. Paper presented at the 14th annual conference of the Society for Social Work Research, San Francisco, CA.

Benjamini, Y., and Y. Hochberg. 1995. Controlling the false discovery rate: A practical and powerful approach to multiple testing. *Journal of the Royal Statistical Society* 57:289–300.

Bluthenthal, R. N., A. H. Kral, J. Lorvick, and J. K. Watters. 1997. Impact of law enforcement on syringe exchange programs: A look at Oakland and San Francisco. *Medical Anthropology* 18:61–83.

Brouwer, K. C., P. Case, R. Ramos, C. Magis-Rodriguez, J. Bucardo, T. L. Patterson, and S. A. Strathdee. 2006. Trends in production, trafficking and consumption of methamphetamine and cocaine in Mexico. *Substance Use and Misuse* 41:707–27.

Brouwer, K. C., R. Lozada, W. A. Cornelius, M. Firestone Cruz, C. Magis-Rodriguez, M. L. Zuniga de Nuncio, and S. A. Strathdee. 2009. Deportation along the U.S.–Mexico border: Its relation to drug use patterns and accessing care. *Journal of Immigrant and Minority Health* 11:1–6.

Brouwer, K. C., R. Lozada, J. R. Weeks, C. Magis-Rodriguez, M. Firestone, and S. A. Strathdee. 2012. Intra-urban mobility and its potential impact on the spread of blood-borne infections among drug injectors in Tijuana, Mexico. *Substance Use and Misuse* 47:244–53.

Brouwer, K. C., S. A. Strathdee, C. Magis-Rodriguez, E. Bravo-Garcia, C. Gayet, T. L. Patterson, S. M. Bertozzi, and R. S. Hogg. 2006. Estimated numbers of men and women infected with HIV/AIDS in Tijuana, Mexico. *Journal of Urban Health* 83:299–307.

Bucardo, J., K. C. Brouwer, C. Magis-Rodriguez, R. Ramos, M. Fraga, S. G. Perez, T. L. Patterson, and S. A. Strathdee. 2005. Historical trends in the production and consumption of illicit drugs in Mexico: Implications for the prevention of blood borne infections. *Drug and Alcohol Dependence* 79:281–93.

Castro, M. C., and B. H. Singer. 2006. A new approach to account for multiple and dependent tests in local statistics of spatial association: Controlling the false discovery rate. *Geographical Analysis* 38:180–208.

Centro Nacional para la Prevención y el Control del VIH/SIDA (CENSIDA). 2010. *HIV/AIDS in Mexico 2010*. Mexico City: CENSIDA.

Cooper, H. L., D. C. Des Jarlais, Z. Ross, B. Tempalski, B. Bossak, and S. R. Friedman. 2011. Spatial access to syringe exchange programs and pharmacies selling over-the-counter syringes as predictors of drug injectors' use of sterile syringes. *American Journal of Public Health* 101:1118–25.

Cruz, M. F., A. Mantsios, R. Ramos, P. Case, K. C. Brouwer, M. E. Ramos, W. D. Fraga, C. A. Latkin, C. L. Miller, and S. A. Strathdee. 2007. A qualitative exploration of gender in the context of injection drug use in two US–Mexico border cities. *AIDS and Behavior* 11 (2): 253–62.

Davey-Rothwell, M. A., and C. A. Latkin. 2007. Gender differences in social network influence among injection drug users: Perceived norms and needle sharing. *Journal of Urban Health* 84:691–703.

Dutilleul, P. 1993. Modifying the *t* test for assessing the correlation between two spatial processes. *Biometrics* 49:305–14.

Feldman, H. W., and P. Biernacki. 1998. The ethnography of needle sharing among intravenous drug users and implications for public policies and intervention strategies. *NIDA Research Monographs* 80:28–39.

Friedman, S. R., T. Perlis, and D. C. Des Jarlais. 2001. Laws prohibiting over-the-counter syringe sales to injection drug users: Relations to population density, HIV prevalence, and HIV incidence. *American Journal of Public Health* 91:791–93.

Friedman, S. R., D. Rossi, and N. Braine. 2009. Theorizing "big events" as a potential risk environment for drug use, drug-related harm and HIV epidemic outbreaks. *International Journal of Drug Policy* 20:283–91.

Frost, S. D., K. C. Brouwer, M. A. Firestone Cruz, R. Ramos, M. E. Ramos, R. M. Lozada, C. Magis-Rodriguez, and S. A. Strathdee. 2006. Respondent-driven sampling of injection drug users in two U.S.–Mexico border cities: Recruitment dynamics and impact on estimates of HIV and syphilis prevalence. *Journal of Urban Health* 83:83–97.

Heckathorn, D. D. 1997. Respondent-driven sampling: A new approach to the study of hidden populations. *Social Problems* 44:174–99.

Iniguez-Stevens, E., K. C. Brouwer, R. S. Hogg, T. L. Patterson, R. Lozada, C. Magis-Rodriguez, J. P. Elder, R. M. Viani, and S. A. Strathdee. 2009. Estimating the 2006 prevalence of HIV by gender and risk groups in Tijuana, Mexico. *Gaceta Medica de Mexico* 145:189–95.

Instituto Nacional de Estadística y Geografía [INEGI]. 2000. *XII censo general de población y vivienda 2000* [XII general census of population and housing 2000]. Mexico City, Mexico: INEGI.

———. 2010. *Censo de poblacion y vivienda 2010: Estados Unidos Mexicanos resultados preliminares*. Mexico City, Mexico: INEGI.

Kim, A. A., F. Malele, R. Kaiser, N. Mama, T. Kinkela, J. C. Mantshumba, M. Hynes, et al. 2009. HIV infection among internally displaced women and women residing in river populations along the Congo River, Democratic Republic of Congo. *AIDS and Behavior* 13: 914–20.

Knoblauch, R. L., M. T. Pietrucha, and M. Nitzburg. 1996. Field studies of pedestrian walking speed and start-up time. *Transportation Research Record, Pedestrian and Bicycle Research* 1538:27–38.

Koester, S. K. 1994. The context of risk: Ethnographic contributions to the study of drug use and HIV. *NIDA Research Monographs* 143:202–17.

Magis-Rodriguez, C., K. C. Brouwer, S. Morales, C. Gayet, R. Lozada, R. Ortiz-Mondragon, E. P. Ricketts, and S. A. Strathdee. 2005. HIV prevalence and correlates of receptive needle sharing among injection drug users in the Mexican–U.S. border city of Tijuana. *Journal of Psychoactive Drugs* 37:333–39.

Marshall, B. D., T. Kerr, J. A. Shoveller, J. S. Montaner, and E. Wood. 2009. Structural factors associated with an increased risk of HIV and sexually transmitted infection transmission among street-involved youth. *BMC Public Health* 9:7.

Marx, M. A., B. Crape, R. S. Brookmeyer, B. Junge, C. Latkin, D. Vlahov, and S. A. Strathdee. 2000. Trends in crime and the introduction of a needle exchange program. *American Journal of Public Health* 90:1933–36.

Monitoring the AIDS Pandemic (MAP) Network. 2004. *AIDS in Asia: Face the facts.* Washington, DC: MAP.

Moyer, L. B., K. C. Brouwer, S. K. Brodine, R. Ramos, R. Lozada, M. F. Cruz, C. Magis-Rodriguez, and S. A. Strathdee. 2008. Barriers and missed opportunities to HIV testing among injection drug users in two Mexico–US border cities. *Drug and Alcohol Review* 27:39–45.

Negron, S. 2008. Baghdad, Mexico. *Texas Monthly* 37 (1):60–64.

Rangel, T. F., J. A. F. Diniz-Filho, and L. M. Bini. 2010. SAM: A comprehensive application for spatial analysis in macroecology. *Ecography* 33:46–50.

Rhodes, T., M. Singer, P. Bourgois, S. R. Friedman, and S. A. Strathdee. 2005. The social structural production of HIV risk among injecting drug users. *Social Science & Medicine* 61:1026–44.

Rhodes, T., G. V. Stimson, N. Crofts, A. Ball, K. Dehne, and L. Khodakevich. 1999. Drug injecting, rapid HIV spread, and the "risk environment": Implications for assessment and response. *AIDS* 13 (Suppl. A): S259–69.

Rockwell, R., D. C. Des Jarlais, S. R. Friedman, T. E. Perlis, and D. Paone. 1999. Geographic proximity, policy and utilization of syringe exchange programmes. *AIDS Care* 11:437–42.

Sarang, A., T. Rhodes, and L. Platt. 2008. Access to syringes in three Russian cities: Implications for syringe distribution and coverage. *International Journal of Drug Policy* 19 (Suppl. 1): S25–36.

Simmonds, L., and R. Coomber. 2009. Injecting drug users: A stigmatised and stigmatising population. *International Journal of Drug Policy* 20:121–30.

Small, W., T. Rhodes, E. Wood, and T. Kerr. 2007. Public injection settings in Vancouver: Physical environment, social context and risk. *International Journal of Drug Policy* 18:27–36.

Strathdee, S. A., R. Lozada, V. D. Ojeda, R. A. Pollini, K. C. Brouwer, A. Vera, W. Cornelius, L. Nguyen, C. Magis-Rodriguez, and T. L. Patterson. 2008. Differential effects of migration and deportation on HIV infection among male and female injection drug users in Tijuana, Mexico. *PLoS One* 3:e2690.

Strathdee, S. A., R. Lozada, R. A. Pollini, K. C. Brouwer, A. Mantsios, D. A. Abramovitz, T. Rhodes, et al. 2008. Individual, social, and environmental influences associated with HIV infection among injection drug users in Tijuana, Mexico. *Journal of Acquired Immune Deficiency Syndromes* 47:369–76.

Strathdee, S. A., and C. Magis-Rodriguez. 2008. Mexico's evolving HIV epidemic. *Journal of the American Medical Association* 300:571–73.

Strathdee, S. A., M. M. Philbin, S. J. Semple, M. Pu, P. Orozovich, G. Martinez, R. Lozada, et al. 2008. Correlates of injection drug use among female sex workers in two Mexico–U.S. border cities. *Drug and Alcohol Dependence* 92:132–40.

Tempalski, B., R. Friedman, M. Keem, H. Cooper, and S. R. Friedman. 2007. NIMBY localism and national inequitable exclusion alliances: The case of syringe exchange programs in the United States. *Geoforum* 38:1250–63.

Tempalski, B., and H. McQuie. 2009. Drugscapes and the role of place and space in injection drug use-related HIV risk environments. *International Journal of Drug Policy* 20:4–13.

UNAIDS/World Health Organization. 2010. *Report on the global AIDS epidemic | 2010.* Geneva, Switzerland: UNAIDS/World Health Organization.

Unger, J. B., M. D. Kipke, C. J. De Rosa, J. Hyde, A. Ritt-Olson, and S. Montgomery. 2006. Needle-sharing among young IV drug users and their social network members: The influence of the injection partner's characteristics on HIV risk behavior. *Addictive Behaviors* 31:1607–18.

Veldhuizen, S., K. Urbanoski, and J. Cairney. 2007. Geographical variation in the prevalence of problematic substance use in Canada. *Canadian Journal of Psychiatry* 52:426–33.

Volkmann, T., R. Lozada, C. M. Anderson, T. L. Patterson, A. Vera, and S. A. Strathdee. 2011. Factors associated with drug-related harms related to policing in Tijuana, Mexico. *Harm Reduction Journal* 8:7.

Weeks, J. R., P. Jankowski, and J. Stoler. 2011. Who's knocking at the door? New data on undocumented immigrants to the United States. *Population, Space and Place* 17:1–26.

Welch, W. M. 2009. Tijuana off-limits to U.S. Marines. *USA Today* 21 January:3A.

Wylie, J. L., L. Shah, and A. Jolly. 2007. Incorporating geographic settings into a social network analysis of injection drug use and bloodborne pathogen prevalence. *Health Place* 13:617–28.

Structural Violence and Women's Vulnerability to HIV/AIDS in India: Understanding Through a "Grief Model" Framework

Vandana Wadhwa

Department of Geography and Environment, Boston University

HIV/AIDS remains one of India's major health concerns today, with 2.27 million people affected by the disease. Ineffective/inappropriate policy stances at the incipient stages of the epidemic are primarily to blame. The Indian government's slow progression through stages of denial and stalling to final acceptance for comprehensive action is analogous to the psychiatric model of dealing with grief. This qualitative study, based on in-depth interviews of key informants from urban health posts serving four slums in Delhi and Hyderabad cities, explores AIDS awareness and attitudes in the community and HIV/AIDS policy efficacy. Findings reveal (i) a largely reactive policy response creating a circular relationship between policy, prevalence, and awareness, where policies often create local patterns of HIV/AIDS occurrence and awareness, which then inform next steps; (ii) the existence of institutional and socioeconomic barriers (poverty, underdevelopment, lack of transparency, taboo, and stigma), which can be conceptually framed as "structural violence." The article concludes that the government's "Grief Model" policy response is another frame of reference through which structural violence can be understood.

艾滋病毒/艾滋病仍然是印度当今的主要健康问题之一，有二百二十七万人受该疾病的影响。这主要归咎于在疫情的早期阶段政府所持的无效/不适当的政策立场。印度政府从拒绝和拖延到最后同意采取全面行动的这几个阶段进程缓慢，类似于精神科处理悲伤过程的模式。这个定性研究，基于对在德里和海得拉巴市四个贫民窟供职的，来自城市卫生服务业的关键知情人的深入访谈，探讨社区对艾滋病的认识和态度，和艾滋病毒/艾滋病的政策效力。结果显示：(i) 主要响应的政策反应，在政策，患病率，和认识之间产生了一个循环关系，在其中，政策往往创建艾滋病毒/艾滋病发生和认识的本地模式，然后通知下一个步骤；(ii) 机构和社会经济障碍的存在（贫穷，不发达，缺乏透明度，禁忌，羞耻），可以被概括为"结构性暴力。"本文认为政府的"悲伤模型"的政策响应是另一个参考的框架，结构性暴力通过它得以理解。*关键词：悲伤模式，卫生政策，艾滋病毒/艾滋病，印度，结构性暴力。*

El HIV/SIDA se mantiene como una de las mayores preocupaciones actuales de la India, que registra 2.27 millones de personas afectadas por la enfermedad. Primariamente, se le echa la culpa a las posturas políticas inefectivas e inapropiadas durante las etapas incipientes de la epidemia. El lento avance del gobierno indio, desde las etapas de denegación y evasivas hasta la aceptación final de una acción comprensiva, es análogo al modelo psiquiátrico para lidiar con la pena. Este estudio cualitativo, basado en entrevistas a profundidad con informantes claves de unidades de salud urbana que sirven a cuatro barriadas tuguriales de las ciudades de Delhi y Hyderabad, explora la conciencia y actitudes que se tiene sobre el SIDA en las comunidades, y la eficacia de las políticas sobre HIV/SIDA. Los descubrimientos revelan, (i) una política de respuesta grandemente reactiva que crea una relación circular entre política, prevalencia y conciencia, donde las políticas a menudo crean patrones locales de ocurrencia y conciencia del HIV/SIDA, que luego informa las siguientes etapas; y (ii) la existencia de barreras institucionales y socioeconómicas (pobreza, subdesarrollo, falta de transparencia, tabú y estigma), que pueden ser conceptualmente enmarcadas como "violencia estructural." El artículo concluye que la política de respuesta del "Modelo de Pena" del gobierno es otro marco de referencia a través del cual se puede entender la violencia estructural.

In her seminal work, Kübler-Ross ([1969] 1997) posited that human reactions to death and dying included stages of denial, anger, bargaining, depression, and acceptance, which might overlap, or occur nonsequentially. In morbid irony, the Indian government has displayed many of these same stages in its policy stances toward the HIV/AIDS epidemic. I use the analogy drawn from the Kübler-Ross grief model to examine state response to the HIV/AIDS epidemic in India. Adding evidence from a qualitative field study conducted in 2006, I link the data to existing theorizations of vulnerability and structural barriers through an exercise in grounded theory (see Glaser and Strauss 1967). In conclusion, I posit that the preceding grief model constitutes an additional frame through which the concept of structural violence can be perceived and understood.

HIV/AIDS in India: Women's Vulnerability

India has the third-highest prevalence of HIV/AIDS in the world, with approximately 2.27 million persons living with HIV/AIDS (PHLA) in 2008. Women currently make up 39 percent of this population but are increasingly in positions of greater vulnerability (National AIDS Control Organization [NACO] 2010; United Nations General Assembly Special Session [UNGASS] 2010). Women's vulnerability to HIV/AIDS in the global south is impacted by factors contextual to place, political economy, and culture: underdevelopment and poverty, unequal gender relations, disempowerment, harmful social practices and mores, stigma and taboo, and global and local political-economic paradigms, including structural adjustment, urban location, mobility, and employment patterns (Craddock 2001; Gould 2005; Kalipeni 2008; Jongsthapongpanth and Bagchi-Sen 2010; for overviews and compilations, see Parker, Easton, and Klein 2000; Bates et al. 2004a, 2004b; Kalipeni, Oppong, and Zerai 2007; Kalipeni, Flynn, and Pope 2009; Pope, White, and Malow 2009; Pope and Kalipeni 2010).

Women in India face similar vulnerability contexts of male preference, early marriage, restrictions on education and autonomy, exposure to sexual abuse, violence and sex trafficking, and ramifications of a neoliberal political economy (Bharat, Aggleton, and Tyrer 2001; Rao Gupta 2002; Verma and Roy 2002; Pradhan and Sundar 2006; D. Ghosh 2007; J. Ghosh and Olson 2007; Eliot 2009; Kishor and Gupta 2009; Dutt 2011). Participant interviews from the 2006 field study revealed that such gendered vulnerability stemmed from lack of autonomy, awareness, and access to relevant information and health care; low investment in education; taboo; and conditions of poverty (J. Ghosh, Wadhwa, and Kalipeni 2009; Wadhwa, Ghosh, and Kalipeni 2010).

HIV/AIDS Policy Response: The Grief Model and Feedback Loop

Next, I describe the state's policy response, as reflected in the phases of the National AIDS Control Program (NACP) and mirroring the Kübler-Ross ([1969] 1997) grief model. Also demonstrated is the complex, circular relationship among HIV prevalence, awareness, and policy in the study cities, where HIV/AIDS policies affected local patterns of prevalence and awareness, which in turn inform the next steps. Hyderabad and Delhi were studied because then-available statistics showed contrasting HIV/AIDS prevalence and awareness rates; Hyderabad had higher HIV prevalence rates than Delhi but the reverse trend in awareness (see J. Ghosh, Wadhwa, and Kalipeni 2009).

1986–1992: Denial, Blame, and Anger

HIV was first diagnosed in 1986 in India's southern state of Tamil Nadu, but the Indian government's initial response was akin to denial (see the Kübler-Ross [1969] 1997 grief model). Hiding behind dominant Indian cultural tropes of sexual modesty, morality, and monogamy, the governmental labeled HIV/AIDS a "foreign disease." The "blaming" subphase that precedes "anger" quickly followed, and "high-risk groups" (HRGs) such as commercial sex workers, professional blood donors, and "foreigners" were made scapegoats and targeted by law and enforcement agencies (Asthana 1996; Ramasubban 1998; Kadiyala and Barnett 2004).

As a result, the epidemic progressed virtually unchecked through the early 1990s, albeit spatially differentiated. Lack of cohesive policy also kept awareness levels low even in high-prevalence states such as Andhra Pradesh. Only 23 percent of ever-married women in Andhra Pradesh's neighboring surveyed state of Tamil Nadu had "heard of AIDS" in 1992–1993. Conversely, despite low HIV prevalence, Delhi's status as the national capital resulted in greater exposure of government-sponsored HIV/AIDS-related media messages—awareness rates here were relatively higher at 35.8 percent (International Institute for Population Studies [IIPS] 1995; Kadiyala and Barnett 2004).

NACP Phase I (1992–1999): Bargaining

The National AIDS Committee was formed within the Ministry of Health and Family Welfare (MoHFW) by 1987, but NACO, the nodal agency for national, cohesive HIV/AIDS policy, was not constituted until

1992. NACO's first mandate was the implementation of NACP I, which unfortunately largely ignored the epidemic's underlying socioeconomic issues and focused primarily on surveillance and other "targeted" (read targeting) interventions (Ramasubban 1998; NACO 2010). This is suggestive of the bargaining phase, often a delaying tactic of dealing with smaller immediacies to avoid facing reality (Kübler-Ross [1969] 1997). Within some years, however, NACO realized the need to address all social groups and began some measure of awareness campaigns (Asthana 1996; NACO 2010).

These campaigns appeared to have some impact; in urban areas of both Delhi and Andhra Pradesh, proportions of ever-married women having heard of AIDS rose sharply to 81 percent by 1999 (IIPS and ORC Macro 2000). These efforts were too late and limited, however, as Delhi became a "highly vulnerable state" by 2000 (HIV prevalence ≥ 5 percent among HRGs), and Andhra Pradesh entered a statewide generalized epidemic (≥1 percent HIV prevalence rate at antenatal clinics [ANCs]; Kadiyala and Barnett 2004; Andhra Pradesh State AIDS Control Society [APSACS] 2005; Delhi State AIDS Control Society [DSACS] 2005). Interestingly, an "after-the-fact" awareness was also reported in high-prevalence states like Andhra Pradesh (IIPS and ORC Macro 2000), supported by a statistically significant positive correlation between rates of HIV prevalence and AIDS awareness (Wadhwa 2006).

NACP Phase II (1999–2006): Acceptance

In response to rising prevalence rates among HRGs in major cities, awareness campaigns were further ramped up during NACP II. Hyderabad received particular attention, as it had become a Category-A district (HIV prevalence rates ≥ 1 percent in ANCs and ≥ 5 percent among HRGs; Mitra 2004; APSACS 2005). As a result, AIDS awareness rates climbed further to 88 percent in Delhi, 87 percent in urban Andhra Pradesh, and an even higher 89 percent in Hyderabad city by 2006, although comprehensive awareness was only 32 percent in Hyderabad and 49 percent in Delhi (IIPS and ORC Macro 2008). The mature stage of acceptance, wherein practical solutions are finally sought (Kübler-Ross [1969] 1997), became more visible in the latter part of NACP II, at least outwardly. Focus shifted to behavior change, PLHA and nongovernmental organization partnerships, youth AIDS education, free provision of antiretroviral therapy (ART) in high-prevalence states and Delhi through "selected hospitals," establishment of Integrated Counselling and Testing Centres (ICTCs), strengthened surveillance, and recognition of

HIV/AIDS as a social issue (Ramasubban 1998; NACO 2010).

Thus, as government reaction to the epidemic cycled through denial and blame to bargaining to acceptance, precious time was lost on not addressing the disease with required immediacy and comprehensiveness. Response was often reactive rather than proactive, and the more progressive policy responses by NACO were too little, too late, as also evidenced in the analysis. As demonstrated in the concluding section, this slow evolution to a mature stage is tantamount to structural violence.

Methodology

The qualitative field study on which this article is based was conducted in Delhi and Hyderabad in July and August 2006. Thirty-two women of reproductive age (fifteen to forty-nine years) residing in slums were asked in-depth questions regarding HIV/AIDS and socioeconomic awareness aspects to explore issues of gendered vulnerability. Slum selection was stratified across relative socioeconomic and health-access situations. To maximize understanding, semistructured interviews of four key informants (KIs) from urban health posts (UHPs) serving each selected slum were also conducted. The current work draws only on the insights provided by the KIs (for participants' perspectives, see J. Ghosh et al. 2009; Wadhwa et al. 2010).

KI recruitment did not include or exclude on the basis of position or title, except that they had to be health staff. This ensured anonymity for the small staff and diverse representation. KIs were asked about their perception of AIDS awareness and attitudes among their communities, their own knowledge and training, and impact of various HIV/AIDS policies on the ground. Informed consent was obtained for conducting and taping the interviews. Median interview length was approximately two hours, conducted in Hindi and English in Delhi and in English and Telugu (translator used) in Hyderabad. These were later transcribed (verbatim in sections—language not denaturalized) and coded into emergent themes that form the basis of the analysis. Interviews were buttressed by field observation.

The study adhered to the Ethical Guidelines for Social Science Research in Health (National Committee for Ethics in Social Science Research in Health [see NCESSRH] 2000). Additionally, study protocol and description were provided to obtain permission from the health department of the relevant municipal corporations (MCs), as UHPs fall under MC jurisdiction. An ethical dilemma occurred during the KI consent process, however; both informants from one city expressed discomfort, offering that they desired to be heard yet feared reprisal. This was despite blanket permission for conducting the study in broad zones rather than specific UHPs or slums and the fact that people from one slum utilize multiple UHPs per their convenience, both of which protect against UHP identification. Therefore, care was taken to explain that anonymity would be ensured by eschewing even pseudonyms and using labels designated in random order (cities X and Y, slums A, B, C, D). Thus, quotes by a KI from city X, slum A would be ascribed to KI,X/A. Consent was then double-checked in case participants felt pressured because I had MC permission. It also merits mention that a study update was attempted in August 2010 without success. This is also addressed as a structural barrier later.

Findings and Analysis

The following findings do not and cannot claim to represent a universality of perceptions and experiences across slums in India. They do, however, facilitate an understanding of contextual, structural barriers, which, when unaddressed by policymakers, result in gendered vulnerability to structural violence.

Economic and Institutional Barriers

Previously published findings from this study revealed that the foremost economic barrier of poverty created structural impediments through lack of autonomy, lack of access to health care and education, and of time to access information due to the everyday grind of making ends meet (J. Ghosh, Wadhwa, and Kalipeni 2009; Wadhwa, Ghosh, and Kalipeni 2010). Underdevelopment and the overall political economy of the state greatly compound gendered vulnerability. India's participation in structural adjustment programs under the International Monetary Fund kept state health care expenditure depressed, averaging only 0.9 percent of gross domestic product from 1975 through 2004, only half the average 2 percent spent by most developing countries (Ekstrand, Garbus, and Marseille 2003; World Health Organization 2006).

By the government's own admission, health care for the urban poor has been largely neglected, and private health expenditure is significant for already stressed slum populations. Quality also remains moot; UHPs have traditionally suffered from resource and accessibility issues—in fact, only 32 percent of slum residents in Delhi and 20 percent in Hyderabad uti-

lized public health resources, citing such issues (Gupta, Arnold, and Lhungdim 2009; MoHFW 2010). Absolute numbers of users are still high, however, because slum populations exceed 3 million in Delhi and 2 million in Hyderabad (Government of NCT, Delhi 2004; Greater Hyderabad MC 2006). The UHPs I visited catered to populations of 10,000 to 50,000, with allowed staff strengths between four and nine (MoHFW 2010), but all were short-staffed.

The National Reproductive and Child Health (RCH) Program Phase I (1997–2005) brought reproductive and sexual health services to UHPs, but it was not until RCH II was launched in 2005 that HIV/AIDS-related services were included in their purview (MoHFW 2010; NACO 2010). UHP staff is thus usually the first and often only point of contact for slum communities regarding reproductive and sexual health, including HIV/AIDS. Despite this, most staff had inadequate training on HIV/AIDS issues, with the maximum being a three-day workshop for the head medical officer.

> [Nurses] etc. have three hours training on AIDS—they are constrained by time and resource, so do not want more training. *Baki* [other patients] they refer to hospitals—usually under skin specialty—[designated hospitals] have HIV/AIDS counter. (KI,X/A)

The "HIV/AIDS counter" this participant referred to is a single window at already understaffed public hospitals lacking in resources, which typically further referred patients to other HIV/AIDS-specific facilities or departments. The state's low investment in HIV/AIDS training for these primary health contacts, and the fragmented and inadequate nature of HIV/AIDS care on all fronts is clear from the field:

> We refer people with problems [*sic*-reference was to HIV/AIDS] to hospitals and their testing centers. (KI,X/B)

> As it is, patients hesitate to come forward—then they have to be referred further—they don't have time, or, they know me and are comfortable here but don't want to reveal [possible or confirmed diagnosis of HIV], so disease goes untreated too long. (KI,Y/C)

Additionally, despite NACO's ambitious launch of the free ART program (NACO 2010):

> Treatment is not really available—ART exists—NACO does provide it, but it is limited. We have heard that the doctors and nurses in the major hospitals that keep it, hoard it for themselves—they are so fearful they will catch it [HIV infection] themselves. Here, mostly we just tell people who have it [HIV+ status and/or AIDS] to follow good diet and exercise, and take vitamins. (KI,X/A)

There is a shortage of drugs [ART] because of government policy—my unit [UHP] here does major procedures [e.g., tubectomies and caesarean sections], but ART are not available—those are available only at major hospitals. (KI,Y/C)

There is no treatment available at a basic level, not even counseling. (KI,Y/D)

This nonavailability of resources like ART is not surprising; at the time, Delhi had only three ART centers and Hyderabad only one (APSACS 2005; DSACS 2005). This can have serious ramifications; for example:

I get all the pregnant women tested—consent and revelation [to partner] is necessary, so usually if the woman is [HIV] positive, we do the husband's test also. There was recently a case where the woman was pregnant, and both parents were [HIV] positive. I recommended termination [abortion]. (KI,Y/C)

When asked if that was standard advice or policy, and if ART options for prevention of parent-to-child transmission (PPTCT) were available, this interviewee clarified:

I know chances of PTCT are only 30 to 33 percent, but I still give that advice to HIV-positive women, given they agree, and they have no income and other care arrangements—tell me, what is their option? They can't get ART—so who will take care of the child after parents are gone? The child might be HIV-positive too, and even if he isn't, there is so much stigma. (KI,Y/C)

This position seems extreme, but perhaps the KI's (and clients') perspectives are (unfortunately) understandable, given the unavailability of ART and general attitudes regarding HIV/AIDS (see the next section). During NACP II, only twenty-two PPTCT centers had been established in Delhi and ninety-three in the entire state of Andhra Pradesh (APSACS 2005; DSACS 2005). PPTCT did not become an NACO priority area until NACP III began in 2006—nationwide, a mere 7.5 percent of HIV-positive mothers received Nevirapine prophylaxis in 2006, although some gains have been made since then (UNGASS 2010).

As addressed by one informant:

There is no comprehensive care or point of contact for HIV care, PPTCT advice is a referral to [private] doctor. (KI,Y/D)

The disjuncture between policy and reality is reflected here; the traditional top-down nature of health policy formulation in India (see Ramasubban 1998) prevents flow of information from below in the state's own health agencies. A more bottom-up approach would have given NACO greater insight regarding PPTCT-related issues, including facility location, which they are only beginning to address now (see UNGASS 2010).

Testing is also a problem; only 28 and 209 ICTCs served the millions in Delhi and the entire state of Andhra Pradesh, respectively (APSACS 2005; DSACS 2005). Lack of knowledge regarding location of these centers, prohibitive costs, and psychological elements presented barriers:

Every cluster [slum] should know its local testing and health center—for example, I went to one of the [health] camps [which had an HIV/AIDS stall], and asked where they were from. They named the local testing center, and even I didn't know about it. (KI,X/A)

Testing kits should be available at basic levels [UHPs] so patients can go to known person—things should start from bottom, not top—any doubts can be removed quickly. (KI,Y/C)

KI,Y/D revealed that high cost of testing was also a barrier:

Only antenatal mothers are tested free—others cannot get services.

The hegemonic system, where action rests on the decision of a few, was also apparent when this study was conducted in 2006. Due to stonewalling by one branch of one of the MCs, another branch had to be approached for permission to conduct the study. The latter's support is highly appreciated because it facilitated greater insight on gendered vulnerability to HIV/AIDS in slums. Creation of impediments when none are warranted is also clear, however. The ethical dilemma outlined earlier regarding KI consent also raises issues regarding top-down, hegemonic policy processes. Additionally, attempting to update the study for this article in August 2010 yielded only obstruction. Promises of support were made by MC officials but never materialized, nor did any explanations, despite phone calls and e-mails. Such obstacles can only hinder greater understanding of realities on the ground. Important policy actions were taken by NACO since 2006 as NACP Phase III was launched, and it would be useful to know whether they have made any impact on KI perceptions of policy efficacy. The preceding situations underscore the need for reducing the opacity and top-down nature of decision-making entities.

Sociocultural Barriers

As mentioned earlier, government efforts toward raising awareness have been somewhat successful, although comprehensive knowledge of HIV/AIDS is still low, and lower still among lower wealth quintiles (IIPS and ORC Macro 2008). As demonstrated in previously published findings from this field study (see J. Ghosh, Wadhwa, and Kalipeni 2009; Wadhwa, Ghosh, and Kalipeni 2010) and later, taboo and stigma are the primary barriers against accessing information, testing, and treatment. The overarching patriarchal nature of Indian society also prevents women from safe sexual practices, ranging from accessing relevant information to negotiating safe sex, even within marriage (Rao Gupta 2002; Verma and Roy 2002; Ekstrand, Garbus, and Marseille 2003; J. Ghosh and Olson 2007).

There is maybe 85 percent awareness about spread and what AIDS is, in [slum A], but otherwise not much. Not much knowledge about MTCT and other ways [of HIV transmission, and], where to go for treatment or testing. (KI,X/A)

Government awareness campaigns have been successful. Awareness is very high, [that HIV] comes from unprotected sex and injections—[they] don't know treatment aspects [and] that life can be prolonged. (KI,Y/C)

[The] community knows all about it—so they say, but don't know all facts—they think there is no treatment for it, or about PTCT, or perhaps prevention. (KI,Y/D)

Taboo and stigma related to a sexually transmitted disease works in more insidiously damaging ways as well:

Many people ... don't like to show they know about it; don't like to hear about it because it is a "dirty" thing ... family talks don't happen and people don't disclose their status even to family so hazard of lack of treatment increases ... family members are more accepting now, but society still discriminates, particularly in low status communities—this is despite knowing you can't get it by casual contact—but inside [their hearts], they can't accept that AIDS might not get them too. ... People are dying more due to hesitation to get treatment. (KI,Y/D)

KI,X/A confirmed similar attitudes:

We do [health] camps, and put up one stall for AIDS etc., with pamphlets and pictures—there is good attendance from the slums. But messages cover overall reproductive health, too—if it were only AIDS, no one would come. Also, prevention is stressed: messages of faithfulness in marriage—other messages are there [condom usage], but not really welcome.

This is despite the fact that for many women in such settings, the greatest risk factor for HIV infection is monogamous relationships (Verma and Roy 2002). The lower status of women in the Indian social structure as seen in the literature cited earlier also puts women at risk:

Some prevention methods known but ... if wives suspect husbands, Health Post tells them to use condoms, but the women feel they can't ask [the husband]—society pressure. ... Condom decision [is] by men only and women can't say no to sex. (KI,X/A)

The men don't use condoms—just not interested—even women don't take, say husband is not interested—women cannot refuse him since they are financially and psychologically dependent—husbands don't allow the use of condoms. (KI,Y/C)

The obvious point is that infrastructural investments are inadequate on their own—the need to address the sociocultural barriers that allow HIV/AIDS to present a threat is immense.

Structural Violence and the Grief Model: A New Frame

In the tradition of grounded theory (Glaser and Strauss 1967), the preceding analysis can be linked directly to existing and interrelated conceptualizations of structural barriers, powerlessness or vulnerability, and structural violence (see reviews in Parker, Easton, and Klein 2000; Kalipeni, Oppong, and Zerai 2007; White, Pope, and Malow 2009; Wadhwa, Ghosh, and Kalipeni 2010).

I support the argument that structural barriers are not themselves direct causes of vulnerability but symptomatic of larger systemic inequities—an idea best encapsulated by the conceptual framework of structural violence. Galtung (1969, 171, 190) simply but eloquently qualified, "Any effort to explore structural violence will lead to awareness of asymmetric conflict, between parties highly unequal in capabilities." Dominant themes in the preceding analysis demonstrate exactly such power–resource inequities in society and the political economy, pointing to the commission of structural violence, particularly against women, and the resultant suffering.

Farmer (1996) asked:

How might we discern the nature of structural violence and explore its contribution to human suffering? Can we devise an analytic model, one with explanatory and predictive power, for understanding suffering?... [The] analysis must, first, be geographically broad ... historically deep ... [and have] ... simultaneous consideration of various social "axes" ... to discern a *political economy of brutality*. (274, italics added)

It is here that I assert the explanatory power of the grief model in elucidating the political economy of brutality, by providing a new frame through which to understand and indeed, discern structural violence. I add the dimension of time to Farmer's geography, history, and social axes. The original individualized grief model posits that when faced with a terminal diagnosis, a person's inability to reach acceptance in adequate time can cause great harm and suffering to the individual and perhaps his or her immediate circle (Kübler-Ross [1969] 1997). Similarly, if the state or its policymaking entities fail to expediently transition to the final stage of acceptance, harm and suffering are caused, of a magnitude that qualifies as structural violence.

The initial commission of structural violence by the state within the context of the HIV/AIDS epidemic in India was its inaction, allowing its unchecked progression in the first decade. Moreover, field evidence presented earlier demonstrates that the state's evolution to acceptance during NACP Phase II (1999–2006) was at best superficial, visible in the continued lack of adequate infrastructure and care and failure to address sociocultural barriers such as taboo and stigma. Hegemonic policy formulation processes and impediments to research are also indicative of structural violence, as these actions efface agency and voice. In effect, the state's transition from first diagnosis to some manner of acceptance by the end of NACP II has taken a long and painful twenty years, providing a measure of the full scope of structural violence.

Epilogue

NACP Phase III began in 2006. A midterm review shows that tremendous gains have been made on all issues mentioned in this study (NACO 2010), including that of encouraging bottom-up processes (see "Assessment of ART Centres: Clients' and Providers' Perspectives" in NACO 2010; UNGASS 2010). Although several targets were not reached and challenges remain (UNGASS 2010), the report is good on paper, but how is it on the ground? This question leads to the next steps in research, exploring whether "acceptance" has finally been reached in full measure or whether structural violence persists in these or other

forms more than twenty years after the epidemic began. The state might be wise to remember:

Structural violence is itself situated in historically constituted political and economic systems—systems in which diverse political processes and policies ... not only create the dynamic of the epidemic, but also provide what is potentially the most effective source of intervention in order to curb its impact. (Parker, Easton, and Klein 2000, S23)

Acknowledgments

This study was made possible by an Association of American Geographers (AAG) Research Grant (2006) and Regional Development and Planning Specialty Group (AAG) Travel Grant—my sincerest thanks to these bodies. Thanks also to Dr. Baleshwar Thakur, Dr. Kalpana Markandey, Dr. P. P. Singh, Dr. Sathyavathi, and Mr. S. V. Bhikkaji for support in the field. I am indebted to the participants and KIs for sharing their invaluable perspectives and to the two anonymous reviewers who provided immensely constructive feedback to help refine this work.

References

Andhra Pradesh State AIDS Control Society. 2005. *Facts, figures and response to HIV/AIDS in Andhra Pradesh*. Hyderabad, India: APSACS, Population Foundation of India (PFI), and Population Reference Bureau (PRB). http://www.prb.org/pdf06/FactsFiguresResponse_HIVAIDS_AndraPradesh.pdf (last accessed 7 August 2011).

Asthana, S. 1996. AIDS-related policies, legislation and programme implementation in India. *Health Policy and Planning* 11 (2): 184–97.

Bates, I., C. Fenton, J. Gruber, and D. Lalloo. 2004a. Vulnerability to malaria, tuberculosis, and HIV/AIDS infection and disease: Part 1. Determinants operating at individual and household level. *Lancet Infectious Disease* 4:267–77.

———. 2004b. Vulnerability to malaria, tuberculosis, and HIV/AIDS infection and disease: Part 2. Determinants operating at environmental and institutional level. *Lancet Infectious Disease* 4:368–75.

Bharat, S., P. Aggleton, and P. Tyrer. 2001. *India: HIV and AIDS-related discrimination, stigma and denial*. Geneva, Switzerland: UNAIDS and Populations Council. http://data.unaids.org/publications/IRC-pub02/jc587-india_en.pdf (last accessed 1 August 2011).

Craddock, S. 2001. Scales of justice: Women, inequity, and AIDS in east Africa. In *Geographies of women's health*, ed. I. Dyck, S. MacLafferty, and N. D. Lewis, 41–60. London and New York: Routledge.

Delhi State AIDS Control Society. 2005. *HIV/AIDS in Delhi: Meeting the challenge*. New Delhi, India: DSACS, PFI and PRB. http://www.prb.org/pdf06/

HIVAIDS_Delhi_MeetingtheChallenge.pdf (last accessed 1 August 2011).

Dutt, S. 2011. Factors influencing willingness to comply to HIV/AIDS prevention measures by female college students in Kolkata, India. Unpublished dissertation, Department of Geography, Kansas State University, Manhattan, KS.

Ekstrand, M., L. Garbus, and E. Marseille. 2003. *HIV/AIDS in India*. San Francisco: University of California. http://ari.ucsf.edu/programs/policy/countries/India.pdf (last accessed 1 August 2011).

Eliot, E. 2009. Spatial dimension and emergence of the HIV/AIDS epidemic in India. In *Indian health landscapes under globalization*, ed. A. Vaguet, 33–60. New Delhi, India: Manohar (publication of French Research Institutes of India).

Farmer, P. 1996. On suffering and structural violence: A view from below. *Daedalus* 125 (1): 261–83.

Galtung, J. 1969. Violence, peace, and peace research. *Journal of Peace Research* 6 (3): 167–91.

Ghosh, D. 2007. Predicting vulnerability of Indian women to domestic violence incident. *Research and Practice in Social Sciences* 3 (1): 48–72.

Ghosh, J., and B. Olson. 2007. HIV/AIDS in South Africa and India: Understanding the vulnerability factors. In *City society and planning*. Vol. 2, ed. B. Thakur, G. Pomeroy, C. Cusack, and S. Thakur, 162–80. New Delhi, India: Concept.

Ghosh, J., V. Wadhwa, and E. Kalipeni. 2009. Vulnerability to HIV/AIDS among women of reproductive age in the slums of Delhi and Hyderabad. *Social Science & Medicine* 68 (4): 638–42.

Glaser, B. G., and A. L. Strauss. 1967. *The discovery of grounded theory: Strategies for qualitative research*. Chicago: Aldine.

Gould, W. T. S. 2005. Vulnerability and HIV/AIDS in Africa: From demography to development. *Population, Space and Place* 11:473–84.

Government of NCT, Delhi. 2004. Socio-economic profile of Delhi, 2003–04. http://delhiplanning.nic.in/Socioeco profiles/finalsocioecoprofile.pdf (last accessed 1 August 2011).

Greater Hyderabad MC. 2006. Hyderabad—City development plan. http://www.ghmc.gov.in/cdp/default.asp (last accessed 1 August 2011).

Gupta, K., F. Arnold, and H. Lhungdim. 2009. *Health and living conditions in eight Indian cities*. National Family Health Survey (NFHS-3), India, 2005–06. Mumbai, India: IIPS. http://www.nfhsindia.org/urban_health_report_for_website_18sep09.pdf (last accessed 1 August 2011).

International Institute for Population Studies. 1995. *National Family Health Survey, India 1992–93*. Bombay, India: IIPS. http://www.nfhsindia.org/pub_nfhs-1.shtml (last accessed 1 August 2011).

International Institute for Population Studies and ORC Macro. 2000. *National Family Health Survey, India 1998–99: State reports Delhi and Andhra Pradesh*. Mumbai, India: IIPS. http://www.nfhsindia.org/pub_nfhs-2.shtml (last accessed 1 August 2011).

———. 2008. *National Family Health Survey, India 2005–06: State reports Delhi and Andhra Pradesh*. Mumbai, India IIPS. http://www.nfhsindia.org/report.shtml (last accessed 1 August 2011).

Jongsthapongpanth, A., and S. Bagchi-Sen. 2010. Spatial and sex differences in AIDS mortality in Chiang Rai, Thailand. *Health & Place* 16:1084–93.

Kadiyala, S., and T. Barnett. 2004. AIDS in India: Disaster in the making. *Economic and Political Weekly* 39 (19): 1888–92.

Kalipeni, E. 2008. HIV/AIDS in women: Stigma and gender empowerment in Africa. *Future HIV Therapy* 2 (2): 147–53.

Kalipeni, E., K. C. Flynn, and C. Pope. 2009. *Strong women, dangerous times: Gender and HIV/AIDS in Africa*. New York: Nova Science.

Kalipeni, E., J. R. Oppong, and A. Zerai., eds. 2007. HIV/AIDS in Africa: Gender, agency and empowerment. *Social Science and Medicine* 64 (5): 1015–50.

Kishor, S., and K. Gupta. 2009. *Gender equality and women's empowerment in India*. National Family Health Survey (NFHS-3), India, 2005–06. Mumbai, India: IIPS. http://www.nfhsindia.org/a_subject_report_gender_for_website.pdf (last accessed 1 August 2011).

Kübler-Ross, E. [1969] 1997. *On death and dying*. London and New York: Routledge.

Ministry of Health and Family Welfare. 2010. *National urban health mission*. New Delhi: Government of India. http://mohfw.nic.in/NRHM/Documents/Urban_Health/UH_Framework_Final.pdf (last accessed 1 August 2011).

Mitra, P. 2004. India at the crossroads: Battling the HIV/AIDS epidemic. *The Washington Quarterly* 27 (4): 95–107.

National AIDS Control Organization. 2010. *Annual report 2009–10*. New Delhi: Ministry of Health and Family Welfare. http://www.nacoonline.org/upload/AR%202009–10/NACO_AR_English%20corrected.pdf (last accessed 1 August 2011).

National Committee for Ethics in Social Science Research in Health. 2000. *Ethical guidelines for social science research in health*. Mumbai: National Committee for Ethics in Social Science Research in Health, Centre for Enquiry into Health and Allied Themes. http://www.cehat.org/publications/ethical2.html (last accessed 10 February 2012).

Parker, R. G., D. Easton, and C. H. Klein. 2000. Structural barriers and facilitators in HIV prevention: A review of international research. *AIDS* 14 (S1): S22–S32.

Pope, C., and E. Kalipeni. 2010. Introduction to special issue on international geographies of HIV/AIDS. *GeoJournal*. Published electronically. DOI 10.1007/s10708-010-9346-x

Pope, C., R. White, and R. Malow. 2009. *HIV/AIDS: Global frontiers in prevention /intervention*. London and New York: Routledge.

Pradhan, B. K., and R. Sundar. 2006. *Gender impact of HIV/AIDS in India*. Geneva, Switzerland: United Nations Development Programme. http://data.undp.org.in/hivreport/Gender.pdf (last accessed 25 July 2011).

Ramasubban, R. 1998. HIV/AIDS in India: Gulf between rhetoric and reality. *Economic & Political Weekly* 33 (45): 2865–72.

Rao Gupta, G. 2002. How men's power over women fuels the HIV epidemic (editorial). *BMJ* 324:183–84.

UNGASS. 2010. India country progress report. http://www.unaids.org/en/dataanalysis/monitoringcountryprogress/

2010progressreportssubmittedbycountries / india _ 2010 _ country_progress_report_en.pdf (last accessed 5 August 2011).

Verma, R., and T. Roy. 2002. HIV risk behavior and the socio-cultural environment in India. In *Living with the AIDS virus: The epidemic and the response in India*, ed. S. Panda, A. Chatterjee, and A. Abdul-Quader, 77–90. Thousand Oaks, CA: Sage.

Wadhwa, V. 2006. Spatial patterns of AIDS awareness among Indian women: What variables are at play? Paper presented at the annual conference of the Association of American Geographers, Chicago, IL.

Wadhwa, V., J. Ghosh, and E. Kalipeni. 2010. Vulnerability to HIV/AIDS among female slum youth in Delhi and Hyderabad, India. *GeoJournal*. Published electronically. DOI: 10.1007/s10708–010-9359–5.

White, R. T., C. Pope, and R. Malow, 269–77. 2009. HIV, public health, and social justice: Reflections on the ethics and politics of health care. In *HIV/AIDS: Global frontiers in prevention/intervention*, ed. C. Pope, R. White, and R. Malow, 269–77. London and New York: Routledge.

World Health Organization. 2006. Tough choices: Investing in health for development—Experiences from national follow-up to the Commission on Macroeconomic and Health. Electronic Annex C. http://www.who.int/macrohealth/documents/Electronic_Annex_C.pdf (last accessed 28 July 2011).

U.S. Migration, Translocality, and the Acceleration of the Nutrition Transition in Mexico

Fernando Riosmena,* Reanne Frank,[†] Ilana Redstone Akresh,[‡] and Rhiannon A. Kroeger[†]

*Department of Geography and Population Program, University of Colorado at Boulder
[†]Department of Sociology, The Ohio State University
[‡]Department of Sociology, University of Illinois at Urbana–Champaign

Migrant flows are generally accompanied by extensive social, economic, and cultural links between origins and destinations, transforming the former's community life, livelihoods, and local practices. Previous studies have found a positive association between these translocal ties and better child health and nutrition. We contend that focusing on children only provides a partial view of a larger process affecting community health, accelerating the nutrition transition in particular. We use a Mexican nationally representative survey with socioeconomic, anthropometric, and biomarker measures, matched to municipal-level migration intensity and marginalization measures from the Mexican 2000 Census to study the association between adult body mass and community migration intensity. Our findings from multilevel models suggest a significant and positive relationship between community-level migration intensity and the individual risk of being overweight and obese, with significant differences by gender and with remittance intensity playing a preponderant role.

移民流动一般都伴随着在起源和目的地之间的，广泛的社会，经济和文化的联系，它改变着起源地的社会生活，生计，和当地的做法。以前的研究已经发现，在这些跨本地的联系和儿童健康和营养状况的改善之间存在一个正相关。我们认为，把注意力集中于儿童，只提供了影响社区卫生的大过程中的一个局部视图，特别是加速了营养的过渡。我们使用墨西哥全国代表性的调查，以及与墨西哥 2000 年人口普查的市级迁移强度和边缘化指标相匹配的社会经济，人体和生物标志物的指标，来研究成年体重和社区迁移强度之间的关联。我们的多层模型研究结果表明，在社区级的迁移强度和超重和肥胖的个人风险之间，存在显著的正相关，并随性别而显著不同，而且汇款强度在此关系中发挥着主导作用。关键词：国际移民，墨西哥，营养过渡，肥胖，跨本地。

Los flujos de migrantes van generalmente acompañados de extensos vínculos sociales, económicos y culturales entre orígenes y destinos, que transforman la vida de comunidad, medios de vida y prácticas locales en el primero de estos. En estudios anteriores se ha detectado una asociación positiva entre estos lazos translocales y una mejor salud y nutrición infantil. Nuestra posición es que si el estudio es enfocado solamente sobre niños apenas se podrá dar una visión parcial de un proceso más grande que afecta la salud de la comunidad, acelerando en particular la transición nutricional. Utilizamos un estudio mejicano nacionalmente representativo con mediciones socioeconómicas, antropométricas y biométricas, pareadas con la intensidad migratoria a nivel municipal y mediciones de marginalidad del Censo Mejicano del 2000, para estudiar la asociación entre la masa corporal adulta y la intensidad de migración de la comunidad. Nuestros hallazgos, obtenidos de modelos de niveles múltiples, sugieren una relación significativa y positiva entre la intensidad de la migración a nivel de comunidad y el riesgo individual de estar pasado de peso y obeso, con diferencias significativas por género, y con la intensidad de las remesas como factor que juega un papel preponderante.

Translocal ties originating from international migration processes transform sending areas in profound ways (Jones 1998; Levitt 1998). Health is no exception. Previous studies have focused on the positive effects of migration on infant and child health (Kana'iaupuni and Donato 1999; Frank and Hummer 2002). Migration appears to be beneficial in these places primarily as the money transferred or brought back by migrants, generally known as *remittances*, helps ameliorate the poverty conditions responsible for poor health and nutrition. A focus on infant and child health, however, illustrates only one side of what we contend is a larger process in which the pecuniary and nonpecuniary exchanges associated with migration might be influencing community health by accelerating the nutrition transition in sending communities. The nutrition transition refers to an increased availability of high-fat and processed foods and altered home cooking practices (Popkin 2001).

Given that these changes translate into significant weight gains, the objective of this study is to understand whether the exchanges that are part of the international migration process are associated with increases in adult body mass index (BMI). We use multilevel techniques and nationally representative anthropometric data from Mexico to study whether the probability that an adult is overweight or obese is associated with the migration intensity of his or her municipality of residence.

We consider two mechanisms for this process: (1) the money remitted and brought back by migrants (Jones 1998) might allow households to afford a higher caloric intake, and (2) the transnational and translocal circulation of people and ideas (Levitt 1998) might change food, portion, and body size preferences. This second set of processes might be associated with an overall decline in health and a higher BMI due to dietary and lifestyle changes adopted during immigrant tenure in the United States (Akresh 2007), a particularly problematic trend given the health problems known to be associated with obesity (Monteverde et al. 2010).

We also hypothesize that this acceleration should occur more rapidly in groups and places in which the nutrition transition is in its earlier stages. Given the rapid but gendered and spatially uneven pace of the epidemiological and nutrition transitions across Mexico, with women and people in large cities experiencing the transition earlier than men and rural residents (Rivera et al. 2002), we expect the relationship between migration and obesity to be especially strong among these latter groups. As such, our research questions are as follows:

1. Is the migration intensity of a municipality (measured as the level of remittances and the level of return migration) associated with an increased likelihood of overweight or obesity, net of other factors?
2. Does the association between overweight or obesity and municipality level of remittances persist after controlling for the municipality level of return migration?
3. Do the relationships tested in the first two hypotheses differ between men and women and between people living in less and more populated areas?

Our findings support a general relationship between the level of migration of a community and the likelihood of overweight or obesity but only partially support some of our more specific assertions. Overweight and obesity are indeed higher for people in places with higher migration intensities (net of a series of relevant controls). Moreover, translocal connections associated with U.S. migration are strongly correlated with male obesity, consistent with the idea that these links might accelerate the nutrition transition for individuals who would otherwise experience it more slowly.

This association, however, seems to be mostly a function of the direct and indirect income effects of remittance intensity rather than of migrant return. These findings suggest that higher BMIs in migrant-sending communities are due mostly to increased caloric intake allowed by reduced budget constraints or to decreased caloric expense. The higher relevance of remittances relative to return migration also suggests that the acceleration of the transition is not due to the fact that places with high migration intensity have a different set of preferences in terms of food or body size compared to similar places with lower migration intensities, nor is it explained by the fact that return migrants themselves have higher BMIs after experiencing the aforementioned dietary changes during their tenure in the United States. Lastly, we find significant differences in the effect of migration on overweight and obesity across less and more urban spaces but only before controlling for the socioeconomic status of a community. We can thus only cautiously conclude that the acceleration of the nutrition transition is quicker in less urban areas provided that these same migration-related translocal links were an important determinant of the socioeconomic status of those communities.

Migration and Health in Sending Communities

Several studies have demonstrated that remittance receipt, migration experience, or both at the household and community scales are significantly associated with lower odds of low birth weight and infant mortality (Kana'iaupuni and Donato 1999; Frank and Hummer 2002; Frank 2005; Hildebrandt et al. 2005; McKenzie 2006; Hamilton, Villarreal, and Hummer 2009). By focusing on child health outcomes, however, these studies have provided only a partial view of the mechanisms through which migration can influence community health. We posit that several types of exchanges associated with international migration might accelerate the nutrition transition in sending communities in Mexico not only by improving child nutrition but also by increasing adult BMI, another typical by-product of this transition (Popkin 2001). We describe two categories of mechanisms producing these results.

Economic Mechanisms

Remittances from migration have a direct income effect on the households receiving them, lowering food deprivation and thus increasing total caloric consumption. In rural areas, this additional income might also be used to purchase equipment and tools that allow households to move from labor- to capital-intensive agricultural methods (Taylor and Lopez-Feldman 2010, which may reduce caloric expense). As these flows generate nontrivial multiplicative effects in local economies (Taylor et al. 1996), remittances can also have indirect income effects, influencing the nutrition of nonmigrant households as well. A community-wide measure of remittance rates, like the one we use in our analyses, would capture both direct and indirect effects.

Noneconomic Mechanisms

Serving sizes and food preferences can also be altered by migration. Translocal, transnational ties might help diffuse nutritional norms and eating habits prevalent in the United States, where fruit and vegetable consumption has decreased and intake of refined sugars has been on the rise (Levi et al. 2009). Immigrants generally adopt these habits, especially as their tenure in the host society increases (Akresh 2007), resulting in increased overweight and obesity (Antecol and Bedard 2006; Akresh 2007).

As migrants bring these practices back to their communities of origin, they might subsequently influence the lifestyles and preferences of nonmigrants (Levitt 1998). One of the previously mentioned studies on child health has shown evidence suggesting that migration alters sociocultural aspects of community life in sending areas (e.g., health knowledge), which in turn lowers the risk of poor infant health outcomes (Hamilton, Villarreal, and Hummer 2009). Likewise, ideas about portion size, food preferences, and body type could be different in communities with high levels of return migration over and beyond income effects. We thus hypothesize that these altered preferences and practices would be captured by a positive association between measures of return migration at a larger scale, such as the municipality, and body mass (even after controlling for remittance intensity).

The Nutrition Transition in Mexico: Gendered Patterns and Variation across Place

The nutrition transition is well under way in Mexico (Rivera et al. 2002; Rivera et al. 2004; Barquera et al. 2009) and has maintained a rapid pace in the past decade. As a result, Mexico has one of the highest levels of obesity in the world, at 32 percent (World Health Organization 2009), 37 percent for women and 24 percent for men (Barquera et al. 2009). Not only is this level high and increasing at a rapid rate in recent years, but the excess mortality attributable to obesity is also higher in Mexico than in the United States (Monteverde et al. 2010).

Although these changes are certainly occurring independent of the particular brand of translocal ties created by international migration, we assert that migration-related exchanges are *accelerating* the nutrition transition. This association should manifest as higher BMI levels in communities with more established migratory traditions. Furthermore, given the gendered and spatially uneven nature of the transition, we should expect this association to be particularly evident among groups that have traditionally experienced the transition more slowly, such as men and people living in rural areas (Rivera et al. 2002; Rivera et al. 2004).

Alternative Explanations: Development and Globalization-Related Changes

Health and nutrition conditions in a community could be influenced by other forms of translocal exchanges of ideas and norms about health, food, and body size not associated with the international migration process per se but linked to the development and globalization processes believed to drive the nutrition and epidemiological transitions (Chaput and Tremblay 2009; Huneault, Mathieu, and Tremblay 2011). These processes could also be influencing international migration (Sassen 1988; Stark and Bloom 1985), thus potentially creating spurious relationships between community migration and health if one does not control for the level of development and globalization of a community. We take these impacts into account by controlling for local levels of development and foreign direct investment, the latter as a proxy for capital penetration processes (Sassen 1988; Massey and Espinosa 1997) in addition to other relevant correlates of health. Again, we do not argue that the influence of migration predates or is more relevant than broader development and globalization processes; our claim is that migration accelerates the transition, not that it initiates it.

Data and Methods

We use data from the Mexican Health Survey (hereafter, ENSA, its Spanish acronym). The ENSA was conducted by the National Institute of Public Health between September 1999 and March 2000 and is a nationally representative, multistage sample of the Mexican population with a 97 percent participation rate. The data are representative at the state and urban and rural levels, yielding 45,756 households (Barquera et al. 2008). The sampling procedure began with the random selection of fourteen counties in each state. Within each county, five basic geostatistical areas (analogous to a census tract) were selected. Seven households in three different blocks within each of these areas were randomly selected. One child, adolescent, and adult (aged twenty and over) was selected in each household (Barquera et al. 2007; Barquera et al. 2008; for more details, see Valespino et al. 2003).

We use data from the adult samples, which contain socioeconomic, health, and anthropometric information on each respondent. Height and weight, measured in light clothing and without shoes, were recorded to the nearest 5 mm and 0.1 kg, respectively (Valespino et al. 2003). Overweight and obesity are categorized using BMI cutoffs created for adults twenty years old and older (http://apps.nccd.cdc.gov/dnpabmi). BMIs that fall in the range between 25.0 and 29.9 kg/m^2 are categorized as overweight, and BMIs that are 30.0 kg/m^2 and above are categorized as obese. By these definitions, 39 percent of the sample is classified as overweight and 24 percent is considered obese (see Table 1). Approximately 7 percent of the adult sample was excluded because of missing or implausible values on the weight and height variables.

We matched the adult sample of the ENSA with municipality-level indexes constructed using 2000 Mexican census data and published by the National Population Council (hereafter, CONAPO; see http://www.conapo.gob.mx). Although it is unlikely that the mechanisms through which translocal exchanges might affect health as laid out earlier operate at a higher scale than the municipal, it is more likely that they operate at a finer scale; for instance, in an area composed by several dwellings and public places, bound by different types of interlocal relations and located within part of a locality (i.e., township; Fussell and Massey 2004) or spanning part of several localities. As such, the municipal scale and zoning used here should capture the average influence of translocal exchanges from migration on health from different communities plus spatial lags related to the influence of these on surrounding towns in the same municipality. Although greater flexibility in the spatial scale of our contextual information would be desirable to test for the robustness of our model specifications to potential scale and, especially, zoning biases (Kwan 2009), this is not possible, as the survey only includes municipal identifiers and census measures (which come from a long-form sample) are likewise only available for municipalities.

First, we included an index of migration-related patterns that came from the International Migration Supplement of the 2000 census. The migration index was constructed using principal components analysis (PCA) on the percentage of households in the municipality (1) receiving remittances, (2) with at least one member emigrating to the United States between 1995 and 1999, (3) with at least one member returning from the United States between 1995 and 1999, and (4) with at least one member emigrating to and returning from the United States between 1995 and 1999. This normalized index, with mean around zero and a standard deviation close to 1 (see Table 1), is the main measure we use in models shown later. We also present results using these measures separately, with a focus on remittance receipt and return migration in an attempt to distinguish income effects from those associated with the degree of transnational connections in the community.

We also control for various community-level socioeconomic characteristics to avoid confounding the impact of the migration intensity level with those of the development and globalization forces as discussed in the previous section. For this purpose, we use an index of marginalization from CONAPO, also based on 2000 census data, using PCA, and composed of the proportion of households in the municipality (1) with dirt floors, (2) without indoor plumbing or a toilet, (3) without electricity, (4) without access to piped water, and (5) with more than two people per room, as well as the proportion of adults in the municipality (6) who are illiterate, (7) who have not completed primary education, and (8) who earn less than twice the minimum wage. Finally, also at the *municipio* level, we include a measure of the proportion of the active labor force that is employed in the *maquiladora* sector (a factory run by a foreign company exporting virtually all of its

Table 1. Weighted means (and standard deviations)

Means	Total	Men	Women
Outcome variables			
Overweight	0.39	0.41	0.36
Obese	0.24	0.19	0.28
Individual-level attributes			
Female	0.53	—	—
Age	37.37 (14.95)	37.63 (15.80)	37.14 (14.50)
Union status [single]			
Married	0.57	0.57	0.57
Cohabiting	0.15	0.15	0.16
Divorced or widowed	0.08	0.04	0.11
Education [college or more]			
High school	0.19	0.19	0.20
Secondary	0.23	0.25	0.21
Primary or less	0.48	0.43	0.53
Household in large urban area	0.64	0.64	0.64
Municipality scale indicators			
Migration measures			
Migration index	−0.29 (0.68)	−0.29 (0.65)	−0.28 (0.70)
Percentage remitting	4.18 (4.84)	4.15 (4.70)	4.22 (4.90)
Percentage return migrants	0.83 (1.15)	0.81 (1.09)	0.84 (1.17)
Marginalization index	−1.13 (0.89)	−1.12 (0.90)	−1.14 (0.88)
Percentage rural municipalities	23.12 (27.81)	23.45 (27.98)	22.82 (27.73)
Proportion *maquiladoras*	0.03 (.08)	0.03 (.08)	0.03 (.08)
Unweighted N	39,843	12,395	27,448

production), our measure of capital penetration and globalization.

We further include measures of the urbanization level of the community, which are particularly relevant for our third research objective. We classified households as more urban if their locality has more than 15,000 inhabitants. Localities are smaller units than municipalities and are generally used to classify the level of urbanization of a place in the Mexican context. Although rural localities in Mexico are defined as those with fewer than 2,500 inhabitants, the 15,000 cutoff is the only classification available in the ENSA. To further consider the heterogeneity of communities below the cutoff in terms of their level of urbanization, we also control for the percentage of the municipality's population living in localities with fewer than 2,500 inhabitants.

Given our use of data at both the individual and municipal levels and our interest in reliably estimating the effects and significance of cross-level interactions, we use hierarchical linear modeling (HLM) in our analysis. Our main approach is to assess the independent effect of community-level migration intensity (and that of its different components) on an individual's risk of being overweight and obese. In addition to the variables previously described, we include controls for socioeco-

nomic characteristics at the individual and household levels, all of which are listed in Tables 2 and 3. All models were estimated using HLM 6.0.7.

Findings

Tables 2 and 3 show results of our multilevel multinomial regression models predicting overweight and obesity (both relative to being underweight or normal) for women and men separately. We present four sets of models in each of the tables, each with a different migration indicator or set of migration indicators (all of them also include our full set of controls). In Model I, we include the global migration intensity index. Models II and III include the remittance intensity and circulation components entered separately, and Model IV includes both of them entered simultaneously.

Our results generally show a positive association between the municipal level of migration and the likelihood that an individual is classified as obese or—to a lesser extent—overweight. Looking at Model I in Tables 2 and 3, the coefficients of the migration intensity index are statistically significant and nontrivial in magnitude for obesity for both sexes, implying that women have 11.3 percent (i.e., $100 \cdot [\exp\{0.107\} - 1]$)

higher odds of being obese and 10.0 percent (i.e., 100 · [exp{0.095} – 1]) higher odds of being overweight for each unit change (i.e., a standard deviation) in the migration intensity index. The equivalent figure for obesity among men is more than two times higher than that of women at 26.5 percent (i.e., 100 · [exp{0.235} –1]). Although these coefficients are substantial and significant after controlling for potential confounders such as our local measure of foreign investment, note that this measure (i.e., the percentage of manufacturing workers in the municipality working in a *maquiladora*) is significantly associated with obesity for both women and men.

The notion that migration might be accelerating the nutrition transition for men in particular is also reflected in the implied effects of our schooling variable. The education gradients for women in Table 2 imply that women with more schooling are less likely to be classified as overweight or obese, whereas the results in Table 3 imply that men with more schooling are

more likely to be overweight and obese. As the correlation between obesity and education changes from positive to negative over the evolution of the nutrition transition (e.g., Smith and Goldman 2007), these results suggest that men have yet to experience many of these changes, an inference confirmed by the fact that male obesity levels are lower than those of females (see Table 1).

The association between BMI levels and the migration intensity index seems to be mostly explained by the additional income that migration generates and less by other types of translocal links related to it. Although both remittance intensity (Model II) and return migration (Model III) indicators are positively associated with the likelihood of obesity among men and overweight among women, the return migration indicator loses statistical significance when we add remittance intensity, which does not decrease in magnitude and retains almost most of its statistical power (see Model IV in Tables 2 and 3). For women, the odds of being

Table 2. Coefficients from multinomial multilevel logistic regression of overweight/obesity risk among Mexican women

Variable	Model I Overweight	Model I Obese	Model II Overweight	Model II Obese	Model III Overweight	Model III Obese	Model IV Overweight	Model IV Obese
Individual level								
Age	0.025***	0.037***	0.026***	0.037***	0.026***	0.037***	0.026***	0.037***
Union status (single)								
Married	0.633***	0.691***	0.633***	0.691***	0.633***	0.691***	0.633***	0.691***
Cohabiting	0.488***	0.497***	0.487***	0.497***	0.484***	0.493***	0.487***	0.497***
Divorced/widowed	0.262**	0.136	0.261**	0.136	0.261**	0.135	0.261**	0.135
Education (college)								
High school/tech	0.244*	0.371**	0.244*	0.371**	0.244*	0.371**	0.243*	0.370**
Secondary	0.418***	0.611***	0.419***	0.612***	0.420***	0.613***	0.419***	0.611***
Primary or less	0.453***	0.907***	0.455***	0.908***	0.457***	0.909***	0.455***	0.908***
Household in large urban area	0.123†	0.114	0.120†	0.113	0.114†	0.103	0.118†	0.108
Municipality scale								
Migration measures								
Migration index	0.095*	0.107*						
Percentage remittances			0.012*	0.014*			0.019*	0.031**
Percentage return migrants					0.034	0.028	−0.041	−0.094*
Marginalization index	−0.146*	−0.320**	−0.151*	−0.321**	−0.173*	−0.359**	−0.153*	−0.326**
Percentage rural	−0.001	−0.002	−0.000	−0.002	0.001	−0.000	−0.000	−0.002
Proportion *maquiladoras*	0.342	1.268**	0.359	1.284**	0.285	1.240**	0.457	1.508**
Intercept	−1.944***	−3.159***	−2.028***	−3.252***	−2.046	−3.284	−2.030***	−3.256***
u_0	0.05	0.16	0.05	0.16	0.05	0.16	0.05	0.16
Df	316	316	316	316	316	316	315	315
N individual level (unweighted)	27,448		27,448		27,448		27,448	
N community level	321		321		321		321	

†p < 0.10.
*p < 0.05.
**p < 0.01.
***p < 0.001.

Table 3. Coefficients from multinomial multilevel logistic regression of overweight and obesity risk among Mexican men

Variable	Model I		Model II		Model III		Model IV	
	Overweight	Obese	Overweight	Obese	Overweight	Obese	Overweight	Obese
Individual level								
Age	0.015***	0.024***	0.015***	0.024***	0.015***	0.024***	0.015***	0.024***
Union status (single)								
Married	0.689***	0.822***	0.690***	0.822***	0.690***	0.825***	0.690***	0.823***
Cohabiting	0.454***	0.587***	0.451***	0.586***	0.452***	0.578***	0.451***	0.587***
Divorced/widowed	0.317†	0.326	0.316†	0.326	0.317†	0.325	0.316†	
Education (college)								
High school/tech	−0.369**	−0.222	−0.369**	−0.222	−0.370**	−0.222	−0.370**	−0.222
Secondary	−0.252*	−0.338*	−0.251*	−0.336*	−0.251*	−0.335*	−0.251*	−0.336*
Primary or less	−0.430***	−0.455**	−0.428***	−0.453**	−0.429***	−0.450**	−0.429***	−0.452**
Household in large urban area	0.223**	0.277*	0.216*	0.270*	0.215*	0.259*	0.216*	0.265*
Municipality scale								
Migration measures								
Migration index	0.044	0.235**						
Percentage remittances			0.001	0.031**			0.004	0.049**
Percentage return migrants					0.001	0.087*	−0.015	−0.100
Marginalization index	−0.205*	−0.363**	−0.224*	0.365**	0.230*	−0.419**	−0.225*	−0.374**
Percentage rural	0.001	−0.000	0.001	−0.000	0.002	0.003	0.001	−0.000
Proportion *maquiladoras*	0.286	0.926*	0.303	0.973*	0.306	0.768†	0.342	1.232**
Intercept	−1.024***	−2.506***	−1.077***	−2.700***	−1.082***	−2.749***	−1.078***	−2.710***
u_0	0.10	0.18	0.10	0.18	0.10	0.19	0.10	0.18
df	316	316	316	316	316	316	315	315
N individual level (unweighted)	12,395		12,395		12,395		12,395	
N community level	321		321		321		321	

†$p < 0.10$.
*$p < 0.05$.
**$p < 0.01$.
***$p < 0.001$.

classified as overweight or obese relative to underweight or normal rise by 9.5 and 16.0 percent when the municipality's remittance intensity increases by one standard deviation (i.e., ~4.8 percent). For men, again, these effects are only significant for obesity: For them, an additional standard deviation in municipal remittance intensity levels increases the odds of obesity by 26.5 percent. These results suggest that most of the association between the level of return migration from a municipality and BMI levels as shown in Model III is likely an artifact of the former's correlation with remittance levels. They also confirm the notion that migration more profoundly influences obesity than overweight for both men and women.[1]

To test whether the migration indicators are associated with a particularly dramatic acceleration of the nutrition transition in less urban places, Model I in Table 4 shows results of models similar to those of Tables 2 and 3, adding interactions between the level of urbanization of the locality in which the household is located and our migration intensity index. Although these models include the same controls as Tables 2 and 3, for the sake of brevity, we report only the coefficients of the urbanization, migration, and marginalization variables along with the interactions. As the interactions are not significant for either men or women in the prediction of obesity or overweight, there is no prima facie evidence that migration is accelerating the nutrition transition at a faster pace in less urban areas.

The migration–BMI relationship could be mediated in part by the socioeconomic status of a municipality, captured in the marginalization index. In less urban municipalities where migration rates are high and remittances represent a large portion of pecuniary flows, the level of marginalization would likely be much higher without these financial links. Model II in Table 4 illustrates this by showing results of a model similar to Model I but in which we did not include the marginalization index. The migration–urban place interaction in this case is significant for males in predicting overweight

Table 4. Results of models interacting the migration intensity index with the level of urbanization of the locality of residence, with community-level socioeconomic status (first two columns) and without (second two columns)

	Women				Men			
	Model I		Model II		Model I		Model II	
Variable	Overweight	Obese	Overweight	Obese	Overweight	Obese	Overweight	Obese
Household in urban area	0.107†	0.113	0.217***	0.367***	0.181*	0.316**	0.337***	0.639***
Municipality scale								
Migration index	0.102*	0.097†	0.128**	0.145*	0.080	0.189**	0.110†	0.237**
Marginalization index	−0.154***	−0.358***	NOT INCL	NOT INCL	−0.178**	−0.382***	NOT INCL	NOT INCL
Cross-level interaction								
Household in large urban area • Migration index	−0.068	−0.029	−0.187*	−0.286**	−0.150	0.196	−0.270*	−0.391
N individual level	27,448		27,448		12,395		12,395	
N community level	321		321		321		321	

†p < 0.10.
*p < 0.05.
**p < 0.01.
***p < 0.001.

and for females in predicting overweight and obesity, all with the expected sign (negative, implying that the effect of the migration intensity index is weaker in large urban areas, as expected). We regard this as weak evidence of a differential effect of migration in accelerating the epidemiological transition.

Discussion

Our findings support the notion that community-level migration processes influence health-related behaviors, specifically observed in BMI levels, presumably accelerating the nutrition transition in sending areas. These influences operate mostly through direct and indirect income effects from remittance flows, which lower budget constraints and allow households to increase their caloric intake and potentially reduce their caloric expense. Although the eating habits and body mass of adult migrants themselves are altered (generally for the worse) through diet and lifestyle changes made while in the United States, we do not find conclusive evidence that their transnational links rapidly diffuse new notions about health and nutrition throughout high-migration communities net of the remittance effect.

Our findings also support the notion that the translocal links associated with U.S. migration might have more far-reaching implications for men, who otherwise tend to experience the nutrition transition more slowly than women. Given that the evidence supports the importance of remittances and not of other forms of

migration-related exchanges, remittances might be disproportionately increasing the caloric consumption of men (those returning from the United States and nonmigrants alike). Additionally, men in rural areas might benefit more from the transition to capital-intensive agricultural methods than women do, at least in those places where men do more of the agricultural work. As a result, men who would otherwise be overweight are more likely to become obese.

Finally, our results show ambiguous support for the notion that the nutrition transition accelerates faster in less urban areas as a result of migration-related exchanges. If the marginalization level itself in rural areas is affected by these translocal exchanges, the links could be indirectly accelerating the nutrition transition, especially for overweight people and for men in general, at a particularly rapid pace in less urban areas. Our results provide weak support for this notion, only holding if we assume that the stronger relationship between migration intensity and overweight and obesity in rural communities operates through migration's impact on the community's level of marginalization.

Overall, our results confirm the importance of remittances in changing migrants' places of origin, but the particular effects we examine are negative: The changes associated with migration set the stage for weight-related health conditions that could continue to increase in prevalence and worsen in severity as time goes on. Our findings continue to highlight the importance of diffusing nutritional awareness throughout sending communities so that, as households enjoy the physically less demanding lifestyle that remittances support, they do not see a corresponding decline in health.

Acknowledgments

This research was supported in part by Grant R24-HD058484 from the Eunice Kennedy Shriver National Institute of Child Health and Human Development (NICHD) awarded to The Ohio State University Initiative in Population Research. We also acknowledge administrative and computing support from the NICHD-funded Population Center at the University of Colorado. We thank Nancy Mann for her helpful editing suggestions and Pablo Ibarraran for sharing the maquiladora data used in the article.

Note

1. Unfortunately, with cross-sectional data, we cannot observe individual change. Thus, a shift from overweight to obese, for example, and a shift from normal weight to obese are observationally equivalent in our study.

References

Akresh, I. R. 2007. Dietary assimilation and health among Hispanic immigrants to the United States. *Journal of Health and Social Behavior* 48 (4): 404–17.

Antecol, H., and K. Bedard. 2006. Unhealthy assimilation: Why do immigrants converge to American health status levels? *Demography* 43 (2): 337–60.

Barquera, S., C. Carrión, I. Campos, J. Espinosa, J. Rivera, and G. Olaiz-Fernandez. 2007. Methodology of the fasting sub-sample from the Mexican health survey, 2000. *Salud Pública* 49:S421–S426.

Barquera, S., R. A. Durazo-Arvizu, A. Luke, G. Cao, and R. S. Cooper. 2008. Hypertension in Mexico and among Mexican Americans: Prevalence and treatment patterns. *Journal of Human Hypertension* 22 (9): 617–26.

Barquera, S., L. Hernández-Barrera, I. Campos-Nonato, J. Espinosa, M. Flores, A. B. J, and J. A. Rivera. 2009. Energy and nutrient consumption in adults: Analysis of the Mexican National Health and Nutrition Survey 2006. *Salud Publica De Mexico* 51 (Suppl. 4): S562–S573.

Chaput, J. P., and A. Tremblay. 2009. Obesity and physical inactivity: The relevance of reconsidering the notion of sedentariness. *Obesity Facts* 2 (4): 249–54.

Frank, R. 2005. International migration and infant health in Mexico. *Journal of Immigrant Health* 7 (1): 11–22.

Frank, R., and R. A. Hummer. 2002. The other side of the paradox: The risk of low birth weight among infants of migrant and nonmigrant households within Mexico. *International Migration Review* 36 (3): 746–65.

Fussell, E., and D. S. Massey. 2004. The limits to cumulative causation: International migration from Mexican urban areas. *Demography* 41 (1): 151–71.

Hamilton, E. R., A. Villarreal, and R. A. Hummer. 2009. Mother's, household, and community U.S. migration experience and infant mortality in rural and urban Mexico. *Population Research and Policy Review* 28 (2): 123–42.

Hildebrandt, N., D. J. McKenzie, G. Esquivel, and E. Schargrodsky. 2005. The effects of migration on child health in Mexico [with comments]. *Economia* 6 (1): 257–89.

Huneault, L., M. È. Mathieu, and A. Tremblay. 2011. Globalization and modernization: An obesogenic combination. *Obesity Reviews* 12 (5): e64–e72.

Jones, R. C. 1998. Remittances and inequality: A question of migration stage and geographic scale. *Economic Geography* 74 (1): 8–25.

Kana'iaupuni, S. M., and K. M. Donato. 1999. Migradollars and mortality: The effects of migration on infant survival in Mexico. *Demography* 36 (3): 339–53.

Kwan, M.-P. 2009. From place-based to people-based exposure measures. *Social Science & Medicine* 69 (9): 1311–13.

Levi, J., S. Vinter, L. Richardson, R. St. Laurent, and L. M. Segaml. 2009. *F as in fat: How obesity policies are failing in America: Trust for America's health.* Princeton, NJ: Robert Wood Johnson Foundation.

Levitt, P. 1998. Social remittances: Migration driven local-level forms of cultural diffusion. *International Migration Review* 32 (4): 926–48.

Massey, D. S., and K. E. Espinosa. 1997. What's driving Mexico–U.S. migration? A theoretical, empirical, and policy analysis. *American Journal of Sociology* 102 (4): 939–99.

McKenzie, D. J. 2006. Beyond remittances: The effects of migration on Mexican households. In *International migration, remittances and the brain drain*, ed. M. Schiff and C. Özden, 123–47. Washington, DC: World Bank.

Monteverde, M., K. Noronha, A. Palloni, and B. Novak. 2010. Obesity and excess mortality among the elderly in the United States and Mexico. *Demography* 47 (1): 79–96.

Popkin, B. M. 2001. Nutrition in transition: The changing global nutrition challenge. *Asia Pacific Journal of Clinical Nutrition* 10:13–18.

Rivera, J. A., S. Barquera, F. Campirano, I. Campos, M. Safdie, and V. Tovar. 2002. Epidemiological and nutritional transition in Mexico: Rapid increase of non-communicable chronic diseases and obesity. *Public Health Nutrition* 5 (1a): 113–22.

Rivera, J. A., S. Barquera, T. González-Cossío, G. Olaiz, and J. Sepúlveda. 2004. Nutrition transition in Mexico and in other Latin American countries. *Nutrition Reviews* 62 (Suppl. 2): S149–S157.

Sassen, S. 1988. *The mobility of labor and capital: A study in international investment and labor flow.* Cambridge, UK: Cambridge University Press.

Smith, K. V., and N. Goldman. 2007. Socioeconomic differences in health among older adults in Mexico. *Social Science & Medicine* 65 (7): 1372–85.

Stark, O., and D. E. Bloom. 1985. The new economics of labor migration. *American Economic Review* 75 (2): 173–78.

Taylor, J. E., J. Arango, G. Hugo, A. Kouaouci, D. S. Massey, and A. Pellegrino. 1996. International migration and community development. *Population Index* 62 (3): 397–418.

Taylor, J. E., and A. Lopez-Feldman. 2010. Does migration make rural households more productive? Evidence from Mexico. *Journal of Development Studies* 46 (1): 68–90.

Valespino, J. L., G. Olaiz, M. de la Paz Lopez, L. Mendoza, O. Palma, O. Velazquez, R. Tapia, and J. Sepulveda. 2003. *Encuesta Nacional de Salud 2000. Tomo IL Vivienda,*

Población y Utilización de Servicios de Salud. Cuernavaca, Morelos, Mexico: Instituto Nacional de Salud Pública.

World Health Organization. 2009. *World Health Organization statistical information system.* http://apps.who.int/whosis/data/Search.jsp (last accessed 10 November 2009).

BOOK REVIEW ESSAY

The Geography of Life and Death: Deeper, Broader, and Much More Complex

Melinda Meade

Department of Geography, The University of North Carolina at Chapel Hill

Malaria in South Asia: Eradication and Resurgence During the Second Half of the Twentieth Century. Rais Akhtar, Ashok K. Dutt, and Vandana Wadhwa, eds. New York: Springer, 2010. xxxv and 241 pp., figures, tables, color plates, glossary, and index. $129.00 cloth (ISBN 978-90-481-3357-4).

Forgotten People, Forgotten Diseases: The Neglected Tropical Diseases and Their Impact on Global Health and Development. Peter J. Hotez. Washington, DC: ASM Press, 2008. xix and 215 pp., figures, tables, photos, color plates, appendix, chapter notes, and index. $68.74 cloth (ISBN 978-1-55581-440-3).

Geospatial Analysis of Environmental Health. Juliana A. Maantay and Sara McLafferty. New York: Springer, 2011. xiv and 498 pp., maps. $129.00 cloth (ISBN 978-94-007-0328-5).

Space, Place, and Mental Health. Sarah Curtis. Burlington, VT: Ashgate, 2010. ii and 299 pp., boxes, figures, and index. $124.95 cloth (ISBN 978-0-7546-7331-6).

Geographies of Obesity: Environmental Understandings of the Obesity Epidemic. Jamie Pearce and Karen Witten, eds. Burlington, VT: Ashgate, 2010. xxii and 331 pp., figures, maps, tables, and index. $124.95 cloth (ISBN 978-0-7546-7619-5).

Medical Geography. 3rd rev. ed. Melinda S. Meade and Michael Emch. New York: Guilford, 2010. xii and 498 pp., color plates, figures, maps, tables, and index. $59.66 cloth (ISBN 978-1-60623-016-9).

Geographies of Health: An Introduction. 2nd rev. ed. Anthony C. Gatrell and Susan J. Elliot. Malden, MA: Wiley-Blackwell, 2009. xx and 282 pp., figures, maps, tables, and index. $110.95 cloth (ISBN 978-1-4051-7576-0); $44.95 paper (ISBN 978-1-4051-7575-3).

An Introduction to the Geography of Health. Peter Anthamatten and Helen Hazen. London and New York: Routledge, 2011. xx and 273 pp. $140.00 cloth (ISBN 978-0-415-49805-0); $52.95 paper (ISBN 978-0-415-49806-7).

A Companion to Health and Medical Geography. Tim Brown, Sara McLafferty, and Graham Moon, eds. Malden, MA: Wiley-Blackwell, 2010. xxii and 610 pp., diagrams, maps, photos, notes, and index. $199.95 cloth (ISBN 978-1-4051-7003-1).

Global health—whether taken as the health of the total Earth ecosystem, the health of cities and their social and economic components, or the health of diverse human populations—has changed in more ways over the past half-century than could ever have been imagined. These major transformations include demographic dynamics, especially involving mobility and age structure; massive and accelerating urbanization; changes in climate, land cover, and water resources; new understandings of genetics at the molecular level and manipulative technologies; the arrival of the digital age with its advances in computer technology and data acquisition; the emergence or reemergence of infectious agents, diffusion of new infectious diseases, and near-eradication of some old ones; unprecedented stresses on mental health; new forms and greater prevalence of drug addiction; and developing epidemics of cancers, diabetes, and autoimmune diseases.

There have been so many changes in the need for and the provision of types and magnitude of health care that entirely new settlement systems of vulnerability and assisted care have developed. Among regions and countries, the spatial patterns of health inequalities at every scale have become more pronounced and undeniable. "Geography is destiny," said the physicians and epidemiologists who produced the *Dartmouth Atlas of Health Care* (Dartmouth Medical School, Center for the Evaluative Clinical Sciences 1998) that traced the patterns of medical procedures, expenses, and outcomes among hospital service areas across the nation. How long you live, your quality of life, and how you die evidently depend to a remarkable degree on where you live. The rather urgent need for public education amidst transformative changes and the bewilderment of a daily onslaught of specialized research reports seems obvious. Yet one thing notably lacking in the cacophony about global conditions and locally lived life today has been a cogent public voice of geography.

Forty years ago I was on a plane from New York to Hong Kong when the man sitting next to me asked what a young woman alone was doing going to Asia. Fully expecting incomprehension and some comment about knowing capitals, I simply stated that I was a medical geographer on my way to Malaysia to do my doctoral research. He immediately replied, "I read the *Geography of Life and Death* by that guy, Stamp, I think. Are you going to study conditions for infectious diseases like malaria in the countryside or the increase of heart disease and cancer as all those million-person cities develop everywhere?" Two years later on a plane returning home, I responded to a similar question by telling the man next to me that I had just defended my geography dissertation on changing disease ecology and land development for agricultural settlement in the rainforest of Malaysia. He casually responded, "I read the *Geography of Life and Death*. You wanted to know where the diseases are more or less, and why they are there and

changing the way they are, what the people there are doing." Really. I remember those two random conversations clearly (confirmed by my field journal), startled as I was to realize how people cared about the subject and the potential of geographic research to improve health and save lives. Apparently L. Dudley Stamp's (1964) slim volume had found a niche among frequent fliers that those of us in geographic education can only envy today.

The changes in global health and their relevance to each individual still fascinates, but geographers had left it to others to do the synthesizing and explanation. Garrett's (1994) bestselling *The Coming Plague: Newly Emerging Diseases in a World Out of Balance* informed millions, including many geography students, about the consequences of population growth, migration, deforestation, and developing megacities totally unequipped to provide adequate shelter, water, and sanitation services. Steingraber's (2010) updated *Living Downstream* educated millions about the production of carcinogenic chemical stews that no life on Earth can now escape; about the research complexities of multiple scales and small numbers; and about the expressions of common humanity fearing and confronting cancer. In other recent work, read by a smaller public because of the taboo nature of her subject, George (2008) explained the developing health crisis of billions of people, in burgeoning urban slums as well as densely populated countrysides, without means of sanitary disposal of human waste.

Although geography still does not offer the public such integrative, high-impact works, the geographic study of life and death is now growing dynamically, flourishing in methodologies, technologies, theories, and perspectives. This essay, with only a few backward glances for context, aims to present the contributions, limitations, and research perspectives in health and medical geography (HMG), connecting each to the other as well as the larger discipline. Reviews of history, discussions of various dualisms, and debates about international variations of the field are readily available.[1] This review uses the intellectual frameworks of the discipline's development as posited in the December 2010 centennial issue of *The Annals*, especially those of Kobayashi (2010) and Zimmerer (2010), to structure the developments and perspectives of HMG and lay out the context for ongoing responses to current global health issues.

The specialty of medical geography in the United States was constituted dualistically in 1974 by the Association of American Geographers (AAG) Committee on Medical Geography and Health Care, which in 1979 became one of the founding Specialty Groups. Yet little progress was made in conceptually linking disease ecology and health care. Over time, health service interests gravitated toward social geography with its developing theoretical and structural approaches. Deriving from the original dualism, however, every topic within medical geography that was not health care was referred to as "disease ecology." The varied interests under that umbrella included modeling disease diffusion, spatially analyzing disease patterns and causalities, and understanding the regional cultural ecology of disease etiology. Each of these subfields developed along separate theoretical and methodological tracks.

The cultural ecology of disease developed in tandem with medical anthropology, sharing foundations in human ecology and field research in developing countries. The eloquent genius of René Dubos presented a guiding, transformative vision for microbiology and public health and an inspiration for health as human ecology.[2] Following Zimmerer's (2010) framework, those Sauerian Berkeley School concepts of anthropogenic landscape evolution were first applied to historical development, as of food crops, storage, and nutrition. Four of his six core topics—coupled human–environment interactions; land use and cover change science; environmental hazards, risk, and vulnerability science; and environmental history landscapes and ideas—directly relate to foundational and continuing studies of the cultural ecology of disease. Combined with ecological concepts, such as feedback relationships and nonequilibrium theory, landscape evolution explained disease emergence and how transformations along the settlement continuum from village to megacity profoundly altered human health conditions. Chicago School concepts of natural hazards and management policy were folded into the impact of artificial water bodies on vectored diseases including malaria or schistosomiasis. John Hunter, collaborating with medical anthropologist Charles Hughes (Hughes and Hunter, 1970), wrote a seminal article on disease and development (e.g., the impact of roads, dams, and urban growth) that inspired studies of population shifts and land-use changes and their impact on settlement and economic geography (Meade 1976, 1977,1978). Land use and cover change science has allowed new field studies to be placed in a broader perspective and also facilitates greater comparative study. Environmental hazards are now studied from swimming pools and arthropod-borne viruses to cholera and dysentery in areas of flooded and embanked land to malaria following settlers along the Amazon's new logging roads. Moreover, climate change is coming to the forefront as it affects human health through drought and flooding, which in turn impact water supplies for hygiene and sanitation. Upstream policy and management repercussions are increasingly addressed at the macroscale as political ecology (Zimmerer's final pair of core topics), but that promising avenue has so far been poorly developed.

Akhtar, Dutt, and Wadhwa have tried in *Malaria in South Asia*, editing and writing a book that explores theoretical and conceptual implications of dynamic human–environment systems about what has been learned from the resurgence of malaria across South Asia. The history of the eradication efforts of the 1960s and 1970s is presented, followed by an ecological overview of malaria's resurgence. Research perspectives on control possibilities using new integrated intervention strategies are presented for several countries. Wadhwa and Dutt's own research addresses the urban ecology behind the emergence of malaria (a quintessentially rural disease) in the cities of India, where an Anopheles mosquito species became uniquely adapted to breed in the construction pits and other water-holding cavities associated with rapid urban development. The lessons they draw from the dynamics of malaria amidst global programs, new drugs, local cultural practices, socioeconomic constraints, and monsoonal variations, are used to address the need for heightened surveillance, effective measures of prevention and control, and the wider availability of curatives.

A book for readers interested in an accessible historical overview of malaria is Shah's (2011) *The Fever*; it offers an in-depth look at the evolution, genetics, control efforts, and resistance of this scourge but adds little conceptually or geographically. A more satisfying read, and an invaluable primer for geographers interested in development and health conditions of the poor but knowing little about parasitic, infectious disease, is Hotez's *Forgotten People, Forgotten Diseases*, a compact but clearly illustrated book on forgotten people and diseases. He brings neglected tropical diseases (NTDs) out of their limbo of unfunded study, genetic ignorance, and pharmaceutical unprofitability and places them at the center of rural poverty, violent upheavals, hunger, and despair. With support from the Bill and Melinda Gates Foundation and Jeffrey Sach's Earth Institute and Millennium Village project, he offers a wide readership the opportunity to "understand the science behind many of the world's afflictions," and provides scientists knowledgeable about infectious diseases an understanding of their impact on the world's poorest people. He describes NTDs as being ancient, mostly unseen in cities, exhibiting low mortality but a high disease burden from disabling, stigmatizing conditions with links to rural poverty that are not appreciated by policymakers. Hotez moves succinctly through a core group of thirteen terrible infections, their sources, and their tolls. He begins with intestinal worm infections of blood loss and iron deficiency; schistosomiasis; the filarial diseases of river blindness; lymphatic filariasis (elephantiasis, spreading in cities); and dracunculiasis (guinea worm). Then come discussions of the protozoan diseases of leishmaniasis, Chagas disease, and African sleeping sickness, followed by the bacterial diseases of trachoma, buruli ulcer, and leprosy. Biology is basic. Most of the substance is concerned with the effects of disturbance, as in former Sudan, on vaccination and eradication efforts or the impact of ineffective garbage collection on rats, stray dogs, and mosquitoes in the *favelas*. He raises hopeful prospects through interdisciplinary research and technology transfer, as in cell line–based vaccines, which might lead to successful efforts to inoculate against rabies, which still kills tens of thousands annually in developing countries.

The failure to develop new theorizations has limited the development of the cultural ecology of disease and its conceptual relevance for cognate disciplines. In fact, those of us researching disease have not clearly communicated with scholars working in our discipline's cultural ecology sphere, nor have we connected with biogeographers and landscape ecologists. There is no shortage of recent ecological research using spatial methodologies, but so far this work has not yielded theoretical contributions to disease ecology. And important challenges coming out of evolutionary biology have not yet been addressed. Ewald (2002) discovered that disease agents do not necessarily become less virulent as human hosts become more resistant, moving over time toward commensalisms, as has been assumed; instead, agents can become more virulent when conditions for their transmission are made easier. Thus, dengue fever in Asian cities, teeming with rural migrants and supersaturated with *Aedes aegypti* vectors, developed a lethal hemorrhagic form; and tuberculosis has reemerged from Russian prisons that is resistant to every treatment. How can research be designed to address the coming disease ecology of, and model new interventions for, explosively growing African cities without sanitation or Indian slums of extreme density facing floods, among other hazards?

The progression from cultural ecology to political ecology of disease to preeminent social environmental context is the movement in intellectual space from the human–environment to the nature–society perspective. Nature–society geography (Zimmerer 2010, p. 1083) is the fourth field in *The Annals'* matrix that also encompasses human and physical geography as well as GIScience and methodology. The core of the human–environment tradition mostly coincides with physical geography space, whereas the core of nature–society focus mostly coincides with human geography space (the two streams of HMG). Anthropologists, however, have been more successful than HMG in constructing the necessary conceptual bridges. Farmer (1992, 1999, 2003), through a series of books—proceeding from early HIV in Haiti through tuberculosis to poverty, women's impoverishment, and international power relations among and within societies—achieved major impact as a public scholar. Using his experience in the field, this medical anthropologist has made both conceptual and practical connections across scales from the beliefs and suffering of individuals in their households to global processes of economy, power, and their institutional expressions. Moran, whose *Human Adaptability* (1979, 2008) brought systems theory into human ecology and established the foundation of ecological anthropology, has pioneered epistemology through more than twenty books. His latest (Moran 2010) offers an interdisciplinary bridge between "environmental social science" and "sustainability science," which explicitly uses a spatial framework to approach and promote more unified theory and methodology.

Kwan's (2010) analysis of the trends and themes of method-oriented scholarship in *The Annals* not only lays out the trajectory of HMG with amazing precision; it also parallels its advance through nature–society into environmental health as well as people and place. The history of cartography and disease analysis, as it developed through a series of ever more sophisticated atlases, is recounted in the volumes noted later. Quantitative models of the analysis of flows, movements, and diffusion of people and infectious agents; of spatial structures of health service and network analysis, locational analysis, and optimization models for provision of health care; and of the patterns and associations of spatial clusters have followed the course of Kwan's progression. Many of these analyses, especially those concerned with the structures and movements of disease diffusion and the associations of their occurrence, were expressed in maps. These have been described, interpreted, and beautifully presented by both Cliff, Haggett, and Smallman-Raynor (2004) and Koch (2005, 2011).

It is GIScience and geographical visualizations, however, that enabled geographers to develop new spatial techniques for studying causality, impact, and intervention in environmental health. The modifiable areal unit problem that geographers have long grappled with had been expressed as a kind of chastity girdle in epidemiology: The "ecological fallacy" of asserting causal connections across scale had resulted in a narrow vision of rigidly pure statistical methodology that could not address broader questions of social and environmental etiology or impact. Now, geographic information systems (GIS) employing digital spatial data at multiple scales is allowing "neighborhood analysis" to separate composition and context of populations as well as natural and social environmental associations. Cromley and McLafferty (2012) have updated their exposition of this transformation, bringing geographic modeling of environmental hazard, vector-borne disease, infectious diffusion, health service location, and public participation GIS to spatial data sets for public and community health. Current methodological extension using qualitative methods and narrative analysis is also allowing justice, feminist, and subjective notions of the world to be included in addressing both etiology and outcomes.

Juliana Maantay and Sara McLafferty have brought their expertise to this interdisciplinary study by editing and contributing to what will be a new landmark work, *Geospatial Analysis of Environmental Health*. Twenty-four diverse chapters of original research by fifty-one contributors offer something of quality to instruct and inspire every geographer. First, environmental health issues of air, water, toxic waste pollution, and GIS modes of analysis are discussed, enhanced by environmental justice, residential histories, and exposure assumptions. Next, a diverse series of topical case studies examines impacts ranging from suburban West Nile Virus, brownfields, asthma, municipal solid waste, housing quality, and racial disparities in North America to diarrhea in Bangladesh and the inequities of war in Nigeria. The third section focuses on investigating environmental health: methods for detection and measurement of distance decay, spatial regression and cluster analysis, and the use of maximum entropy models in permanent hazard surveillance. Satellite imagery and ground-based monitoring data are highlighted in one chapter and time–space modeling of poverty determinants of acute respiratory infections in another.

Such spatial epidemiology illustrates the possibilities of using geospatial methods in research on etiology and uses in surveillance and assessment. Yet even the good research described in this book does not rise to the level of driving the research ideas or understandings of etiology. A more sophisticated theoretical framework of disease ecology is needed to guide the study of environmental health, because the geneticization of all noncommunicable disease research, with its concomitant focus and expenditure on treatment, is now both consuming funding and allowing policy to ignore precaution and prevention.

The overlapping phases and varied perspectives of Kobayashi's (2010) review of "people, place, and region" seem especially apposite to HMG. Determinism yielded to antideterministic cultural field research, which opened the perspective to ecological interaction and the importance of human agency in disease creation. Disease ecology as spatial analysis increasingly utilized quantitative methods, an "inventory" perhaps of what was where and associated with what, while not "looking for laws." As the study of health services developed, however, economic geography's "midlevel models" of spatial flows and areal connections came to dominate research: location–allocation modeling of patients' use of medical facilities; optimal routing of ambulances and distribution of emergency response centers; central-place analyses; and, later, adapting Wolpert's framework for decision making, amenity attractions in the location of medical practitioners. Over time, the efficiencies of locational analysis became mere algorithms for hospital marketing and government planning. As inequities in the provision of health services to meet needs became increasingly obvious, the humanist and Marxist approaches came to seem especially apropos. HMGers questioned assumptions as the magnitudes of ethnic and regional variations in life expectancy, infant mortality, and survival of heart attacks or common cancers were recognized. As foreign national health care systems and American municipal hospitals became privatized, poststructuralist perspectives and more qualitative methods were applied to study globalized flows of health care providers and their acceptance. Clearly, society had constructed significant differences in health outcomes. Field research had acquainted HMG with traditional medical systems and alternative healing approaches that could now be more meaningfully considered anew. Therapeutic landscapes, so powerful in the mind, became more widely recognized and even constructed to promote health.

Interest in directly promoting health has now united social–theoretical and humanistic approaches in health geography. Kearns and Moon (2002) identified three distinguishing themes that are quite congruent with

Kobayashi's review: the emergence of "place" as a framework for understanding health; the application of social-theoretical positions; and efforts to develop "critical" geographies of health. People are not just "observations"; places are not just "containers" in which to put observations, such as counties for mapping; qualitative methods are best for addressing interests of well-being and justice. In-depth presentations addressing disabilities, mental health, the frailties of old age and other previously ignored vulnerabilities are now appearing in cohesive geographical studies. The remarkable collection of books recently published as Ashgate's Geography of Health series evinces such consistently high quality that choosing a few titles to present here was difficult. The two reviewed here (Curtis; Pearce and Witten) clearly explain the concepts and issues of this dimension of HMG for those new to the field and provide depth, substance, and critical analysis for those with greater familiarity.

Mental illness studies have a long but very shallow history in HMG. The location of asylums and psychiatric services and patients in particular has been framed by distance decay and other economic geography models since the 1820s. The classic question has been whether the blighted conditions and degrading circumstances of certain urban neighborhoods were responsible for increasing mental illness among residents or whether the inferiority of housing and other services in those neighborhoods attracted the mentally ill to settle there. Dear and Wolch's (1987) *Landscapes of Despair*, which traced the deinstitutionalization of mentally ill people into communities where the intended care seldom materialized, has for decades been the major influence on our geographic understanding. Only recently have geographers like Parr (2008) fully brought the communities and spaces of mental illness into HMG, challenging the way in which social spaces can be envisioned to include and empower rather than to isolate and exclude people. Viewed as "participatory space," she moves mental health through social geography to the promotion of wellness.

The importance of *Space, Place, and Mental Health* has emerged from fragmented research that discussed relationships and concepts seemingly shrouded in the fog of jargon. In this book, Curtis has cleared the view by writing a masterful overview of perspectives, concepts, terms, and cross-disciplinary connections. Her presentation covers a range of theories, concepts, measurement issues, and interpretations on the ways in which our built social environment relates to mental illness and well-being. She explains how the social capital of bonding, trust, reciprocity, and cohesive activity are expressed spatially and link to psychological health and how social contexts such as ethnic density, gender, and population attributes can support psychological resilience even in the face of socioeconomic disadvantage. Curtis also relates the mind–body–environment complex to positive, real, or imaginary therapeutic landscapes as well as their converse, those containing elements of hazard and deprivation, through conceptualization of risk and its paths in social space through the life-course experience. There are many theoretical challenges to comprehending how "material features" (e.g., natural landscapes), social attributes (inclusion, cohesion), and symbolic dimensions (appreciation of environmental variables, identity, self-esteem) interact and how individuals respond. That understanding might be necessary to build sustainable societies for mental health. Biophilia is mentioned in regard to the significance of natural landscapes, and even garden therapy; but in both this book and Parr's, the singular lack of any consideration of pets in the space of mental health is striking. The benefits of social connections and the biological benefits of exercise from walking a dog for example, or even the therapeutic value through oxytocin production of pet companionship or simple visitation (Olmert 2010) seems deliberately left out.

Obesity is a predominant public health concern today for the simple reason that as a global pandemic it will soon surpass tobacco in malevolent impacts on health. Diabetes and its consequences for disability, amputation, and blindness are well known to the public. Too soon, obese children will grow up to die young from cardiopulmonary disease, kidney failure, and the basic lack of functionality needed to maintain employment or simply run a home. Genes cannot explain the current explosive epidemic. Pearce and Witten's *Geographies of Obesity* is an impressive, lucid exposition of how HMG can explain changing spatial distributions and contribute toward effective interventions. Environment is defined as including "all factors that are external to the individual including the social, political, economic, built or biophysical spheres" (p. 5), and is addressed at various geographic scales. Agricultural systems and global markets are clearly important but by themselves do not explain fat, youthful urban populations even in countries of hunger and malnutrition. This book also examines the built and social environments as well as the habits and activities of individuals in groups. Pearce and Witten then set about "theorizing pathways" through which these environmental factors interact with people in their daily lives. The usual energy balance of food consumption and activity expenditure becomes an inquiry into neighborhood availability issues, urban designs of parks and green spaces, school-based activities, and the distribution of employment opportunities. *Obesogenic* is a term that is immediately understandable but takes some getting used to before the full range of determinants and macrolevel changes of the population health burden can be appreciated. Commentaries on fast food outlets, school lunch provision, lack of inner-city retail outlets for fresh produce, the need to revive urban gardening public space, and other changes in childhood activities and our passive sedentary lives in front of computer and television screens today fill the airwaves and abound like junk in magazines. In this expressive and effective health geography, these authors successfully connect the pathways between environments and behaviors and lay out policy responses, multiscalar interventions, and research agendas aimed at understanding obesity.

The four remaining titles under review here tackle the challenge of comprehensively surveying and assessing the entire spectrum of geographical engagement with human health. They include the newest editions of two foundational textbooks (Meade and Emch; Gatrell and Elliot); a new, more abbreviated, simplified text (Anthamatten and Hazen); and a major guide and research anthology that covers all of HMG.

Medical Geography is concerned with explaining the etiology of local and international patterns of health and mortality through the global demographic and cultural forces of land use and changing habitat, urbanization, space–time exposure, and social vulnerability. Although its basic orientation is that of cultural and political ecology, involving both infectious and noncommunicable diseases in natural as well as social environments, it aims to be comprehensive in covering the dimensions and approaches to HMG. Concepts and methods from social geography and perspectives of humanism have been included in several chapters, but if I were rewriting it now I would add a separate chapter on "wellness geography."

Geographies of Health: An Introduction quickly dismisses positivist approaches as it aims to present the methodological issues and shifting terrain of a subdiscipline that "has moved on" from disease ecology. The second section's social-interactive studies on inequalities in health outcomes associated with lifestyles, countries, provision and utilization of health services (rationing, efficiency, equity, need), and migration show the power of this text's ideas and perspective.

Its environmental health section, however, addresses unidirectional health effects through irrelevant correlations in graphs and maps of inappropriate scale. This edition is more international in scope, written without unnecessary jargon, and centered on the health of people in society.

An Introduction to the Geography of Health strives to be comprehensive while simplifying, dividing the text into three main approaches: ecological, social, and spatial. Each approach is presented in a few chapters that succinctly cover theory, concepts, and specialized terminology, accompanied by case studies that exemplify the approach. Large maps and boxes are used generously to illustrate concepts, but little space is left for conceptual elaboration or topical substance. What at first seems to be a fresh, exciting juxtaposition of ideas soon devolves into a frustrating cramming of lists on unrelated topics. Social concepts are best developed, and those of the "retheorization" of both culture and of the body are explained. Neither author, however, seems to have read the other's book sections because there is no cross-referencing. An extended attack on the reproductive-rights injustices of China's population policy, for instance, fails to mention impacts on infant, child, and maternal mortality; nutritional status; and health. The concluding presentation on ten current disease eradication efforts and the societal and political considerations underlying such efforts, however, is effectively integrated.

A Companion to Health and Medical Geography is a truly remarkable editorial achievement that recognizes the significance of HMG contributions to the discipline as well as the specialty. (Editor's note: This work was reviewed in the March 2011 issue of *The Annals*.) Following an introduction to the subdiscipline and its approaches and debates, this compendium offers sections on disease studies and methodologies; health and well-being, including landscapes of therapy, despair, and vulnerabilities; health inequalities; and health care and caring. The two latter sections, especially, incorporate all that the previous approaches and perspectives have contributed to their study. In their introductory chapter, the editors emphasize that they seek not to "reify" any particular "field of vision" but to showcase the breadth and depth of an exciting and relevant area of study. Unquestionably, the *Companion* achieves that goal.

The geography of life and death is much deeper and more complicated than it used to be. Human life is not only a state of being alive; it is a process of living on the surface of the Earth as a sentient social being. The dimensions of that for health cannot be comprehended by any single paradigm or approached by any specific methodology. Disease ecology used to assume, but not study, higher sociopolitical structures of causation. Except for the provision of health services, however, mental health, addictive behaviors, disabilities, and social injustice were not only largely ignored but unimagined in social space. Spatial analysis was useful for addressing problems but contributed little etiological insight without theoretical context. Social–theoretical analyses of power and control have been needed to advance understanding. The creation of social difference and identity in space, however, are so made *of* society that a nonsocial environment barely exists in place. When it is considered at all, the "natural" environment is an external influence and not dynamically and interactively created in its particulars by us human animals who are being biologically selected, as Dubos put it, by our own culture. But mosquitoes and the pathogens they transmit *really* exist, whether we imagine or construct them or not. Whether it is the eradication of lead poisoning and malaria or the creation of inclusive, supportive places for our aging population to thrive, the promotion of human health, as Hunter challengingly put it, needs us all.

HMG is a microcosm of geography. All of the discipline's bridging of physical and social sciences, its theoretical struggles and insights, its methodological developments for spatial analysis, its integrative immersion in place, and its dimensions of humanism in identity and landscape are expressed in its study. If we can get past the singular correctness of our own theories and communicate with each other about space and the multiple dimensions of place, we might be able to educate the public again. I end my work, and this review, believing more than ever in the power of geographic research to improve health and save lives.

Key Words: cultural ecology, disease ecology, health and medical geography, spatial analysis, wellness promotion.

Notes

1. The history, development dualisms, and arguments of HMG are variously but thoroughly covered in each of the final four books reviewed in this essay. A short European view is presented in the opening chapters of Akhtar and Izhar's (2010) *festschrift* for the Belgian medical geographer Yola Verhasselt; histories of European national developments had earlier been presented by McGlashan and Blunden (1983) in a *festschrift* for the British medical geographer, Andrew Learmonth. The work of Bar-

rett (2000) on the history of the idea of medical geography before 1900 is considered definitive. Mayer's (1982, 1996, 2000) series of research position papers summarizes the scholarship of the late twentieth century. Kearns and Moon (2002) provide the most thoughtful review of the ideas and their evolution in health geography. In a tour de force, Cliff and Haggett (1988) present the accomplishments of geographers in the analytical mapping of disease. The broader history of disease cartography and geographic visualization has been thoroughly and beautifully described by Koch (2005, 2011).

2. Dubos (1901–1982), one of the most influential intellectuals of the second half of the twentieth century, deserves to be read by the new generation of geographers and not just for his works (Dubos 1959, 1965, 1968) that have most influenced the study of human health. In his journey from bacteriologist (tuberculosis) to biographer (of Pasteur) to philosopher of science and humanist husbandry of the Earth, this French-born American scholar wrote more than twenty books and won almost every major prize, among them the Lasker Prize, Pulitzer Prize for nonfiction, Arches of Science Award, and the Tyler (environmental) Prize. His vision of states of health or disease as an expression of success or failure in adaptively responding to challenges inspired a renaissance in the approach to studying infectious disease as well as recognition of the need for broad cultural and environmental research on human biological changes resulting from civilization, urbanization, and degradation of the environment. His *New York Times* obituary on 21 February 1982 particularly noted his "profound humanity and the study of man's harm to himself through environmental pollution."

References

Akhtar, R., and N. Izhar, eds. 2010. *Global medical geography: Essays in honour of Professor Yola Verhasselt*. New Delhi, India: Rawat.

Barrett, F. A. 2000. *Disease and geography: The history of an idea*. Geography Monograph No. 23. Toronto: York University, Atkinson College.

Cliff, A., and P. Haggett. 1988. *Atlas of disease distribution: Analytic approaches to epidemiological data*. Oxford, UK: Blackwell.

Cliff, A., P. Haggett, and M. Smallman-Raynor. 2004. *World atlas of epidemic disease*. London: Arnold.

Cromley, E. K., and S. L. McLafferty. 2012. *GIS and public health*. 2nd rev. ed. New York: Guilford.

Dartmouth Medical School, Center for the Evaluative Clinical Sciences. 1998. *The Dartmouth atlas of health care*. Chicago: American Hospital Association.

Dear, M. J., and J. R. Wolch. 1987. *Landscapes of despair*. Princeton, NJ: Princeton University Press.

Dubos, R. 1959. *Mirage of health: Utopias, progress, and biological change*. New Brunswick, NJ: Rutgers University Press.

———. 1965. *Man adapting*. New Haven, CT: Yale University Press.

———. 1968. *So human an animal: How we are shaped by surroundings and events*. New York: Scribners.

Ewald, P. W. 2000. *Plague time: The new germ theory of disease*. New York: Free Press.

Eyles, J., and A. Williams, eds. 2008. *Sense of place, health, and quality of life.* Burlington, VT: Ashgate.

Farmer, P. 1992. *AIDS and accusation: Haiti and the geography of blame.* Berkeley: University of California Press.

———. 1999. *Infections and inequalities: The modern plagues.* Berkeley: University of California Press.

———. 2003. *Pathologies of power: Health, human rights, and the new war on the poor.* Berkeley: University of California Press.

Garrett, L. 1994. *The coming plague: Newly emerging diseases in a world out of balance.* New York: Farrar, Straus & Giroux.

George, R. 2008. *The big necessity: The unmentionable world of human waste and why it matters.* New York: Holt.

Hughes, C. C., and J. M. Hunter. 1970. Disease and development in Africa. *Social Science and Medicine* 35:443–93.

Kearns, R., and G. Moon. 2002. From medical to health geography: Novelty, place and theory after a decade of change. *Progress in Human Geography* 25:605–25.

Kobayashi, A. 2010. People, place, and region: 100 years of human geography in *The Annals. Annals of the Association of American Geographers* 100:1095–1106.

Koch, T. 2005. *Cartographies of disease: Maps, mapping, and medicine.* Redlands, CA: ESRI Press.

———. 2011. *Disease maps: Epidemics on the ground.* Chicago: University of Chicago Press.

Kwan, M.-P. 2010. A century of method-oriented scholarship in *The Annals. Annals of the Association of American Geographers* 100:1060–75.

Mayer, J. D. 1982. Relations between the two traditions of medical geography: Health systems planning and geographic epidemiology. *Progress in Human Geography* 6:216–30.

———. 1996. The political ecology of disease as one new focus for medical geography. *Progress in Human Geography* 20:441–56.

———. 2000. Geography, ecology, and emerging infectious diseases. *Social Science and Medicine* 50:937–52.

McGlashan, N. D., and J. R. Blunden, eds. 1983. *Geographical aspects of health: Essays in honour of Andrew Learmonth.* New York: Academic Press.

Meade, M. S. 1976. Land development and human health in West Malaysia. *Annals of the Association of American Geographers* 66:428–39.

———. 1977. Medical geography as human ecology: The dimension of population movement. *Geographical Review* 67:379–93.

———. 1978. Community health and changing hazards in a voluntary agricultural resettlement. *Social Science and Medicine* 12:95–102.

Moran, E. F. 1979. *Human adaptability: An introduction to ecological anthropology.* North Scituate, MA: Duxbury.

———. 2008. *Human adaptability: An introduction to ecological anthropology.* 3rd rev. ed. Boulder, CO: Westview.

———. 2010. *Environmental social science: Human–environmental interactions and sustainability.* Malden, MA: Wiley-Blackwell.

Olmert, M. D. 2010. *Made for each other: The biology of the human–animal bond.* Cambridge, MA: Da Capo Press.

Parr, H. 2008. *Mental health and social space: Towards inclusionary geographies?* Malden, MA: Wiley-Blackwell.

Shah, S. 2011. *The fever: How malaria has ruled humankind for 500,000 years.* New York: Farrar, Straus, & Giroux.

Stamp, L. D. 1964. *The geography of life and death.* Ithaca, NY: Cornell University Press.

Steingraber, S. 2010. *Living downstream: An ecologist's personal investigation of cancer and the environment.* Cambridge, MA: Da Capo Press.

Zimmerer, K. S. 2010. Retrospective on nature–society geography: Tracing trajectories (1911–2010) and reflecting on translations. *Annals of the Association of American Geographers* 100:1076–94.

Index

Page numbers in **bold** type refer to figures
Page numbers in *italic* type refer to tables
Page numbers followed by 'n' refer to notes

T - #0011 - 071024 - C368 - 276/216/20 [22] - CB - 9780415870016 - Gloss Lamination